www.brookscole.com

www.brookscole.com is the World Wide Web site for Thomson Brooks/Cole and is your direct source to dozens of online resources.

At *www.brookscole.com* you can find out about supplements, demonstration software, and student resources. You can also send email to many of our authors and preview new publications and exciting new technologies.

www.brookscole.com
Changing the way the world learns®

NATURAL HAZARDS AND DISASTERS

Donald Hyndman
University of Montana

David Hyndman
Michigan State University

Australia • Canada • Mexico • Singapore • Spain • United Kingdom • United States

Natural Hazards and Disasters
Donald Hyndman and David Hyndman

Earth Science Editor: Keith Dodson
Assistant Editor: Carol Benedict
Editorial Assistant: Megan Asmus
Technology Project Manager: Ericka Yeoman-Saler
Marketing Manager: Jennifer Somerville
Marketing Assistant: Michele Colella
Advertising Project Manager: Kelley McAllister
Project Manager, Editorial Production: Hal Humphrey
Art Director: Vernon Boes
Print/Media Buyer: Karen Hunt
Permissions Editor: Kiely Sisk
Production Service: Penmarin Books
Text Designer: Adriane Bosworth

Art Editor: Connie Hathaway
Copy Editor: Steven Summerlight
Illustrator: new pieces by Accurate Art
Cover Designer: Denise Davidson
Cover Printer: Phoenix Color Corp
Compositor: G&S Typesetters
Printer: Courier Corporation/Kendallville
Cover Images: Hurricane Felix, Virginia Beach: © Corbis/Annie Griffiths Belt; Taiwan toppled building: Jack P. Moehle, University of California, Berkeley; Tornado: © Corbis/Jim Zuckerman; La Conchita slide: Robert Schuster photo, USGS; Grand Forks flood: Richard Larson, University of North Dakota; Mount St. Helens: © Corbis/Gary Braasch.

Printed in the United States of America
2 3 4 5 6 7 09 08 07 06 05

For more information about our products, contact us at:
Thomson Learning Academic Resource Center
1-800-423-0563
For permission to use material from this text or product, submit a request online at **http://www.thomsonrights.com.**
Any additional questions about permissions can be submitted by email to **thomsonrights@thomson.com.**

ISBN 0-534-99760-0

Thomson Higher Education
10 Davis Drive
Belmont, CA 94002-3098
USA

Asia (including India)
Thomson Learning
5 Shenton Way
#01-01 UIC Building
Singapore 068808

Australia/New Zealand
Thomson Learning Australia
102 Dodds Street
Southbank, Victoria 3006
Australia

Canada
Thomson Nelson
1120 Birchmount Road
Toronto, Ontario M1K 5G4
Canada

UK/Europe/Middle East/Africa
Thomson Learning
High Holborn House
50/51 Bedford Row
London WC1R 4LR
United Kingdom

Latin America
Thomson Learning
Seneca, 53
Colonia Polanco
11560 Mexico
D.F. Mexico

Spain (including Portugal)
Thomson Paraninfo
Calle Magallanes, 25
28015 Madrid, Spain

TO SHIRLEY AND TERESA
for their endless encouragement and patience.

DONALD HYNDMAN is a recently retired Professor in the Department of Geology at the University of Montana, where he has taught courses in natural hazards, regional geology, igneous and metamorphic petrology, volcanology, and advanced igneous petrology. He continues to teach courses in natural hazards. Donald is co-originator and co-author of six books in the Roadside Geology series and one on the geology of the Pacific Northwest, and is also the author of a textbook on igneous and metamorphic petrology. His B.S. in Geological Engineering is from the University of British Columbia and his Ph.D. in Geology is from the University of California, Berkeley. He has received the Distinguished Teaching Award and the Distinguished Scholar Award, both given by the University of Montana.

DAVID HYNDMAN is an Associate Professor in the Department of Geological Sciences at Michigan State University, where he teaches courses in natural hazards, environmental geology, physical geology, and advanced hydrogeology. His B.S. in Hydrology and Water Resources is from the University of Arizona and his M.S. in Applied Earth Sciences and Ph.D. in Geological and Environmental Sciences are from Stanford University. David is an Associate Editor for the journals *Water Resources Research* and *Ground Water*, and was selected as a Lilly Teaching Fellow and for the Ronald Wilson Teaching Award. For 2002 he was the Darcy Distinguished Lecturer.

BRIEF CONTENTS

CONTENTS

LIVING WITH NATURE

*"The further you are from the last disaster,
the closer you are to the next."*

Why We Wrote This Book

In teaching large introductory environmental and physical geology courses for many years—and, more recently, natural hazards—it has become clear to us that topics involving natural hazards are among the most interesting for the students. Thus, we realize that employing this thematic focus can stimulate students to learn basic scientific concepts, to understand how science relates to their everyday lives, and to see how such knowledge can be used to help mitigate both physical and financial harm.

Natural hazards and disasters courses should achieve higher enrollments, have more interested students, and be more interesting and engaging than those taught in a traditional environmental or physical geology framework. A common trend is to emphasize the hazards portions of physical and environmental geology texts while spending less time on subjects that do not engage the students. Students who have previously had little interest in science can be awakened with a new curiosity about the Earth and the processes that dramatically alter it. Science majors experience a heightened interest, with expanded and clarified understanding of natural processes. In response to years of student feedback and discussions with colleagues, we have reshaped our courses to focus on natural hazards.

Students who take a natural hazards course greatly improve their knowledge of the dynamic Earth processes that will affect them throughout their lives. They should be able to make educated choices about where to build houses, business offices, or engineering projects. Perhaps some of those who take this course will become government officials or policy makers who can change some of the current culture that contributes to major losses from natural disasters.

Undergraduate college students, including nonscience majors, should find the writing clear and stimulating. Our emphasis is to provide them a basis for understanding important hazard-related processes and concepts. This book encourages students to grasp the fundamentals while still appreciating that most issues have complexities that are beyond the current state of scientific knowledge and involve societal aspects beyond the realm of science. Students not majoring in the geosciences may find motivation to continue studies in related areas and to share these experiences with others.

Natural Hazards and Society

Events in nature cause increasing property damage as populations grow and more people move into settings that they do not realize are hazardous. Much of the resultant costs fall on federal and state governments and therefore eventually back to taxpayers. The balance generally falls on local residents who can ill afford the damages. A strong awareness of natural hazards can help people avoid becoming either victims or part of the problem.

A recent secretary general of the United Nations aptly stated, "The term *natural disaster* is a misnomer. In reality, human behavior transforms natural hazards into what should be called 'unnatural disasters.'" Natural events become hazards only when people place themselves in harm's way. An enormous event in a thinly populated area is not much of a hazard. For example, the major November 2002 earthquake on the Denali Fault in central Alaska caused only one broken arm and negligible property damage because of the region's sparse population. A moderate-sized earthquake in December 2003 near Paso Robles, California, killed two people. In contrast, a similar-sized earthquake in the same month in Iran caused more than 26,000 casualties because a large population lives near an active fault in weak mud-brick buildings. Disaster mitigation efforts need to consider people's living conditions and the factors that influence where they live.

Various economic and behavioral factors influence societal decisions. Economic factors include costs to people, communities, and businesses, as well as the loss of potential profits to various individuals or groups. Behavioral factors include a desire to live in certain environments and perceived property rights by individuals, corporations, and the general public. If an owner's rights affect those of a neighbor or those of society in general, whose rights are diminished?

Today's ethics compel us to help those in need, not only emotionally and physically, but also financially. In response to natural disasters, individuals and government programs often help those who have suffered financial losses, even when those people made choices that predestined their loss. Some are even repeat offenders in the same location; we as a society need to better deal with such behavior.

It is important to understand the basis and limitations of hazard forecasts; unfortunately, few people do. Even fewer people have any idea of how to prevent natural disasters or mitigate their effects. We cover these topics in order to help students deal more effectively with future natural events.

People acquire insurance to protect against catastrophic losses that they cannot afford. Because risk is proportional to the probability of an event times the cost of such an event, we touch on the factors that determine the basis for assessing insurance premiums for potential natural hazard losses. As local populations increase, more people move to hazardous areas; thus there is an increased need for understanding natural hazards and their physical and financial consequences.

The expectation that we can control nature through technological change stands in contrast to the fact that natural processes will ultimately prevail. We can choose to live *with* nature or we can try to fight it. Unfortunately, people who choose to live in hazardous locations tend to blame either "nature on the rampage" or others for permitting them to live there. People do not often make such poor choices willfully but through their lack of awareness of natural processes. Even when they are aware of an extraordinary event that has affected someone else, they somehow believe "It won't happen to me." These themes are revisited throughout the book, ensuring that our discussion of principles and processes is frequently related to societal behavior and attitudes.

Our Approach

Our focus here is on Earth and atmospheric hazards that appear rapidly, often without significant warning. Throughout the book, we emphasize interrelationships between hazards, such as the fact that building dams on rivers often leads to greater coastal erosion. Similarly, wildfires generally make slopes more susceptible to floods, landslides, and mudflows. We attempt to provide balanced coverage of natural hazards across North America. Toward the end of the course, our students sometimes ask, "So where in North America is a safe place to live?" We often reply that you can choose hazards that you are willing to deal with and live in a specific site or building that you know will minimize impact of that hazard.

Extensive illustrations and "Case in Point" examples bring reality to the discussion of principles and processes. These cases tie the process-based discussions to individual cases and integrate relationships between them. They emphasize the natural processes and human factors that affect disaster outcomes. Illustrative cases are interwoven with topics as they are presented.

The book is up-to-date and clearly organized, with most of its content derived from current scientific literature and from our own personal experience. It is packed with relevant content on natural hazards, the processes that control them, and means of avoiding catastrophes. Numerous excellent and informative color photographs, many of them our own, illustrate concepts in a manner that is not possible without color. The diagrams are clear, straightforward, and instructive.

The text begins with an overview of the dynamic environment in which we live and the variability of natural processes, emphasizing the fact that most daily events are small and generally inconsequential. Larger events are less frequent, though most people understand that they can happen. Fortunately, giant events are infrequent; regrettably, most people are not even aware that such events can happen. People often decide on their residence or business location based on a desire to live and work in scenic environments without understanding the hazards around them. Once they realize the risks, they often compound the hazards by attempting to modify the environment. Students who read this book should be able to avoid such decisions.

This book emphasizes Earth and atmospheric hazards that appear suddenly or rapidly, often without significant warning. It includes chapters on dangers generated internally, including earthquakes, tsunami, and volcanic eruptions. Society has little control over the occurrence of such events but can mitigate their impacts through a deeper understanding that can afford more enlightened choices. The landslides section addresses hazards influenced by both in-ground factors and weather, a topic that forms the basis for many of the following chapters. A chapter on sinkholes, subsidence, and swelling soils addresses other destructive in-ground hazards that we can to some extent mitigate, and which are often subtle yet highly destructive.

The following hazard topics depend on an understanding of the dynamic variations in climate and weather, so we begin with a chapter to provide that background and an overview of global warming. Chapters on streams and floods begin with the characteristics and behavior of streams and how human interaction affects both a stream and people around it. Chapters follow on wave and beach processes, hurricanes and Nor'easters, thunderstorms and tornadoes, and wildfires. The final chapters discuss asteroid impacts and future concerns related to natural hazards. Appendixes present the geological time scale and a brief discussion of the nature of rocks and minerals, primarily as background for some of the physical hazards.

Acknowledgments

We are grateful to many people for assistance in gathering material for this book, far too many to individually list here. However, we especially appreciate help from the following colleagues.

- For editing and suggested additions: Dr. Dave Alt (University of Montana, emeritus); Ted Anderson; Shirley Hyndman; Teresa Hyndman; Dr. Duncan Sibley, Dr. Kaz Fujita, and Dr. Tom Vogel (Michigan State University).

- For information and photos on specific sites:
Dr. Maurizio Bonardi, Instituto per lo Studio della Dinamica Delle Grandi Masse, Venice, Italy; Dr. Scott Burns, Portland State University, Oregon; Dr. Susan Cannon, US Geological Survey (USGS), Denver; Helen Delano, geologist, Pennsylvania Department of Conservation and Natural Resources, Harrisburg; Dr. Kazuya Fujita, Michigan State University; Robert Glancy, National Oceanic and Atmospheric Administration, Boulder, Colorado; Bruce Heise, geologist, National Park Service, Denver; Christina Heliker, USGS, Hawaii Volcanoes National Park; Agnes Hiramoto, University of Hawaii Press, Honolulu; Dr. Roy Hyndman, Pacific Geoscience Center, Victoria, British Columbia; Dr. Matthew Larsen, USGS, Guaynabo, Puerto Rico; William Lund, senior scientist, Utah Geological Survey; Robert Larson, former Los Angeles County engineering geologist; Larry Schock, civil engineer, Montana Department of Natural Resources and Conservation; David Noe, chief of engineering geology, Colorado Geological Survey, Denver; Gregory Ohlmacher, engineering geologist, Kansas Geological Survey; Dr. Byron Olson; Dr. Orrin Pilkey, Duke University; Dr. George Plafker, earthquake hazards team, USGS, Menlo Park, California; Dr. Perry Rahn, professor emeritus, South Dakota School of Mines and Technology; Dr. Robert Schuster, USGS landslide hazards group (retired); Dr. Robert Tilling, volcano hazards team, USGS, Menlo Park, California; and Dr. Robert Webb, USGS, Tucson, Arizona.

- For providing us with original photos: Professor Jean-Pierre Bardet, University of Southern California; Michael Bohlander; Karl Christians, state coordinator, Montana Department of Natural Resources, Helena; Dr. John Clague, Simon Fraser University, Burnaby, British Columbia; Wendy Davidson, American Institute of Professional Geologists; James Van Dongen, New Hampshire Office of Emergency Management, Concord; Alexander Feldman, Shannon & Wilson, Inc., Seattle, geotechnical engineers; Dr. Manny Gabbet, University of Montana; Gary Goreham, North Dakota State University; Dr. Carl Hobbs, Virginia Institute of Marine Science; Norm Hughes, senior photographer, California Department of Water Resources; Dr. Robert Jarrett, Paleohydrology and Climate Change, USGS, Lakewood, Colorado; Dr. Randall Jibson, USGS, Denver; Peter Judge, Commonwealth of Massachusetts, Framingham; Dr. Ben Kennedy, USGS; John Kiefer, Kentucky Geological Survey, Lexington; William Kochanov, Pennsylvania Geological Survey, Middletown; Dr. Vladmir Kostoglodov, Instituto de Geofisica, UNAM, Mexico, DF; Dr. Ian Lange, University of Montana (emeritus); Dr. Matt Larsen, Caribbean District Chief, USGS; Richard Larson, University of North Dakota, Grand Forks; Robert A. Larson, GeoArchives, Sylmar, California; Dr. Stephen Leatherman, Director of the International Hurricane Research Center, Florida International University; Dr. George Leung, Physics, University of Massachusetts, Dartmouth; Tom Lisle, U.S. Forest Service, Arcata, California; Chris Mack, coastal engineer, U.S. Army Corps of Engineers, Charleston, South Carolina; Dr. Mark McCourt, North Dakota State University, Fargo; Dr. Marli Miller, University of Oregon, Eugene; Mark Milligan, Utah Geological Survey; Dr. Donald Murray, Missoula, Montana; National Geoscience Database, Iran; Jan Pardell, Longview, Washington; Kent Petersen, KLP Consulting Engineers, Englewood, Colorado; Dr. Fred Phillips, New Mexico Institute of Mining and Technology, Socorro; Dr. Perry Rahn, South Dakota School of Mines and Technology, Rapid City; Joel Schiff, *Meteorite International Quarterly*, Auckland, New Zealand; Dr. Peter Scholle, New Mexico Bureau of Geology and Mineral Resources, Socorro; Dr. Don Schwert, North Dakota State University, Fargo; Dr. David B. Scott, Dalhousie University, Nova Scotia; Dr. Thomas Scott, Florida Geological Survey; Hank Shiffman, Mountain View, California; U.S. Army Corps of Engineers, Galveston, Texas; Susan Snyder, Bancroft Library, University of California, Berkeley; Jill Sommer, Pacific Tsunami Museum, Hilo, Hawaii; Thomas Spittler, California Geological Survey; Keith Stoffel, Spokane, Washington; Washington State Department of Transportation; Dr. Robert Weidman, emeritus, University of Montana; Dr. Gary Weissmann, Michigan State University; Peter Wohlgemuth, U.S. Forest Service, Riverside, California; and Dr. Leslie Youd, Brigham Young University, Provo, Utah.

- For providing access to the excavations of the Minoan culture at Akrotiri, Santorini: Dr. Vlachopoulos, head archaeologist, Greece;

- For assisting our exploration of the restricted excavations at Pompeii, Italy: the site's chief archaeologist.

- For logistical help: Roberto Caudullo, Catania, Italy; Brian Collins, University of Montana; and Keith Dodson, Brooks/Cole's Earth sciences textbook editor.

In addition, we thank the following reviewers of the text for their insights and helpful comments: Ihsan S. Al-Aasm, University of Windsor; Jennifer Coombs, Northeastern University; Jim Hibbard, North Carolina State University; Mary Leech, Stanford University; David Evans, California State University, Sacramento; Wang-Ping Chen, University of Illinois at Urbana-Champaign; Stephen Nelson, Tulane University; Alan Lester, University of Colorado at Boulder; Luther Strayer, California State University, East Bay; Katherine Clancy, University of Maryland; and James Harrell, University of Toledo.

Donald Hyndman and David Hyndman
February 2005

Downtown Grand Forks, North Dakota, was flooded for many weeks in 1997 when the Red River of the North backed up behind ice jams as the river thawed upstream to the south. Details are provided in Chapter 11.

NATURAL HAZARDS AND DISASTERS

Catastrophic Events in Nature

Those who cannot remember the past are condemned to repeat it.

—GEORGE SANTAYANA, 1905 (SPANISH PHILOSOPHER)

Living in Harm's Way

Scientists and engineers understand much about the natural world but less than many people suppose. In addition, policy makers place too much reliance on the ability of scientists and engineers to control nature. In most cases, we need to learn to live with natural events instead of trying to control them. Far too many efforts to avoid impending catastrophe fail, and too commonly without any redeeming benefit. People continue to learn that tampering with nature often makes matters worse.

Textbooks commonly emphasize that everyday processes can produce large effects over the enormous length of Earth history. Of course that is true, but the implication is that those great effects generally arrive slowly, not as sudden major events. However, most geological processes that

1

have significant effects move in fits and starts. Not much happens for a long time; then there is a sudden event or a pulse of rapid activity before the system again becomes quiet. Such abrupt events produce large effects that can be disastrous if they affect people.

We see that most streams run clear throughout the year but then become muddy during floods. That mud reflects renewed erosion during the few days or weeks of high water. Catastrophic floods do far more damage and move more material than all of the intervening floods put together. Soil moves slowly downslope by creep, but occasionally a huge part of a slope may slide. Mountains rise, sometimes slowly but commonly by sudden movements within the Earth. During an earthquake, a mountain can abruptly rise as much as several meters above the adjacent valley.

Catastrophes come in all sizes, some so immense that they alter the course of history. Large or sometimes gigantic events are a part of nature. Although we all know that catastrophes take place, many scientists have been slow to embrace the importance of catastrophic events. In the context of natural hazards, that view is beginning to change.

Many geological processes are potentially hazardous. We know that streams flood, if not every year, then every two or three years. A moderate flood that spills over the floodplain does not often create a **disaster.** Only when a major flood causes significant damage, either to life or property, is it considered a disaster or **catastrophe.** For a natural event to become a catastrophe, it must kill or injure large numbers of people or cause major property damage. Thus, catastrophic events must happen in populated areas.

A catastrophe often requires that several inherently unlikely events happen in the same place at the same time; they conspire to create an incredible result. So it seems unfortunate that few textbooks mention the equally obvious truth that inherently improbable events are ultimately inevitable. If it can happen, it eventually will—Murphy's Law.

Large numbers of people live and work in notoriously dangerous places. They crowd up to the flanks of large volcanoes for the views or because volcanic ash degrades into richly productive soil. They build homes at the bases or tops of large cliffs for a scenic view, not realizing that large portions of the cliffs can give way in landslides or rockfalls. Large floodplains attract people because they provide good agricultural soil, inexpensive land, and natural transportation corridors (see opening photograph). For understandable reasons, such people are living in the wrong place (▶ Figure 1-1).

We read and hear reports of disasters in every country, but the nature of the disaster differs from place to place. The hazard may be the same, but the consequences depend heavily on the wealth and education of the populace. Disasters cost money and lives wherever they occur, but developed countries almost always lose more money while underdeveloped countries lose more lives.

More people also crowd into dangerous areas for frivolous reasons. They long to live along edges of sea cliffs where they can enjoy the ocean view, or they want to live

▶**FIGURE 1-1.** This four-year-old house near Zion National Park in southern Utah was built near the base of a steep rocky slope capped by a sandstone cliff. Early one morning in October 2001, the owner awoke with a start as a giant boulder 4.5 meters (almost 15 feet) across crashed into his living room and bedroom, narrowly missing his head.

▶**FIGURE 1-2.** Orting, Washington, with spectacular views of Mount Rainier, is built on a giant, ancient mudflow from the volcano. If mudflows happened in the past, they almost certainly will happen again.

along the beach to experience the ocean more intimately. Others build beside rivers because they are picturesque or seem so calmly soothing. Far too many people build houses in the woods because they enjoy the seclusion and scenic nature of this setting. Some experts who are concerned with natural catastrophes say that such people have chosen to live in an "idiot zone." People who deliberately choose to live in hazardous areas might as well choose to park their cars on a railroad track. The hazardous event may not arrive today or tomorrow, but it is just a matter of time. (▶See Figure 1-2.)

Avoiding Natural Disasters

The problem of avoiding natural disasters is like the problem drivers face in avoiding collisions with trains. They can do nothing to prevent trains, so they must look and listen. Catastrophic natural hazards are much harder to avoid than passing freight trains; we may not recognize the signs of imminent catastrophes because these events are infrequent. So many centuries may pass between eruptions of a large volcano that most people forget that it is active. Many people live so long on a valley floor without seeing a big flood that they forget it is a floodplain. The great disaster of a century ago is long forgotten so people move into the path of disaster without a thought for the tragic sequel that will follow on some unknowable future date.

People who watch Earth processes proceed at their normal and unexciting pace rarely pause to imagine what might happen if that slow pace were suddenly punctuated with a major event. The fisherman enjoying an afternoon pursuing trout in a quiet stream can hardly imagine how a 100-year flood might transform the scene. Someone strolling along a beach on a placid summer morning can hardly imagine what will happen in a few months when a hurricane comes ashore there. Most of us live in ignorance of what may happen.

Land Use Planning

Land use planning offers an obvious way to avoid many natural catastrophes. Ideally, we should reserve land along major active faults for parks and natural areas. We should also limit housing and industrial development on floodplains to minimize damage from floods and along the coast to minimize hurricane losses. Limiting building on active volcanoes can minimize the hazards associated with eruptions.

It is hard, however, to impose land use restrictions in many areas because such imposition tends to come too late. Many hazardous areas are already heavily populated, perhaps even saturated with people. Many people want to live as close as they can to a coast or a river and resent being told that they cannot; they oppose any attempt at land use restrictions because they feel it infringes on their property rights. Almost any attempt to regulate land use in the public interest is likely to ignite intense political and legal opposition.

Pointing the Blame Finger

People tend to blame all the wrong entities for natural disasters. Some ancient journalistic tradition seems to require that the news media interpret a great flood as "nature on a rampage," while insurance companies consider it an "act of God." We have also seen floods blamed on mayors, county commissioners, governors, members of Congress, and Republicans or Democrats in general. Agencies such as the U.S. Forest Service, the Bureau of Land Management, the Bureau of Reclamation, and the Soil Conservation Service all get their fair share of potshots. Only the Corps of Engineers seems able to stay out of the line of fire. Many people believe that the Corps of Engineers prevents floods with levees that wall floodwaters out of large floodplain areas. In fact, levees do not prevent floods; they merely displace them from one part of the floodplain to another. People in the areas to which floods are displaced then demand more levees to protect their property. The Corps is generally praised for constructing "protection" structures or replenishing beaches in spite of the fact that these features may do serious long-term damage, may not last long, and are paid for by public funds. Those funds, of course, are provided by the taxes that most people complain about.

Of course, any river has flooded thousands, perhaps millions, of times. Its routine floods became natural disasters only after large numbers of people settled on its floodplain. All floodplains are subject to flooding—which is why we call them **floodplains.** Maps are available in many areas that show exactly what parts of the floodplain will be submerged during floods of different sizes. The only mystery is when those floods will happen. The pressure for floodplain development is intense, incessant, and generally involves significant profits to individuals or corporations. Decisions often become political.

Predicting Catastrophe

Despair comes easily to those who would predict that a natural disaster will occur on a particular date. So far, there have been few well-documented cases of prediction, and even those may have involved some luck. Use of the same techniques in similar circumstances has involved false alarms. The complexities of most natural situations make it unlikely that reliable disaster predictions will become possible anytime soon. Are hazard predictions really possible?

Ask a stockbroker where the market is going, and you will probably hear that it will continue to go wherever it has been going during recent weeks. Ask a scientist to predict an event, and he or she will probably look to see what has happened in the geologically recent past and predict more of the same. Most predictions of any kind are based on linear projections of past experience. That undoubtedly explains why so many people lose money in the stock market.

Statistical predictions are simply a refinement of past-recorded experiences. They are typically expressed as recurrence intervals that relate to the probability that a particular natural event will happen within a certain period of time. Flip a coin, and the **chance** that it will come up heads will be 50 percent. Flip it again, and the chance is again 50 percent. If it comes up heads five times in a row, the next flip still has a 50 percent chance of coming up heads. So it goes

with earthquakes, floods, and many other kinds of apparently random natural events.

Fifty-year floods are likely to happen sometime in the next several decades, but not at intervals of fifty years. The chance that anyone's favorite fishing stream will stage a fifty-year flood this year and every year is 1 in 50, regardless of what it may have done during the last fifty years. Such a flood last year does not indicate that there will not be one this year or next. The past history of a fault may indicate that it is likely to produce an earthquake of a certain size once every 100 years—on **average.** The likelihood of an earthquake along an active fault is certainly much greater than it would be for a location somewhere well away from such a fault. Nothing in the statistical reasoning suggests, however, that it will produce those earthquakes on a regular schedule, just that the probability that it will produce one this or any other year is approximately 1 in 100.

It is not safe to assume that all natural events occur quite as randomly as floods or tosses of a coin. In some situations, past events create a contingency that influences future events. It is certainly true, for example, that sudden movement on a fault causes an earthquake. But the same movement also changes the stress on other parts of the fault and probably on other faults in the region, so the next earthquake will likely differ considerably from the last. Similar problems arise with many other types of destructive natural events.

The usual procedure to estimate the recurrence interval of a particular kind of natural event is to plot a graph of each event size versus the time interval between sequential individual events. Such plots often make curved lines that cannot be reliably extrapolated to larger events that might lurk in the future (▶ see Figure 1-3). Plotting the same data on a logarithmic scale often leads to a straight-line graph that can be extrapolated to values larger than those in the historic record. Whether the extrapolation produces a reasonable result is another question.

Some natural events involve disruption of a temporary "equilibrium" between opposing influences. Unstable slopes, for example, may hang precariously for thousands of years, held there by friction along the slip surface until some small perturbation such as water soaking in from a large rainstorm sets them loose. Similarly, the opposite sides of a fault may stick until the continuing stress along them finally tears them loose, triggering an earthquake. A bulge may form on a volcano as molten magma slowly rises into it, then the bulge collapses as the volcano erupts. The behavior of these natural systems is somewhat analogous to a piece of fabric or plastic wrap that remains intact as it stretches until it suddenly tears.

It is impossible in our current state of knowledge to predict natural events, even if we understand in a general way what controls them. We have no way of knowing how firm the natural restraints on a landslide, fault, or volcano may be. We also cannot know what changes are occurring at depth. Small events commonly fall in the category of the

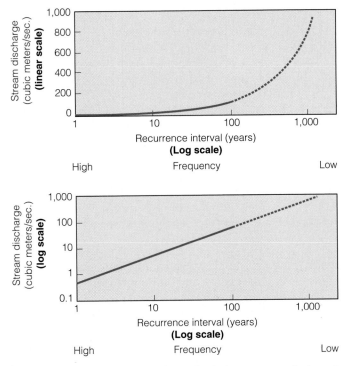

FIGURE 1-3. On a graph of magnitude (e.g., stream discharge) versus the frequency of such magnitude, a logarithmic plot is often equivalent to a straight-line graph that can then be extrapolated to larger values. Also note that small magnitude events tend to be frequent, whereas large magnitude events tend to be infrequent.

random noise of chaos theory. We may be confident that the landslide or fault will eventually move or that the volcano will erupt. And we can reasonably understand what those events will involve when they finally happen. But it is generally impossible to predict exactly when they will happen, and that seems likely to remain so for a long time.

Considering the number of variables involved in most natural events and the uncertainties surrounding each, it is not surprising that we have not had much success in predicting major catastrophes. Nevertheless, the people who make it their business to understand natural disasters have learned enough about them to provide some guidance to people who are at risk. They cannot predict exactly when an event will occur. However, based on past experience, they can often forecast, with an approximate percentage probability, that a hazardous event will occur in a certain area within the next several decades or hundreds of years. They can forecast that there will be a large earthquake in the San Francisco Bay region over the next several decades. They can forecast that Mount Shasta will likely erupt sometime in the next few hundred years. In many cases, their advice can greatly reduce the danger to lives and property. Their efforts have not been in vain. Once a volcano shows signs of reawakening, volcanologists can better predict the specific eruption timing based on precursor events that occur before the main eruption.

Randomness in Natural Events

Many people have sought to find predictable cycles in natural events. Even most recurrent events are generally not really cyclic. Too many variables control the behavior of natural events. Even with **cyclic events,** overlapping cycles would make the resultant extremes noncyclic (▶ see Figure 1-4). That would affect the predictability of the event. So far as anyone can tell, most events, large and small, occur at seemingly random and essentially unpredictable intervals. The calendar does not predict them.

Most natural disasters happen when a number of unrelated variables happen to fall in such a way that they reinforce each other (▶ Figure 1-4). If the high water of a hurricane storm surge happens to arrive at the coast during the daily high tide, the two reinforce each other to produce a much higher storm surge. If this occurs on a section of coast that happens to have a large population, then the situation becomes much worse. Such a coincidence caused the catastrophic hurricane that killed 8,000 people in Galveston, Texas, in 1900. Bad luck prevailed.

Are different kinds of events somehow related? Is one type a consequence of another? **Plate tectonics,** or the slow movement of the huge outer layers of the Earth colliding or sliding past one another, clearly explains the driving forces behind volcanic eruptions and earthquakes. Heavy or prolonged rainfall can cause a flood or a landslide (▶ Figure 1-5). Are some events unrelated? If an earthquake happens at the time of a volcanic eruption, did the eruption cause the earthquake or did the earthquake cause the eruption—or neither? Or did the earthquake not **cause** the eruption but merely trigger the final eruption? Geologists studying the stirrings of Mount St. Helens, Washington, in 1980 before its catastrophic eruption, monitored swarms of earthquakes and decided that most recorded the movements of rising magma as it squeezed upward, expanding

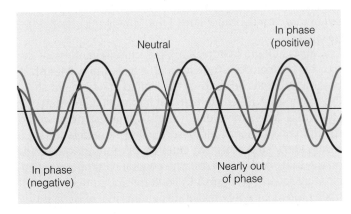

FIGURE 1-4. Unrelated cycles with different wavelengths may overlap so that their high points are *in phase*—that is, they occur at the same time. Or they may be *out of phase,* in which case they tend to cancel out one another.

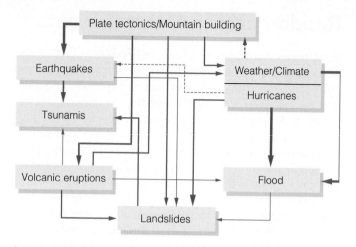

FIGURE 1-5. This flowchart indicates interactions between natural hazards. The bolder arrows indicate stronger influences. Can you think of others?

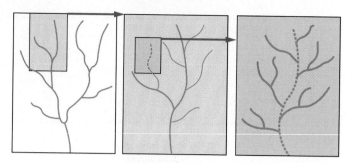

FIGURE 1-6. The branching of streams is fractal. The general branching of patterns looks similar regardless of scale—from a less-detailed map on the left to the most-detailed map on the right.

the volcano. What about the bigger earthquake at the instant the volcano exploded the morning of May 18? The expanding bulge over the rising magma collapsed in a huge landslide. Certainly neither the landslide nor the earthquake caused the formation of molten magma. Once formed, neither caused the magma to rise, but did they trigger the final eruption? If so, which one triggered the other—the earthquake, the landslide, or the eruption? Were the events directly related—that is, did one cause the others? What about the tragic mudflows that immediately followed? It was a beautiful clear morning before the eruption. Was the falling ash hot enough to melt snow on the volcano, or did the eruption cause it to rain? One or more of these possibilities could be true in different cases.

Even seemingly unrelated events may **overlap** to amplify an effect; some events are directly related to others—formed as a direct consequence of another event. Could any of the arrows in Figure 1-5 be reversed? Given all of the interlocking possibilities, the variability, and the uncertainties, we could call Figure 1-5 a "chaos net" for natural hazards.

Many geologic features look the same regardless of the size of the feature, a quality that makes them **fractal.** A broadly generalized map of the United States might show the Mississippi River with no tributaries smaller than the Ohio and Missouri rivers. A more detailed map would show many smaller tributaries. An even more detailed map would show still more. The number of tributaries depends on the scale of the map, but the general pattern looks similar at all scales (▶Figure 1-6). Patterns apparent on a small scale quite commonly resemble patterns that exist on much larger scales that we cannot so easily perceive.

Geologic events may also be fractal. They are most numerous on a small scale and become less common at larger

scales (Sidebar 1-1). We see great numbers of small events, many fewer large events, and only a rare giant event. Grains of sand falling on a pile form a slope at their angle of repose, sometimes slightly steeper, sometimes slightly gentler. We don't know whether adding another grain will cause an avalanche and, if so, how big it will be (▶Figure 1-7). It is often rewarding to study the small events because they may well be scale models of their infrequent larger counterparts that may occur in future. They also provide insight into huge events that occurred in the distant past but are larger than any seen in historic time; we may find evidence of these big events if we search.

Hazard and Risk

A **hazard** exists where a natural event is likely to harm people or property. A natural event in a thinly populated area can hardly pose a major hazard. For example, the Borah Peak earthquake that struck central Idaho in 1983 was severe but posed little hazard because it happened in a region with few people or buildings. The eruption of St. Helens in 1980 caused few fatalities and remarkably little property damage simply because few people lived in the area surrounding the mountain. On the other hand, a similar eruption of Vesuvius, on the outskirts of Naples, Italy, could kill or injure hundreds of thousands of people and cause property damage beyond reckoning.

Risk is essentially a hazard considered in the light of its recurrence interval and expected costs. The greater the haz-

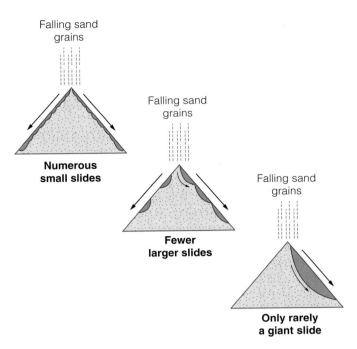

FIGURE 1-7. Sand grains falling on a pile produce numerous small avalanches, fewer larger avalanches, and only rarely a giant avalanche.

Sidebar 1-2

Insurance costs are *actuarial:* They are based on past experience. For insurance, a "hazard" is a condition that increases the severity or frequency of a loss.

Risk α [probability of occurrence] × [cost of the probable loss from the event]

(Recall that α means "proportional to.")

ard and the shorter its recurrence interval, the greater the risk (Sidebar 1-2). The concept is most widely used in the insurance industry in establishing premium rates for policies. People buy property **insurance** to shield themselves from a major loss that they cannot afford. In most cases, the company can estimate the cost of a hazard event to a useful degree of accuracy, but its recurrence interval is hardly better than an inspired guess. The history of experience with a given natural hazard in any area of North America is typically less than 200 years. Large events come around, on average, only every few decades or every few hundred years, or even more rarely. Estimating risk for these events becomes a perilous exercise likely to lose the company large amounts of money. In some cases, most notably floods, the hazard and its recurrence interval are both firmly enough established to support a rational estimate of risk. But the amount of risk, the potential cost to the company, can be so large that a catastrophic event would put the company out of business. Such a case explains why private insurance companies are not eager to offer disaster policies.

Consider the case of Tokyo. This enormous city is subject to devastating earthquakes that for more than 500 years came at intervals of close to seventy years. The risk steadily increased as the population grew during those years and the strain across the fault zone grew. This is a situation in which both hazard and recurrence intervals were well enough known to make it possible to assess risk with some degree of confidence. The last major earthquake ravaged Tokyo in 1923, so everyone involved awaited 1993 with considerable consternation. More than ten years later, no large earthquake has occurred. However, earthquake risk is not purely random; as the time since the last earthquake increases, so does the risk. Obviously, the recurrence interval does not predict events at equal intervals, in spite of the 500-year Japanese historical record.

The uncertainties of estimating risk make it impossible for private insurance companies to offer affordable policies to protect against many kinds of natural disasters. As a result, insurance is generally available for events that present relatively little risk, mainly those with more or less dependably long recurrence intervals. The difficulty of obtaining policies from private insurers for certain types of natural hazards has inspired a variety of governmental programs. Earthquake insurance is available in areas such as Texas where the likelihood of an earthquake is low. California, where the risks and expected costs are much higher, required insurance companies to provide earthquake coverage if they were to offer any type of insurance in the state. As a result, the companies now make insurance available through the California Earthquake Authority, a consortium of companies. Similarly, most hurricane-prone southeastern states have mandated insurance pools that provide property insurance where individual private companies are unwilling to provide such coverage.

Insurance for some natural hazards is simply not available. Landslides, most mudflows, and ground settling or swelling are too risky for companies. The large number of variables makes the risk too uncertain; it is too expensive to estimate the different risks for the relatively small areas involved.

The United States government is involved in many aspects of natural hazards. It conducts and sponsors research into the nature and behavior of many kinds of natural disasters. It attempts to find ways to predict hazardous events and mitigate the damage and loss of life they cause. The governmental programs are split between several agencies.

The U.S. Geological Survey (USGS) is heavily involved in earthquake and volcano research, as well as in studying stream behavior and monitoring stream flow. The National Weather Service monitors rainfall and uses this and the USGS data to try to predict storms and floods.

The Federal Emergency Management Agency (FEMA) was created in 1979, primarily to bring order to the chaos of relief efforts that seemed invariably to emerge after natural disasters. After the hugely destructive midwestern floods of 1993, it has increasingly emphasized hazard reduction. Rather than pay victims to rebuild in their original unsafe locations, such as floodplains, the agency now focuses on relocating them. Passage of the Disaster Mitigation Act in 2000 signals greater emphasis on identifying and assessing risks before natural disasters strike and taking steps to minimize potential losses. The act funds programs for hazard mitigation and disaster relief through FEMA, the U.S. Forest Service, and the Bureau of Land Management.

To determine the level of risk and to estimate potential losses from earthquakes, federal agencies such as FEMA use a computer system called HAZUS (Hazard United States). It integrates a group of interdependent modules that include potential hazard, inventory of the hazard, direct damage, induced damage, direct economic and social losses, and indirect losses. The system may be used as well for hazards other than earthquakes.

Population and Social Pressures

Increasing world populations in urban and coastal settings results in more people occupying land that is subject to major natural events. In effect, people place themselves in the path of unusual but catastrophic events. Economic centers of society are increasingly concentrated in larger urban centers. As a result, those urban centers tend to expand into areas that were previously considered undesirable, including those with greater exposure to natural hazards.

On average, there has been no increase in the frequency of earthquakes or volcanic eruptions. However, there has been a large increase in dollar losses per year, especially in **developed countries.** In **underdeveloped countries,** there are increasing numbers of deaths from natural disasters. In both developed and underdeveloped countries, population growth fosters the increases, with more people living in less suitable and more dangerous areas.

Catastrophic events are natural and expected, but the most common human reaction to a current or potential catastrophe is to modify the natural system in an attempt to control nature. The **mitigation** effort typically seeks to avoid or eliminate the hazard through engineering. Such efforts require financing, either from the government or by individuals or groups that are likely to be affected. Less commonly but more appropriately, mitigation requires changes in human behavior. Change is commonly much less expensive and more permanent than the necessary engineering work. We cannot change the behavior of the natural system because we cannot change natural laws. Individually and as a society, we need to learn to live with nature, not try to control it. In recent years, governmental agencies have begun to learn this lesson, generally through their own mistakes. In a few places along the Mississippi and Sacramento

rivers, for example, levees are being moved back away from the river to permit the river to spread out on its floodplain during future floods.

If, through lack of forethought, you find yourself in a hazardous location, *what can you do about it?* You might build a river levee to protect your land. Or you might build a rock wall into the ocean-side surf to stop sand from leaving your beach and undercutting the hill on which your house is built.

If you do any of these things, however, *you merely transfer the problem elsewhere, to someone else, or to a later point in time.* For example, if you build a levee to prevent a river from spreading over a floodplain and damaging your property, the flood level past the levee will be higher than it would have been without the levee. Constricting river flow with the levee will also back up the floodwater, potentially causing flooding of your upstream neighbor's property. Deeper water also flows faster past your levee, so it may cause more erosion of your downstream neighbor's riverbanks. As in the stock market, individual stocks go up and down. If you make money because you bought a stock when its price was low and sold it when its price was high, then you effectively bought it from someone else who lost money. In the stock market, over the short term, the best we can do, from a selfish point of view, is to shift disasters to our neighbors. The same is true in tampering with nature.

The scale of some natural catastrophes that have affected the Earth, and will do so again, is almost too large to fathom. Examples include catastrophic failure of the flanks of oceanic volcanoes or the impact of large asteroids. For these, reality is more awesome than fiction. Yet each is so well documented in the geological record that we need to be aware of the potential for such future extreme events.

Most people do not realize the inherent danger of an unusual occurrence, or they believe that they will not be affected in their lifetime because such events occur infrequently. That inference incorrectly assumes that the probability of another severe event is lower for a considerable length of time after a major event.

Developers, companies, and even governments often aggravate hazards by allowing or even encouraging people to move into hazardous areas. Many developers and private individuals view restrictive zoning as infringements on their rights to do as they wish with their land. Developers, real estate agents, and some companies are reluctant to admit the hazards that may affect a property for fear of lessening its value and scaring off potential clients. Many local governments consider news of hazards bad for growth and business. They shun restrictive zoning or minimize the possible dangers for fear of inhibiting improvements in their tax base. As in many other venues, different groups have different objectives (▶ Figure 1-8). Some are most concerned with economics, others with safety, still others with the environment.

Benjamin Franklin, back in 1776, noted that "an ounce of prevention is worth a pound of cure." And so it is with natural hazards. Unfortunately, this is easier said than done.

▶**FIGURE 1-8.** Some developers seem unconcerned with the hazards that may affect the property they sell. High spring runoff floods this proposed development site in Missoula, Montana.

It is hard to persuade people to expend time and money to prevent a disaster that does not appear to be imminent and, in their minds, might never happen.

In 1997, FEMA's associate director for mitigation argued that policies concerning hazards in the United States are driven by politics and capitalism. In spite of all that has been learned about specific hazards, those who are interested in mitigating the effects of hazards must accept and use this reality. Policy makers must understand that mitigating hazards is wise both politically and economically. They need to be convinced, for example, that an improved quality of life or new recreational opportunities will lead to a stronger tax base. Similarly, corporations are driven by sales, profits, and return on equity. Although mitigation programs cost money, if companies can be convinced that safety standards will save money overall, then they will adopt them. They must understand that the costs of a disaster include not only physical damages, some of which may be uninsured, but also interruption of business activity, loss of market, and emotional stress on employees who will be less productive for a protracted period. Many of those employees will also have personal losses.

The Role of Public Education

Much is now known about natural hazards and the negative impacts they have on people and their property. It would seem obvious that any logical person would avoid such potential impacts or at least modify their behavior or their property to minimize such impacts. However, most people are not knowledgeable of potential hazards, and human nature is not always rational. Until someone has a personal experience or knows someone who has such an experience,

most people subconsciously believe "It won't happen here" or "It won't happen to me." Even knowledgeable scientists who are aware of the hazards, the odds of their occurrence, and the costs of an event do not always act appropriately. Compounding the problem is the lack of tools to reliably predict the specific location and timing of many natural hazards.

Unfortunately, a person who has not been adversely affected in a major way is much less likely to take specific steps to reduce the effects of a potential hazard. Migration of the population toward the Gulf and Atlantic coasts accelerated in the last half of the twentieth century and still continues. Most of those people, including developers and builders, are not familiar with coastal storms. Even where a hazard is apparent, people are slow to respond. Is it likely to happen? Will I have a major loss? Can I do anything to reduce the loss? How much time will it take and how much will it cost? Who else has experienced such a hazard? People are motivated to act if they think it is their own idea.

Several federal agencies have programs to foster public awareness and education. The Emergency Management Institute—in cooperation with FEMA, NOAA, USGS, and other agencies—provides courses and workshops to educate the public and governmental officials. Some state emergency management agencies, in partnership with FEMA and other federal agencies, provide workshops, reports, and informational materials on specific natural hazards.

Given the hesitation of many local governments to publicize natural hazards in their jurisdictions, people need to educate themselves. Being aware of the types of hazards in certain regions allows people to find evidence for their past occurrence. It also prepares them to seek relevant literature and ask the appropriate questions of knowledgeable authorities.

Some people are receptive to making changes in the face of potential hazards. Some are not. The distinction depends partly on several factors: knowledge, experience, and whether they feel vulnerable. A middle-age person whose house was badly damaged in the 1989 Loma Prieta earthquake is likely to either move to a less earthquake-prone area or to live in a house that is well-braced for earthquake resistance. A similar person losing his home to a landslide is more likely to live away from a steep slope. The best window of opportunity for effective reduction of hazard is immediately following a disaster of the same type. Studies show that the window of opportunity is short—generally, not more than two or three months.

Successful public education programs such as some of those on earthquake hazards in parts of California and presented by the U.S. Geological Survey have shown that information must come from various credible sources and be presented in nontechnical terms that spell out specific steps that people can take. Broadcast messages can be helpful, but they should be accompanied by written material that people can refer to. Discussion among groups of people that would be affected can help them understand the hazard and to act on this information. If people think the risk is plausible, they tend to seek additional reliable information to validate what they have heard. And the range of additional sources must be those that different groups of people trust. Some people will believe other scientists; others will believe only structural engineers. Some will seek out information online. Successful education programs must include specialists and should adapt the material to the different interests of specific groups such as homeowners, renters, and corporations. Overall, natural hazard education depends on tailoring the message clearly to different audiences using nontechnical language. It must convey not only the nature of potential events but also show that certain relatively simple and inexpensive actions can substantially reduce potential losses.

Unfortunately, politics also enters the equation. Disaster assistance continues to be provided without a large cost-sharing component from states and local organizations. Thus, local governments continue to lobby Congress for funds to pay for losses but lack incentive to do much about the causes. FEMA is charged with rendering assistance following disasters; it continues to provide funds for victims of earthquakes, floods, hurricanes, and other hazards.

It remains reactive to disasters, as it should be, but is only beginning to be proactive in eliminating the causes of future disasters. Congress continues to fund multimillion-dollar Army Corps of Engineers projects to build levees along rivers and replenish sand on beaches. The Small Business Administration disaster loan program continues to subsidize credit to finance rebuilding in hazardous locations. The federal tax code also subsidizes building in both safe and hazardous sites. Real estate developers benefit from tax deductions, and ownership costs such as mortgage interest and taxes can be deducted from income taxes. A part of uninsured "casualty losses" can still be deducted from the disaster victim's income taxes. Such policies encourage future damages from natural hazards.

Natural hazards and disasters can be fascinating and even exciting for those who study them. Just don't be on the receiving end! The following chapters address the main natural hazards: earthquakes and volcanic eruptions; extremes of weather, including hurricanes; floods, landslides, tsunamis, wildfires, and asteroid impacts. For each, we examine the nature of the hazard, the factors that influence it, the dangers associated with the hazard, and methods of forecasting or predicting such events. Examples are used to clarify and illustrate the broad discussions. Background chapters or appendixes are provided on plate tectonics, characteristics of common rocks, and the geological time scale.

This book is organized around individual types of hazards, with chapters that emphasize understanding of each hazard, its characteristics, the conditions that promote it, the environments in which we find it, its magnitudes and probability of occurrence, and its effects. Examples are mentioned in the context of these discussions. Where more extensive discussions can be used to illustrate important points, we have included "Case in Point" examples. These cases integrate such aspects in more detail, emphasize both dominant and secondary underlying causes, and how different and often unrelated influences affect the result. Most examples are provided from North America because of their familiarity to more of our readers. Additional examples are from Italy, another modern industrialized country where the large population density leads to greater public impact and losses. Does Italy provide a glimpse of the future for parts of North America as populations grow near hazardous areas?

KEY POINTS

✓ Humans need to learn to live with some natural events rather than trying to control them. **Review p. 1.**

✓ Many natural processes that we see are slow and gradual, but occasional sudden or dramatic events can be hazardous to humans. **Review pp. 1–2.**

✓ There are numerous small events, fewer larger events, and only rarely a giant event. We are familiar with the common small events but do not expect the giant events that can create major catastrophes because they come along so infrequently. **Review pp. 2, 6–7; Sidebar 1-1.**

✓ A large event becomes a disaster or catastrophe only when it affects people or their property. Large natural events have always occurred but did not become disasters until people placed themselves in harm's way. **Review pp. 2–3, 6.**

✓ Developed countries lose large amounts of money in a disaster; underdeveloped countries lose large numbers of lives. **Review pp. 2, 8.**

✓ We tend to blame others for disasters, but we tend to place ourselves in harm's way. **Review pp. 2–4.**

✓ Statistical predictions or recurrence intervals are average expectations based on past experience. **Review pp. 4–5; Figure 1-3.**

✓ Events are often neither cyclic nor completely random. Overlapping influences of multiple factors can lead to the extraordinarily large events that often become disasters. **Review pp. 5–6; Figure 1-4.**

✓ Different types of natural hazards often interact or influence one another. **Review pp. 5–6; Figure 1-5.**

✓ Many natural features and processes are fractal— that is, they have similarities across a broad range of sizes. **Review p. 6; Figures 1-6 and 1-7.**

✓ Greater risk is proportional to the probability of occurrence and the cost from such an occurrence. **Review pp. 6–7; Sidebar 1-2.**

✓ Mitigation involves efforts to avoid disasters rather than merely dealing with the resulting damages. **Review pp. 8–9.**

✓ Erecting a barrier to some hazard will typically transfer the hazard to another location or to a later point in time. **Review p. 8.**

✓ Most people believe that a disaster will not happen to them. They need to be educated about natural processes and how to learn to live with and avoid the hazards around us. **Review pp. 8–10.**

IMPORTANT WORDS AND CONCEPTS

Terms

average, p. 4
catastrophe, p. 2
cause, p. 5
chance, p. 4
cyclic events, p. 5
developed countries, p. 8
disaster, p. 2
fractal, p. 6
hazard, p. 2
insurance, p. 7
mitigation, p. 8
overlap, p. 6
plate tectonics, p. 5
risk, p. 6
underdeveloped countries, p. 8

QUESTIONS FOR REVIEW

1. Is the geological landscape controlled by gradual and unrelenting processes or intermittent large events with little action in between? Provide an example to illustrate.

2. Why do people live in geologically dangerous areas?

3. Contrast the general nature of catastrophic losses in developed countries versus poor countries.

4. If people should not live in especially dangerous areas, what beneficial use is there for those areas?

5. When people or governmental agencies try to restrict or control the activities of nature, what is the *general* result?

6. Some natural disasters happen when the equilibrium of a system is disrupted. What are some examples?

7. Why are most natural events not perfectly cyclic?

8. Give an example of a fractal system.

9. When an insurance company decides on the cost of an insurance policy for a natural hazard, what are the two main deciding factors?

10. What are the main reasons for the ever-increasing costs of catastrophic events?

FURTHER READINGS

Assess your understanding of this chapter's topics with additional quizzing and conceptual-based problems at:

 http://earthscience.brookscole.com/hyndman.

CHAPTER 2

Jennifer Tidwell photo.

The Himalaya Mountains are the highest in the world because of the immense plate tectonic forces that push India northward into Eurasia.

PLATE TECTONICS AND PHYSICAL HAZARDS

Development of the Theory

Back in the 1960s, the main debate about tectonics revolved around a concept called **continental drift.** In 1912, Alfred Wegener proposed that North and South America had separated from Europe and Africa, widening the Atlantic Ocean in the process. He suggested that Earth's spin caused the continents to drift through the oceanic crust, forming mountains along their leading edges. To some, the evidence for continental drift seemed compelling. There is a striking similarity in the shapes of the continents along the two sides of the Atlantic Ocean, as viewed on a map or globe (▶ Figure 2-1), suggesting they had once been together. The match is especially good if we use the real edge of the continents, including the shallowly submerged continental shelves. Continued work showed that ancient rocks, their fossils, and their mountain ranges also matched well across the Atlantic Ocean. This analysis is similar to what you would use when you put a jigsaw puzzle together; the pieces fit and the patterns match across the reconnected pieces.

Other lines of evidence supported the continental drift hypothesis. Exposed surfaces of ancient rocks in the southern parts of Australia, South America, India, and Africa showed grooves carved by immense areas of continental glaciers. The grooves showed that glaciers with embedded rocks at their base might have moved from Antarctica into India, eastern South America, and Australia (▶ Figure 2-2). The rocks had clearly been buried under glacial ice, yet many of these areas now have warm to tropical climates. In addition, the remains of fossils that formed in warm climates such as coal with fossil impressions of tropical leaves, the distinctive fossil fern *Glossopteris*, and coral reefs were found in areas such as Antarctica and present-day arctic areas.

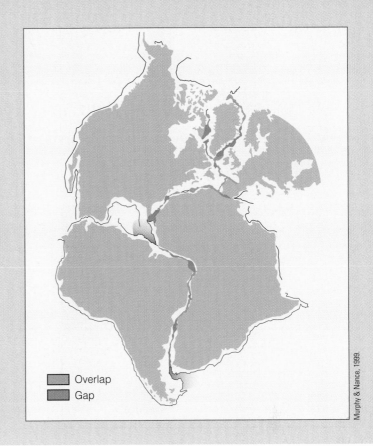

Overlap
Gap

Murphy & Nance, 1999.

▶ **FIGURE 2-1.** Before continental drift a few hundred million years ago, the continents were clustered together as a giant "supercontinent" that has been called Pangaea. The Atlantic Ocean had not yet opened. The pale blue fringes on the continents are continental shelves, which are part of the continents. The areas of overlap and gap (in red and darker blue) are small.

Many geologists and geophysicists disagreed with the theory of continental drift; they demanded a mechanism whereby such continental movement could be accomplished. English geophysicist Harold Jeffreys argued that the ocean floor rocks were far too strong to permit the continents to plow through them. As it turns out, Wegener's assumption that the continents plowed through the ocean floor was incorrect; however, many other aspects of Wegener's hypothesis were correct.

A variety of points of view led to an understanding of how the continents were separating. Oceanographers from Woods Hole Oceanographic Institute in Massachusetts, who were measuring ocean depths from all over the Atlantic Ocean in the late 1940s and 1950s, found an immense mountain range or ridge down the center of the ocean and extending for its full length (▶ the seafloor topography is shown in Figure 2-3). Later, scientists recognized that most earthquakes in the Atlantic Ocean were concentrated in that central rift.

In 1960, Harry Hess of Princeton University conjectured that the ocean floors acted as giant conveyor belts carrying the continents. New oceanic crust welled up at the oceanic rifts, spread away, and finally sank into the deep oceanic trenches along the edges of some continents, a process later called **seafloor spreading.** Hess calculated the spreading rate to be approximately 2.5 centimeters (1 inch) per year across the Mid-Atlantic Ridge. If correct, the whole Atlantic Ocean floor would be created in only 180 million years or so.

Confirmation of seafloor spreading finally came in the mid-1960s through work on the magnetic properties of rocks of the ocean floor. We are all aware that the Earth has a **magnetic field** because a magnetized compass needle points toward the north magnetic pole. Slow convection movements in the Earth's molten nickel-iron outer core are believed to generate that magnetic field (▶ Figure 2-4). That magnetic field reverses its north–south orientation every 10,000 to several million years (every 600,000 years on average) because of changes in those currents.

Additional confirmation of a key aspect of spreading ocean floors came from drilling by the research vessel *Glomar Challenger* (▶ Figure 2-5). The ages of basalts and sediments dredged and drilled from the ocean floor showed that those near the Mid-Atlantic Ridge were young and had only a thin coating of sediment, both results in contradiction to the prevailing notion that the ocean floor was extremely old. The ridge had a prominent valley or rift down its center. In contrast, rocks from deep parts of the ocean floor far from the ridge were consistently much older.

The ocean floor consists of basalt that erupted at the mid-ocean ridge and solidified from molten magma. Iron atoms crystallizing in the magma orient themselves like tiny compass needles, pointing toward the north magnetic pole so the rock is slightly magnetized with an orientation like the compass needle. When the magnetic field reverses, that reversed magnetism is frozen into the magmatic rocks solidified at the time (▶ Figure 2-6).

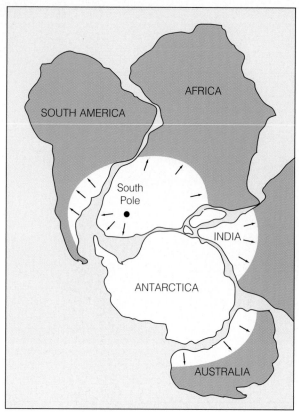

(a)

▶FIGURE 2-2. **(a)** Continental masses of the southern hemisphere appear to have been parts of a supercontinent 300 million years ago in which a continental ice sheet centered on Antarctica spread outward to cover adjacent parts of South America, Africa, India, and Australia. **(b)** After separation, the continents migrated to their current positions. **(c)** The inset photo shows glacial grooves like those found in those continents.

(c)

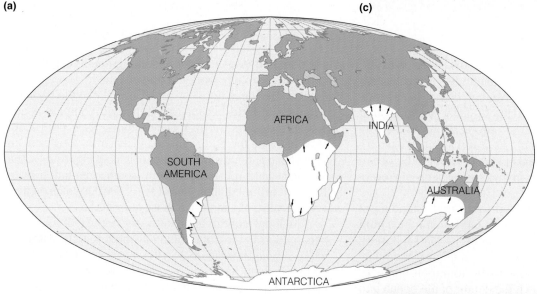

(b)

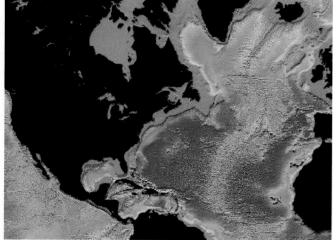

▶FIGURE 2-3. In this seafloor topographical map for the world oceans, shallow depths at the oceanic ridges are shown in orange to yellow, deeper water off the ridge crests are in green, and deep ocean is blue. Shallow continental shelves are shown in red.

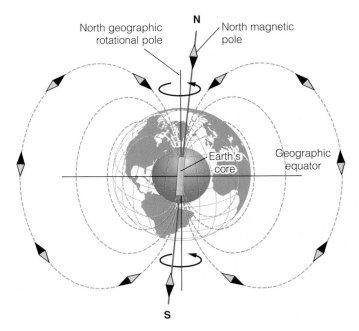

FIGURE 2-4. Earth's magnetic field is shaped as if there were a huge bar magnet in the Earth's core. Instead of a magnet, Earth's rotation is thought to cause currents in the liquid outer core. Those currents cause a magnetic field similar to the way in which power plants generate electricity when steam or falling water rotates an electrical conductor in a magnetic field.

FIGURE 2-5. The *Glomar Challenger* did the early drilling and dredging of the ocean floor that contributed to plate tectonic theory.

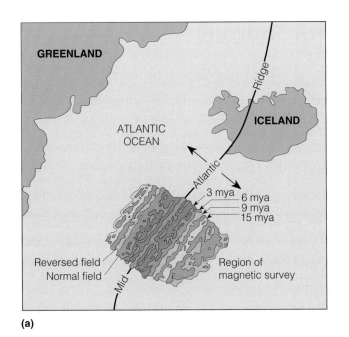

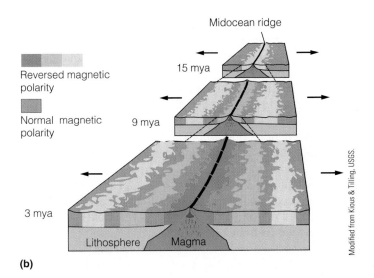

FIGURE 2-6. (a) The magnetic polarity, or orientation, across the Mid-Atlantic ridge shows a symmetrical pattern as shown in this regional survey near Iceland (a similar nature of stripes exists all along the spreading center). Basalt lava erupting today records the current northward-oriented magnetism; basalt lavas that erupted 3 million years ago (mya) recorded the reversed, southward-oriented magnetic field at that time. The south-pointed magnetism in those rocks is largely cancelled out by the present-day north-pointing magnetic field so the ocean floor shows alternating strong (north-pointing) and weak (south-pointing) magnetism in the rocks. **(b)** This schematic diagram shows how the solidified basalt moves away from the spreading center, carrying with it the normal (northward) or reversed (southward) magnetism.

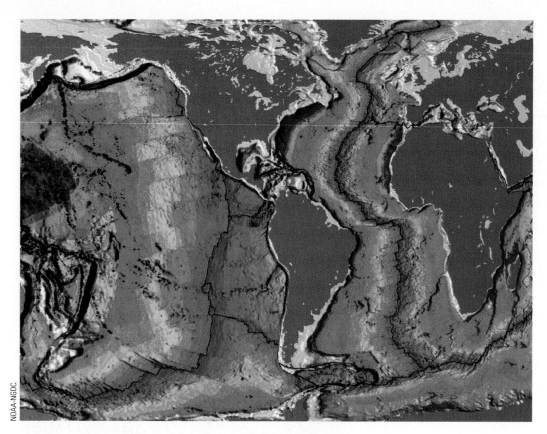

NOAA-NGDC.

▶**FIGURE 2-7.** Ocean floor ages are determined by their magnetic patterns. Red colors at the oceanic spreading ridges grade to yellow at 48 million years ago, to green 68 million years ago, and to dark blue some 155 million years ago.

British oceanographers Frederick Vine and Drummond Matthews, studying the magnetic properties of ocean-floor rocks in the early 1960s, showed a striped pattern parallel to the oceanic ridge. Some of the stripes were strongly magnetic; adjacent stripes were weakly magnetic. They realized that the magnetism is stronger where the rocks solidified while Earth's magnetism was oriented parallel to the present-day north magnetic pole. Where the rock magnetism was pointing toward the south magnetic pole, the recorded magnetism was weak—it was partly cancelled by the present-day magnetic field. The Earth's magnetic field imposed a pattern of **magnetic stripes** as the basalt solidified at the ridge, because the magnetic field reversed from time to time. As the ridge spread apart, ocean floor formed under alternating periods of north- versus south-oriented magnetism to create the matching striped pattern on opposite sides of the ridge (Figure 2-6).

The continents are actually part of continent–ocean plates that are moving in unison. Those plates pull apart at the Mid-Atlantic Ridge to make the ocean floor wider. In the Pacific Ocean, the plates pull apart at the East Pacific Rise; their oldest edges (some 180 million years old) sink in the deep ocean **trenches** near the western Pacific continental margins as discussed below (▶Figures 2-7 and 2-8).

As the lithospheric plates drifted over the asthenosphere, they collected to form one giant supercontinent, called **Pangaea,** 225 million years ago (▶Figure 2-9). The con-vection currents that drive plate motion continued to move deep in the asthenosphere, so the continents again began to drift apart. Pangaea began to break up, and the plates slowly moved the continents into their current positions.

Earth Structure

We now know that the Earth has a stiff outer rind that is 60 to 200 kilometers (47 to 124 miles) thick called the **lithosphere.** It is colder and therefore more rigid than the easily deformed **asthenosphere** that lies beneath it (Figures 2-8 and 2-10). Continental lithosphere includes silica-rich crust from the surface to an average depth of 30 to 50 kilometers (see Appendix 2 for detailed rock compositions). Oceanic lithosphere, some 60 kilometers thick, includes a 7-kilometer thick crust. Continental crust is largely composed of high silica-content minerals, which give it the lightest density (2.7 g/cm^3) of the major regions on Earth (▶Figure 2-10). (See Appendix 1 for detailed descriptions of rocks.) Oceanic crust has a higher density (3 g/cm^3) because it contains more iron- and magnesium-rich minerals.

As shown in the inset to Figure 2-10, the low-density continental crust is thicker and stands higher than the higher-density oceanic crust. The concept of **isostacy** or buoyancy explains this elevation difference. Although the Earth's mantle is not liquid, its high temperature (above 450°C or 810°F)

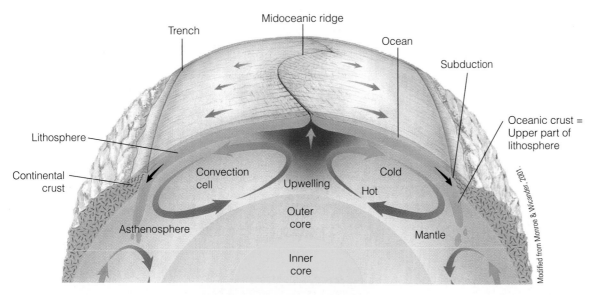

FIGURE 2-8. A generalized cross-section through the Earth shows its main concentric layers. The more rigid lithosphere moves slowly over the less rigid asthenosphere, which is thought to circulate slowly by convection. The lithosphere pulls apart at ridges and sinks at trenches.

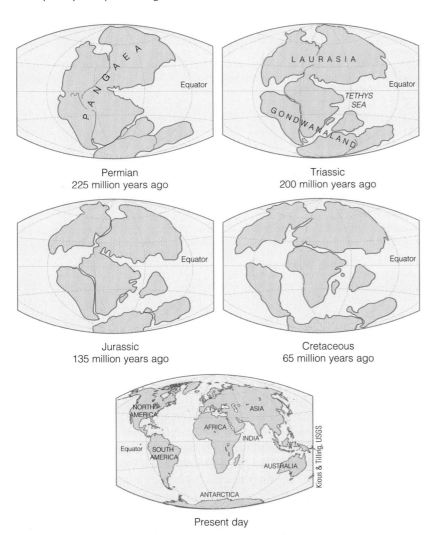

Permian
225 million years ago

Triassic
200 million years ago

Jurassic
135 million years ago

Cretaceous
65 million years ago

Present day

FIGURE 2-9. The supercontinent Pangaea broke up into individual continents starting approximately 225 million years ago.

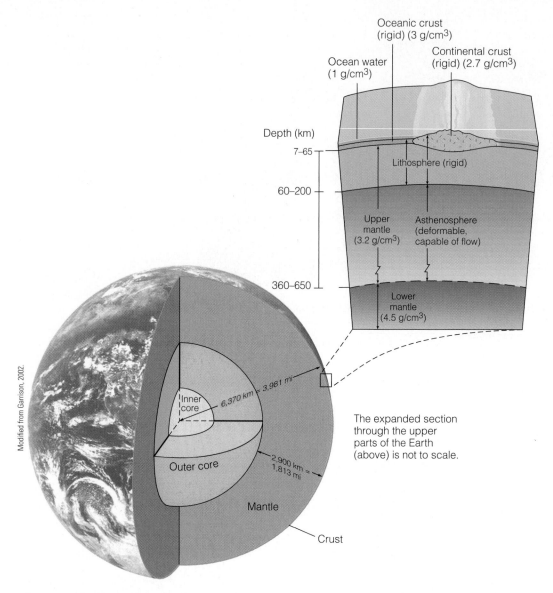

Ocean water
(1 g/cm³)

Oceanic crust
(rigid) (3 g/cm³)

Continental crust
(rigid) (2.7 g/cm³)

Depth (km)

7–65

60–200

360–650

Lithosphere (rigid)

Upper
mantle
(3.2 g/cm³)

Asthenosphere
(deformable,
capable of flow)

Lower
mantle
(4.5 g/cm³)

The expanded section
through the upper
parts of the Earth
(above) is not to scale.

Modified from Garrison, 2002.

Inner
core

6,370 km = 3,981 mi

Outer core

2,900 km =
1,813 mi

Mantle

Crust

▶**FIGURE 2-10.** A slice into the Earth shows a solid inner core and a liquid outer core, both composed of nickel-iron. Peridotite of the Earth's mantle makes up most of the volume of the Earth. The Earth's crust, on which we live, is as thin as a line at this scale.

permits it to flow slowly as if it were a viscous liquid. A floating solid object will displace a liquid of the same mass. As a result, the percentage of a material immersed in the liquid can be calculated from the density of the floating solid divided by the density of the liquid. For example, when water freezes, it expands; it thus has a lower density (ice density is 0.90 g/cm³ relative to liquid water at 1.0 g/cm³). Thus, 92 percent of an ice cube or iceberg will be underwater; similarly, approximately 84 percent of a mountain range of continental rocks (2.7 g/cm³) will submerge into the mantle (3.2 g/cm³) as a deep mountain root (▶ Figure 2-11). If the

weight of an extremely large glacier is added to a continent, the weight will do the same thing: The crust and upper mantle will slowly sink into the deeper mantle.

A map of Earth's topography clearly shows that the continents stand high relative to the ocean basins because of isostacy. Thin lithosphere of the ocean basins stands low; the continents with their thick lithosphere sink deep into the asthenosphere and float high. The thickest parts of the continental lithosphere sink deepest and stand highest as major mountain ranges such as the Rockies, Alps, and Himalayas (▶ Figures 2-12 and 2-13).

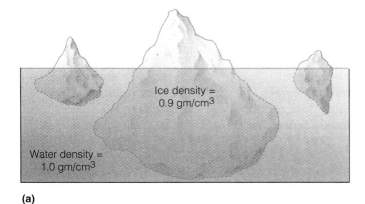

Ice density =
0.9 gm/cm³

Water density =
1.0 gm/cm³

(a)

Ocean crust
(3.0 g/cm³)

Continental crust
(2.7 g/cm³)

Mountain

Mountain root

Mantle
(3.2 g/cm³)

Thompson & Turk, 1998.

(b)

▶**FIGURE 2-11.** **(a)** An iceberg sinks 90 percent of its mass into water because the ice (90 percent of the density of water) displaces an equivalent mass of water (1.0 g/cm³). **(b)** Similarly, the load of a thick mass of a continental mountain range sinks the continental crust (2.7 g/cm³) into the underlying denser mantle (3.2 g/cm³) to provide a "mountain root."

We do not have direct observations of crustal thickness, thus scientists measure the gravitational attraction of the Earth (greater over denser rocks) and analyze seismic waves as they propagate away from the locations of earthquakes to provide indirect evidence of the changing density and velocity of subsurface materials. The boundary between the Earth's crust and mantle has been identified as a difference in density that we call the **Mohorovičic Discontinuity,** or Moho. Deeper in the mantle, the next major change in material properties occurs at the boundary between the strong and rigid lithosphere and the weak and deformable asthenosphere. This was identified as a zone of low velocity from the analysis of seismic waves; we call it the **low-velocity zone** (Figure 2-13).

The lithosphere is not continuous like the rind on a melon. It is broken into about a dozen or so large **plates** and another dozen or so much smaller plates. Even though

NOAA.

▶**FIGURE 2-12.** This shaded relief map shows the continents standing high. Mountain ranges in red tones concentrate at converging margins. Light blue ridges in oceans are generally spreading centers, and trenches are visible at Pacific continental margins where oceanic crust is subducting below continental crust.

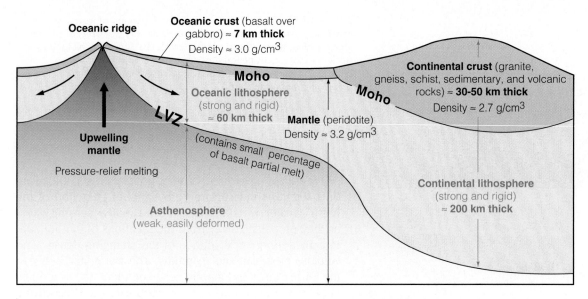

▶**FIGURE 2-13.** Earth's lithosphere and asthenosphere are distinguished by their strength: strong or weak. Earth's crust and mantle are distinguished on the basis of composition: basalt or peridotite. These rock types are described in Appendix 2. Moho = Mohoroviçic Discontinuity = boundary between crust and mantle. LVZ = low-velocity zone, or region of low seismic velocity because it contains a small percentage of partial melt.

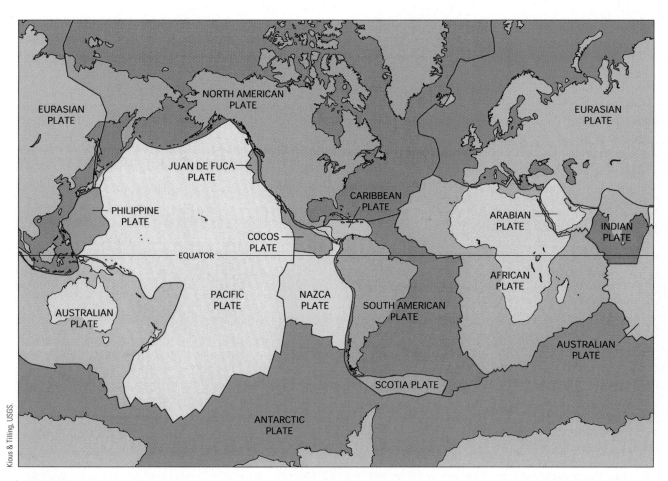

▶**FIGURE 2-14.** Most large lithospheric plates consist of both continental and oceanic areas. Although the Pacific Plate is largely oceanic, even it contains a part of the South Island of New Zealand.

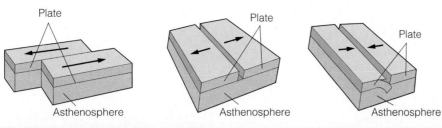

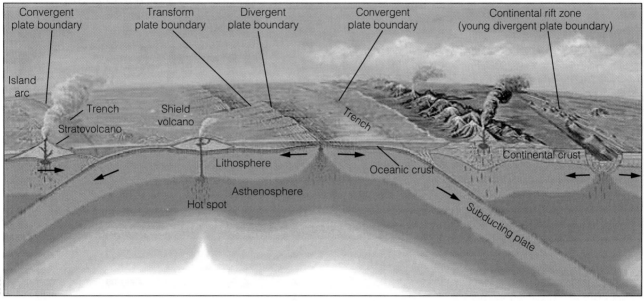

▶**FIGURE 2-15.** This three-dimensional view shows the different types of lithospheric plate boundaries: convergent, divergent, and transform.

they are uneven in size and irregular in shape, the plates fit neatly together in a mosaic that covers the entire surface of the Earth. The plates do not correspond to continent versus ocean areas; most plates consist of a combination of the two (▶Figure 2-14). Even the Pacific Plate, which is mostly ocean, includes a narrow slice of western California and part of New Zealand.

As part of the overall unifying concept, the current theory of **plate tectonics,** the Earth's plates move at an average rate of up to 11 centimeters (4.2 inches) per year. Many move in a more-or-less east–west direction, but not everywhere. Some separate, others slide under or over or past one another, and still others collide (▶Figure 2-15). In some cases, their encounters are head on; in others, the collisions are more or less oblique. Plates move away from each other at divergent boundaries, most commonly at mid-ocean spreading centers. Plates move toward each other at convergent boundaries, which results in subduction zones in cases where one or both of the plates are oceanic lithosphere. When two continental plates collide, neither side is dense enough to be subducted into the asthenosphere so the two sides may often crumple into a thick mass of low-density continental material. This type of convergent

margin is where the largest mountain ranges on Earth such as the Himalayas are built. The remaining category of plate interactions is where two plates slide past each other at a **transform fault** such as the San Andreas Fault.

Most of Earth's earthquake and volcanic activity occurs along or near moving plate boundaries (▶Figures 2-15, 2-16, and 2-17). Most of the collision boundaries between oceanic and continental plates form subduction zones along the Pacific coasts of North and South America, Asia, Indonesia, and New Zealand. Collisions between continents are best expressed in the high mountain belts extending across southern Europe and Asia (Figure 2-12). Most rapidly spreading boundaries follow oceanic ridges. In some cases, slowly spreading boundaries rift continents apart.

In some places, different types of plate boundaries join at **triple junctions.** For example, the Cascadia subduction zone off the Washington and Oregon coast joins both the San Andreas transform fault and the Mendocino transform fault at the Mendocino triple junction just off the northern California coast. The north end of the same subduction zone joins both the Juan de Fuca spreading ridge and the Queen Charlotte transform fault at a triple junction just off the north end of Vancouver Island (▶Figure 2-18).

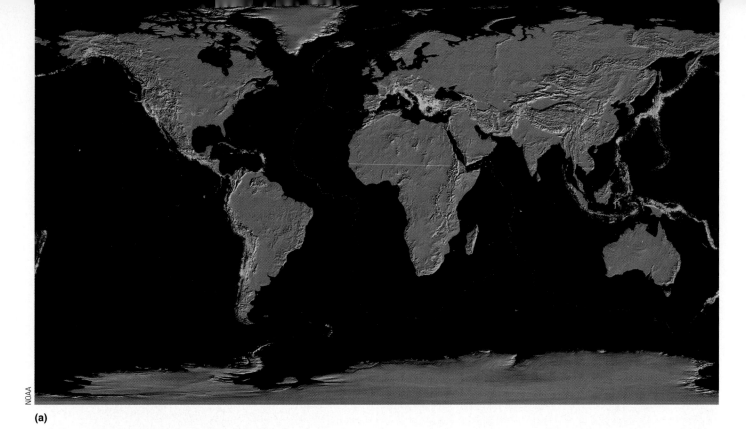

NOAA

(a)

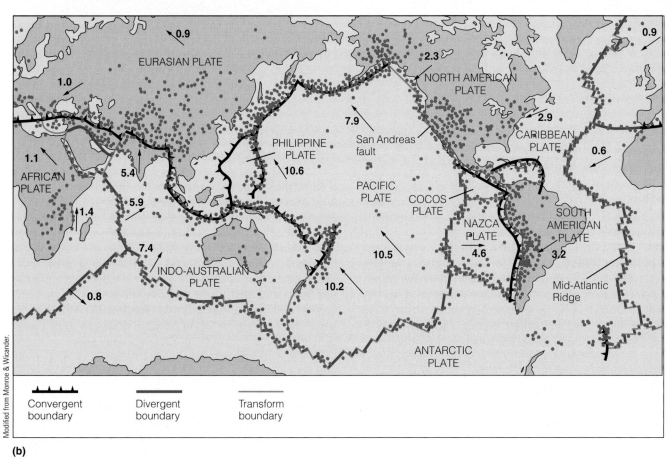

Modified from Monroe & Wicander.

EURASIAN PLATE

0.9

1.0

2.3

NORTH AMERICAN PLATE

0.9

1.1

AFRICAN PLATE

5.4

PHILIPPINE PLATE

7.9

San Andreas fault

2.9

CARIBBEAN PLATE

0.6

5.9

10.6

PACIFIC PLATE

COCOS PLATE

1.4

7.4

INDO-AUSTRALIAN PLATE

NAZCA PLATE

SOUTH AMERICAN PLATE

3.2

10.5

4.6

0.8

10.2

Mid-Atlantic Ridge

ANTARCTIC PLATE

▲┴┴┴ Convergent boundary

━━━ Divergent boundary

━━━ Transform boundary

(b)

▶ **FIGURE 2-16.** **(a)** Most earthquakes are concentrated along major tectonic plate boundaries, especially subduction zones and transform faults, with fewer along spreading ridges. **(b)** General direction and velocities of plate movement (compared with hotspots that are inferred to be anchored in the deep mantle), in centimeters per year, are shown with arrows.

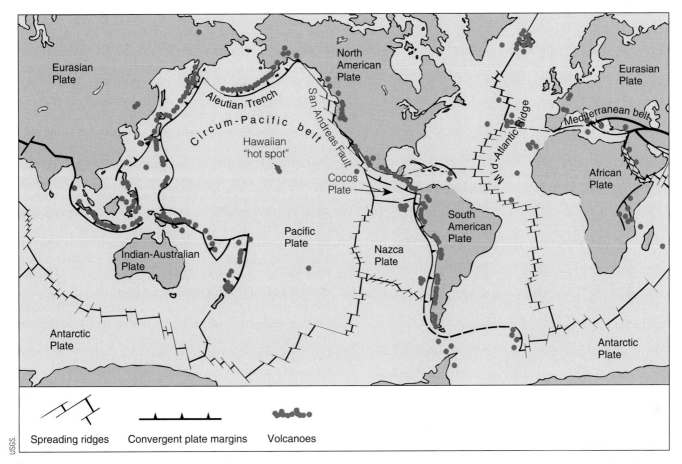

Spreading ridges   Convergent plate margins   Volcanoes

USGS.

▶**FIGURE 2-17.** Most volcanic activity also occurs along plate tectonic boundaries. Eruptions tend to be concentrated along the continental side of subduction zones and along divergent boundaries such as rifts and midocean ridges.

Where Plates Pull Apart (Spreading Zones or Divergent Boundaries)

Boundaries where plates pull apart make a system of more or less connected oceanic ridges that wind through the ocean basins like the seams on a baseball. Iceland is the only place where that ridge system rises above sea level; elsewhere, it is submerged to an average depth of a few thousand meters.

A broad valley doglegs from south to north across Iceland. The hills east of it are on the Eurasian Plate that extends all the way east to the Pacific Ocean; the hills to the west are on the North American Plate, which extends west to the Pacific Ocean. Repeated surveys over a period of decades have shown that the valley is growing wider at a rate of several centimeters per year. The movement is the result of the North American and Eurasian plates pulling away from each other, making the Atlantic Ocean grow wider at this same rate.

Iceland's long recorded history shows that a broad fissure opens in the floor of the central valley of Iceland every 200 to 300 years. It erupts a large basalt lava flow that covers as much as several thousand square kilometers. The last fissure opened about the time of the American Revolution, and another such event could happen almost anytime. Fortunately, the sparse population of the region limits the potential for a great natural disaster.

It now seems clear that similar events happen fairly regularly all along the oceanic ridge system. They are the source of the basalt lava flows that cover the entire ocean floor, roughly two-thirds of the Earth's surface, to an average depth of several kilometers. The hot basalt magma rises to the surface where it comes in contact with water and rapidly cools to form pillow-shaped blobs of lava, each with an outer solid rind that encases still molten magma. As the plate moves away from the spreading center, it cools, shrinks, and thus increases in density. This explains why the hot spreading centers stand high on the subsea topography. New ocean floor continuously moves away from the crests of the oceanic ridges as the oceans grow wider at an average rate of several centimeters every year (▶ Figures 2-16 and 2-19). The frequent earthquakes and volcanic eruptions along the oceanic ridges pose no danger to people or any threat of property damage anywhere except in Iceland, where the oceanic ridge rises above sea level.

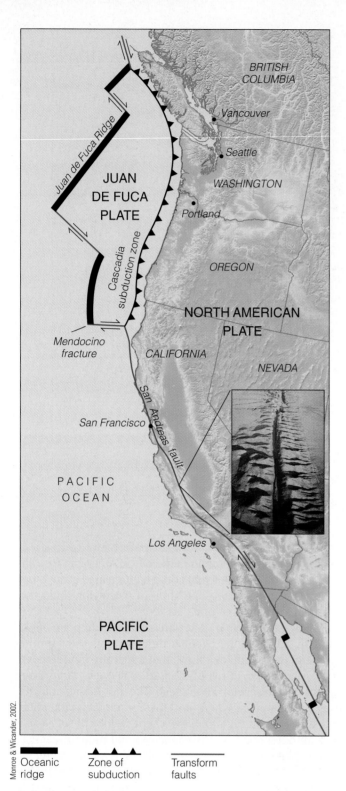

FIGURE 2-18. Three types of plate boundaries are found at the western edge of North America: The San Andreas Fault runs through the western edge of California; the Cascadia subduction zone parallels the coast off Oregon, Washington, and southern British Columbia; and the Juan de Fuca spreading ridge lies farther offshore. Spreading at the Juan de Fuca Ridge carries ocean floor of the Juan de Fuca plate down the Cascadia subduction zone.

Spreading centers in the continents pull apart at much slower rates and do not generally form plate boundaries (▶Figure 2-20). The Rio Grande Rift of New Mexico and the Basin and Range of Nevada and Utah are active North American examples (▶Figure 2-21). The East African Rift zone runs much of the length of that continent (▶Figure 2-22). Fewer, though sometimes large, earthquakes and volcanic eruptions accompany the up and down, "normal fault" movements. Volcanic activity is varied, ranging for example from large rhyolite calderas in the Long Valley Caldera of the Basin and Range of southeastern California and the Valles Caldera of the Rio Grande Rift of New Mexico to small basaltic eruptions at the edges of the **spreading zone.** The Red Sea Rift, at the northeast edge of Africa, is an example where the rift does form a plate boundary (▶Figure 2-23). Continental rifts such as the Rio Grande Rift of New Mexico spread so slowly that they cannot split the continental plate to form new ocean floor.

Most of the magmas that erupt in continental **rift zones** are either perfectly ordinary rhyolite or basalt with little or no intermediate andesite (see pages 27, 31, and Appendix 2). But some of the magmas are peculiar, with high contents of sodium or potassium. Some of the rhyolite ash deposits in the Rio Grande Rift and in the Basin and Range (▶Figure 2-21) provide evidence of extremely large and violent eruptions, giant rhyolite volcano activity. But those events appear to be infrequent and much of the region is sparsely populated, so they do not pose much of a volcanic hazard.

Where Plates Come Together (Convergent Boundaries)

SUBDUCTION ZONES If the Earth generates new oceanic crust at boundaries where plates pull away from each other, then it must destroy old oceanic crust somewhere else (▶Figure 2-15). It swallows old oceanic crust at convergent plate boundaries. If not, our planet would be growing steadily larger at the same rate as new oceanic crust forms. That is clearly not the case.

The idea of two plates colliding is truly horrifying at first thought—the irresistible force finally meets the immovable object. But the Earth solves its dilemma as one plate slides beneath the other and dives into the hot interior of the planet. The plate that sinks is the denser of the two, the one with oceanic crust on its outer surface. It absorbs heat as it sinks into the much hotter rock beneath until it finally loses its identity as cold lithosphere at a depth of several hundred kilometers. That ocean–continent **collision zone** is called a **subduction zone** (▶Figure 2-24).

Where an oceanic plate sinks in a subduction zone, a line of picturesque stratovolcanoes rises inland from the trench (▶Figures 2-25 and 2-26). The process begins at the oceanic spreading ridge where fractures open in the ocean floor. Seawater penetrates into the peridotite of the upper mantle, where the two react to make a greenish rock called

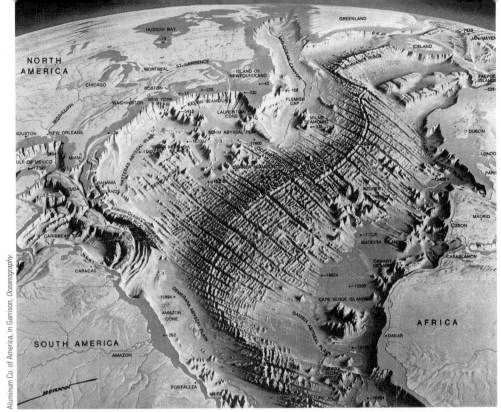

▶**FIGURE 2-19.** The spreading Mid-Atlantic Ridge, fracture zones, and transform faults are dramatically exhibited in this exaggerated topography of the ocean floor.

Aluminum Co. of America, in Garrison, *Oceanography.*

Donald Hyndman photo.

▶**FIGURE 2-20.** This Basin and Range terrain is found southwest of Salt Lake City, Utah.

▶**FIGURE 2-21.** The Basin and Range spreading zone in the western United States is marked, as the name implies, by prominent basins and mountain ranges. Centered in Nevada and western Utah, it gradually decreases in rate of spreading to the north across the Snake River Plain. Its western edge includes the eastern edge of the Sierra Nevada Range in eastern California, and its main western edge is at the Wasatch Front in Utah. An eastern branch includes the Rio Grande Rift of central New Mexico.

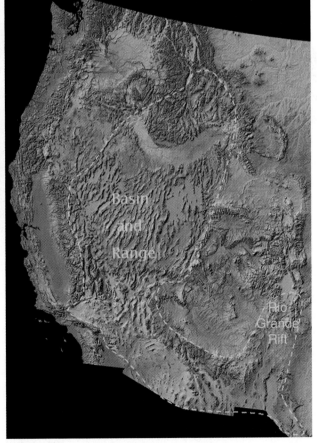

Modified from NOAA.

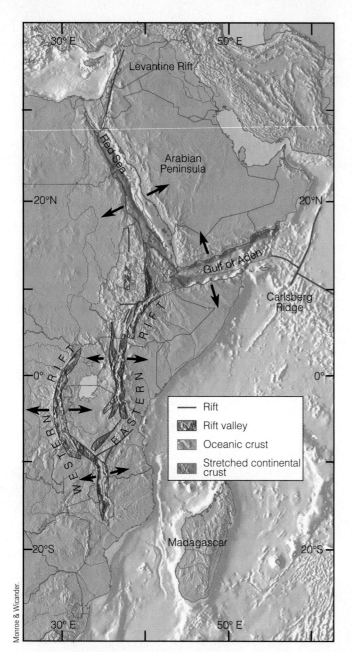

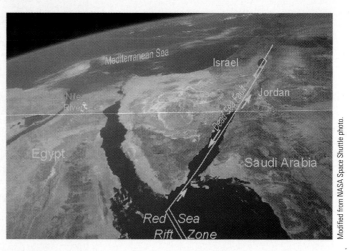

▶**FIGURE 2-23.** The north end of the Red Sea spreading zone connects with the Dead Sea transform fault.

Modified from NASA Space Shuttle photo.

▶**FIGURE 2-22.** The East African Rift Valley spreads the continent apart at rates 100 times as slow as a typical oceanic rift zone. The rift forms one arm of a triple junction, from which the Red Sea and Gulf of Aden form somewhat more rapidly spreading rifts.

Monroe & Wicander.

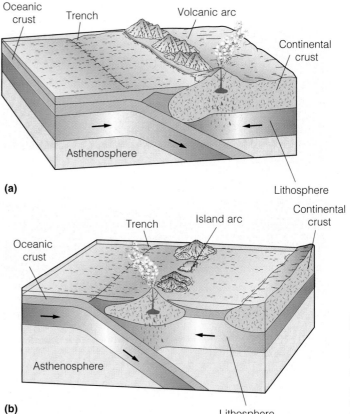

(a)

(b)

Modified from Kious & Tilling, USGS.

▶**FIGURE 2-24.** **(a)** A continental volcanic arc forms on the continent where the oceanic lithosphere descends beneath the continental margin. **(b)** An oceanic island arc forms on oceanic lithosphere where the lithosphere descends beneath more oceanic lithosphere. In both cases, earthquakes are generated in the subduction zone where the overriding lithosphere sticks against the descending lithosphere and then suddenly slips.

Steve Schilling photo, USGS.

▶ **FIGURE 2-25.** Mount St. Helens in foreground and Mount Adams, behind, are two of the picturesque active volcanoes that lie inland from the Cascadia subduction zone.

serpentinite that contains a lot of water. That altered ocean floor eventually sinks through an oceanic trench and descends into the upper mantle, where the serpentinite heats up, breaks down, releases its water, and reverts back to peridotite. The water rises into the overlying mantle, which it partially melts to make basalt magma, that rises. If the basalt passes through continental crust, it can heat and melt some of those rocks to make rhyolite magma. The basalt and rhyolite may erupt separately or mix in any proportion to form andesite and related rocks, the common volcanic rocks in stratovolcanoes. The High Cascades volcanoes in the Pacific Northwest are a good example; they lie inland from an oceanic trench (▶ Figure 2-26).

In some cases, an oceanic plate descends beneath another section of oceanic plate attached to a continent (▶ Figure 2-24b). The same melting process described above generates a line of basalt volcanoes because there is no overlying continental crust to melt and form rhyolite.

Volcanoes above a subducting slab present major hazards to people who live on or near them and to their property. It is hard to reduce the risks associated with these hazards because volcanoes are so scenic and the volcanic rocks break down into rich soils that support and attract large populations. They are prominent all around the Pacific basin and in Italy and Greece, where the African Plate collides with Europe.

The sinking slab of lithosphere also generates many earthquakes, both shallow and deep. Earth's largest earthquakes are generated along subduction zones; some of these cause major natural catastrophes. Somewhat smaller earthquakes occur in the overlying continental plate between the oceanic trench and the line of stratovolcanoes.

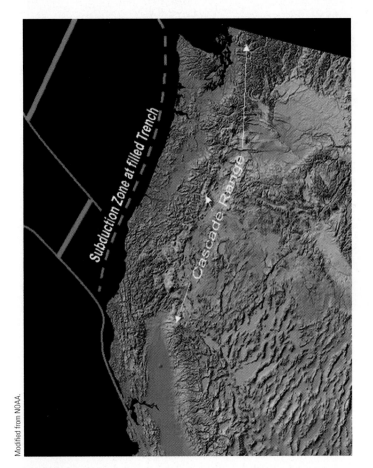

Modified from NOAA.

▶ **FIGURE 2-26.** The Cascade volcanic chain forms a prominent line of peaks parallel to the oceanic trench and 100 to 200 kilometers inland.

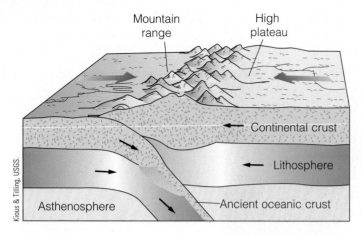

▶**FIGURE 2-27.** Collision of two continental plates generally occurs after subduction of oceanic crust. The older, colder, denser plate may continue to sink, or the two may merely crumple and thicken. Collision promotes thickening of the combined lithospheres and growth of high mountain ranges.

▶**FIGURE 2-28.** The Himalayas, which are the highest mountains on any continent, were created by collision between the Indian and Eurasian plates.

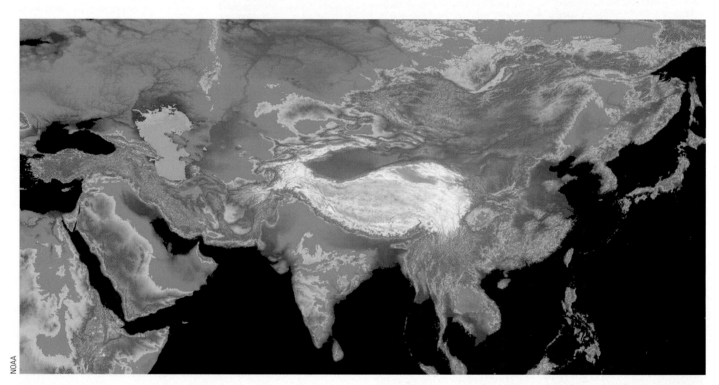

▶**FIGURE 2-29.** Collision of the Indian Plate with the Eurasian Plate has thickened the continental lithosphere and continues to push up the Himalayas, the high-elevation area shown in white.

COLLISION OF TWO CONTINENTS Where two continental plates collide, the results can be catastrophic. Neither plate sinks; high mountains such as Mount Everest are pushed up in fits and starts, accompanied by large earthquakes (▶Figures 2-27 to 2-29). Earthquakes regularly kill thousands of people from the collisions between India and Asia during the formation of the Himalayas and between the Arabian Plate and Asia during the formation of the Caucasus. These earthquakes are distributed across a wide area because of the thick stiff crust in these mountain ranges (▶Figures 2-29 and 2-30).

Subduction before collision

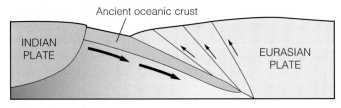

Early collision

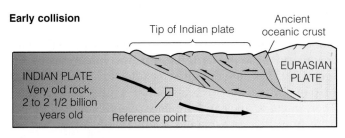

After

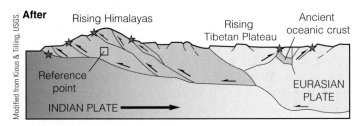

▶**FIGURE 2-30.** Collision between two continental plates deforms the edge of both plates, causing frequent earthquakes (red stars) in the broad zone of collision.

Modified from Kious & Tilling, USGS.

Where Plates Slide Past Each Other (Transform Faults)

In some places, plates simply slide past each other without pulling apart or colliding. Those are called transform plate boundaries or **transform faults.** Some of them offset the midoceanic ridges. Because the ridges are spreading zones, the plates move away from the ridges. As shown in ▶Figure 2-31, the section of the fault between the offset ends of the spreading ridge shows significant relative movement; spreading occurs in the opposite direction from the two offset ridge sections. However, note that there is no relative movement across the same fault beyond the spreading ridges as long as the two adjacent ridge segments are spreading at the same rate and are moving in the same direction. Note also that the offset between the two ridge segments does not indicate the direction of relative movement on the transform fault.

Transform faults generate significant earthquakes without causing casualties because no one lives anywhere along this ridge except in Iceland. On continents it is a different story. The San Andreas Fault system in California is a well known continental example. The North Anatolian Fault in Turkey is another that is even more deadly. The San Andreas Fault is the dominant member of a swarm of more or less parallel faults that move horizontally. Together, they have moved a large western slice of California, part of the Pacific Plate, north at least 500 kilometers so far. If that trend

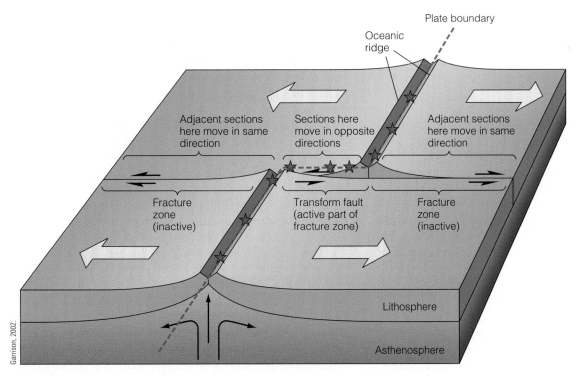

Garrison, 2002.

▶**FIGURE 2-31.** In this perspective view of an oceanic spreading center, earthquakes (stars) occur along spreading ridges and on transform faults offsetting the ridge.

▶**FIGURE 2-32.** Interstate 280 and the heavily populated area south of San Francisco border the San Andreas Fault zone, marked here by San Andreas Lake and Crystal Springs Reservoir.

continues for millions of years, they will eventually move that slice far enough north to lodge it against Alaska.

Transform plate boundaries typically generate large numbers of earthquakes, a few of which are catastrophic. A sudden movement along the San Andreas Fault caused the devastating San Francisco earthquake of 1906 with its large toll of casualties and property damage. The San Andreas system of faults passes through the metropolitan areas south from San Francisco (▶Figure 2-32) and just east of Los Angeles. Both areas are home to millions of people, all of whom live at risk of major earthquakes that could cause enormous numbers of casualties and substantial property damage with little or no warning. Even moderate-size earthquakes in 1971 and 1994 near Los Angeles, in 1989 near San Francisco, and in 2003 near Paso Robles killed some people. The absolutely genuine threat of such sudden havoc in a still larger event inspires much public consternation as well as major scientific efforts to find ways to predict large earthquakes.

For reasons that remain mostly unclear, some transform plate boundaries also associate with volcanic activity. Several large volcanic fields have erupted along the San Andreas system of faults during the last 15 or so million years. One of those, the Clear Lake volcanic field north of San Francisco, erupted recently enough to suggest that it may still be capable of further eruptions.

Hotspot Volcanoes

Some volcanoes erupt in places remote from any plate boundary; most appear to be **hotspot volcanoes** (▶Figure 2-33). These are the surface expression of hot columns of partially molten rock anchored in the deep mantle. Their origin is unclear, but many scientists infer that they arise from the core–mantle boundary. Because they are anchored deep in the Earth, lithospheric plates drift over them, and the hotspots burn a track in the moving overlying plate. Typical hotspot volcanoes therefore erupt at the active end of a long chain of extinct volcanoes that become progressively older with increasing distance from the active volcano. Mauna Loa and Kilauea, for example, erupt at the eastern end of the Hawaiian Islands, a chain of extinct volcanoes that become older westward toward Midway Island. Beyond Midway, the Hawaiian-Emperor chain doglegs to a more northerly course. It continues as a long series of defunct volcanoes that are now submerged. They form seamounts to the west end of the Aleutian Islands west of Alaska (▶Figure 2-34).

Plumes of abnormally hot but solid rock rising within the Earth's mantle begin to melt as the rock pressure on them drops. Wherever peridotite of the asthenosphere partially melts, it releases basalt magma that fuels a volcano on the surface. If the hotspot is under the ocean floor, the basalt magma erupts as basalt lava. If the hot basalt magma rises under continental rocks, it partially melts those rocks to form rhyolite magma; that magma often produces violent eruptions of ash.

That rising column or plume of hot rock appears to remain fixed in its place as one of Earth's plates moves over it. As the Earth's lithosphere moves over the asthenosphere beneath, the active volcano moves off its deep root in the asthenosphere and dies out while another volcano grows in its old place above the mantle plume. The result is a line of extinct volcanoes that become progressively younger toward the active volcano directly above the mantle plume (▶Figure 2-35). Mauna Loa and Kilauea in the Pacific Ocean mark the active end of the Hawaiian hotspot track (Figure 2-34). Reunion volcano in the Indian Ocean is the active end of a hotspot track, which extends north to near Bombay on the west coast of India. So far as anyone knows, this may continue almost indefinitely as the hotspot track of dead volcanoes lengthens (Figure 2-34). Eventually the volcanoes and the plate carrying them slide into a subduction zone and disappear.

Hotspot volcanoes, being anchored deep in the Earth's mantle and leaving a burning track in the overlying litho-

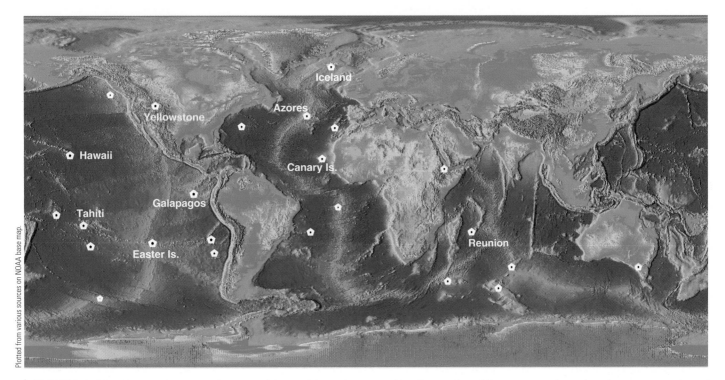

▶**FIGURE 2-33.** The main hotspot volcanoes of the world are shown as white dots. A few of the more prominent locations are named. Note that most, like Hawaii, are in ocean basins, at the ends of hotspot tracks.

sphere, leave a clear record of the direction and rate of movement of the lithospheric plates. Remnants of ancient hotspot volcanoes show the direction of movement in the same way that a saw blade traces the direction of movement of the board being sawn. The ages of those old volcanoes provide the rate of movement of the lithospheric plate (Figure 2-34). The assumption, of course, is that the mantle containing the hotspot is not itself moving. Comparison of different hotspots suggests that this is at least approximately valid.

Basalt magma evolved from mantle plumes also drives continental hotspot volcanoes. The basalt magma rises through continental crust, which it partially melts to make rhyolite magma. The melting temperature of basalt is more than 300°C above that of rhyolite, so a small amount of molten basalt can melt a large volume of rhyolite. The molten rhyolite rises in large volumes, which may erupt explosively through giant rhyolite calderas such as the Yellowstone volcano in Yellowstone National Park, the Long Valley Caldera in eastern California, and Taupo Caldera in New Zealand. The Snake River Plain of southern Idaho is probably the best example of a continental hotspot track.

Along the Snake River Plain, a series of extinct **resurgent calderas** that began to erupt some 14 million years ago tracks generally east and northeast in southern Idaho. They become progressively younger northeastward (▶ Figure 2-36). They are a continental hotspot track that leads from

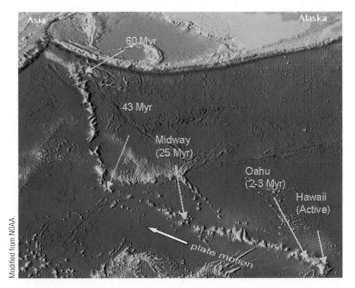

▶**FIGURE 2-34.** The topographic image of the Hawaiian-Emperor chain of volcanoes clearly shows the movement of the crust over the hotspot that is currently below the big island of Hawaii, where there are active volcanoes. Two to three million years ago, the part of the Pacific Plate below Oahu was over the same hotspot. The approximate rate and direction of plate motion can be calculated using the common belief that the hotspot is nearly fixed in space through time. The distance between two locations of known ages divided by the time (age difference) indicates the rate of movement.

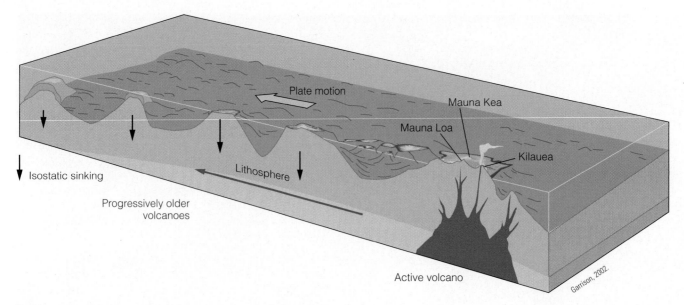

Plate motion

Mauna Kea

Mauna Loa

Kilauea

Isostatic sinking

Lithosphere

Progressively older
volcanoes

Active volcano

Garrison, 2002.

▶**FIGURE 2-35.** The lithospheric plate, moving across a stationary hotspot in the Earth's mantle (to the left in this diagram), leaves a track of old volcanoes. The active volcanoes are over the hotspot.

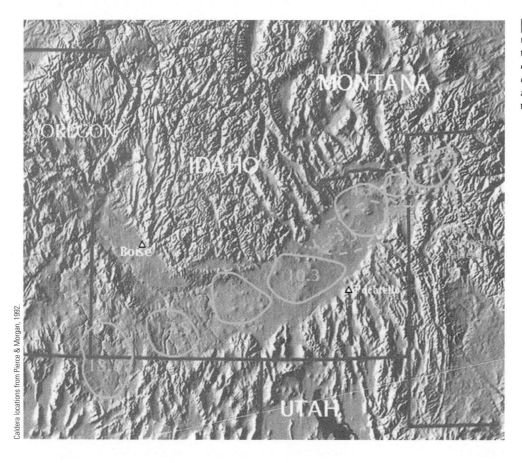

Caldera locations from Pierce & Morgan, 1992.

MONTANA

OREGON

IDAHO

Boise

10.3

Pocatello

UTAH

▶**FIGURE 2-36.** This shaded relief map of the Snake River Plain shows the outlines of ancient resurgent calderas leading northeast to the present-day Yellowstone caldera. Caldera ages are shown in millions of years before the present.

its western end near the border between Idaho and Oregon to the Yellowstone resurgent caldera at its active northeastern end in northwestern Wyoming within Yellowstone National Park.

Summary

The modern theory of plate tectonics is supported by a large mass of data collected over the last century. The development of this theory is a good example of how the **scientific method** works (see Box 2-1). Initially Alfred Wegener, like many others, noted that the coastlines matched across the Atlantic Ocean, suggesting that they were once together. To test this initial hypothesis, Wegener went further; he searched for connections between other aspects of geology across the Atlantic: mountain ranges, rock formations and their ages, and fossil life forms. With confirmation of such former connections, he hypothesized that continents that were once together had drifted apart through oceanic crust. As we now know, his hypothesis of continental drift was correct, but his proposed mechanism of the drift process was incorrect.

Many scientists rejected Wegener's whole hypothesis because they could show that his proposed mechanism was not physically possible. Others who were willing to consider other possibilities eventually came up with another possible mechanism: the ocean floor separating at the mid-Atlantic Ridge, spreading apart, and moving with the continents. Still others made additional tests of the concept of plate tectonics, leading to its confirmation. Modern data continue to support the concept that plates move and substantiate the mechanism of new oceanic plate generation at the midocean ridges and destruction of this same plate at subduction zones. As a result, Wegener's initial hypothesis of continental drift was modified to be a hypothesis that is the foundation for the modern theory of plate tectonics. This theory is a fundamental foundation for the geosciences and is important for understanding why and where we have a variety of major geologic hazards such as earthquakes and volcanic eruptions.

KEY POINTS

✓ A dozen or so nearly rigid lithospheric plates make up the outer 60 to 200 kilometers of the Earth. They slowly slide past, collide with, or spread apart from each other. **Review pp. 20–30; Figures 2-14 and 2-16b.**

✓ Much of the tectonic action, in the form of earthquakes and volcanic eruptions, occurs near the boundaries between the lithospheric plates. **Review pp. 22–28; Figures 2-16 to 2-26.**

✓ Continental drift was proposed by matching shapes of the continental margins on both sides of the Atlantic Ocean, as well as the rock types, deformation styles, and fossil life forms. The fact that plates move was confirmed initially by matching alternating magnetic stripes on opposite sides of the oceanic spreading ridges. **Review pp. 12–16; Figures 2-1 and 2-9.**

✓ The concept of isostacy explains why the lower-density continental rocks stand higher than the denser ocean-floor rocks and sink deeper into the underlying mantle. This behavior is analogous to ice (lower density) floating higher in water (higher density). **Review pp. 16–19; Figures 2-11 and 2-12.**

✓ Spreading ridges or rifts mark the locations of earthquakes along faults that slip up and down and may erupt lava flows. **Review pp. 23–24; Figures 2-19 to 2-23.**

✓ Subduction zones, where ocean floors slide beneath continents or beneath other slabs of oceanic crust, are the locus of major earthquakes and cause volcanoes to erupt and mountains to grow on the overriding plate. **Review pp. 24, 26, 27; Figures 2-24 to 2-26.**

✓ Continent–continent collision zones, where two continental plates collide, are regions with major earthquakes and the tallest mountain ranges on Earth. **Review p. 28; Figures 2-27 to 2-30.**

✓ Transform faults, where two lithospheric plates slide laterally past one another, are areas of major earthquakes where they cross continents. **Review pp. 29–30; Figures 2-31 and 2-32.**

✓ Hotspots form chains of volcanoes within individual plates rather than near plate boundaries. They grow as a trailing track of progressively older extinct volcanoes because lithosphere is moving over hotspots fixed in the underlying Earth's asthenosphere. **Review pp. 30–33; Figures 2-33 to 2-36.**

✓ The scientific method involves developing tentative hypotheses that are tested by new observations and experiments, which can lead to confirmation or rejection. **Review p. 33.**

IMPORTANT WORDS AND CONCEPTS

Terms

asthenosphere, p. 16
collision zone, p. 24
continental drift, p. 12
hotspot volcanoes, p. 30
hypothesis, p. 33
isostacy, p. 16
lithosphere, p. 16
low-velocity zone, p. 19
magnetic field, p. 13
magnetic stripes, p. 16
Mohoroviçic Discontinuity, p. 19
Pangaea, p. 16

plates, p. 19
plate tectonics, p. 21
plumes, p. 30
resurgent calderas, p. 31
rift zones, p. 24
scientific method, p. 33
seafloor spreading, p. 13
spreading zone, p. 24
subduction zone, p. 24
theory, p. 33
transform fault, p. 29
trenches, p. 16
triple junctions, p. 21

QUESTIONS FOR REVIEW

1. What direction is the Pacific Plate currently moving, based on Figure 2-34?

2. Before people understood plate tectonics, what evidence led some scientists to believe in continental drift?

3. If the coastlines across the Atlantic Ocean are spreading apart, why isn't the Atlantic Ocean deepest in its center?

4. What evidence confirmed seafloor spreading? Be brief but to the point.

5. Distinguish between Earth's crust, lithosphere, asthenosphere, and mantle.

6. What are the main types of lithospheric plate boundaries in terms of relative motions? Provide a real example of each (by name or location).

7. What does oceanic lithosphere consist of and how thick is it?

8. Along which type of lithospheric plate boundary are large earthquakes common?

9. Along which type(s) of lithospheric plate boundary are andesite stratovolcanoes common? Provide an example.

10. Why does oceanic lithosphere almost always sink beneath continental lithosphere at such convergent zones?

FURTHER STUDY

Assess your understanding of this chapter's topics with additional quizzing and conceptual-based problems at:

 http://earthscience.brookscole.com/hyndman.

This house near Redwood Grove, in Santa Cruz, California, slid off its concrete foundation during the Loma Prieta Earthquake on October 17, 1989.

EARTHQUAKES AND THEIR DAMAGES
Shaking Ground, Collapsing Buildings

"We only get a few small quakes here, nothing to worry about. They have some pretty big ones around San Francisco and Los Angeles, but we don't get them." We heard this remark recently in Santa Rosa, some 70 kilometers (43.5 miles) north of San Francisco. Unfortunately, the speaker lacked some critical information. It is true that the main San Andreas Fault lies 40 kilometers to the west, far enough to diminish somewhat the shaking from an earthquake there. In fact, the 1906 San Francisco earthquake wrecked Santa Rosa before almost anyone now living in the area was even born. Most current residents moved to the area in the last few decades, so even their relatives have no knowledge of the event. In addition, Santa Rosa lies on the Hayward-Rodgers Creek Fault, a major strand of the overall San Andreas system that is currently considered more dangerous than the main San Andreas Fault in the San Francisco Bay area. Earthquake seismologists estimate that there is a

The teams were warming up and the crowd was settling into its seats in Candlestick Park on the southern edge of San Francisco. Game Three of the World Series between the San Francisco Giants and the Oakland Athletics was about to begin. At 5:04 P.M. on October 17, 1989, the sudden shock of an earthquake jarred everyone to a stop. That was the P wave arriving. Ten seconds later, the shaking suddenly intensified enough to knock a few people off their feet. That was the S wave arriving. The ten-second interval between the P and S waves indicated that the earthquake was some 80 kilometers away. Then the light towers began to sway and the entire country experienced the Loma Prieta earthquake on television. By morning, it was clear that the San Andreas Fault had moved near Loma Prieta in the Santa Cruz Mountains 80 kilometers southeast of the stadium. The magnitude was 6.9.

Meanwhile, cars swerved and traffic stopped on a 2.25-mile stretch of double-decked section of Interstate 880 across the bay in Oakland. Some drivers first thought they had flat tires. Initial excitement turned to terror for those on the lower deck, where chunks of concrete popped out of the support columns as the upper deck collapsed onto their vehicles. It seemed a miracle to rescuers that only forty-two motorists died (▶Figure 3-1). Overall, it was most fortunate that the earthquake happened at the beginning of the World Series in San Francisco. Most people in the San Francisco Bay area who were not at the game were home watching it on television. Freeways that would normally be filled with rush-hour traffic were nearly empty. Otherwise, many more people would have been killed.

▶**FIGURE 3-1.** Severe shaking of the double-deck Interstate 880 freeway in Oakland sheared off heavily reinforced concrete supports, and much of the upper deck collapsed onto the lower deck. Note that the heavy column in midphoto, for example, failed at the level of the lower deck where the two parts were joined during construction.

Howard G. Wilshire photo, USGS.

Soft mud along the edge of San Francisco Bay amplified the ground motion under the freeway by a factor of 5 to 8 despite the 90-kilometer distance from the epicenter (▶Figure 3-2). It was especially unfortunate that the vibration frequency of the sediments reverberated with the vibration frequency of the elevated freeway (2 to 4 lateral cycles per second) because that greatly amplified the damage.

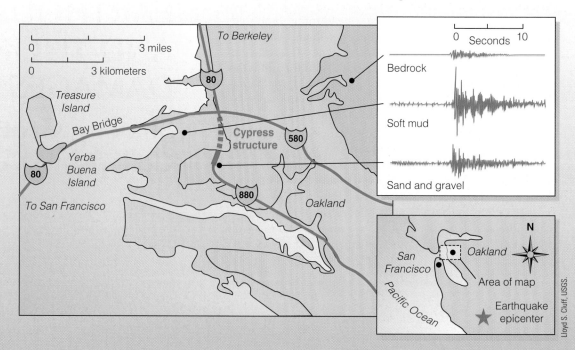

▶**FIGURE 3-2.** The Cypress section of Interstate 880 in Oakland collapsed during the 1989 Loma Prieta earthquake. Seismographs show that the shaking was much stronger on mud than on bedrock.

Lloyd S. Cluff, USGS.

Santa Cruz, less than 16 kilometers southwest of the epicenter, was badly damaged, as were neighboring towns. Landslides closed highways in the coast ranges north of Santa Cruz. Buildings in the Marina district were especially vulnerable because much of that district on the edge of San Francisco Bay stands on fill that amplified ground motion (▶Figures 3-3 and 3-4). The water-saturated sediments of the landfill turned to a mushy fluid, causing settling that broke gas lines and set fires. It also broke water mains, forcing firefighters to string hoses to pump seawater from the bay.

Though the Loma Prieta earthquake was not by any means the "Big One" that the people of the Bay Area have come to dread, collapsing structures did kill sixty-two people and injured 3,757. Some 12,000 people were displaced from their homes. The earthquake destroyed 963 homes and inflicted $6 billion in property damage ($8.7 billion in 2002 dollars). Damage to the San Francisco–Oakland Bay Bridge closed it for a month (▶Figure 3-5). The collapsed I-880 freeway slowed traffic along the East Bay freeway for years until a single-level freeway was completed to replace the fallen multilevel structure.

The fault slippage that caused the Loma Prieta earthquake began at a depth of 18 kilometers, where rocks west of the fault moved 1.9 meters (6.2 feet) north and 1.3 meters up. The offset did not break the surface or even the upper 6 kilometers of the crust.

The U.S. Geological Survey had convened a Working Group on California Earthquake Probabilities. It had already identified the fault segment that caused the Loma Prieta earthquake as one likely to produce an earthquake of magnitude 6.5 or greater in the thirty years after 1988. As with many other earthquakes, this one in the Santa Cruz Mountains segment of the fault filled a **seismic gap,** an area along the fault that had not seen recent earthquakes. Seismographs did not detect any precursor events that might have warned of an imminent earthquake.

C. Meyer, USGS.

▶**FIGURE 3-3.** Major fires broke out in the San Francisco Marina district after the 1989 Loma Prieta earthquake.

J. K. Nakata photo, USGS.

▶**FIGURE 3-4.** This apartment building, constructed on the artificial fill of the Marina district of San Francisco, collapsed during the Loma Prieta earthquake in 1989. The car parked next to the building did not fare well, but people inside the building and above the first floor probably survived. Note that the second-floor balcony behind the car is now at street level.

Edgar V. Leyendecker photo, USGS.

▶**FIGURE 3-5.** Violent shaking of the San Francisco–Oakland Bay Bridge sheared off the array of heavy bolts securing one section of the westbound upper roadway and dropped it onto the lower roadway. This view is westward from near the east end of the bridge. Note the skid marks on the roadway in lower right. The western half of the bridge, a suspension design, was not damaged.

62 percent chance of at least one earthquake of magnitude 6.7 or larger in this area before the year 2032. That is the size earthquake that caused severe damage in the Los Angeles area in 1994. People are also apparently unaware that even though they occasionally feel small earthquakes in an earthquake-prone area, larger ones are likely and a very large earthquake is quite possible.

Innumerable tiny earthquakes, especially in western California, cause no more ground movement than a passing truck and generally go unnoticed. Somewhat larger earthquakes shake the ground and make people sit up and take notice but still do not do much damage. It is the large events that wreck havoc, collapsing buildings and highways and killing people.

Before 1989, earthquakes in the United States cost an average of $230 million per year. Costs escalate as more people move into more dangerous areas and as property values rise. Some authorities now expect future losses to average more than $4.4 billion per year, with 75 percent of that in California. Roughly one-third of California's earthquake damage will probably occur with a few large earthquakes in Los Angeles County. A large proportion of the remainder will probably occur in the San Francisco Bay area. California gets far more than its share of North American earthquakes because of its location along the boundary between the Pacific and North American tectonic plates. That part of the plate boundary is marked by the San Andreas Fault, one of the Earth's largest and most active transform faults.

If people did not live near that fault, its sudden shifts would not create a problem, but cities and towns grew in those areas for reasons that had nothing to do with movements in the Earth. Now with millions of people living in hazardous environments, society is beginning to realize that we have a worsening problem, both for many individuals and for society as a whole. In order to deal with the hazards, we need to know more about what creates each hazard and how to deal with it.

Earthquakes and Earthquake Waves

People experience a major earthquake differently whether they are near the epicenter, indoors or out, on a rocky hill, or on soft mud. Hit a solid rock with a hammer and it rings—that is, it vibrates with a high frequency. Hit a soft material and you hear a thud or nothing at all—a rather low vibration frequency. Soft wet mud in an earthquake sloshes back and forth like a bowl of jelly.

The first event is the arrival of the P waves, which come as a sudden jolt. People indoors wonder for a moment whether a truck just hit the house. Then comes a brief interval of quiet while the cat heads under the bed and plaster dust sifts down from the cracks in the ceiling.

Next come the S waves, an oscillating motion that makes it hard to stand. Chimneys snap off and fall through the floors to the basement. Streets and sidewalks twist and turn. Buildings jarred by the earlier P waves now distort and may collapse.

Finally, the surface waves arrive—a long series of rolling motions almost like broad swells on the ocean. When these waves finally arrive, the P and S waves have already done their worst. Surface waves find buildings of all kinds loosened and weakened, ready for the final blow.

Inertia tends to keep people and loose furniture in place as ground motion yanks the building back and forth beneath them. Shattering windows spray glass shrapnel as plaster falls from the ceiling. If the building is weak or the ground loose, it may collapse.

People outdoors are much safer than those indoors because no roof is overhead to come down around their ears. A car parked next to a building provides little safety from falling debris (▶ Figure 3-6). Glass and other debris falling from nearby buildings, however, might make it advisable to run for open ground. Electric wires would certainly fall to an accompaniment of great sparks that are likely to start fires. Buckled streets, heaps of fallen rubble, and broken water mains would hamper firefighters' efforts. Release of gas or other petroleum products in port areas such as Los Angeles could spark a firestorm.

The days after a major earthquake typically bring epidemics of the diseases of impure water as broken sewer mains leak contaminants into broken water mains. Decomposing bodies that are buried in rubble also contribute to the spread of disease. It is not uncommon for many people to die of disease following a major earthquake especially in the poorest countries. Meanwhile, the fires continue to burn and the expenses mount.

Observant people have noticed for centuries that many earthquakes arrive as a distinct series of shakings that feel different. We will consider only these waves, ignoring the more complex records of internal refractions of waves as they pass between different Earth layers. Those complications do not much affect the damage that earthquakes inflict because the direct waves are significantly stronger.

Frequency and Wavelength

To describe the vibrations of earthquake waves, we use a variety of terms (▶ Figure 3-7). The time for one complete cycle between successive wave peaks to pass is the **period;** the distance between wave crests is the **wavelength;** and the amount of positive or negative wave motion is the **amplitude.** The number of peaks per second is the **frequency** in cycles per second or Hertz (Hz). P and S waves generally cause vibrations in the frequency range between one and thirty cycles per second (1–30 Hz), a high frequency for earthquake waves. Surface waves generally cause vibrations at much lower frequencies, which dissipate less

▶ **FIGURE 3-6.** These cars were crushed by falling bricks during the moderate-sized March 4, 2001, earthquake in Olympia, Washington.

rapidly than those associated with body waves. That is why they commonly damage tall buildings at distances as great as 100 kilometers from the epicenter. Tall buildings are most vulnerable to the vibrations at lower frequencies.

Because the swaying of a tall building is similar to that of a pendulum swinging upside down, you can simulate it by dangling a weight at the end of a string and moving the hand holding the string. For example, if you move your hand back and forth 30 centimeters (1 foot) each second, the weight at the end of a 30-centimeter-long string will cause a large swing back and forth. This swing is in resonance with the oscillation period of the pendulum (see Box 3-1). On the other hand, if you move your hand back and forth three times per second, the weight will hardly move. Similarly, at three times per second, the weight will hardly move if the

string is much longer than 30 centimeters. A tall building sways back and forth more slowly than a short one, so low-frequency earthquake vibrations are more likely to damage tall buildings. See also Sidebar 3-2 (page 61), for the sway of buildings of different heights during an earthquake.

Seismographs

We record the shaking of earthquake waves on a device called a **seismograph.** When recording seismographs finally came into use during the early part of the twentieth century, it became possible to see those different shaking motions as a series of distinctive oscillations that arrive in a predictable sequence. Imagine the seismograph as an extremely sensitive mechanical ear clamped firmly to the ground, constantly listening for noises from the depths. It is essentially the geologist's stethoscope.

We normally stand firmly planted on the solid Earth to watch things move. Seismographs are the answer to the question of how do we stand still to watch the Earth move? They are simple instruments—at least in principle. Reduced to their raw essentials, seismographs designed to record horizontal motions consist of a heavy weight suspended from a wire attached to a rigid column that is firmly anchored to the ground (▶ Figures 3-8, 3-9, and 3-10). Those designed for vertical motions have a heavy weight suspended from a weak spring. In both cases, everything moves with the

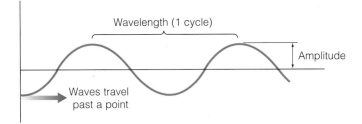

▶ **FIGURE 3-7.** *Wavelength* is the distance between successive wave crests. *Wave frequency* is the number of wave crests or cycles per second. *Period* is the time for one cycle to pass.

BOX 3.1

Movement of a Pendulum

The back-and-forth oscillation of a pendulum depends only on the length of the swinging object. Specifically, the period (*P*) of the pendulum, or total time for a back-and-forth movement, is equal to the square root of the pendulum length (*L*).

$$P = \sqrt{L}$$

Note also that the earthquake wave frequency times its wavelength equals the wave velocity (▶ Figure 3-7). For example, a wave frequency of two cycles per second times a wavelength of 3 kilometers equals a velocity of 6 kilometers per second. That is comparable to the velocity of typical P waves in the continental crust.

USGS.

▶**FIGURE 3-8.** Although many seismograph stations now record earthquake waves digitally, a recording drum seismograph like this one is especially useful for visualizing the nature of earthquake waves: their amplitude, wavelength, and frequency of vibration. Different seismographs are used to measure, for example, horizontal versus vertical motion and small versus large earthquakes.

earthquake motion except the suspended weight, which pretty much stays still because of its inertia. Inertia requires a motionless object to stay motionless unless some outside force causes it to move. Because the suspended weight is so weakly coupled to the Earth that accelerates during slip on the fault, it hardly moves. Actually, its movement is not synchronous with movement of the shaking Earth.

The first seismograph used in the United States was at the University of California, Berkeley, in 1887. It used a pen attached to the suspended weight to make a record on a sheet of paper that was attached to the moving ground. Most modern seismographs work on the same basic principle but detect and record ground motion electronically.

Just as you hear the snap and feel the vibration when you bend a stick until it breaks, the Earth vibrates back and forth at the frequency of a low rumble when it breaks. We think of rocks as brittle solids, but rocks are elastic, like a spring. They break if stretched too far. The highest earthquake frequencies are twenty or thirty cycles per second, but people hear sounds with frequencies as low as fifteen cycles per second, so the earthquake frequencies we hear are between fifteen and thirty cycles per second.

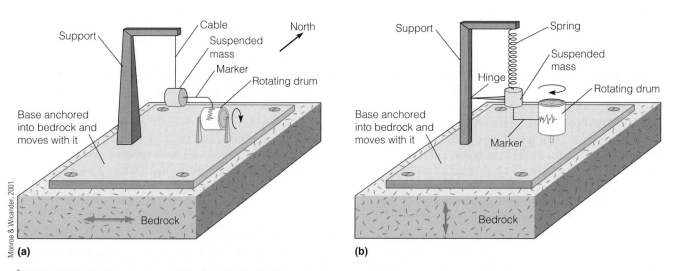

Monroe & Wicander, 2001.

▶**FIGURE 3-9.** These seismographs measure **(a)** horizontal motion and **(b)** vertical motion.

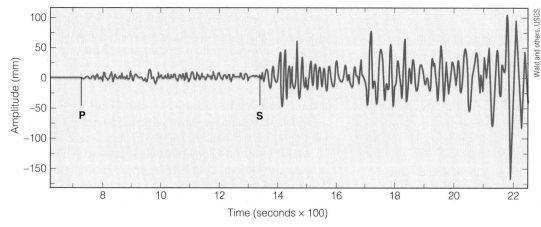

Wald and others, USGS.

▶**FIGURE 3-10.** A seismogram for the 1906 San Francisco earthquake shows the main seismic waves—P, S, and surface waves.

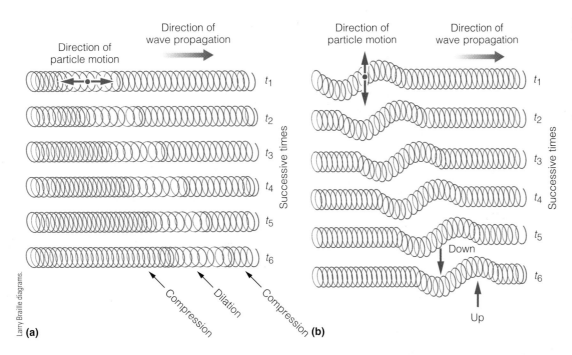

Direction of particle motion

Direction of wave propagation

Successive times

t_1

t_2

t_3

t_4

t_5

t_6

Compression

Dilation

Compression

Larry Braille diagrams.

(a)

Direction of particle motion

Direction of wave propagation

Successive times

t_1

t_2

t_3

t_4

t_5

Down

t_6

Up

(b)

You first feel the **P waves,** the *primary* or compressional waves (▶Figures 3-10 and 3-11a). They consist of a train of compressions and expansions that travel roughly 8 kilometers per second in the dense, less compressible rocks of the upper mantle, and 5 to 6 kilometers per second in the less dense continental crust. People sometimes hear the low rumbling of the P waves of an earthquake before the arrival of the stronger and more destructive *secondary* (S) waves and surface waves. Sound waves are also compressional and closely comparable to P waves, but they travel through the air at only 0.34 kilometers per second.

S waves (secondary or shear waves) are slower than P waves and arrive at the seismograph shortly after them. They move with an up-and-down wriggling motion like that of a rhythmically shaking rope (Figure 3-11b). S waves travel at speeds of 4.5 kilometers per second in the upper mantle and 3.5 kilometers per second in the crust. These shear waves do not travel through liquids, so their absence from an otherwise normal seismograph record suggests the existence of molten magma at depth. Their wriggling motion and larger wave height or amplitude make them more destructive than P waves.

P waves and S waves are both called *body waves* because they travel through the Earth instead of along its surface. Because the velocity of earthquake waves in rocks increases with depth, they follow paths within the Earth that are slightly concave upward toward the Earth's surface.

Surface waves travel along the Earth's surface and fade downward. They are the last to arrive at a seismograph partly because they are slower than the body waves, 2 to 4.5 kilometers per second. For distant earthquakes, they follow a longer path along the surface instead of through the Earth. Surface waves include Love and Rayleigh waves, which move in perpendicular planes. Love waves move

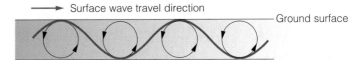

▶**FIGURE 3-12.** In the rolling motion of surface waves, individual particles at the Earth's surface move in a circular motion. This motion is found in Rayleigh waves.

from side to side. Rayleigh waves move up and down in a motion that somewhat resembles swells on the ocean (▶Figure 3-12).

Surface waves generally involve the greatest ground motion, so their trace on a seismogram shows the largest amplitudes. Thus, they cause a large proportion of all earthquake damage. Deep fault movement does not cause nearly as much movement of the ground as do fault movements near the surface.

Faults with different orientations and directions of movement (▶Figure 3-13) generate various patterns of motion. Specialized seismographs are designed to measure those various directions of earthquake vibrations—north to south, east to west, and vertical motions. Knowing the directions of ground motion makes it possible to infer the direction of fault movement from the seismograph records.

Faults and Earthquakes

Faults are simply fractures in the Earth's crust along which rocks on one side of the break move past those on the other. Faults come in a wide range of sizes measured according to

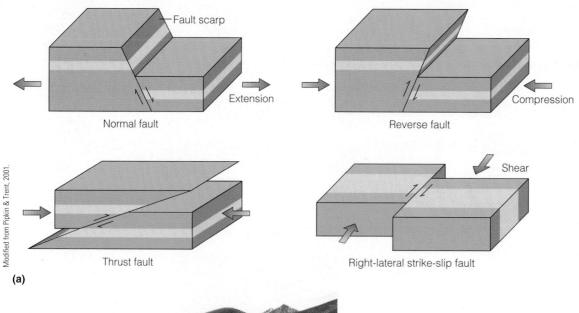

FIGURE 3-13. (a) These are the basic types of faults. **(b)** A normal fault near Challis, Idaho, that moved in the 1993 earthquake.

those first announce their presence when they cause an earthquake. Earthquakes are common in the mountainous western parts of the country, where the rocks are deformed into complex patterns of faults and folds.

Basic types of faults form under differently oriented stresses (Figure 3-13). **Normal faults** move on a steeply inclined surface. Rocks above the fault surface slip down and over the rocks beneath the fault. Normal faults move during crustal extension. **Reverse faults** move rocks on the upper side of the fault up and over those below. **Thrust faults** are similar to reverse faults, but the fault surface is more gently inclined. Rocks above the fault move up and over those below, stacking the same section of rocks over the rocks below. Reverse and thrust faults move during crustal compression. **Strike-slip** faults move horizontally as rocks on one side of the fault slip laterally past those on the other side. If the far side of the fault moves to the right as in Figure 3-13, it is a right-lateral fault. If it moved in the opposite direction, it would be a left-lateral fault.

Causes of Earthquakes

After the great San Francisco earthquake of 1906, the governor of California appointed a commission to find the cause of earthquakes, which was then a complete mystery. The director was Andrew C. Lawson, a geologist of great distinction and one of the most colorful personalities in the history of California. Lawson and his students at UC Berkeley had recognized the San Andreas Fault and mapped large parts if it but had no idea that it could cause earthquakes. Members of the earthquake commission found numerous places where roads, fences, and other structures had broken during the 1906 earthquake just where they crossed the San Andreas Fault. In every case, the side west of the fault had

the amount of displacement along the fractures. The rocks west of the San Andreas Fault of California have moved, over several million years, at least 450 kilometers north of where they started. Thousands of other faults have moved less than 1 kilometer.

Some faults produce earthquakes when they move; others produce almost none. Some faults have not moved for such a long time that we consider them dead; others are clearly still active and potentially capable of causing earthquakes.

Active faults are rare in regions such as the American Midwest, where the continental crust has been stable for hundreds of millions of years. Such stable regions contain many faults that geologists have yet to recognize. Some of

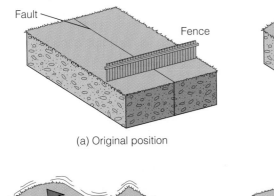

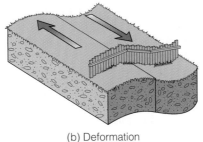

(a) Original position

(b) Deformation

▶**FIGURE 3-14.** Elastic rebound theory: Rocks near a fault are slowly bent elastically until the fault breaks during an earthquake; the rocks on each side then slip past each other, relieving the stress. Distortion of the Earth's crust actually extends over tens of kilometers outward from the main fault.

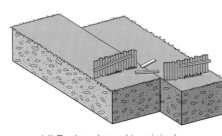

Pipkin & Trent, 2001.

(c) Rupture and release of energy

(d) Rocks rebound to original undeformed shape

moved north as much as 7 meters. That led to the theory of how fault movement causes earthquakes.

The earthquake commission concluded that the opposite sides of the fault had stuck for many years. The Earth's crust had continued to move, with the rocks on either side of the fault bending instead of slipping. Anyone who has ever watched a big sheet of plate glass waver as it was carried from the truck to the window will have no trouble with the idea that other rigid materials bend. As the rocks on opposite sides of the fault bend, they accumulate stress in exactly the way of a drawn bow. When the stuck segment of the fault finally slips, the bent rocks on opposite sides straighten with a sudden snap that causes the earthquake. The earthquake explanation, the **elastic rebound theory,** is that elastic strain in the rocks builds up as the rocks slowly bend. That strain is then released when the rocks suddenly snap back in the seismic wave that we call an earthquake. Imagine the cut edges of two blocks of sponge rubber catching as you try to slide them past one another. The rubber will stretch where the edges catch until they suddenly break loose and move (▶ Figure 3-14). The theory has withstood so many rigorous tests that geologists now consider it valid.

We use the term ***stress*** to refer to the forces imposed on a rock and ***strain*** to refer to the change in shape of the rock in response to the imposed stress. We now know from laboratory experiments that rocks deform in broadly consistent ways. Stress to simulate greater forces in one direction is added by compressing the rock from its ends. Such a differential compression generally breaks the rock in a diagonal orientation (▶ see Figure 3-15). The results help us understand the behavior of faults and earthquakes.

Deformation experiments show that most rocks near the Earth's surface, where they are cold and not under much pressure of overlying rocks, deform elastically when affected by small forces; they revert to their former shape when the force is relieved. The larger the stress applied, the greater the strain. In response to large stresses, rocks are brittle; that is, they fracture or break. Under other conditions, such as deep in the Earth where they are hot and under high pressure imposed by the overlying load of rocks, rocks are more

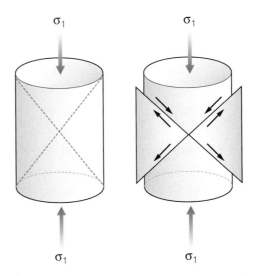

▶**FIGURE 3-15.** Deformation of a cylinder of rock by compression from the top and bottom breaks the rock on diagonal shear planes. Shear is generally on one plane only, as shown by the red line. σ_1 is the maximum principle stress.

likely to be plastic or *ductile;* they permanently change shape or flow when forces are applied.

A typical rock will deform elastically under low stress, then change to plastic behavior at higher stress, and ultimately succumb to brittle failure (▶see Figure 3-16). Along a real fault in nature, differential plate motions apply stresses continuously. Because those plate motions do not stop, elastic deformation progresses to plastic deformation within meters to kilometers of the fault. The fault finally ruptures in an earthquake.

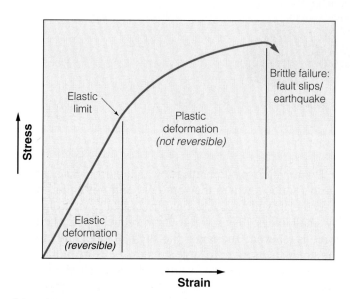

▶**FIGURE 3-16.** With increasing stress, a rock deforms elastically, then plastically, before ultimately failing or breaking. A completely brittle rock fails at the elastic limit.

Locating Earthquakes

One of the first questions that seismograph records can help answer is "Where did it happen?" The answer lies most obviously in the time interval between the arrivals of the P and S waves. Imagine them as two cars that start at the same place and at the same time, one going 100 kilometers per hour, the other at 90 kilometers per hour. The faster car gets farther and farther ahead with time. An observer who knows the speeds of the two cars could determine how far they are from their starting point simply by timing the interval between their passage. In exactly the same way, the time interval between the arrivals of the P and S waves reveals the approximate distance between the seismograph and the place where the earthquake struck (▶Figures 3-17 and 3-18).

But the earthquake could have happened anywhere on the perimeter of a circle drawn with the seismograph at its center and the distance to the earthquake as its radius. Seismograph records do not easily yield information about direction. The first step to a solution for that dilemma is to draw another circle on the same map using the data from a different seismograph station. The two circles will intersect at two different points, either of which could have been the location of the earthquake. The last step to a solution is to draw a third circle using the data from a third seismograph station. The three circles will intersect at only one location, and that is where the earthquake struck (Figure 3-18). Actually, because the earthquake waves travel at slightly different velocities through different rocks on their way to the seismograph, their apparent distances are slightly different and the circles intersect in a small triangle of error.

The magnitude of the earthquake can be estimated using an *earthquake nomograph,* on which a straight line is

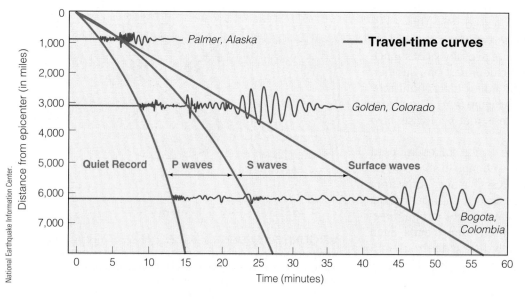

▶**FIGURE 3-17.** The difference in arrival time between the P and S waves reveals the distance from a seismograph to an earthquake. This plot shows records from seismographs at different distances from a single earthquake. The later arriving wave labeled L waves are one type of surface wave.

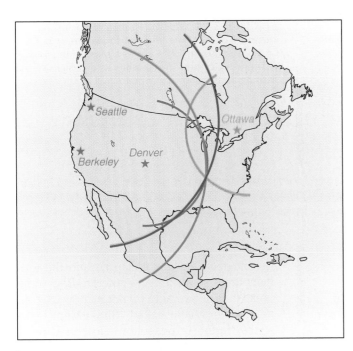

▶**FIGURE 3-18.** Three circles of distance to the earthquake drawn from at least three seismograph stations locate the earthquake on a map.

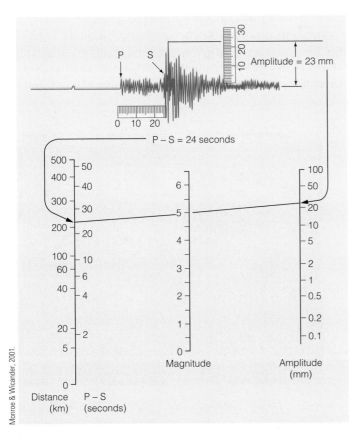

▶**FIGURE 3-19.** A nomograph chart uses the distance from the earthquake (P – S time in seconds) and the S-wave amplitude (in mm) to determine the earthquake magnitude.

plotted between the P–S time and the S-wave amplitude (▶Figure 3-19). This line intersects the central line at the approximate magnitude of the earthquake.

The map location of an earthquake is called the **epicenter** (▶Figure 3-20). It is the point directly above the **focus,** or hypocenter, which is the place where the fault first moved. Complete movement on the fault extends, of course, well beyond the epicenter.

In practice, seismograph stations communicate the basic data to a central clearing house that locates the earthquake, evaluates its magnitude, and issues a bulletin to report when and where it happened. The bulletin is often the first news of the event. That is why we so often find the news media reporting an earthquake before any information arrives from the scene.

Earthquake Size and Characteristics

Mercalli Intensity Scale

Another question that comes to mind when people feel an earthquake, or see the wild scribbling of a seismograph recording its ground motion, is "How big is it?" The Mercalli Intensity Scale, used before invention of the seismograph,

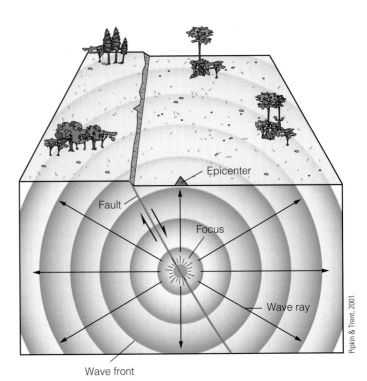

▶**FIGURE 3-20.** The epicenter of an earthquake is the point on the Earth's surface directly above the focus where the earthquake originated.

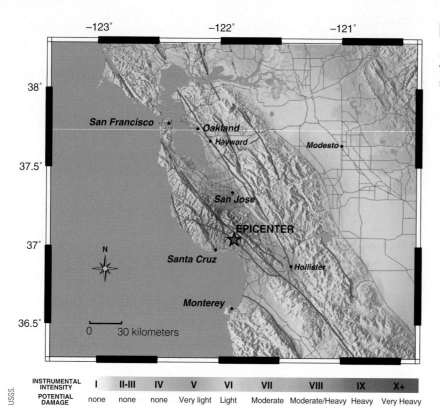

INSTRUMENTAL INTENSITY	I	II-III	IV	V	VI	VII	VIII	IX	X+
POTENTIAL DAMAGE	none	none	none	Very light	Light	Moderate	Moderate/Heavy	Heavy	Very Heavy

USGS.

▶**FIGURE 3-21.** This TriNet map of earthquake intensities of the Loma Prieta earthquake near Santa Cruz, California, in 1989, shows a Mercalli Intensity VIII at the epicenter. The epicenter as determined from seismographs is shown as a star.

provides the first and simplest way to answer that question. Its roots go back hundreds of years.

After the great Lisbon earthquake of 1755, the archbishop of Portugal sent a letter to every parish priest in the country asking him to report the type and severity of damage in his parish. Then the archbishop had the replies assembled into a map that clearly displayed the pattern of damage in the country. Jesuit priests have been prominent in the study of earthquakes ever since.

Italian seismologist Giuseppe Mercalli formalized the system of reporting in 1902 with his development of the Mercalli Intensity Scale. It is based on how strongly people feel the shaking and the severity of the damage it causes. The original Mercalli Scale was later modified to adapt it to construction practices in the United States. The Modified Mercalli Intensity Scale is still in use. The U.S. Geological Survey sends postcard or email questionnaires to people it considers qualified observers who live in the area of an earthquake and then assembles the returns into an earthquake intensity map, the higher Roman numerals recording greater intensities (▶ Figure 3-21). Note that the intensity distribution around the earthquake is not circular. In the case of the Loma Prieta earthquake, the zones of greater intensity are, as expected, elongate parallel to the San Andreas Fault. Greater intensities also exist over San Francisco Bay because its loose, wet muds amplify the shaking. If you look carefully at Figure 3-21, you will see areas of higher intensity in the soft sediment of the Marina district at the north edge of San Francisco, in the soft mud area of Oakland where the

double-deck freeway collapsed and the low-lying areas south of San Francisco Bay. Some areas such as Southern California now have an automated system called "**ShakeMaps**" (described further in the section below on "Ground Motion and Failure During Earthquakes") that uses the new TriNet seismograph grid to map the degree of shaking over a large area within a five-minute period after the earthquake (▶ Figure 3-22). Such real-time maps can help send emergency-response teams to areas that have likely suffered the greatest damage.

Mercalli Intensity Scale maps reflect both the subjective observations of people who felt the earthquake and an objective description of the level of damage. They typically show the strongest intensities in areas near the epicenter

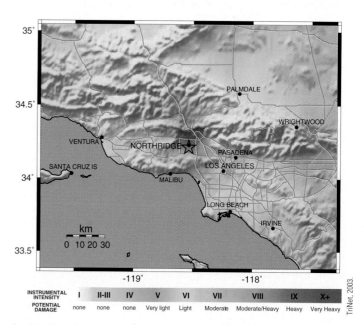

INSTRUMENTAL INTENSITY	I	II-III	IV	V	VI	VII	VIII	IX	X+
POTENTIAL DAMAGE	none	none	none	Very light	Light	Moderate	Moderate/Heavy	Heavy	Very Heavy

▶**FIGURE 3-22.** This TriNet ShakeMap shows the distribution of shaking during the 1994 Northridge earthquake. This one was created after the earthquake because the system was not available at the time. The system is now operational in southern California.

and areas where ground conditions favor the strongest shaking. They are especially useful in land use planning because they predict the pattern of shaking in future earthquakes along the same fault.

Richter Magnitude Scale

Suppose you were standing on the shore of a lake on a perfectly still evening admiring the flawless reflection of a mountain on the opposite shore. Then a ripple arrives, momentarily marring the reflection. Did a minnow jump nearby, or did a deranged elephant take a flying leap into the distant opposite shore of the lake? Nothing in the ripple as you would see it could answer that question. You also need to know how far it traveled and spread before you saw it.

Useful as it is, the Mercalli Intensity Scale does not help answer some of the basic questions. It tells us nothing about events such as the big 1983 Borah Peak earthquake that shook a region of central Idaho that is almost devoid of people or buildings. Nor does it help answer the vexing question of whether a minnow jumping nearby or a diving elephant across the lake caused the ripple that disturbed the reflection. That is the problem that Charles Richter of the California Institute of Technology addressed when he first devised a new earthquake magnitude scale in 1935.

Richter developed an empirical scale based on the maximum amplitude of earthquake waves measured on a seismograph of a specific type, the Wood-Anderson seismograph. Wave amplitude decreases with distance, so Richter contrived the magnitude scale as though the seismograph were 100 kilometers from the epicenter.

The squiggles the seismograph writes on its records, the **seismograms,** vary greatly in amplitude for earthquakes of different sizes. To make that variation more manageable, Richter chose to use a logarithmic scale to compare earthquakes of different sizes. At a given distance from the earthquake, an amplitude 10 times as great on a seismograph records a magnitude difference of 1.0—an earthquake of magnitude 5 sends the seismograph needle swinging 10 times as high as one of magnitude 4 (▶ Figure 3-23).

The **Richter Magnitude Scale** is simple in principle and easy to use, but actual practice leads to all sorts of complications, which have inspired many modifications. Seismographs, like buildings and people, sense shaking at different frequencies. Tall buildings, for example, sway back and forth more slowly than short ones—they have longer periods of oscillation. P waves, S waves, and surface waves have different amplitudes and different periods. With this variability in earthquake waves, Richter chose to use as the standard for his "local" magnitude, M_L, waves with periods, or back-and-forth sway times, of 0.1 to 3.0 seconds.

The Richter magnitude is now known as M_L. The number quoted in the news media is generally the surface-wave magnitude, M_S, as described below. Seismologists use different magnitude scales, specifically based on the amplitudes of surface waves, P waves, or S waves, or the moment magnitude (see below). Larger earthquakes have longer periods, so different seismographs are used to measure short- and long-period wave motion.

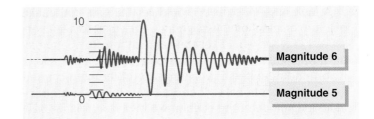

▶**FIGURE 3-23.** An earthquake of magnitude 6 registers with 10 times the amplitude as an earthquake of magnitude 5 from the same location and on the same seismograph. That difference is an increase in 1 on the Richter scale. The horizontal axis on these seismograms is time, and the vertical axis is the ground motion recorded.

Distant earthquakes travel through the Earth's interior at higher velocities and frequencies. To work with distant earthquakes, Beno Gutenberg and Charles Richter developed two more-specific magnitude scales in 1954.

M_S is the surface-wave magnitude, which is calculated in a similar manner to that described for M_L. Surface waves with a period of twenty seconds or so generally provide the largest amplitudes on seismograms. Special seismographs record earthquake waves with such long periods. M_B, the body-wave magnitude, is measured from the amplitudes of P waves. P waves move through the Earth's liquid outer core, but surface waves do not.

Large-Earthquake Characteristics

If the M_L magnitude is above 6.5, the strongest earthquake oscillations, with their lower frequency, may lie below the frequency range of the seismograph. This may cause "saturation" of earthquake records, which occurs when the ground below the seismograph is still going in one direction while the seismograph pendulum, which swings at a higher frequency, has begun to swing back the other way. Then the seismograph does not record the maximum amplitude. So the Richter Magnitude becomes progressively less accurate above M_L 6.5, and a different scale such as the moment magnitude scale becomes more appropriate.

Moment Magnitude (M_W) is essentially a measure of the total energy expended during the earthquake. It is determined from long-period waves taken from broadband seismic records that are controlled by three major factors that affect the energy expended in breaking the rocks. The moment magnitude calculation depends on the seismic moment, which is determined from the shear strength of the displaced rocks multiplied by both the surface area of earthquake rupture and the average slip distance on the fault. The largest of these variables and the one most easily measured is the offset or slip distance.

A fault offset of 1 meter would generate an earthquake of approximate moment magnitude 6.5, whereas a fault offset of 13 meters would generate an earthquake of approximate moment magnitude 9 for typical rupture thicknesses. If you find a fault with a measurable offset that occurred in a single earthquake, then you can infer the approximate magnitude of the earthquake it caused (Figure 3-27, page 50). An earthquake of magnitude 6 indicates ground motion or seismograph swing 10 times as large as that for an earthquake of magnitude 5, but the amount of energy released in the earthquakes differs by a factor of 33 (see Sidebar 3-1).

Below magnitude M_L of 6 or 6.5, the various measures of magnitude differ little, but above that the differences increase with increase in magnitude (▶ Figure 3-24). But moment magnitude 9 is approximately equivalent to M_L of 7.6, body-wave magnitude of 8, and surface-wave magnitude of 8.4.

Earthquake Frequency

In 1954, Gutenberg and Richter worked out the relationship between frequency of occurrence of a certain size of earthquake and its magnitude. Small earthquakes are far more numerous than large earthquakes, and giant earthquakes are extremely rare. Most of the total energy release for a fault occurs in the few largest earthquakes.

A plot of the frequency of earthquakes on a log scale versus their magnitude can often be fit with a straight line, the Gutenberg-Richter frequency–magnitude relationship, as shown in Figure 3-24. It seems likely that earthquakes in a region with numerous faults will follow such a distribution

up to the maximum size earthquake for the area. That maximum size depends on the strength of the rocks and the rate of deformation. The practical upper limit is magnitude 10.

An increase in magnitude above 6 does not cause much stronger shaking so much as it increases the area and total time of shaking. Structures weakened or cracked in the first few seconds of an earthquake are commonly destroyed with continued shaking of no greater violence. Earthquakes of magnitude 5 generally last only two to five seconds, those of magnitude 7 from twenty to thirty seconds, and those of magnitude 8 almost fifty seconds (▶ see Table 3-1).

Ground Acceleration, Shaking Time, and Displacement

Sometimes it helps to think of ground motion during an earthquake as a matter of acceleration, that is, the strength of the shaking. **Acceleration** is normally designated as some proportion of the acceleration of gravity (g); 1.0 g is the acceleration felt by a freely falling body such as what you feel when you step off a diving board. Most earthquake accelerations are less than 1.0 g; a few are more. A famous photograph taken after the San Francisco earthquake of 1906 shows a statue of the eminent nineteenth-century scientist Louis Agassiz stuck headfirst in a courtyard on the campus of Stanford University, its feet in the air (▶ Figure 3-25). Perhaps the statue was tossed off its pedestal at a moment when the vertical ground acceleration was greater than 1 g. It is commonplace after a strong earthquake to find boulders tossed a meter or more from one place to another without having rolled.

The duration of strong shaking in an earthquake depends on the size of the earthquake, including the amount of offset and the length of the total break or rupture. The time that the ground moves in one direction during an earthquake, before the oscillation moves back in the other direction, is similar to the time of initial fault slip in one direction. The total duration of motion is longer because the ground oscillates back and forth. As a result, waves arrive at the seismograph over a period of time. P, S, and surface waves travel at different velocities, so they arrive at the seismograph at different times.

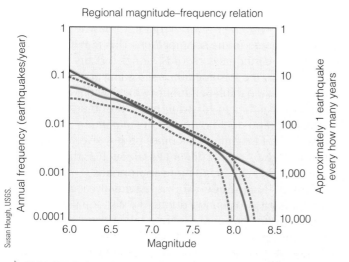

Regional magnitude–frequency relation

▶FIGURE 3-24. This graph plots the Gutenberg-Richter frequency–magnitude relationship for the San Francisco Bay region. The logarithm of the annual frequency of earthquakes plotted against their Richter magnitude generally plots as a straight line, or nearly so (red line). The curved line provides the best fit to the data, which mostly plot between the dashed lines. Different faults plot as somewhat different lines.

Table 3.1 Comparison of Richter Magnitude and Mercalli Intensity Scales

Richter Magnitude (M_L)	Mercalli Intensity (Approx.) at Epicenter	Effect on People and Buildings	Max. Acceleration (Approx. % of g)[a]	Maximum Velocity of Back-and-Forth Shaking (cm/sec.)	Time of Shaking[a] Near Source (sec.)	Displacement or Offset	Surface Rupture Length (km)
<2	I–II	Not felt by most people.	0.1–0.2				
~3	III	Felt indoors by some people.	<1.4	<1	0–2		
~4	IV–V	Felt by most people; dishes rattle, some break.	1.4–9	1–8	0–2		
~5	VI–VII[b]	Felt by all; many windows and some masonry cracks or falls.	9–34	8–31	2–5	~1 cm	1
~6	VIII–IX[c]	People frightened; most chimneys fall; major damage to poorly built structures.	34–124	31–116	10–15	60–140? cm	~8
~7	X–XI	People panic; most masonry structures and bridges destroyed.	>124	<116	20–30	~2 m	50–80
~8	XII	Nearly total damage to masonry structures; major damage to bridges, dams; rails bent.	>124	>150	>30	4.1 m	200–400
~9+	>XII	Nearly total destruction; people see ground surface move in waves; objects thrown into air.	>124		>80	13 m	>1,200

[a]g = 980 cm/sec². Measures the strength of shaking. Note that maximum acceleration above magnitude 6 or 7 does not increase much but duration of shaking does. Also note that a magnitude 6 earthquake shakes 10–15 seconds, providing only a short time to evacuate, if any time at all. A magnitude 7 earthquake provides more time to evacuate, but evacuation is harder to do so with accelerations approaching 1 g.
For example,
- the magnitude 6.5 San Fernando Valley earthquake (1971) lasted 12 seconds,
- the magnitude 7.0 Loma Prieta earthquake (1989) lasted 15 seconds,
- the magnitude 8.25 San Francisco earthquake (1906) lasted 45–60 seconds, and
- the magnitude 7.9 Fort Tejon earthquake (1957) lasted 1–3 minutes.

[b]Magnitude 5 = Intensity VI in the western United States but Intensity VII in the eastern United States because differences in bedrock transmit earthquake waves differently. Bedrock in the western United States slows down earthquake waves more than the older, colder, more rigid, and less compressible basement rocks in the eastern United States.
[c]Magnitude 6 = Intensity VII in the western United States but Intensity IX–XII in the eastern United States. See also Figure 4-57, page 94.

USGS image.

▶ **FIGURE 3-25.** After the 1906 San Francisco earthquake, Louis Agassiz's head was firmly planted in the sidewalk at Stanford University.

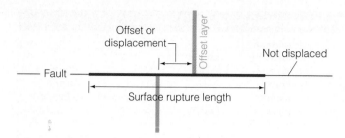

FIGURE 3-26. This diagram shows displacement and surface rupture length on a fault. On either side of the rupture, the fault does not break or offset.

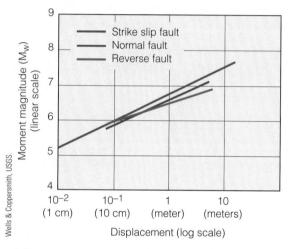

FIGURE 3-27. This graph shows the relationship between the maximum fault offset (see Figure 3-26) during earthquakes on all types of faults to the magnitude of the earthquake.

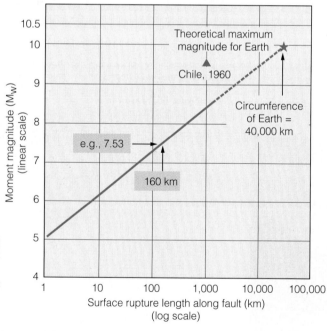

FIGURE 3-28. The relationship between the surface rupture length (see Figure 3-26) on all types of faults is graphed here to the magnitude of the earthquake.

Small **offsets** of a fault release small amounts of energy and generate small earthquakes. If the length of fault and the area of crustal rocks it breaks is large, then it will cause a large earthquake. Because the relationships are consistent, it is possible to estimate the magnitude of an ancient earthquake from the total **surface rupture length** (Figures 3-26, 3-27, and 3-28).

Ground Motion and Failure During Earthquakes

Earthquake waves radiating outward from an earthquake source show a significant decrease in violence of shaking, especially in bedrock and firmly packed soil. On reaching an area of loose, water-saturated soils such as old lakebed clay or artificial fill at the edge of bays, earthquake waves are strongly amplified to accelerations many times greater than nearby waves in bedrock (see the comparison in Figure 3-2). The violence of that shaking depends on the frequency of earthquake oscillation of the earthquake waves compared with the frequency of the ground oscillation. The lower frequency oscillations of surface waves often correspond to the natural frequency of oscillation of loose, water-saturated ground. Recently developed computer-generated maps of ground motion called ShakeMaps show the distribution of maximum acceleration and maximum ground velocity for many potential earthquakes; they can be used to infer the likely level of damage (see, e.g., Figure 3-22).

Liquefaction

Some soils that ordinarily seem perfectly stable become almost liquid when shaken and then solidify when the shaking stops. Many soft sediment deposits consist of extremely loose sand or silt grains with water-filled pore spaces between them. An earthquake can shake these deposits down to a much tighter grain packing, expelling water from the pore spaces. The escaping water carries sediment along as it rapidly flows to the surface, creating sand boils and mud volcanoes that are typically a few meters across and several centimeters high (Figure 3-30). This process of **liquefaction** can be fatal to buildings (Figures 3-31, 3-32, and 3-33). Examples of liquefaction include damage to apartment buildings in the Marina district of San Francisco in the 1989 Loma Prieta earthquake; the raised marine terrace of Anchorage, Alaska, in the 1964 earthquake; low-lying harbor areas of Kobe, Japan, in 1995; and parts of Izmit, Turkey, in 1999 (Figures 3-32 and 3-33). In Anchorage, 100 kilometers from the epicenter, the ground began to shake fifteen seconds later and continued for seventy-two seconds. Clays liquefied in the Turnagain Heights district causing bluffs up to 22 meters high to collapse along 2.8 kilometers of coastline. The swiftly flowing liquid clay carried away many modern frame houses that were as much as 300 meters inland. The

On September 19, 1985, the oceanic lithosphere sinking through the trench off the west coast of Mexico suddenly slipped in an area 350 kilometers from Mexico City. The resulting earthquake struck with a magnitude of 7.9. Slippage along almost 200 kilometers of the fault filled a seismic gap that had worried seismologists for more than a decade (see also Figure 4-2, page 67). Large aftershocks two and five days after the main earthquake registered magnitudes 7.5 and 7.3.

Accelerations of 16 percent of g near the coast decreased with distance to 4 percent of g in Mexico City. Most of the older parts of Mexico City are built on clays and sands deposited in the bed of Lake Texcoco. These materials amplified the shaking to cause severe damage to 500 buildings despite their distance from the epicenter. Meanwhile, the buildings of the National University of Mexico stood undamaged on the bedrock hills around the old lakebed.

Many of the damaged buildings were six to sixteen stories tall, a height that vibrates in resonance with the one- to two-second frequencies of the ground shaking. Shorter buildings were generally not structurally damaged. Nor was the thirty-seven-story Latin American Tower, which was built in the 1950s. Its vibration period of 3.7 seconds was not in resonance with the ground vibration.

Many of the damaged buildings were poorly designed. Some had weak first stories that collapsed (▶Figure 3-29). Adjacent buildings of different heights swayed at different frequencies, banging into one another until one or both collapsed (see also Figures 3-53 to 3-55, page 60). Tall buildings with setbacks, to smaller width at a higher floor, commonly failed at the setback. Total property damage was $6.7 billion. Roughly 9,000 people died in collapsing buildings in Mexico City, and 30,000 were injured.

▶**FIGURE 3-29.** The collapse of this seven-story office building in Mexico City in 1985 was caused by weak first-story failure.

Christopher Arnold, Building Systems photo.

$1.74 billion in property damage included roads, bridges, railroad tracks, and harbor facilities.

Shaking of the 1971 San Fernando earthquake near Los Angeles caused liquefaction of the upper face of the Van Norman Dam, nearly causing its failure just upstream from the homes of tens of thousands of people. On June 16, 1964, a magnitude 7.4 earthquake in Niigata, Japan, caused liquefaction and settling of approximately one-third of the city. A magnitude 7.4 earthquake in Izmit, Turkey, had a similar effect. In both cases, strongly constructed apartment buildings in some areas remained intact but toppled over on their sides (Figures 3-31 and 3-32).

USGS photo.

▶**FIGURE 3-30.** This sand boil was formed during the 1989 Loma Prieta earthquake in California.

National Geophysical Data Center, NOAA.

▶**FIGURE 3-31.** These strong buildings in Niigata, Japan, fell over when the sediments below them liquefied during the 1964 earthquake.

Quick Clays

Some clays are deposited in cold salt water, with a charge of positive ions on the surface of each microscopic flake of clay. Flakes with the same charges make them mutually repellent, so they stack like cards on edge. If this type of clay deposit is raised to a level where it is no longer beneath salt water, fresh water rinses the positive ions out of the stack, removing the repellent charges between clay flakes and leaving the stack unstable and susceptible to collapse. The deposit is likely to run like water when an earthquake jars it. Then it settles into a deposit with the clay flakes lying flat, as they normally do. The new deposit is stable, and will not run again. Quick clay flows are most common on northern coasts such as those of eastern Canada and Scandinavia.

Landslides

Earthquakes often trigger landslides: The effect is intuitively obvious. If you place a pile of loose dry sand on a table, then sharply whack the side of the table, some of the sand will immediately slide down the pile. In nature, if sand or soil in the ground is saturated with water, the quick back-and-forth acceleration has a pumping effect on the water between the grains. Water is forced into spaces between the grains with each pulse of the earthquake. This sudden increase in the water pressure in the pore spaces effectively pushes the grains apart and permits the mass to slide downslope.

Mehmet Celebi photo, USGS.

▶**FIGURE 3-32.** Liquefaction caused the toppling of this building during the 1999 Izmit, Turkey, earthquake.

Mehmet Celebi photo, USGS.

▶**FIGURE 3-33.** Liquefaction of the foundation under the left side of this building during the August 17, 1999, earthquake in Izmit, Turkey, caused settling of the left section and destruction of the middle section.

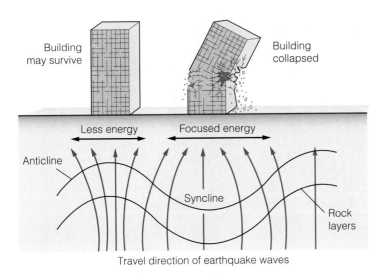

FIGURE 3-34. Earthquake energy is focused in synclines and dissipated over anticlines. Neighboring buildings can respond quite differently over unseen structures at depth.

Effect of Bedrock Structure

Variations in damage after the 1994 Northridge earthquake showed that different orientations of rock contacts at depth can refract earthquake waves with effects much like those of a magnifying glass. Folds in rocks deep below the surface can similarly affect the amount of ground motion and damage at the surface, resulting in focused earthquake wave energy in some areas and diffusion of this energy in others (▶Figure 3-34).

Damage Control

Earthquakes don't kill people—falling buildings and highway structures do. No one suffers much injury from a few seconds of shaking during an earthquake. The main danger is overhead. The Uniform Building Code provides a seismic zonation map of the United States that indicates the level of construction standards required to provide safety for people inside a building (▶Figure 3-35).

Land use planning and building codes are the best defenses against deaths, injuries, and property damage in earthquakes. Zoning should strictly limit development in areas along active faults, situations prone to landslides, or on soft mud or fill (see "Case in Point: Armenian Earthquake, 1988"). Parks and golf courses are good uses for such areas.

Building codes in areas of likely earthquake damage should require structures that are framed in wood, steel, or have appropriately reinforced concrete. They should forbid masonry walls made of brick, concrete blocks, stone, or mud that support roofs (see the discussion of the earthquakes in India in 2001 and Iran in 2003, pages 88 and 90).

CASE IN POINT
Armenian Earthquake, 1988

The Armenian earthquake of December 7, 1988, struck in a zone of continent-to-continent collision, a compressional environment. The magnitude was 6.9. The ground shook for thirty seconds with accelerations as great as 0.8 g. The ground along the fault was offset by as much as 2 meters vertically on the Spitak Fault.

At least 170 schools and kindergartens and eighty-four hospitals were partly or totally destroyed. More than 25,000 people died. Eighty percent of the doctors and nurses in the region were killed or injured. Most of the casualties were in stone buildings.

Almost 95 percent of the buildings framed in reinforced concrete in the town of Leninakan collapsed, generally because they were standing on soft sediment as much as 4 kilometers thick. No failures of similar buildings occurred in the town of Kirovakan, which is on solid bedrock.

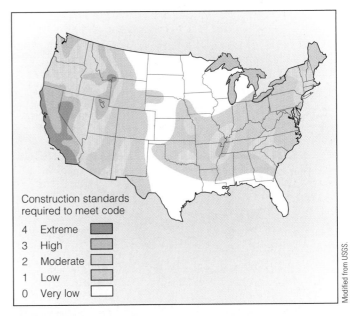

FIGURE 3-35. The Uniform Building Code seismic zone map for the contiguous United States shows the construction required to address various levels of earthquake risk. See text for details.

Load-bearing masonry walls of any kind are likely to shake apart and collapse during an earthquake, dropping heavy roofs on people indoors. Houses can be built to withstand a severe earthquake well enough to minimize the risk to people inside. Even if a properly framed building is severely damaged, it is unlikely to collapse.

▶**FIGURE 3-36.** A major elevated freeway in the low-lying waterfront area of Kobe, Japan, fell over on its side during the 1995 earthquake.

▶**FIGURE 3-38.** The steel reinforcing bars in the rigid concrete of this freeway support column shattered during the 1994 Northridge earthquake. Unsupported, the reinforcing bars buckled and failed.

▶**FIGURE 3-37.** Even elevated freeways that were held together with bolts and welds did not fare well in the 1995 Kobe, Japan, earthquake. This bus managed to stop just before crashing down onto the dropped continuation of the roadway (lower left).

▶**FIGURE 3-39.** Segments of the Interstate 5–California Highway 14 interchange collapsed while under construction in the 1971 San Fernando Valley earthquake. They were rebuilt with the same specifications, not retrofitted, and collapsed again, as shown here, in the 1994 Northridge earthquake.

Reinforced concrete often breaks in large earthquakes, leaving the formerly enclosed reinforcing steel free to buckle and fail (▶Figures 3-36 to 3-39). Prominent recent examples include the collapse of elevated freeways during the earthquake of 1971 in the San Fernando Valley, the Loma Prieta earthquake of 1989, the Northridge earthquake of 1994, and the Kobe, Japan, earthquake of 1995. Reinforced concrete fares much better if it is wrapped in steel to prevent crumbling. Much of the strengthening of freeway overpasses in California during the 1990s involved fitting steel sheaths around reinforced concrete supports. New construction often involves wrapping steel rods around the vertical reinforcing bars in concrete supports.

Houses framed with wood generally have enough flexibility to bend without shattering. The house may bounce off its foundation during an earthquake, but it is unlikely to collapse. Retrofitting to minimize the damage during strong earthquake motion can provide additional protection. Commercial buildings with frames made of steel beams welded or bolted together are generally flexible enough to resist collapse during an earthquake.

Although most well-built wood-frame houses will not collapse during an earthquake, damage may make the house uninhabitable and worthless. Because a house is typically the largest investment for most people, total loss of its value can be financially devastating. In earthquake-prone areas it may be worth buying earthquake insurance even though that insurance can be extremely expensive because homeowners insurance does not cover such damage (recall discussion in Chapter 1). Earthquake insurance can cover only the cost of replacing the house, not the land. The difference is often large, as in the expensive real estate of western California.

Weak floors at any level of a building are a major problem. Garages and storefronts commonly lack the strength to resist major lateral movements, as do commercial buildings with too many unbraced windows on any floor. Lack of diagonal bracing permits lateral movement of the upper floors and the collapse of either individual floors or the whole building (▶ Figures 3-40 to 3-47).

In general, walls of all kinds should be well anchored to floors and the foundation (Figure 3-47). The walls of many houses built before 1935 merely rest on the concrete foundation. The only thing that holds the house in place is its weight. Large earthquakes often shake older houses off their foundations, destroying them (▶ Figure 3-48). Most building codes now require builders to bolt the framing to the foundation—an excellent example of an inexpensive change that can make an enormous improvement. Other recommendations to minimize damage from earthquakes include bolting bookcases and water heaters to walls, and securing chimneys and vents with brackets (Figure 3-47).

Many buildings 100 or more years old have wooden floor beams loosely resting in notches in brick or stone walls (▶ Figure 3-49). Earthquake shaking may pull those beams out of the walls, causing the building to collapse. Some bridge decks and floors of parking garages are not strongly anchored at their ends because they must expand and contract with changes in temperature. A strong earthquake can shake them off their supports (▶ Figure 3-50). Many external walls are loosely attached to the building framework. They may break free during an earthquake and collapse onto

James W. Dewey photo, USGS.

▶**FIGURE 3-40.** Unbraced ground-floor garages of this apartment building in Reseda collapsed on cars in the 1994 Northridge earthquake.

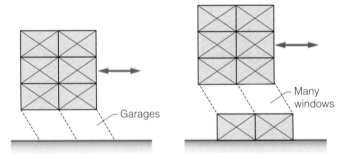

▶**FIGURE 3-41.** A weak ground floor or poorly braced floors at any level are a recipe for lateral failure.

Roger Hutchison photo.

▶**FIGURE 3-42.** This second floor with many windows collapsed in the 1995 Kobe, Japan, earthquake.

EARTHQUAKES AND THEIR DAMAGES **55**

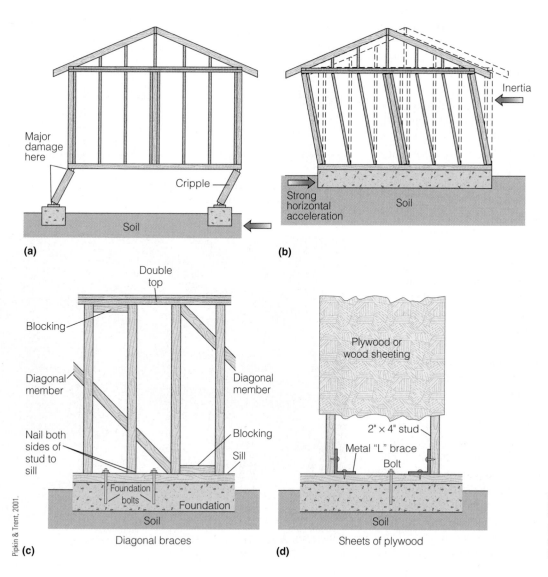

FIGURE 3-43. Diagonal cross braces in walls help keep buildings standing during an earthquake.

Compression of one diagonal

Extension of the other

the street below (▶Figure 3-51). Overhanging parapets are common on old brick or stone buildings. Earthquakes commonly crack them off where the external wall joins the roofline. Even if the building remains standing, falling walls may crush people and cars below (▶Figures 3-6 and 3-52).

Glass is too rigid to fare well during large earthquakes. Shattering glass is one of the most common causes of injuries in earthquakes. Broken glass rained down on the street from the windows of a large department store in downtown San Francisco during the Loma Prieta earthquake of 1989.

Safety glass is now required in the ground floor windows of commercial buildings but not in the upper floors. Glass systems in modern high buildings are designed to accommodate routine sway from wind, and they also perform fairly well during earthquakes. Safety glass similar to that used in cars helps, as does polyester film bonded to the glass.

Earthquakes commonly break gas mains and electric wires, which start fires that firefighters cannot readily fight if the water mains are also broken. Matches or candles are likely to ignite gas in the air. It helps if water and gas mains

Major damage here

Cripple

Soil

(a)

Inertia

Strong horizontal acceleration

Soil

(b)

Double top

Blocking

Diagonal member

Diagonal member

Nail both sides of stud to sill

Blocking

Sill

Foundation bolts

Foundation

Soil

Diagonal braces

(c)

Pipkin & Trent, 2001.

Plywood or wood sheeting

2" × 4" stud

Metal "L" brace

Bolt

Soil

Sheets of plywood

(d)

FIGURE 3-44. Even single-story houses can have similar problems. They can be easily strengthened with diagonal braces or sheets of plywood.

FIGURE 3-45. This U.S. Geological Survey building in Menlo Park, California, near the San Andreas Fault, has been retrofitted with diagonal braces to withstand earthquake shaking. Construction is precast-concrete floor and roof slabs hoisted onto a steel frame.

Donald Hyndman photo.

Mehmet Celebi photo, USGS.

FIGURE 3-46. This modern concrete building in Izmit, Turkey, lacked cross braces or shear walls. It collapsed in the August 17, 1999, earthquake.

are flexible and stoves, refrigerators, and television sets are well anchored to floors or walls.

Taller buildings sway more slowly than their shorter neighbors. Adjacent tall and short buildings tend to bang against each other at the top of the shorter one (▶ Figures 3-53, 3-54, and 3-55). This commonly breaks the taller building at about the level of the top of its shorter neighbor. Buildings should be either firmly attached or

CASE IN POINT
Kobe Earthquake, Japan, 1995

The earthquake that struck Kobe, Japan, on January 17, 1995, had a magnitude of 7.2 (M_w = 6.8). The cause was horizontal slippage along some 40 kilometers of the Nojima Fault, a continental transform fault that is similar to the San Andreas Fault in California. In spite of earthquake design and construction that is among the best in the world, the Kobe earthquake killed more than 5,000 people and caused $355 billion in property damage in this modern city with a population of 1.5 million.

Shaking lasted fifteen seconds but reverberated for 100 seconds in soft sediments around Kobe Bay. It damaged at least 180,000 buildings. One or more floors collapsed at various levels of a building, most commonly at floors weak with too many windows (Figure 3-42). In some cases, the ground and building shook in resonance, causing the top of the building to move in one direction as the ground moved in the opposite direction. That caused the collapse of individual floors at the midlevels of some buildings. Automatic shut-off valves on residential gas lines failed, and broken gas mains fed large fires. Most of the heavy concrete pillars were designed to be strong and rigid, instead of flexible. Many broke, causing collapse of bridges, elevated highways, and rail lines.

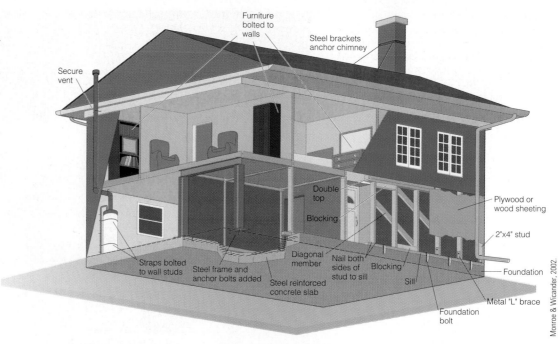

▶**FIGURE 3-47.** Even single-story houses can have problems. They can be easily strengthened with diagonal braces or sheets of plywood.

Furniture bolted to walls

Steel brackets anchor chimney

Secure vent

Double top

Blocking

Plywood or wood sheeting

2"x4" stud

Diagonal member

Nail both sides of stud to sill

Blocking

Foundation

Straps bolted to wall studs

Steel frame and anchor bolts added

Steel reinforced concrete slab

Sill

Metal "L" brace

Foundation bolt

Monroe & Wicander, 2002.

Carl W. Stover photo, USGS.

▶**FIGURE 3-48.** This wood-frame house in Watsonville, California, southeast of Santa Cruz, was shaken off its foundation during the 1989 Loma Prieta earthquake.

Donald Hyndman photo.

▶**FIGURE 3-49.** Floor joists rest loosely in this structural brick wall in a century-old building in Missoula, Montana.

Mehmet Celebi photo, USGS.

▶**FIGURE 3-50.** The precast floor of the Northridge Fashion Center's three-year-old parking garage, which was supported on the ledge of a supporting beam, failed during the 1994 Northridge, California, earthquake.

▶**FIGURE 3-51.** The exterior facade loosely hung on a building framework in the community of Reseda collapsed in the 1994 Northridge earthquake. Directional shaking detached one wall, while leaving the perpendicular wall intact.

stand far enough apart that they do not bash each other during an earthquake.

Most earthquakes last less than a minute. People who find themselves inside a building during an earthquake are well advised to stay there, if for no other reason than because the earthquake will probably end before they can get out. A cursory examination of Table 3-1 will show that for earthquakes of magnitudes of less than 6, the shaking time is so short that by the time you realize what is happening, there is little time to move to safety. For greater magnitudes, there may be enough time but the accelerations are so high that it is hard to stay on your feet. If you do attempt to leave a building, avoid elevators.

The latest guidelines for surviving an earthquake-collapsed building suggest that, you get outside and away from the building if there is time. If not, lie down next to a sturdy object that will not completely flatten if the build-

ing collapses, not under a table or a doorway, as once suggested. Even collapsing concrete leaves triangular spaces next to an object that is heavy and sturdy.

Earthquake Effects on Buildings

Research in response to the 1994 Northridge earthquake included numerical modeling of the effect of a M_w 7.0 earthquake on a building twenty stories high framed in I-beams according to the latest earthquake codes for the Los Angeles area. Over a ground area of 1,000 square kilometers (386 square miles), the model predicted that the first story would move 35 centimeters (13.8 inches) off center, three times what was then considered severe. At the top of the building, the predicted lateral displacement would reach 3.5 meters within 7 seconds, followed by continued swaying of 1.5 meters with a period of 2.5 seconds. The most damaging movement in these models was forward just when the back motion of the ground strikes the building, creating a whiplash effect.

Another simulation showed cracks opening at the welds between vertical columns with a load only 25 percent of the bearing capacity of the column. In some locations, an exterior column that broke during back motion could no longer prevent lateral motion in the first above-ground story, and the building collapsed.

The Federal Emergency Management Agency (FEMA) responded in 2001 with new guidelines for steel construction in areas prone to earthquakes. Most high buildings built in the last thirty years have welded steel frames designed to resist earthquake motions. Many of the problems discovered after the Northridge earthquake were in poor welds, a construction rather than a design issue. Other problems were caused by design failures of the connections between steel beams. Most of the identified problems resulted from the gap between building codes and actual construction,

▶**FIGURE 3-52.** The fourth-story wall and overhanging brick parapet of an unreinforced building in San Francisco collapsed onto the street during the 1989 Loma Prieta earthquake. It crushed five people in their cars.

▶**FIGURE 3-53.** Two buildings of similar height built too closely together swayed at a period close to that of the 1985 Mexico City earthquake. The building on the left hammered the one in the middle, collapsing one story of the taller building.

as has been the case in Mexico City and many other earthquake-prone locations (▶ Figure 3-56).

Building Vibration and Oscillation

Earthquakes shake the ground with frequencies of 0.1 to 30 oscillations per second. Small earthquakes generate a larger proportion of higher frequency vibrations. If the natural oscillation frequency of a building is similar to that of the ground, it may resonate the way a child on a swing pulls in resonance with the oscillations of the swing. In both cases, the resonance greatly amplifies the motion (Sidebar 3-2).

Buildings need to be designed to avoid matching their natural vibration frequency with that of the shaking ground beneath (Sidebar 3-2). Even buildings without weak floors may collapse if their upper floors move in one direction as the ground snaps back in the other.

Base Isolation

Short buildings up to several stories high vibrate at high frequency and do not sway much. In recent years, seismic engineers have tried to minimize the shaking, and therefore the damage, by isolating buildings from the shaking ground. They place the building on thick rubber pads that act like the springs and shock absorbers of a car that isolate us from many of the bumps in a road. This is generally done during initial construction. Historic buildings can be retrofitted with **base isolation** pads after an expensive but sometimes worthwhile process of jacking up the building (▶ Figures 3-57, 3-58, and 3-59).

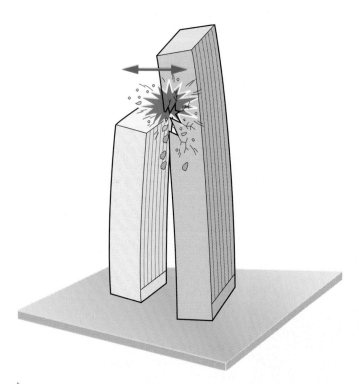

▶**FIGURE 3-54.** Adjacent buildings of quite different heights will sway at different frequencies, so they collide during earthquakes.

▶**FIGURE 3-55.** This flexible parking garage was locked between two shorter rigid buildings. Its upper floors, free to sway, collapsed in the 1985 Mexico City earthquake.

▶**FIGURE 3-56.** This twenty-one-story steel-frame office building completely collapsed during the 1985 Mexico City earthquake. The building of similar height in the upper right background better met the construction code requirements and survived.

Sidebar 3.2

Buildings of different heights sway at different frequencies, as inverted pendulums. The ground also vibrates at different frequencies. Buildings need to avoid shaking at a frequency similar to that of the ground.

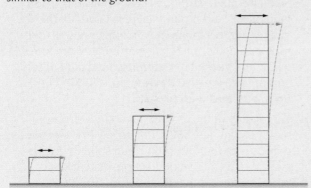

GROUND MOTION

Short building
- Rigid 1- to 2-story building oscillates at 5–10 Hz*
- Shakes back and forth rapidly (high frequency)
- Thus, period* is 1/5 to 1/10 = 0.2–0.1 sec

Midheight building
- 5- to 10- story building oscillates at 0.5–3 Hz
- Shakes back and forth less rapidly (inter-mediate frequency)
- Thus, period is 1/0.5 to 1/3 = 2–0.3 sec

Tall building
- Flexible 20-story building oscillates at ~0.2 Hz
- Sways back and forth slowly (low frequency)
- Thus, period is 1/0 or 5 sec

Therefore:
- A short, rigid building on soft sediment often does okay in an earthquake.**
- Soft sediment vibrates at low frequency (dull thud when hit with a hammer), but building does not.

- A tall, flexible building on bed-rock often survives better in an earthquake.
- Bedrock vibrates at high frequency (rings when hit with a hammer), but building does not.

*Period is the time for a single forward-and-back motion. Hz = Hertz = cycles of back-and-forth motion per second.

**Except for secondary effects such as liquefaction or landslides.

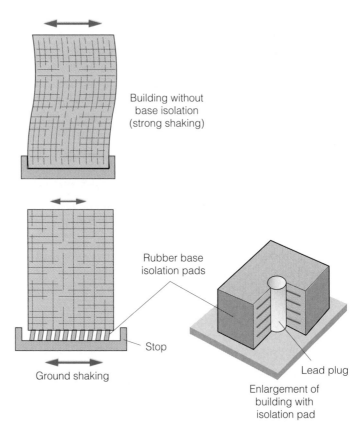

Building without base isolation (strong shaking)

Rubber base isolation pads

Stop

Ground shaking

Lead plug

Enlargement of building with isolation pad

▶**FIGURE 3-57.** Base isolation pads permit a building to shake less than the violent shaking of the ground. They often consist of laminations of hard rubber and steel in a stack 50 centimeters high. A lead plug in the middle helps to dampen the vibrations. Compare Figure 3-58.

▶**FIGURE 3-58.** A black base isolation pad, 30 centimeters thick (black "cube" in photo center), was placed under the drilled-off foundation of the same building.

▶**FIGURE 3-59.** The Salt Lake City, Utah, city and county office building was retrofitted by jacking it up and placing base isolation pads under the foundation to isolate it from the ground.

KEY POINTS

✓ Earthquakes are felt as a series of waves: first the compressional P waves, then the higher amplitude shear-motion S waves, and finally the slower surface waves. **Review p. 41; Figures 3-10 and 3-17.**

✓ Earthquakes are measured using a seismograph, which is essentially a suspended weight that remains relatively motionless as the Earth moves under it. **Review pp. 39–40; Figures 3-8 and 3-9.**

✓ Faults move in both up–down and lateral motions controlled by Earth stresses in different orientations. **Review pp. 39–42; Figure 3-13.**

✓ Earthquakes are caused when stresses in the Earth deform or strain rocks until they finally snap. Small strains may be *elastic* where the stress is relieved and they return to their original shape, *plastic* where the shape change is permanent, or *brittle* where the rocks break in an earthquake. **Review pp. 43–44; Figures 3-14 to 3-16.**

✓ We can estimate the distance to an earthquake using the time between arrival of the S and P waves to a seismograph and then locate the earthquake by plotting the distances from at least three seismographs in different locations. **Review pp. 44–45; Figures 3-17 and 3-18.**

✓ The size of an earthquake as estimated from the degree of damage at various distances from the epicenter is recorded as the Mercalli Intensity; the strength of an earthquake can be determined from its amplitude on a specific type of seismograph as the Richter Magnitude. An in-

crease of one Richter magnitude corresponds to a 10 times increase in ground motion and about a 33 times increase in energy. **Review pp. 45–49; Figures 3-21 to 3-23; Table 3-1; and Sidebar 3-1.**

✓ Small frequent earthquakes are caused by short fault offsets and rupture lengths; larger earthquakes are less frequent with longer offsets and rupture lengths; and giant earthquakes are infrequent and caused by extremely long fault offsets and rupture lengths. **Review pp. 48–50; Figures 3-24 and 3-26 to 3-28.**

✓ The rigidity of the ground has a large effect on building damage. Soft sediments with water-filled pores shake more violently than solid rocks. The ground may fail by liquefaction, clay flake collapse, or landsliding. **Review pp. 50–52.**

✓ People die because things fall on them, not because of the shaking itself. Therefore, weak, rigid buildings and highway overpasses are most dangerous. Flexible, well-built wood-frame houses may be damaged but not collapse on people. **Review pp. 53–55.**

✓ Weak floors are likely to collapse, as are floors without lateral bracing. Walls that are not anchored to floors and the roof can separate and fall. **Review pp. 55–58; Figures 3-40 to 3-45, 3-49 to 3-51.**

✓ Where the frequency of vibration or back-and-forth oscillation is the same in the building as in the ground under it, the shaking is strongly amplified and the building is more likely to fall. **Review pp. 59–62; Sidebar 3-2.**

IMPORTANT WORDS AND CONCEPTS

Terms

acceleration, p. 48
amplitude, p. 38
base isolation, p. 61
elastic rebound theory,
   p. 43
epicenter, p. 45
faults, p. 41
focus, p. 45
frequency, p. 38
liquefaction, p. 50
Mercalli Intensity Scale,
   p. 46
moment magnitude, p. 47
normal fault, p. 42
offsets, p. 50
period, p. 38
P waves, p. 41

reverse fault, p. 42
Richter Magnitude Scale,
   p. 47
seismic gap, p. 37
seismogram, p. 47
seismograph, p. 39
ShakeMaps, p. 46
strain, p. 43
stress, p. 43
strike-slip fault, p. 42
surface rupture length,
   p. 50
surface waves, p. 41
S waves, p. 41
thrust fault, p. 42
wavelength, p. 38

QUESTIONS FOR REVIEW

1. What is the approximate highest frequency of vibration (of back-and-forth shaking) in earthquakes?

2. Which type or types of earthquake waves move through the mantle of the Earth?

3. Which type of earthquake waves shake with the largest amplitudes (largest range of motion)?

4. Where is the safest place to be in an earthquake?

5. What is meant by the *elastic rebound* theory?

6. How is the distance to an earthquake determined?

7. What does the Richter Magnitude Scale depend on?

8. A magnitude 7 earthquake has how much higher ground motion than a magnitude 6 earthquake on a seismogram?

9. In addition to ground offset along a fault and shaking of a building, what types of *ground* failure can lead to severe damage of a building?

10. What kinds of structural materials make dangerously weak walls during an earthquake?

11. What type of wall strengthening is commonly used to prevent a building from being pushed over laterally during an earthquake?

12. Why do the floor or deck beams of parking garages and bridges sometimes fail and fall during an earthquake?

13. Where a tall building is right next to a short building, why is the tall building often damaged? Why does the damage occur and where in the building?

14. What type of feature is commonly used to prevent a building from shaking so much during an earthquake?

15. If a building is to be built on bedrock, is that good or bad? Why?

FURTHER READING

Assess your understanding of this chapter's topics with additional quizzing and conceptual-based problems at:

 http://earthscience.brookscole.com/hyndman.

CHAPTER 4

The Tangshan earthquake of July 27, 1976, leveled the city and killed at least 250,000 people.

EARTHQUAKE PREDICTION AND TECTONIC ENVIRONMENTS

Predicting Earthquakes

As in other aspects of natural hazards, we distinguish between predictions and forecasts. Earthquake **forecasts** involve statements of where and how frequently an event is likely to occur and how large it might be. Earthquake **predictions** involve statements about specifically when and where an earthquake is expected to occur. We now know enough about earthquakes and the tectonic environments that control them to make reasonably reliable forecasts. However, prediction is another matter. Few avenues of research in any subject have attracted more attention and money than the effort to find ways to predict earthquakes, and few have yielded such meager results. No one can now effectively predict when an earthquake will strike. Charles Richter, the developer of the magnitude scale, once remarked that "Only fools, charlatans, and liars predict earthquakes."

We can begin our discussion of earthquake forecasts by noting that most earthquakes will be on plate boundaries. Beyond that, we infer that former earthquakes on a particular boundary provide guidance to future earthquakes. The sense of motion during a future earthquake is dictated by the relative motion across the plate boundary—thrust faults are typically associated with subduction zones and continent–continent collision boundaries, strike-slip faults move along transform boundaries, and normal faults move in spreading zones.

The largest earthquake expected for a particular fault generally depends on the total fault length or the longest segment of the fault that typically ruptures (compare Figure 3-28, page 50). A 1,000-kilometer-long (620 miles) subduction zone such as that off Oregon and Washington or a 1,200-kilometer-long transform fault such as the San Andreas Fault of California, if broken along its full length in one motion, would generate a giant earthquake. Normal faults such as the Wasatch Front of Utah often break in shorter segments, so somewhat smaller earthquakes are to be expected in such areas.

Earthquake Predictions and Consequences

Imagine yourself the governor of a state, perhaps California. Now imagine that your state geological survey has just given you a perfectly believable prediction that an earthquake of magnitude 7.3 will strike within a large city next Tuesday afternoon. What would you do?

If you went on television to announce the prediction, you would run the risk of causing a hysterical public reaction that might cause major physical and economic damage before the date of the predicted earthquake. If you did not announce the prediction, you would stand accused of withholding vital information from the public. Imagine the consequences if you announced the prediction and nothing happened. The problem is a political minefield.

However, the potential benefits in saved lives and minimizing economic damage make short-term predictions a desirable goal. If, for example, a warning of an impending earthquake could be given several minutes before a large event, people could be evacuated from buildings, and bridges and tunnels could be closed, trains stopped, critical facilities prepared, and emergency personnel mobilized.

One main objective of earthquake research has been to provide people with some reasonably reliable warning of an impending earthquake. Those efforts have succeeded in determining the locations for earthquakes at some future time. They have been less successful in finding ways to predict that an earthquake will strike a specific area at a specific time. It is perfectly reasonable to say that earthquakes are far more likely to strike California than North Dakota. However, it is still just as impossible to know exactly when earthquakes will strike California as it is to know in July the dates of next winter's blizzards in North Dakota.

In the current state of knowledge, earthquakes appear to be as inherently unpredictable in the short term as are the oscillations of the stock market.

Earthquake Precursors

One main avenue of research has been to explore phenomena that warn of an imminently impending earthquake. Few of those efforts have proved useful, though that may be changing, as we discuss below.

The fault motions that cause earthquakes happen mainly because plate movements cause stresses in the earth's crust. Plates appear to move at constant speeds, and it seems reasonable to suppose that rock strength in a major fault zone might be constant through time. If so, stresses might accumulate enough to break the stuck fault loose at a predictable time. However, nature is more complex.

It is fairly common to find that a stress applied at a constant rate does not produce results at a constant rate. Broadly speaking, this is because the natural situation is far more complicated than our mental image. More specifically, every time a fault slips, for example, it juxtaposes different rocks of different strengths, thus changing the terms of the problem of predicting when it will slip again. A large fault movement may well change the stress patterns along other faults, which dramatically changes their chances of future movement.

Some hope persists that swarms of minor earthquakes, or **foreshocks,** may warn of major earthquakes if they announce the onset of fault slippage. But they precede only a small minority of large earthquakes in a way that might make them distinguishable from the background of common small earthquakes. Recently, it became possible to precisely determine the focus of large numbers of minute earthquakes. Thus, the changes with time in the detailed distribution of tiny "crackling" movements along a fault may ultimately lead to some kinds of predictions. No one can know until after the commotion is over whether the preceding small earthquake was a foreshock or the main event.

Although there have been many false alarms, there have been a few successes. One of the most notable successful predictions was for the Haicheng earthquake in China on February 4, 1975. In Haicheng, 90 percent of the buildings were destroyed or severely damaged. Were it not for the prediction two days in advance in which 1 million people were ordered to evacuate and remain outside in the winter cold, the magnitude 7.3 earthquake would have killed many tens of thousands to hundreds of thousands of people.

The Haicheng prediction was based mainly on a pronounced foreshock sequence (see the following Case in Point). Other unusual observations included a rise in the ground surface in the area by 2.5 millimeters (0.1 inch) in nine months and unusual changes in the Earth's magnetic field. Many people reported changes in groundwater levels, while others reported peculiar animal behavior.

Haicheng and Tangshan Earthquakes, China, 1975 and 1976

Essentially, the only successful prediction of a major earthquake was the 1975 magnitude 7.3 Haicheng event in northeastern China. Early the afternoon of February 4, 1975, officials warned 3 million people to expect a large earthquake within the next two days and to remain outside to avoid being in their houses in the event of collapse. In spite of the winter cold, people complied, in part because frequent small earthquakes already frightened them. The prediction was based on an increase in small earthquakes, unusual changes in elevation and tilts of the ground near the Jinzhou Fault, changes in groundwater levels, and changes in the Earth's magnetic field. Then there was the strange behavior of animals beginning a few days before the earthquake—chickens flying into trees, dogs barking wildly, animals not entering barns, birds circling without landing, and snakes slithering out of holes in winter and freezing to death. Taken individually, any of these odd events would not be taken as evidence that a large earthquake was imminent. Other predictions of earthquakes

in China, both before and after this event, turned out to be false alarms. However, somehow the Haicheng prediction turned out to be correct. At 7:36 P.M. the same day, a major earthquake collapsed more than 90 percent of the houses. The number of casualties is unknown, but according to official Chinese reports no one died.

Just over a year later, and some 400 kilometers to the southeast, the somewhat larger magnitude 7.6 Tangshan earthquake was completely unexpected. In contrast to the Haicheng event, the Tangshan earthquake on July 27, 1976, showed no precursor activity before a sudden right-lateral shift on a northeast-trending fault leveled the city (▶Figure 4-1). At least 250,000 people died in and around that industrial city with a population of about 1 million. Another half-million were injured. Dozens died in the collapse of walls and heavy roofs of their old adobe and brick homes in Beijing, nearly 150 kilometers to the west.

Hebei Provincial Seismological Bureau; USGS.

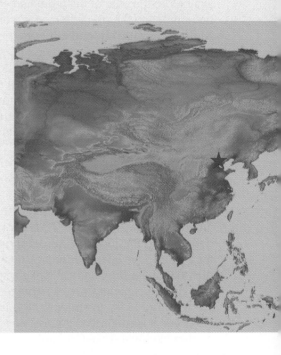

▶**FIGURE 4-1.** The damage was nearly complete during the Tangshan earthquake.

Other large events in the same general region, including the magnitude 7.6 earthquake of July 27, 1976, in Tangshan were not predicted; that event killed 250,000 out of a population of 1 million.

The radioactive gas radon is one step that uranium atoms pass through in their long decay chain, which finally ends with the formation of lead. Radon is one of the rare gases. It forms no chemical compounds and remains in rocks only

because it is mechanically trapped within them. Uranium is a widely distributed element, so most rocks contain small amounts of radon, which escapes along fractures. Radon is responsible for nearly all of the background radiation in well water and presents hazardous long-term exposure in many parts of the world.

Geologists in the former Soviet Union reported during the 1970s that they had observed a strange set of symptoms in the days immediately before an earthquake. These included a rise of a few centimeters (an inch or two) in the surface elevation above the fault, a drop in water table elevations, and a rise in background radioactivity in wells. Evidently, the rocks along the fault were beginning to break. The water table dropped because water was draining into the new fractures, and radon was escaping through the fractures. It all seemed to make sense as a set of precursor events.

Geologists of the U.S. Geological Survey noticed that the surface elevation was rising in the area around Palmdale, California, near Los Angeles. In addition, the groundwater table was dropping while the level of background radiation was rising. Furthermore, Palmdale is on a seismic gap on the San Andreas Fault. Many members of the geologic community began to hold their breaths but there has been no earthquake, at least not yet. Some people now claim that it was a false alarm, that the surveys used to detect changes in ground elevation were incorrect.

There is some hope that forecasts will be possible within periods of less than a year. A team from the University of California at Los Angeles has recently been using an approach that it calls "tail wagging the dog" in which it monitors the constant background seismicity for four symptoms: (1) an increase in the frequency of small earthquakes in a region, (2) a clustering of these quakes in time and space, (3) nearly simultaneous quakes over large parts of a region, and (4) an increase in the ratio of medium magnitude to small magnitude earthquakes for the region. This approach

was apparently successful in forecasting quakes in two fairly large regions in 2003: The San Simon magnitude 6.4 quake occurred in a region of a forecast magnitude 6.4 quake, and the magnitude 8.1 Hokkaido, Japan, quake occurred in a region where the team forecast a quake larger than magnitude 7. At this point, the approach is still being tested. The team's most recent prediction was that a quake of at least magnitude 6.4 would occur in a 12,500-square-mile region of south-central California between January 5 and September 5, 2004. No large earthquakes occurred in this region during the predicted time interval.

Seismic Gaps

Some segments of a major fault may appear to lack historic earthquakes, while others appear to have been more active. A fault segment with few or no historic earthquakes is a **seismic gap** (▶ Figures 4-2 and 4-3). Experience has shown that seismic gaps are far more likely to be the locations of large earthquakes than the apparently more active segments of the same fault. The Loma Prieta earthquake of 1989 filled a seismic gap in the San Andreas Fault that geologists had viewed with deep suspicion for many years and had identified as an area statistically likely to produce an earthquake within a few decades.

Some seismic gaps are fault segments that slip almost continuously, thus relieving stress in the rocks without causing earthquakes. A famous example of this fault behavior

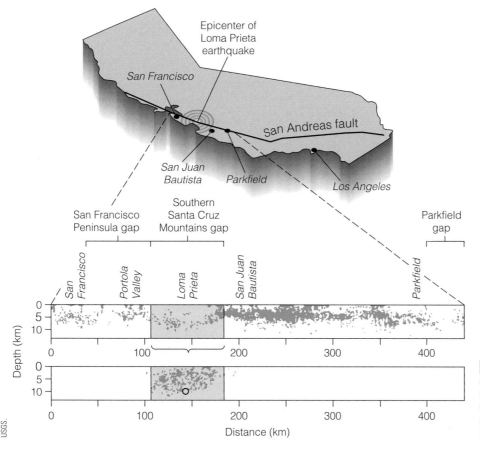

▶ **FIGURE 4-2.** A gap in the historic earthquake pattern along the San Andreas Fault south of San Francisco was filled in 1989 by the Loma Prieta earthquake. The lower plot shows the main shock and aftershocks of the Loma Prieta earthquake.

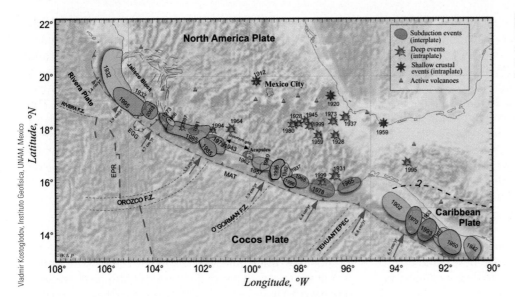

FIGURE 4-3. This map of the Pacific coast of Mexico shows faulted segments of the subduction zone and recent seismic gaps between faulted segments (ovals).

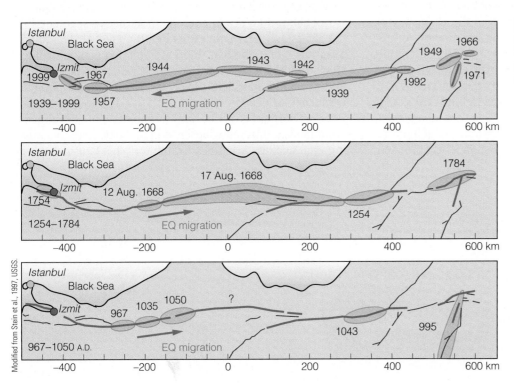

FIGURE 4-4. These three maps show groups of earthquakes on the North Anatolian Fault in Turkey from 967 to 1050, 1254 to 1784, and 1939 to 1999.

is near Hollister, California, south of San Francisco, where the Calaveras Fault splay of the San Andreas system offsets streets year by year. Rocks along that part of the fault are probably not accumulating stress, and thus Hollister seems unlikely to become the epicenter of a major earthquake.

Migrating Earthquakes

In a few places, earthquakes are known to migrate along a major fault. Probably the best example of **migrating earthquakes** is the North Anatolian Fault in Turkey where, between 1939 and 1999, they moved sequentially westward.

Most ruptures along the fault were juxtaposed, though a few left seismic gaps that were later filled with lesser earthquakes (▶ Figure 4-4).

Earthquake Regularity

A few faults fall into a pattern of moving just as regularly as the calendar comes around, but this kind of behavior is so uncommon and the series so unreliable that it has never become a useful method of earthquake prediction.

For nearly 150 years, the San Andreas Fault moved at regular intervals near Parkfield in the central California coast ranges midway between San Francisco and Los Angeles. It generated earthquakes of magnitude 5.5 to 6.0 in 1857, 1881, 1901, 1922, 1934, and 1966—at intervals averaging close to twenty-two years. The last three earthquakes in the sequence, those for which seismograph records exist, were virtually identical. This series prompted seismologists to predict that there was approximately a 95 percent chance that the next earthquake would shake Parkfield between 1985 and 1993. The U.S. Geological Survey installed a wide variety of expensive instruments around the Parkfield area and waited for some action. There were no large earthquakes in this region until September 28, 2004, when a magnitude 6.0 earthquake finally struck.

For more than 500 years, especially strong earthquakes devastated Tokyo, Japan, at intervals of roughly seventy years, most recently in 1923 (▶ Figure 4-5). Many people, especially those in the insurance business, watched the approach of 1993 with extreme discomfort. No earthquake has yet occurred, but it may just be later than expected.

Another pattern that is coming to light is that a major fault may be quite active over many decades and then lie quiet for many more before it again becomes active. In North America, our history may be a bit short to demonstrate such patterns clearly, but in Turkey the historic record is much longer. Beginning in 967 A.D., the North Anatolian Fault moved to cause earthquakes at intervals of seven to forty years, followed by no activity for 204 years. The fault saw more earthquakes from 1254 to 1784, then another long hiatus. Then again in 1939, activity picked up again at intervals from one to twenty-one years and continues to the present day (Figure 4-4).

Paleoseismology

It is commonly possible to recover at least a partial record of prehistoric fault movements. Some basins and ranges in western North America show distinctive white stripes along the lower flanks of the ranges. Many of these mark the fault scars where the range rose during a past earthquake. They are most obvious in the arid Southwest, where vegetation has not obscured the fault scarps.

Trenches dug across segments of an active fault generally expose the fault along with sediments deposited across it and then offset during later movements (▶ Figures 4-6 and 4-7). Older sedimentary layers are offset more because they have experienced a series of fault movements. The amount of offset during a fault movement is generally proportional to the magnitude of the accompanying earthquake (▶ Figures 3-24, page 48; 4-7, 4-8, and 4-9), so trenches also provide a record of the magnitude of prehistoric earthquakes. Similarly, the length of a fault break is proportional to the magnitude of the earthquake. Straight lines surveyed across an active fault gradually become bent before the next earthquake and fault slip. The total amount of offset along this line should approximate the maximum amount of slip on the fault during the next event (Figures 3-26 and 3-27, page 50).

Trenches can also provide evidence of past earthquakes in such features as soil liquefaction, sand boils (Figure 3-30, page 52), and sediment compaction. Radiocarbon dating of organic matter, most commonly charcoal, in the sedimentary layers broken along a fault can reveal the maximum dates of fault movements. However, that is possible only if they happened less than 40,000 years ago, the maximum age for radiocarbon dating method.

Water as a Triggering Mechanism?

Water in the ground is well known to be the greatest culprit in triggering the movement of landslides. Is it possible that water deep in a fault zone could trigger movement of a fault? In a case during the early 1960s, fluids were pumped into an ancient inactive fault zone in Colorado, which apparently triggered a series of earthquakes in an area not noted for earthquakes. When the U.S. Army pumped waste fluids

▶**FIGURE 4-5.** A magnitude 8.2 earthquake in 1923 destroyed Tokyo from an epicenter 90 kilometers away.

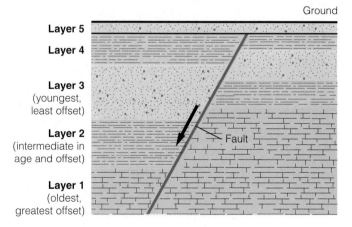

▶**FIGURE 4-6.** This schematic diagram shows a trench dug across a fault to expose the fault and help determine the relative amounts of offset of layers of various ages.

Before faulting

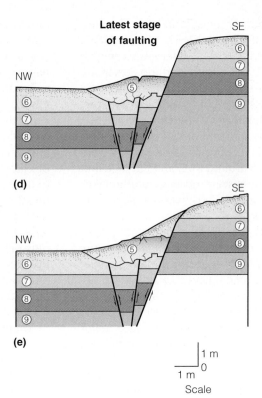

Latest stage of faulting

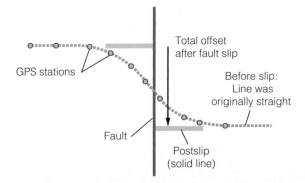

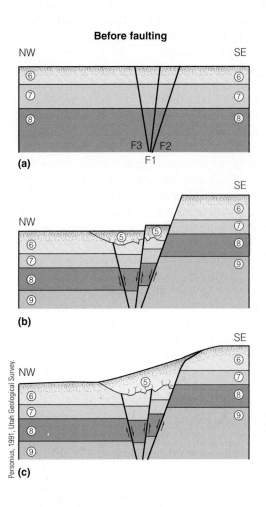

Personius, 1991, Utah Geological Survey.

▶**FIGURE 4-7.** These simplified diagrams show sections of the north wall of a trench across the Wasatch Fault zone at Brigham City, Utah, and represent a sequence of fault movements. Colluvial sediment from the slope (e.g., number 5 in c) covers the fault between earthquakes. The sequence begins with a (before faulting).

1 m

0

1 m

Scale

G. K. Gilbert photo, USGS.

▶**FIGURE 4-8.** This fence near Point Reyes, north of San Francisco, California, was offset approximately 2.6 meters in the 1906 San Francisco earthquake on the San Andreas Fault.

GPS stations

Total offset after fault slip

Before slip: Line was originally straight

Fault

Postslip (solid line)

▶**FIGURE 4-9.** Global positioning system (GPS) stations placed in a straight line across a fault curve as the Earth's crust on one side bends elastically with respect to the crust on the other side.

into a 3,674-meter-deep well at the Rocky Mountain Arsenal northwest of Denver, the area experienced ten to twenty or more earthquakes per month (▶Figure 4-10). When the Army stopped pumping in fluids, the earthquake frequency dropped to fewer than five per month. When pumping began again, so did the earthquakes. Also, the addition of waste under gravity triggered some earthquakes, but injection un-

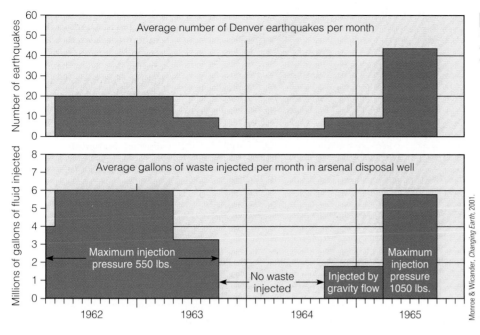

Monroe & Wicander, *Changing Earth*, 2001.

FIGURE 4-10. These two graphs correlate the timing of fluid waste injection into deep wells with earthquakes near Denver, Colorado, in the early 1960s.

der high pressure triggered more than forty per month. The inference is that addition of the fluids increases the pore pressure, which helps to reduce the force between opposite sides of the fault. The situation is analogous to the well-known correlation of water addition as a trigger of landslide movement.

If water can trigger movement of a fault and generation of earthquakes, then might we deliberately inject water into a fault to cause movement and initiate a modest-size earthquake? Some geologists have suggested this approach to preempt occurrence of a giant earthquake. Unfortunately, there are consequences that would require careful consideration before such an action. Consider, for example, the San Andreas Fault near San Francisco:

■ What size earthquake might be generated?

■ Would people be evacuated first? How many and to where? Who pays for the costs of travel, food, and lodging, and for how long?

■ What would be the costs of disruption of local, interstate, and international trade, business, and travel?

■ Would the fault slip immediately? If not, then when?

■ Who is liable for the injuries, deaths, or damages?

■ How many moderate-size earthquakes (e.g., magnitude 6) would it take to forestall a large earthquake (e.g., magnitude 7 or 8)? Would the costs of many smaller earthquakes be less or more than that for one large earthquake?

■ What other consequences might there be? Do we know enough about the behavior of fault movement and earthquakes to be able to reliably predict or control the consequences?

If you were in a position of authority, what would you do?

Early Warning Systems

It is reasonable to entertain hope that a large earthquake at a distance of 100 kilometers or more from vulnerable structures could be detected in time to provide a useful warning. Earthquake waves traveling 4 kilometers per second from the focus of a large earthquake would require twenty-five seconds to reach points 100 kilometers away. That might be just enough time for an early warning system to shut down a critical facility, assuming that the early warning system was close to the epicenter. A consortium of federal, state, university, and private groups began building such a system called TriNet in Southern California in 1997.

Long-Term Forecasts and Risk Maps

Earthquake forecasts for decades or centuries in the future become important in choosing sites for major structures such as large dams, power plants, public buildings, and bridges, and for insurance. Making such forecasts generally requires knowledge of past earthquakes along a fault. Such long-term forecasts are much more reliable than short-term predictions. That makes it possible to estimate the **recurrence interval** for earthquakes of various sizes and statistically estimate the probability that an earthquake of any particular size will strike within a specified period in the future—for example, within the next thirty years.

Earthquake risk maps are based on past activity, both frequency and magnitude (▶ Figures 4-11 to 4-14). Most of the earthquakes that strike the United States concentrate along the San Andreas and related faults in western California. The intermountain seismic zone through central Utah, Yellowstone Park, and western Montana feels more earthquakes than one might expect so far from the Pacific coast,

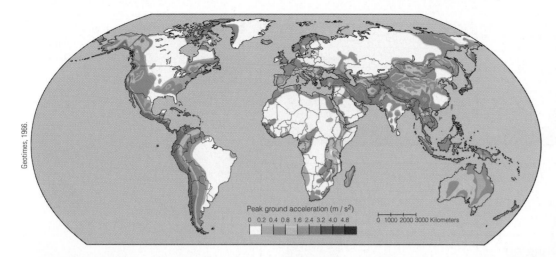

Geotimes, 1966.

▶**FIGURE 4-11.** Global earthquake hazard: Peak accelerations (darker red colors) correspond to regions with strong earthquakes. Note the concentration of strong earthquakes with subduction zones on the Pacific coast of North and South America and in Indonesia, and with collision zones in southern Europe and Asia.

Peak ground acceleration (m / s²)
0 0.2 0.4 0.8 1.6 2.4 3.2 4.0 4.8

0 1000 2000 3000 Kilometers

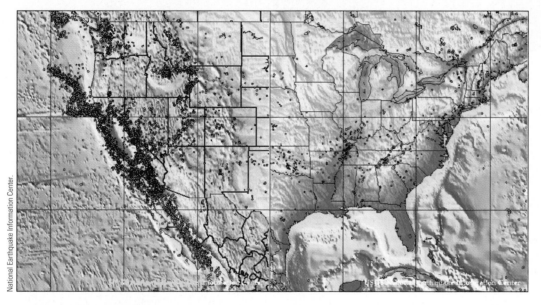

National Earthquake Information Center.

▶**FIGURE 4-12.** In this distribution map of earthquake epicenters in the United States from 1899 to 1990, larger circles represent larger earthquakes.

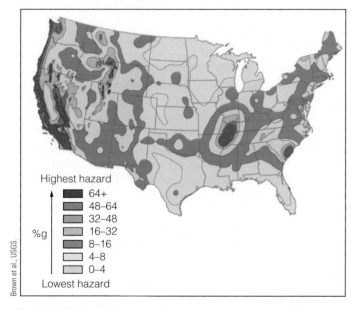

Highest hazard

%g

	64+
	48–64
	32–48
	16–32
	8–16
	4–8
	0–4

Lowest hazard

Brown et al., USGS.

▶**FIGURE 4-13.** This seismic zonation map for the United States shows areas with a 2 percent probability of exceeding the indicated percentage of the acceleration of gravity over the next fifty years. Compare with Figure 4-12, which shows earthquake distributions.

as does the Seattle–Tacoma area of Washington. Most of the earthquakes in Alaska come from the Aleutian oceanic trench, a subduction zone.

Psychics

Psychics, astrologers, crackpots, assorted cranks, and miscellaneous prophets regularly predict earthquakes, never with any success. They may, however, cause great public consternation and serious public and private expense.

The late Iben Browning, an eccentric business consultant with a doctoral degree in biology but no background in seismology or geology proposed an overlap of tidal forces based on the alignment of the sun and moon, along with the locations of recent earthquakes and volcanic eruptions. He then decided that those forces would affect ground movement in a certain broad region. That somehow led him to predict a 50 percent chance of a magnitude 7 earthquake within a few days of December 3, 1990, on the infamous

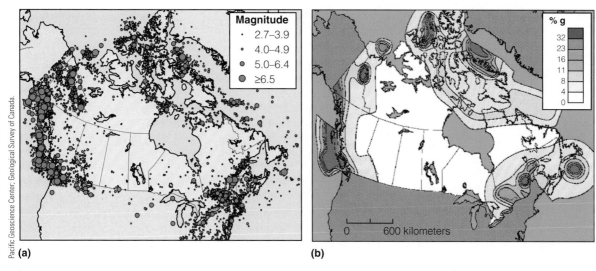

Pacific Geoscience Center, Geological Survey of Canada.

(a) **(b)**

▶**FIGURE 4-14.** These two maps of Canada show **(a)** earthquake epicenters and **(b)** seismic zones with the probability of exceeding the indicated fraction of the acceleration of gravity.

New Madrid fault system in southeastern Missouri. Innumerable people believed him, and the proposal quickly developed into a media frenzy. Seismologists strongly opposed his prediction but to no avail. The news media shamelessly exploited the prediction. Civil officials staged preparedness drills, closed schools, and stationed fire trucks in what they imagined were critical places. Many millions of dollars were expended in wasted preparation. There was, of course, no earthquake.

Surviving Earthquakes

The largest earthquakes do not necessarily kill the most people. Most of the high death counts come from countries notable for poor building construction or unsuitable building sites (see Table 4-1). The death toll from the January 2001 earthquake in San Salvador was the result of both a huge landslide in a prosperous part of the capital and poorly built adobe houses in the poor areas. The much higher death tolls from the January 2001 earthquake in India and the December 2003 earthquake in Iran derived mainly from the collapse of houses poorly built with heavy materials (discussed later in this chapter; see Figures 4-47 to 4-49, page 90). Developed countries tend to be better off in this respect but can still experience a devastating loss of life in an earthquake.

A cursory examination of Table 4-1 also shows that we are not doing much better with time. More than 200 years ago, many tens of thousands of people died in major earthquakes and occasionally hundreds of thousands. Even in the past fifty years, tens of thousands, and occasionally hundreds of thousands, died. Unfortunately, there are now far more people living in crowded conditions and often in poorly constructed buildings (▶ see Figure 4-15).

Donald Hyndman photo.

▶**FIGURE 4-15.** Houses that were damaged in old parts of Athens, Greece, in a magnitude 5.9 earthquake in September 1999 were poorly constructed from local rocks poorly cemented together. The collapsing houses killed 143 people. Would you stay in a hotel with this type of construction? Could you tell it if its walls were covered with plaster or stucco?

The San Andreas Fault

The San Andreas Fault, the dominant earthquake zone in the United States, slices through a 1,200-kilometer length of western California, from just south of the Mexican border to Cape Mendocino in northern California (▶ Figures 4-16, 4-17, and 4-18). It is a continental transform fault. The main sliding boundary marks the relative motion between the

Table 4-1 Some of the Most Catastrophic Earthquakes in Terms of Casualties*

Earthquake	Date	Magnitude (Ms)	Casualties	Probable Cause
Sumatra	Dec. 26, 2004	9.0	>**283,000** (including missing) in a great tsunami caused by the earthquake.	Subduction zone
Iran, Bam	Dec. 26, 2003	6.7	>26,000, most in buildings of mud and brick	Strike-slip motion, continent–continent collision
India, Bhuj	Jan. 26, 2001	7.7	>30,000, most in buildings of mud and brick	Intraplate blind thrust
Turkey, Izmit	Aug. 17, 1999	7.4	>30,000 in poorly built, masonry buildings	Transform fault
Iran	June 20, 1990	7.7	>40,000, most in landslides and collapse of adobe and unreinforced masonry houses	Continent–continent collision
Armenia	Dec. 7, 1988	7.0	25,000, mostly in precast, poorly constructed concrete-frame buildings	Continent–continent collision
China, Tangshan	July 27, 1976	7.6	**250,000**, mostly in collapsed adobe houses	Strike-slip motion
Peru	May 31, 1970	7.8	66,000 in rock slide that destroyed Yungay	Subduction zone
Chile, Chillan	Jan. 24, 1939	7.8	30,000	
India, Quetta	May 31, 1935	7.5	60,000	
Japan, Kwanto	Sept. 1, 1923	8.2	**143,000**, including in great Tokyo fire started by the earthquake	Subduction zone
China, Kansu	Dec. 16, 1920	8.5	**180,000**	Thrust fault (collision)
Italy, Avezzano	Jan. 13, 1915	7	30,000	Normal fault
Italy, Sicily, Messina	Dec. 28, 1908	7.5	120,000	Subduction zone
Ecuador and Colombia	Aug. 16, 1868	?	70,000 total: 40,000 in Ecuador, 30,000 in Colombia	
Japan, Echigo	Dec. 19, 1828		~30,000	
Ecuador, Quito	Feb. 4, 1797	?	40,000	
Italy, Calabria	Feb. 4, 1783		50,000	Subduction zone
Portugal, Lisbon	Nov. 1, 1755		70,000 including in a great tsunami caused by the earthquake	Continent–continent collision?
Northern Persia	June 7, 1755		40,000	Continent–continent collision
India, Calcutta	Oct. 11, 1737		**300,000**	
Italy, Catania	Jan. 11, 1693		60,000	
Caucasia, Shemakha	Nov. 1667		80,000	
China, Shaanxi, Shensi	Jan. 23, 1556		**830,000**, mostly in collapse of homes dug into loess	
Portugal, Lisbon	Jan. 26, 1531		30,000	
Japan, Kamakura	May 20, 1293		30,000	Subduction zone

*Information gathered from various sources.

Table 4-2 The Largest World Earthquakes Since 1900

Earthquake	Date	Magnitude	Cause
Chile	May 22, 1960	9.5	Subduction zone
Anchorage, Alaska	Mar. 28, 1964	9.2	Subduction zone
Andreanof Is., Alaska	Mar. 9, 1957	9.1	Subduction zone
Northern Sumatra	Dec. 26, 2004	9.0	Subduction zone
Kamchatka	Nov. 4, 1952	9.0	Subduction zone

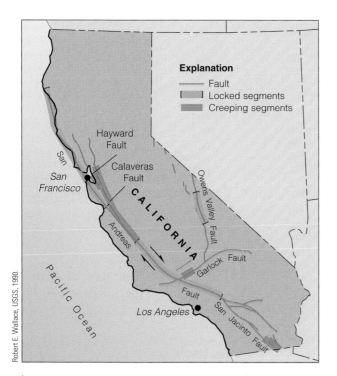

FIGURE 4-16. The San Andreas Fault slices through most of coastal California. The dashed part of the fault, south from Hollister, creeps continuously without significant earthquakes.

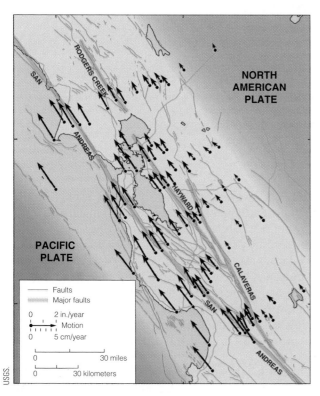

FIGURE 4-17. The San Andreas Fault system is a wide zone that includes nearly the entire San Francisco Bay area. The black arrows are proportional to the rates of ground movement relative to the stable continental interior based on GPS measurements. Energy builds up when there is a differential movement as can be seen across each of the major faults shown in orange.

FIGURE 4-18. The San Andreas Fault shows as a series of straight valleys slicing through the coast ranges in this shaded-relief map of California and western Nevada.

Pacific Plate, which moves northwest, and the North American Plate, which moves slightly south of west. The overall fault is really a zone of more-or-less parallel faults some 50 kilometers wide (Figure 4-17). The San Andreas Fault is simply its main strand. The trace of the San Andreas Fault appears from the air and on topographic maps as lines of narrow valleys (▶Figures 4-18 and 4-19), some of which hold long lakes and marshes (see also Figure 2-32, page 30). In some areas immediately south of San Francisco, developers have filled fault-induced depressions and built subdivisions right across the trace of the fault (▶Figure 4-20). As shown on Figure 4-17, the total motion of the westernmost slice of California moves more than the slices closer to the continental interior. The greatest difference between the arrow lengths across a fault suggests the greatest likelihood of new fault slippage or an earthquake.

The San Andreas Fault results from interaction between the relative northwest movement of the Pacific Plate compared with the North American Plate (▶Figures 4-17 to 4-22). The west side of the northwest-trending fault moves northwestward at an average relative rate of 3.5 centimeters per year or 3.5 meters every 100 years. The rupture length for a magnitude 7 earthquake for that offset would be 50 kilometers; release of all the strain accumulated in 100 years would require a series of such earthquakes along the length

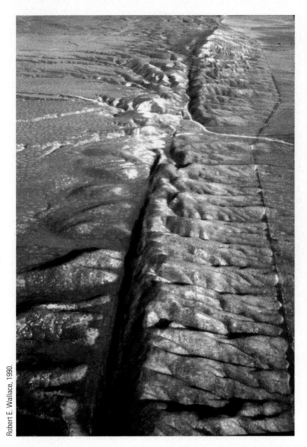

Robert E. Wallace, 1990.

▶**FIGURE 4-19.** The San Andreas makes a long, straight furrow through hills of the Carrizo Plain north of Los Angeles.

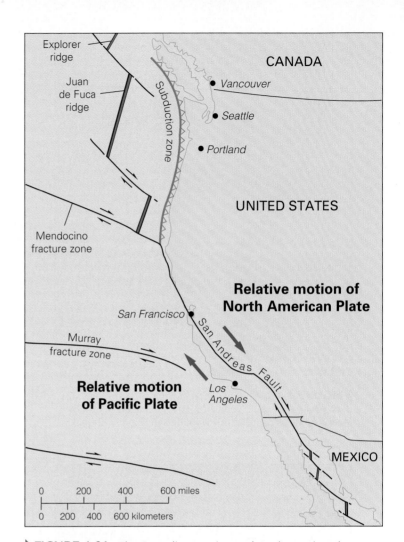

▶**FIGURE 4-21.** The Cascadia oceanic trench to the north and the San Andreas transform fault to the south dominate the Pacific continental margin of the United States.

Robert E. Wallace, USGS.

▶**FIGURE 4-20.** A 1965 air view south along the San Andreas Fault in Daly City and Pacifica, just south of San Francisco, shows how housing developments even then were straddling the fault. Developments now more completely cover the fault zone.

of the fault. However, earthquakes are not evenly distributed in time. Earthquakes frequently occur in clusters separated by periods of relative seismic inactivity.

In the seventy years before the San Francisco earthquake of 1906, earthquakes of magnitude 6 to 7 occurred every ten to fifteen years. No earthquakes greater than magnitude 6 struck the San Francisco Bay area between 1911 and 1979. Four earthquakes of magnitudes 6 to 7, including the one in Loma Prieta, struck the region between 1979 and 1989. We may be in the midst of another cluster of strong earthquakes. U.S. Geological Survey seismologists now estimate a 62 percent chance of an earthquake with a magnitude greater than 6.7 in the San Francisco Bay area before 2032—on the San Andreas, Hayward, or related nearby faults.

The southern end of the San Andreas Fault connects with the northern end of the East Pacific Rise, an oceanic ridge that follows the Gulf of California (Figure 4-21). At its northern end, the fault meets the Cascadia oceanic trench, a col-

▶**FIGURE 4-22.** In this view from the air, streams jog abruptly where they cross the San Andreas Fault in the Carrizo Plain north of Los Angeles. The 1857 Fort Tejon earthquake caused 9.5 meters of this movement.

▶**FIGURE 4-23.** This curb in Hollister, California, has been off-set by creep along the San Andreas Fault.

lision plate boundary, and the Mendocino fracture zone at a triple junction, the join between the San Andreas Fault, the east–west Mendocino Fracture, and the Cascadia Subduction Zone to the north.

A segment of the central part of the San Andreas Fault south of Hollister slips fairly continuously without causing significant earthquakes. In this zone, strain in the fault is released by creep and thus does not accumulate to cause large earthquakes (▶Figures 4-16 and 4-23). Why that segment of the fault slips without causing earthquakes, whereas other segments stick until the rocks break during an earthquake, is not entirely clear. Presumably, the rocks at depth are especially weak, such as might be the case with shale or serpentine; or perhaps water penetrates the fault zone to great depth, making it weak.

In the discussion below, and where not specified otherwise, the earthquake magnitude is generally that based on the surface waves. For comparison between events, all losses are converted to 2002 U.S. dollars.

San Francisco Bay Area Earthquakes

SAN FRANCISCO, 1906 By the late 1800s, people in California had felt enough tremors to know that they were living in earthquake country. Engineers had learned much about the hazards of certain kinds of building designs, but little was actually done to change the way buildings were designed.

The San Francisco earthquake came before dawn on April 18, 1906. It began at 5:12 A.M. with a foreshock that rudely awakened nearly everyone. Then, twenty to twenty-five seconds later, the main shock struck with a magnitude of 7.8. One survivor recalled a rumble and roar like old cannons. Others heard roaring sounds or dull booms. People on the street recalled that the ground rose and fell in waves as the earthquake approached.

Strong shaking lasted forty-five to sixty seconds. Thousands of chimneys snapped off and fell through the houses into their basements. Many brick or stone buildings collapsed into heaps of rubble (▶Figures 4-24, 4-25, 4-26, and 4-27). New skyscrapers with steel frames along Market Street survived with little damage, as did most wood-frame buildings. Building design and materials played a big role in their survival. Unfortunately, many of the wood-frame buildings on filled areas along the edges of San Francisco Bay collapsed (Figures 4-24, 4-25, and 4-26). Others sank as the mud under them compacted. Aftershocks destroyed more buildings weakened in the main event. Buildings on the bedrock hills survived relatively well.

Cooking fires and broken gas and electrical lines sparked fires in many of the wood buildings. Dozens of fires ignited within a half hour, then coalesced into two major fires that spread across much of the city, one north and another south of Market Street. Broken water mains so hampered the fire department that it resorted to dynamite to cut fire lines. Three days of fire destroyed Chinatown, the surviving skyscrapers along Market Street, and the wharves along the edge of San Francisco Bay. More than 28,000 buildings were destroyed, 10.6 square kilometers (4 square miles) of the city (Figures 4-26 and 4-27).

North of San Francisco, the earthquake damaged buildings almost to the Oregon border; to the south, damages reached a third of the way to Los Angeles. The death toll was generally quoted as 700, with nearly 500 in San Francisco. The epidemics that followed killed many more. However, the death toll appears to have been understated at the time, apparently because of anticipated effects on the local economy. In fact, it was probably more than 3,000 from the effects of the earthquake and its aftermath. Damage was $480 million from the earthquake (2002 dollars) and $7 billion from the fire. Some 225,000 people, out of a population

▶**FIGURE 4-24.** The Hibernia Bank building in San Francisco was destroyed by the 1906 earthquake.

W. C. Mendenhall photo, USGS.

▶**FIGURE 4-26.** This view of San Francisco looking down Market Street shows the devastation following the 1906 earthquake and fire.

Courtesy Bancroft Library, University of California, Berkeley.

▶**FIGURE 4-25.** Slumping toppled many buildings during the 1906 earthquake in San Francisco.

NOAA, National Geophysical Data Center photo.

▶**FIGURE 4-27.** These horses were killed by falling debris during the San Francisco 1906 earthquake.

P. E. Hotz Collection, USGS.

of 400,000, lost their homes. Santa Rosa, Healdsburg, and San Jose were severely damaged.

The surface offset extended 430 kilometers from the area east of Santa Cruz to Cape Mendocino (▶Figures 4-28 and 2-18). The maximum horizontal displacement was 2.6 meters near Point Reyes (see broken fence in Figure 4-8), 50 kilometers north of San Francisco. Neither the concept of earthquake magnitude nor seismographs suitable for measuring magnitudes existed in 1906. The size of the area damaged to Mercalli Intensity VII or higher suggests a moment magnitude of approximately 7.8.

LOMA PRIETA, 1989 This magnitude 7.0 quake is described in the introduction to Chapter 3 ("Earthquakes and Their Damages").

THE BIG ONE? The San Andreas Fault began to move along most of its length approximately 16 million years ago and shows no sign of stopping. It seems perfectly reasonable to assume that it has inflicted thousands of earthquakes on the San Francisco Bay area and that thousands more will come. So the people who live in the Bay Area dread the next devastating earthquake, the "Big One." It will

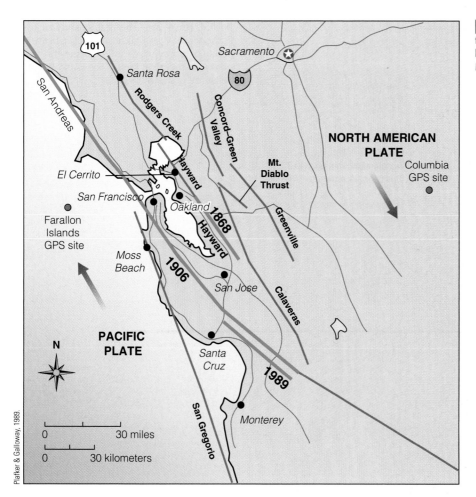

▶**FIGURE 4-28.** This map shows the major faults of the San Francisco Bay area. The 1906 break extended from near Santa Cruz to Cape Mendocino, which is far north of this map.

happen just as surely as the sun will rise tomorrow, but no one knows when or how big.

A certain level of anxiety seems perfectly appropriate. The population and the number of houses and buildings in California have increased enormously since 1906. People build in areas close to or even on top of the fault, right on potential future epicenters (Figure 2-32, page 230; and Figure 4-20). Zoning prevents building in some areas along the fault within San Francisco itself but not in others outside the city limits. All those people and all that development guarantee that the next Big One will be far more devastating than the last big earthquake.

The Gutenberg-Richter frequency versus magnitude relationship suggests that any segment of the San Andreas Fault 100 kilometers long should release energy equivalent to an earthquake of magnitude 6 on average every eight years, one of magnitude 7 on average every sixty years, and one of magnitude 8 on average about every 700 years (see Figure 3-24, page 48). If so, several segments of the fault are overdue. Energy builds up across the fault until an earthquake occurs, so the chance of a large event grows as the time since the last large earthquake increases.

Researchers have calculated different overall slip rates on the fault using different types of data. Using a relatively low slip rate of 2 centimeters per year, strain should accumulate on stuck faults at a rate that would cause an earthquake of moment magnitude M_w 7.4 or 7.5 every 120 to 170 years. Alternatively, there could be six earthquakes of magnitude 6.6 in the same period. Because the actual number has been far fewer, it again seems that we are overdue.

Hayward and Rodgers Creek Faults

The Hayward Fault splays north from the San Andreas Fault south of San Francisco Bay. It trends north along the base of the hills near the east side of the bay through a continuous series of cities, including Hayward, Oakland, Berkeley, and Richmond. The Hayward Fault has not caused a major earthquake since 1868, but many geologists believe it is one of the most dangerous faults in California and may be ready to move (▶Figures 4-17, 4-28, and 4-29).

The Rodgers Creek Fault continues the Hayward Fault trend from the north end of San Francisco Bay and north through Santa Rosa. Along with the San Andreas Fault, these two faults and their southward extension, the Calaveras Fault, pose the greatest hazards in the region, partly because they are likely to cause large earthquakes and partly because they traverse large metropolitan areas with rapidly

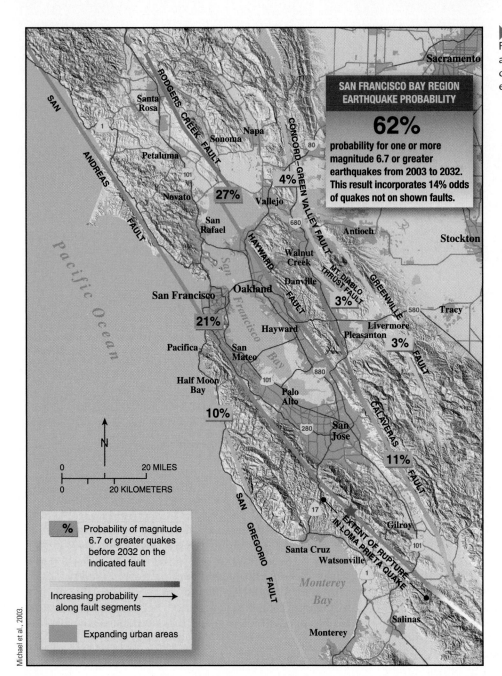

FIGURE 4-29. This map of the San Francisco Bay area faults shows the probability of a magnitude 6.7 or larger earthquake on the main faults. The Loma Prieta epicenter is shown as a star.

growing populations. The risk is not only in the low-lying areas of San Francisco Bay muds, or in the heavily built up cities along the fault trace, but in the precipitous hills above the fault. Those hills are extensively covered with expensive homes, including those at the crest of the range along Skyline Drive where the outer edges of many are propped on spindly looking stilts.

The latest study of earthquake risk by the U.S. Geological Survey and California Geological Survey shows a 62 percent chance of at least one magnitude 6.7 earthquake or an 80 percent chance of at least one magnitude 6 to 6.6 earthquake somewhere in the Bay Area before 2032. The risk on any individual fault is proportionally less (▶Figure 4-29).

An offset of as much as 3 meters on the Hayward Fault would probably cause an earthquake of magnitude 7 that would last twenty or twenty-five seconds. The U.S. Geological Survey estimates that several thousand people might perish and at least 57,000 buildings would be damaged, 11 times as many as in the Loma Prieta earthquake in 1989. The earthquake of magnitude 6.8 that hit Kobe, Japan, in January 1995 killed more than 6,000 people in an equally modern city built on similar ground.

Another recent general assessment by the USGS of the earthquake probabilities in the Bay Area before 2030 suggests a 21 percent probability of a magnitude 6.7 or larger earthquake on the San Andreas Fault, 27 percent on the

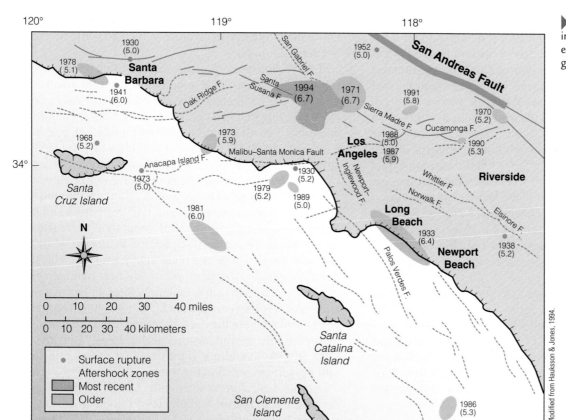

1930
(5.0)

1978
(5.1)

**Santa
Barbara**

1941
(6.0)

1968
(5.2)

34°

*Santa
Cruz Island*

Oak Ridge F.

Santa
Susana F

1994
(6.7)

1973
(5.9)

Malibu–Santa Monica Fault

Anacapa Island F.

1973
(5.0)

1979
(5.2)

1981
(6.0)

San Gabriel F.

1952
(5.0)

1971
(6.7)

1991
(5.8)

Sierra Madre F.

1988
(5.0)
1987
(5.9)

**Los
Angeles**

1930
(5.2)

1989
(5.0)

San Andreas Fault

1970
(5.2)

Cucamonga F.

1990
(5.3)

Riverside

Newport-
Inglewood F.

Whittier F.

Norwalk F.

**Long
Beach**

1933
(6.4)

**Newport
Beach**

Palos Verdes F.

Elsinore F.

1938
(5.2)

N

0   10   20   30   40 miles

0 10 20 30 40 kilometers

*Santa
Catalina
Island*

- Surface rupture
Aftershock zones
Most recent
Older

*San Clemente
Island*

1986
(5.3)

Modified from Hauksson & Jones, 1994.

▶**FIGURE 4-30.** Faults in the Los Angeles area with earthquake magnitudes greater than 4.8 since 1920.

Hayward Fault–Rodgers Creek Fault, and 11 percent on the Calaveras Fault (Figure 4-29). The total probability of at least one magnitude 6.7 earthquake somewhere in the San Francisco Bay area before 2030 is 62 percent or so. All of these faults pose a significant threat.

Los Angeles Area Earthquakes

The San Andreas Fault is 50 kilometers northeast of Los Angeles. A large earthquake on that part of the fault would undoubtedly cause enormous loss of life and property damage. That will almost certainly happen someday. But the prospect of lesser earthquakes on any of the many related faults that cross the Los Angeles basin is a matter of more immediate concern. They are far more dangerous than the big fault because of their proximity to heavily populated areas (▶Figure 4-30).

The northward drag of the Pacific Plate against the continent is slowly crushing the Los Angeles basin, and the sedimentary rocks that fill it, northward at roughly 7 millimeters per year. The sedimentary formations buckle into folds and break along thrust faults, both of which shorten the basin as they move (▶Figure 4-31). The thrust faults are called **blind thrusts** because they do not break the surface. Many of them remain quite unknown until they cause an earthquake. Blind thrusts are quite dangerous.

A total of seventeen earthquakes greater than magnitude 4.8 shook the Los Angeles region between 1920 and

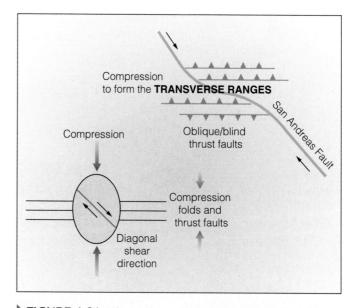

Compression to form the **TRANSVERSE RANGES**

Compression

Oblique/blind thrust faults

San Andreas Fault

Compression folds and thrust faults

Diagonal shear direction

▶**FIGURE 4-31.** These schematic diagrams of the big bend in the San Andreas Fault near Los Angeles show the north–south compression (top) in rocks adjacent to the fault (east–west thrust faults accommodate that compression) and the stresses that cause movement on blind thrust faults in the area (bottom).

1994 (Figure 4-30). Especially dangerous faults include the Sierra Madre–Cucamonga fault system that follows 100 kilometers of the northern edge of the San Fernando and San Gabriel valleys. The blind thrust along the westernmost 19 kilometers of the fault moved in 1971 to cause the San Fernando Valley earthquake. The fault systems beneath downtown Los Angeles include thrust faults that dip down to the northeast. The Santa Monica Mountains fault zone near downtown Los Angeles extends west along the Malibu coast for 90 kilometers. It includes blind thrust faults that do not break the surface and strike-slip faults that do. The Oak Ridge fault system north of the Malibu coast generated the 1994 Northridge earthquake of M_w 6.7. The Palos Verdes thrust fault, along the coast south of downtown Los Angeles, slips approximately 3 millimeters per year; this progressive slip releases accumulating strain that might someday cause an earthquake.

Many California faults are capable of causing earthquakes of magnitudes from 7.2 to 7.6, tens of times more energy than the San Fernando Valley earthquake of 1971 and the Northridge earthquake of 1994. Those future larger earthquakes could cause stronger shaking that would last longer and extend over larger areas than previous California earthquakes. The consequences in densely populated areas would be tragic.

NORTHRIDGE, 1994 The Northridge earthquake struck on January 17, 1994, from an epicenter 20 kilometers southwest of the San Fernando Valley earthquake of similar size in 1971. Like the San Fernando Valley earthquake, the

Northridge earthquake had a magnitude 6.7 M_w. Both accompanied movement on faults in the San Fernando Valley north of Los Angeles.

The Northridge earthquake caused $15.2 billion in property damage. Some 3,000 buildings were closed to all entry; another 7,000 buildings were closed to occupation. Severity of building damage depended on the distance from the epicenter, type of ground under the foundation, and building design. Walls fell from older buildings constructed from structural brick (▶Figure 4-32). Many buildings collapsed because they had weak first floors, most commonly apartments above garages (▶Figure 4-33). Concrete around steel reinforcing bars shattered in concrete overpasses and parking garages, permitting the reinforcing steel to buckle and then collapse (▶Figure 4-34). Such a failure could be prevented if the concrete had been tightly wrapped with steel. Even a carefully designed and nearly new parking garage constructed from flexible materials thought to be able to withstand a strong earthquake failed (▶Figure 4-35). Many welds broke where they attached horizontal beams to vertical columns, permitting the beams to fall (Figure 3-50, page 58). Mobile homes were jarred off their foundations, and broken gas lines ignited fires.

A thrust fault slipped at a depth of 10 kilometers. The offset reached to within 5 kilometers of the surface but did not break it, so the fault is a blind thrust (▶Figures 4-30 and 4-36). The fault, unknown before the earthquake, is now known as the Pico thrust fault. The fault movement raised Northridge 20 centimeters (8 inches); the Santa Susana Mountains north of the San Fernando Valley rose 40 centimeters.

Ground acceleration reached almost 1 g in many areas and approached 2 g, twice the acceleration of gravity, at one site in Tarzana. Local topography or bedrock structure may have amplified ground shaking. Anomalously strong

FEMA photo.

▶**FIGURE 4-32.** Older structural brick buildings were heavily damaged in the 1994 Northridge earthquake.

USGS photo.

▶**FIGURE 4-33.** Many first-floor garages under apartment buildings were not well braced; the buildings then collapsed on the cars below.

FIGURE 4-34. The Northridge earthquake severed gas lines and caused fires.

▶FIGURE 4-35. A nearly new Northridge stadium parking structure collapsed during the 1994 earthquake in spite of being built from flexible materials designed to withstand a strong earthquake.

shaking occurred on some ridge tops and at the bedrock boundaries of local basins. Ground velocity locally exceeded 1 meter per second.

The earthquake occurred at 4:31 A.M. Although sixty-one people died, the toll would probably have been in the thousands had the earthquake happened during the day when people were up and about. Most buildings and parking garages that collapsed were essentially unoccupied (see Figure 3-22, page 46). Freeway collapse occurred at seven locations, and 170 bridges were damaged but only a few unlucky were on the road (▶Figure 4-37). Welded joints cracked in more than 100 steel-frame buildings, making them unsafe.

Damage inflicted during the Northridge earthquake reached Mercalli Intensity IX near the epicenter (see Figure 3-22, page 46). Intensity V effects were noted as far as 120 kilometers to the north, 180 kilometers to the west, and 200 kilometers to the southeast. A few people felt the ground motion as much as 300 to 400 kilometers away from the epicenter.

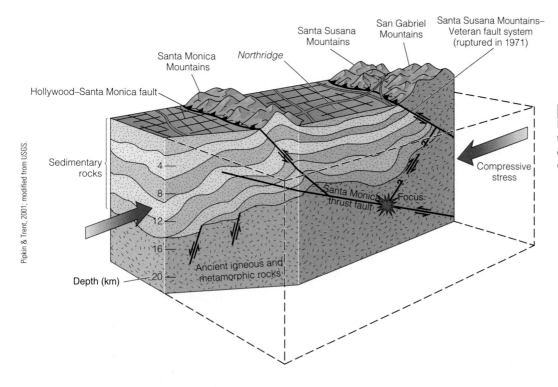

▶FIGURE 4-36. This cutaway block diagram shows the blind thrust fault movement that caused the 1994 Northridge earthquake.

▶**FIGURE 4-37.** This driver had the misfortune to drive off the collapsed freeway in the early morning darkness during the Northridge earthquake.

FORT TEJON, 1857 Movement on the San Andreas Fault 100 kilometers north of Los Angeles caused the Fort Tejon earthquake of January 9, 1857. Two earthquakes of inferred magnitude 6 near Parkfield, northeast of Paso Robles, preceded the main event by a few hours. Some seismologists now interpret those as premonitory foreshocks.

The Fort Tejon earthquake struck long before seismographs were invented. Few people lived in the region at the time, so too few data exist to make a meaningful map. But it is clear that the surface rupture along the trace of the fault extended from near Parkfield 300 kilometers south to Wrightwood, east of Los Angeles. Maximum surface displacement was horizontal, approximately 9.5 meters on the Carrizo Plain in the northern half of the offset. Strong shaking lasted for one to three minutes. These observations lead seismologists to infer a moment magnitude of 7.8, approximately the same as that of the San Francisco earthquake of 1906. Many geologists believe the Fort Tejon earthquake was the largest in the period of recorded California history. Only two people died, and property damage was negligible because the area was then nearly devoid of people or buildings.

Seismic activity has been notably absent along the segment of the San Andreas Fault south of the Fort Tejon slippage of 1857, almost to the Mexican border. Fault segments with such seismic gaps are far more likely to experience an earthquake than fault segments that have recently moved.

THE OTHER BIG ONE? Several moderate earthquakes have shaken Los Angeles and the Transverse Ranges area during the last century. Such a history guarantees that we can expect more (Sidebar 4-1).

Of those listed in the sidebar, only the 1857 and possibly the 1970 earthquakes actually involved movement on the

Sidebar 4-1

Some large earthquakes in the Los Angeles area in the last 150 years:

1994, M_W = 6.7	1978, M_L = 5.9
1991, M_L = 5.8	1971, M_W = 6.7
1990, M_L = 5.3	1970, M_L = 5.2
1989, M_L = 5.0	1967, M_L = 5.9
1988, M_L = 5.0	1933, M_L = 6.4
1987, M_L = 5.9	1930, M_L = 5.2
1979, M_L = 5.2	1857, M_W = 7.9 (Fort Tejon)

Note that following the big earthquake in 1857, there have been many much smaller earthquakes. Recall also that we need 32 or 33 magnitude 7 quakes to expend as much energy as a magnitude 8 quake—and more than 1,000 magnitude 6 quakes for the same energy as the magnitude 8 quake. Smaller quakes do not release much of the pent-up strain on the fault.

San Andreas Fault. The others were associated with movement along other associated faults. Nevertheless, all were driven by northward movement of the Pacific plate west of the San Andreas Fault, including those in the transverse ranges and the blind thrust fault earthquakes of 1971 and 1994 in the San Fernando Valley.

The San Andreas Fault appears to have accumulated a total displacement of 235 kilometers in the approximately 16 million years since it began to move. The fault has been stuck along its big bend, south of Parkfield, since the Fort Tejon earthquake of 1857. More recent earthquakes near the southern San Andreas Fault are associated with blind thrusts, which probably means that at least some of the crustal movement is being taken up in folding and thrust faults near the main fault instead of in slippage along the main fault itself (see Figures 4-30 and 4-36).

Moderate earthquakes similar to the magnitude 6.7 Northridge event of 1994 are likely to shake the Los Angeles basin with an average recurrence interval of less than ten years. Given the overall slip rate in the region, the number of observed moderate earthquakes is fewer than these estimates would lead us to expect. A larger earthquake of M_w 7.5, requiring a rupture length of roughly 160 kilometers, should occur on average once every 300 or so years.

Trenches dug across the San Andreas Fault northeast of Los Angeles exposed sand blows that record sediment liquefaction during nine large earthquakes during the 1,300 years before the Fort Tejon earthquake of 1857. They struck at intervals between 55 and 275 years; the average interval between these events was 160 years. Based on history, the area is almost certainly due for its next major earthquake.

The Gutenberg-Richter frequency-versus-magnitude relation shows that larger earthquakes are far less frequent than small ones, the decrease in numbers being about proportional to the logarithm of the magnitude (Figure 3-24, page 48). The diagram shows the Gutenberg-Richter frequency-versus-magnitude relationship for part of the San Andreas Fault. A magnitude 7.2 earthquake, for example, should occur every 100 years for that part of the fault. However, as in most data, there is some scatter on either side of this line.

Each whole number increase in magnitude corresponds to an increase in energy release of approximately 33 times. Thus, thirty-three magnitude 6 earthquakes would be necessary to equal the total energy release of one magnitude 7 earthquake. And more than 1,000 earthquakes of magnitude 6 would release energy equal to a single earthquake of magnitude 8 ($33 \times 33 = 1,089$). It seems clear that the many magnitude 5 or 6 earthquakes that occur each decade do not significantly relieve the accumulated strain across the fault.

Far too few moderate earthquakes have occurred in the Los Angeles area to relieve the observed amount of strain built up across the San Andreas Fault between 1857 and 1995 (see Sidebar 4-1). Releasing the accumulated strain would have required seventeen moderate earthquakes, but only two (1971 and 1994) have been greater than magnitude 6.7. The reason for the deficiency is unclear. Elsewhere in the world, clusters of destructive earthquakes have occurred at intervals of tens of years. Perhaps several faults that were mechanically linked together ruptured at about the same time to generate a large earthquake. Such combination ruptures appear to have occurred in the Los Angeles area in the past.

A moment magnitude 7.4 earthquake would cause the ground to shake over a much larger area and for a longer duration than during the Northridge earthquake of 1994. The greater area and longer shaking would cause far more casualties and property damage than the Northridge earthquake did.

Other large earthquakes on or near the San Andreas Fault north of Los Angeles include an event of June, 1838, when the San Andreas Fault slipped along a segment more than 60 kilometers long on the San Francisco peninsula. Estimates based on historical data place the magnitude at approximately 7. The Lompoc earthquake struck with a magnitude of 7.3 north of Santa Barbara from an epicenter near the coast in November 1927. The magnitude 7.7 Arvin-Tehachapi earthquake of July 1952 accompanied oblique slip on a steep fault near the south end of the San Joaquin Valley. An earthquake of magnitude 6.5 in May 1983 hit Coalinga, midway between San Francisco and Los Angeles. It was the result of movement along a blind thrust fault in the actively rising Coalinga anticline.

Tectonic Environments of Major Earthquakes

Intraplate, or within-continent, earthquakes with nebulous tectonic context are poorly understood but sometimes devastating. A few examples of devastating earthquakes on such boundaries are illustrative. A recurring theme is that such earthquakes in highly developed countries with modern construction codes and strong enforcement of those codes cause significant damage but fewer deaths. In poor countries with poor quality construction or little enforcement of existing construction codes, such earthquakes result in large death tolls.

Transform Faults

IZMIT EARTHQUAKE: NORTH ANATOLIAN FAULT, TURKEY, 1999
In the eastern Mediterranean region, the Arabian and African plates jam northward against Eurasia at a rate of 1.8 to 2.5 centimeters per year. The North Anatolian Fault trends east over a distance of 900 kilometers. It closely resembles the San Andreas Fault in California and is comparable in length and slip rate (▶ Figure 4-38).

The North Anatolian Fault slipped on August 17, 1999, in the area 100 kilometers east of Istanbul, causing an earthquake of magnitude 7.4. The surface displacement was horizontal, between 1.5 and 2.7 meters along a fault length of 140 kilometers. More than 30,000 people died. Property damage reached $7 billion dollars.

The North Anatolian Fault has caused eleven major earthquakes larger than magnitude 6.7 in the last century (▶ Figure 4-39). A series of six large earthquakes progressed steadily westward between 1939 and 1957. Between 1939 and 1944, the fault ruptured along an incredible 600 kilometers of continuous length. Meanwhile, three other earthquakes occurred near both ends of the fault, beyond the sequence of six. The earthquake of August 17, 1999, filled a seismic gap. Less than three months later, on November 12, the North Anatolian Fault slipped along a length of 43 kilometers in the area immediately east of the previous

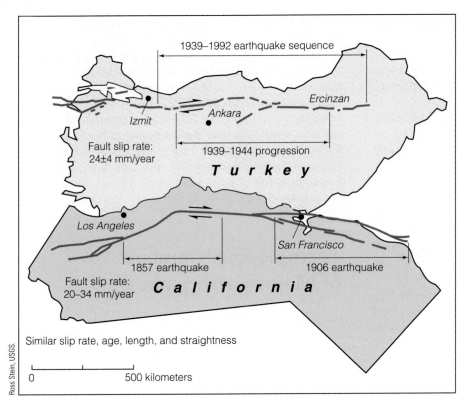

Ross Stein, USGS.

FIGURE 4-38. The North Anatolian Fault of Turkey is comparable to the San Andreas Fault of California in slip rate, age, length, and straightness.

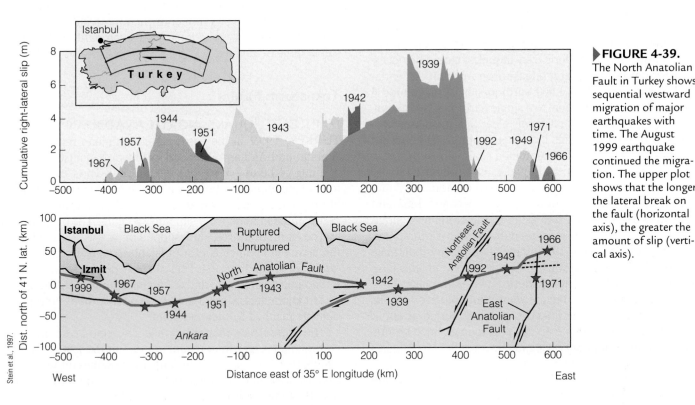

Stein et al., 1997.

FIGURE 4-39. The North Anatolian Fault in Turkey shows sequential westward migration of major earthquakes with time. The August 1999 earthquake continued the migration. The upper plot shows that the longer the lateral break on the fault (horizontal axis), the greater the amount of slip (vertical axis).

movement. That caused an earthquake of magnitude 7.1 that killed 850 people in Duzce. If current trends continue, Istanbul, farther west and only 20 kilometers north of the North Anatolian Fault is next. That is a major concern.

On May 1, 2003, a magnitude 6.4 earthquake struck just south of the east end of the North Anatolian Fault. Collision between the Arabian Plate and the Eurasian Plate caused this earthquake in addition to those on the North Anatolian Fault (see Figures 4-40, 4-41, and 4-42).

Although building codes in Turkey match those of the United States in their anticipation of earthquakes, enforcement is poor and much of the construction is shoddy. The

Mehmet Celebi photo, USGS.

▶**FIGURE 4-40.** This modern, multistory apartment building in Izmit pancaked over to the right. It is unlikely that anyone would have survived this type of collapse.

Mehmet Celebi photo, USGS.

▶**FIGURE 4-42.** This building, which was unfinished at the time of the Izmit earthquake, had its upper floors hanging out over the street. Its heavy superstructure rested on concrete posts with weak links to the foundation and the concrete floors. The floors and foundation also lacked diagonal braces and shear walls that would have prevented lateral shift.

Mehmet Celebi photo, USGS.

▶**FIGURE 4-41.** A group of older apartment buildings in Izmit lies in ruins after the poorly braced lower floors of most of them collapsed. Poor-quality construction contributed.

Mehmet Celebi photo, USGS.

▶**FIGURE 4-43.** Some buildings sank or tilted because of both liquefaction of the ground and heavy construction with upper floors overhanging streets.

tax code favors construction of buildings with second stories that extend over the street because buildings are taxed on the area of the street-level floor (▶ Figures 4-42 and 4-43). Many of these buildings collapsed during the earthquakes. Liquefaction locally caused buildings to tilt or collapse.

Subduction Zones

CHILE, 1960 On May 22, 1960, the largest earthquake on record struck the coast of Chile south of Santiago with a Richter magnitude of 8.5 and a moment magnitude (M_w) of 9.5. The slab sinking through the oceanic trench offshore had broken along a length of 1,000 kilometers, a distance comparable to the entire length of the Cascadia oceanic trench in the Pacific Northwest.

A large foreshock that struck thirty minutes before the main event sent people scurrying into the streets. It saved them from the main shock that collapsed most of their homes. Dozens of big aftershocks ranged up to magnitudes 6 and 7. Approximately 2,000 people died.

Fifteen minutes after the earthquake, three large ocean waves, or tsunami, severely damaged coastal towns in Chile. Fifteen hours after the earthquake, a 10-meter ocean wave struck Hilo, Hawaii (see Chapter 5, Tsunami). Twenty-two hours after the earthquake, the largest wave recorded in 500 years washed onto the coast of Japan. It destroyed thousands of houses and drowned 120 people.

The ground displacement raised some offshore islands in Chile as much as 6 meters. Meanwhile, the coastal mountains dropped 2 meters. That was exactly the pattern of ground movement in the Alaska earthquake of 1964. And it is the pattern that geologists expect when the next big earthquake strikes from the Cascadia collision boundary off the Pacific Northwest.

On March 3, 1985, another earthquake, of magnitude 7.8, struck Santiago, Chile, from an epicenter just offshore. It killed 176 people, injured 2,483, and left 372,532 homeless. Many survived because the earthquake came on a Sunday evening when commercial buildings were closed and people were outside. Building codes enacted in response to past earthquakes required reinforced buildings; these were not seriously damaged. Chile has experienced twenty-five major earthquakes since 1870.

Blind Thrust Faults over an Active Subduction Zone

SEATTLE FAULT, WASHINGTON Every three or four years, the Puget Sound area of northwestern Washington state feels the jolt of a moderate to large earthquake with Richter magnitude 5 to 7. Most of the earthquakes in the Puget Sound area do not involve slip on the collision boundary at the oceanic trench offshore. Instead they accompany movement at shallow depth on faults that trend west or northwest and straddle Puget Sound. A similar set of active faults exists east of the Cascades in Washington and Oregon (▶Figure 4-44).

The Nisqually earthquake struck the Seattle–Tacoma area with a Richter magnitude of 6.8 on February 28, 2001; it was the largest earthquake since the magnitude 7.1

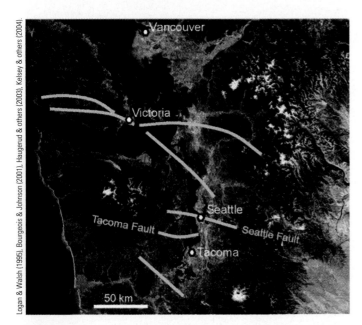

Logan & Walsh (1995), Bourgeois & J. Johnson (2001), Haugerud & others (2003), Kelsey & others (2004).

▶**FIGURE 4-44.** This map of northwest Washington shows the Seattle Fault and related major recent active fault zones in the Seattle area.

Olympia event of April, 1949, and the magnitude 6.5 Seattle event of 1965. The Nisqually earthquake inflicted fairly widespread minor damage as windows broke, cornices fell, and objects of all kinds rattled off their shelves. The shaking continued for thirty to forty-five seconds. Property damage amounted to more than $2 billion, mostly to old buildings on reclaimed and poorly compacted tidelands. One person died, and more than 250 were injured. Although the Seattle area was not severely damaged, scientists there are particularly concerned with the safety of the Alaska Way double-deck elevated highway around the waterfront of Seattle (▶Figure 4-45). Geotechnical studies of the damage and the level of shaking suggest that if the earthquake had lasted ten seconds longer, footings on bay mud near the south end of Alaska Way would have failed and the highway would have collapsed.

So what caused the Nisqually earthquake? It is unclear, but one possibility is bending of the ocean floor as it begins its descent under the continental margin. Another is that parts of the descending slab descend at different rates or different slopes or inclinations than adjacent areas to the north or south. The stress on the slab causes it to break. Alternatively, a northward drag of the ocean floor deformed the continental margin.

Although the magnitude of the Nisqually earthquake matched that of the Northridge earthquake that struck Los Angeles in 1993, it inflicted much less damage and many fewer casualties. The lesser degree of damage was caused by several factors. First, the fault movement that caused the Nisqually earthquake occurred at a depth of 50 kilometers, so no building on the Earth's surface was any closer to the focus. Second, deep earthquakes, though commonly strong, do not generate much surface-wave motion, which is the most damaging. Third, recent bracing of buildings and bridges to better withstand earthquakes helped immensely.

The Seattle Fault is the best known, and perhaps the most dangerous, fault in the region. It trends east through the south end of downtown Seattle, almost through the interchange between Interstate 5 and Interstate 90 (▶Figure 4-46). Seventy kilometers of the fault are mapped; the part that reaches the surface dips steeply down to the south. Studies show that the rocks south of the fault rose 8 meters some 980 years ago, probably accompanying an earthquake of magnitude greater than 7. That movement generated large tsunami waves in the water in Puget Sound and caused landslides into Lake Washington at the eastern edge of downtown Seattle (see Chapter 5, Tsunami). The length of the Seattle Fault suggests that earthquakes of magnitude 7.6 are possible.

Continent–Continent Collision Zones

BAM EARTHQUAKE: IRAN, 2003 Northward collision of the Arabian Plate with the Eurasian Plate creates the Caucasus Mountains, part of a sinuous, more or less east–west collision zone that extends from the Alps in the west to the Himalayas in the east. Convergence is roughly 3 cen-

▶**FIGURES 4-45.** A support beam (left) on the Alaska Way viaduct in Seattle was damaged in the 2001 Nisqually earthquake. The viaduct is in danger of collapse if there should be another large earthquake on the Seattle Fault.

timeters per year. On December 26, 2003, a magnitude 6.7 earthquake struck the city of Bam, Iran (▶ Figure 4-47). The earthquake accompanied right-lateral strike-slip movement on a north-trending fault that moved in the collision zone. The magnitude of the Bam earthquake (M_W 6.5) was slightly less than the magnitude 6.7 M_W Northridge earthquake in

the Los Angeles area of southern California, and both were in the early morning when most people were home in bed, 4:31 A.M. in Northridge and 5:26 A.M. in Bam.

The destruction and death toll in the two earthquakes, however, was dramatically different; sixty-one died in Northridge, but more than 26,000 died in Bam (Figure 4-48). The

▶**FIGURE 4-46.** The Seattle Fault runs east–west through the interchange of I-90 (foreground) and I-5 (middle right) at the south edge of Seattle.

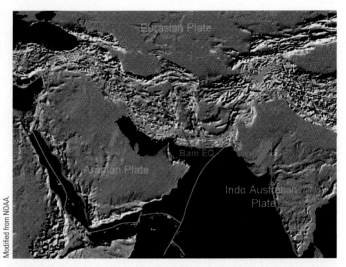

Modified from NOAA.

▶**FIGURE 4-47.** The Bam, Iran, earthquake of 2003, occurred as part of the ongoing northward collision of the Arabian Plate with Asia. That collision pushes up the Caucasus Mountains, northeast of Turkey, and its southeastward extension, the Zagros Mountains of Iran. Plate boundaries are in green.

National Geoscience Database Iran.

▶**FIGURE 4-48.** We often think of casualty numbers in natural hazards as big or small, but those who die in such tragedies are real people from real families. Here people search for relatives among the dead in the 2003 Bam earthquake.

National Geoscience Database Iran.

(a)

Courtesy of Space Imaging.

(b)

▶**FIGURE 4-49.** **(a)** Poor construction materials and quality led to widespread collapse of buildings and the tragic death toll. **(b)** This high-resolution space image of Bam, Iran, shows many intact houses west of the north-south road but complete destruction of most of the houses east of that road.

difference was largely because of the nature of buildings in the two areas. In Northridge, most of the buildings are either of wood-frame construction or reinforced concrete. Some older buildings are bricks held together with mortar; many of these were badly damaged, but few completely collapsed. In contrast, most homes in Bam were built from adobe, poor quality mud bricks, loosely held together by equally poor quality mortar; some 80 percent of the buildings collapsed (▶Figures 4-48 and 4-49). They had little or no reinforcing.

Typical construction in Bam consisted of mud-brick walls supporting steel roof girders of a heavy roof sealed with cement. During the earthquake, the unbraced mud-brick walls crumbled, leading to pancaking of the upper floors and roof. People inside were crushed. The construction style also left few air pockets for survivors.

Blind Thrust Faults Associated with a Continental Collision Zone

BHUJ EARTHQUAKE: INDIA, 2001 On January 26, 2001, an earthquake of magnitude 7.7 struck the Indian city of Bhuj, 560 kilometers north of Bombay in western India. The area is just south of the collision zone where the Indian Plate is colliding north into the southern margin of the Asian Plate. One of many blind thrust faults moved at a depth of 23.6 kilometers. Rolling hills marked the blind thrusts, the largest yet recorded anywhere. Some rivers temporarily reversed their direction of flow where they crossed the fault that caused the earthquake.

More than 30,000 people died, a horrific toll that would have been much higher had it not been Republic Day, when many people were out on the streets celebrating. Many of those killed were caught in narrow streets between collaps-

(a)

(b)

(c)

Courtesy of J.-P. Bardet.

▶FIGURE 4-50. (a) Buildings constructed with adobe collapsed during the Bhuj earthquake. (b) Even buildings with concrete floors and some steel reinforcing bars did not fare well; floors collapsed on each other. (c) Narrow streets were not a safe place or refuge from the collapsing buildings.

ing houses and tall buildings. Some 600,000 were left homeless. Another 1,200 people died in Ahmedebad, a city of 3.6 million, 290 kilometers to the east. Although it is rare to find people alive after four days of burial under rubble, a few survived for six days; two people survived for ten days because they had water.

On paper, India's building codes are similar to those in the United States. Unfortunately, they are merely advisory, not compulsory, and poorly followed. Many buildings are made of mud or brick and collapsed during the shaking (▶Figure 4-50). Severe soil liquefaction occurred over an area of at least 1,000 square kilometers.

Continental Spreading Zones

WASATCH FRONT, UTAH In western North America, the best known area of continental extension and the normal faults it produces is the Basin and Range of Nevada, Utah, and adjacent areas (▶Figures 4-51, 2-20, and 2-21, page 25). This broad region is laced with numerous north-trending faults that separate raised mountain ranges and dropped valleys.

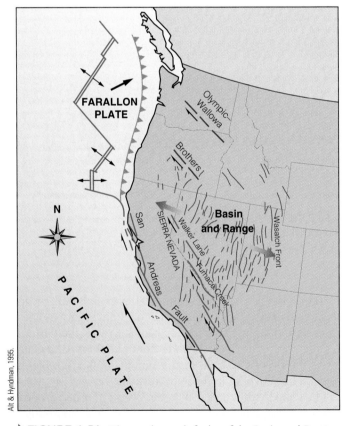

Alt & Hyndman, 1995.

▶FIGURE 4-51. The north–south faults of the Basin and Range of Nevada, western Utah, and adjacent areas occupy a spreading zone accompanying the northwestward drag of the Pacific Ocean floor. That drag also causes shear to form the San Andreas Fault. The western margin of the Basin and Range is marked by the precipitous eastern edge of the Sierra Nevada; the eastern margin is the equally precipitous Wasatch Front at Salt Lake City.

▶**FIGURE 4-52.** A prominent fault scarp crosses the East Bench, out from the base of the Wasatch Front, in a completely built-up part of the Salt Lake City metropolitan area. Arrows mark the fault.

▶**FIGURE 4-54.** In spite of real hazards and educational efforts of the Utah Geological Survey, development continues, even along the scarp itself.

▶**FIGURE 4-53.** A Wasatch Front fault scarp cuts a 19,000-year-old glacial moraine from Cottonwood Canyon near Salt Lake City. This scarp formed during a large earthquake since that time.

The Wasatch Front of central Utah is a high fault scarp that faces west across the Basin and Range and defines its eastern margin (▶Figures 4-51 to 4-54). It is the eastern counterpart to the Sierra Nevada front of California. It overlooks the deserts of Utah in the same way that the Sierra Nevada front overlooks those of Nevada.

Many small earthquakes shake the Wasatch Front, but none of any consequence have been felt since Brigham Young's party founded Salt Lake City in 1847. The Wasatch Front is a fault scarp that separates the broad flat of the Salt Lake basin on the west from the high Wasatch Range to the east.

One way to interpret the modest size of many alluvial fans along the base of the Wasatch Front is to suggest that the fault movement has dropped the valley relative to the Wasatch Range during the geologically recent past, probably within a few million years. That would roughly correspond to the time in which the Sierra Nevada rose.

Radiocarbon dates on sediments offset by fault movement indicate that no movement has happened on the main fault near Brigham City, north of Salt Lake City, for at least 1,400 years. But that is not nearly a long enough time to offer much reassurance to the people of the Salt Lake area. Neither does the extremely fresh appearance of the Wasatch Front. The people of the Salt Lake area should consider themselves at high risk for major earthquakes. It is also not reassuring to see that an obvious west-facing fault scarp crossing the broad East Bench glacial outwash fan at the base of the Wasatch Front is covered with homes (Figure 4-52).

Nor is it reassuring to see an obvious fault scarp that faces west, offsetting glacial moraines in Little Cottonwood Canyon in the southeastern part of Salt Lake City (Figure 4-53). Radiocarbon dates on scraps of charcoal showed that the scarp rose approximately 9,000 years ago. More radiocarbon dates on offset sedimentary layers that are exposed in trenches cut across the fault revealed evidence of movements 5,300, 4,000, 2,500, and 1,300 years ago. The average recurrence interval appears to be 1,350 years with an offset averaging 2 meters during each event. More unsettling is the continued development pressure near, and even within, the fault zone (Figures 4-53 and 4-54).

The Uniform Building Code (Figure 3-35, page 53) places the Salt Lake Valley in seismic zone 3. Some new structures are built to zone 4, the highest standard, and some older ones are being upgraded. In 1989, the historic City and County Building, five floors of stone masonry construction, was reinforced and fitted with base isolation pads (Figures 3-57 and 3-58, pages 61 and 62). The lowest floor was detached from its foundation, jacked up, and placed on rubberized pads more than 30 centimeters thick at a cost of $43.6 million dollars. The building is now designed for an

▶**FIGURE 4-55.** Broad low mounds in a nearly flat area below the Wasatch Front west of Farmington, near Ogden, Utah, mark an area of liquefaction of soft clays laid down in Glacial Lake Bonneville, the ancestor of Great Salt Lake.

earthquake of Richter magnitude 5 or 6. Seismic risk maps indicate a 10 percent probability of such an earthquake within fifty years.

The greatest earthquake hazards are in the Salt Lake Valley with its deep fill of soft sediments and high groundwater levels. Together, these factors may amplify ground motion more than 10 times. The groundwater table is only a few meters below the surface at Salt Lake City. Liquefaction of wet clays would cause loss of bearing capacity and downslope flow (▶Figure 4-55).

Intraplate Earthquakes without Obvious Tectonic Context

Earthquakes occasionally strike without warning in places that lack any recent record of earthquakes and are remote from any plate boundary. Most local people are unaware of any threat. Some isolated earthquakes are enormous, easily capable of causing a major natural catastrophe. Although many geologists have offered tentative explanations for these earthquakes, their causes remain generally obscure.

The New Madrid earthquakes that struck southeastern Missouri in 1811 and 1812 were among the most severe to strike North America during its period of recorded history. If the region had not been so sparsely populated, they would have caused a major catastrophe. A repetition could devastate Memphis, St. Louis, Louisville, Little Rock, and many smaller cities that have older masonry buildings. Another isolated earthquake that struck Charleston, South Carolina, in 1886 was indeed a natural catastrophe. It caused many casualties and heavy property damage. The Charleston event, near the east coast of the United States, is along what has been called a "trailing continental margin." This is not a current plate margin but the margin between the North

American continent and the Atlantic Ocean basin; it was originally a plate margin when the Atlantic Ocean floor began to spread more than 100 million years ago.

NEW MADRID EARTHQUAKES: MISSOURI, 1811–1812
Had it not actually happened in 1811 and 1812, no one would suspect that the southeastern corner of Missouri might become the site of a great earthquake. The American Midwest is so far from any plate boundary that it seems an unlikely place for any earthquake, large or small. Nevertheless, three great earthquakes struck near New Madrid in December 1811, January 1812, and February 1812.

The earthquakes predated seismographs by nearly a century. The region was then too thinly populated to provide data for maps of Mercalli intensities. But toppled chimneys in Ohio, Alabama, and Louisiana suggest intensities of VII in those distant places. The earthquakes caused church bells to ring in Boston. Indeed, the three large earthquakes were felt throughout the eastern United States. The best estimates place their Richter magnitudes at about 7.5, 7.3, and 7.8.

Those three earthquakes were not freak events. The area is seismically active enough that people as far away as St. Louis and Memphis, the nearest big cities, occasionally hear the ground rumble as their dishes and windows rattle. Few of the buildings in either city are designed or built to resist significant earthquakes. A repetition of any of the earthquakes of 1811 and 1812 would cause enormous loss of life and property damage in both cities (▶Figures 4-56 and 4-57).

The Earth's crust in eastern North America transmits earthquake waves more efficiently, with less loss of energy, than the continental crust of the west, which is hotter and more broken along faults. That explains why the area of

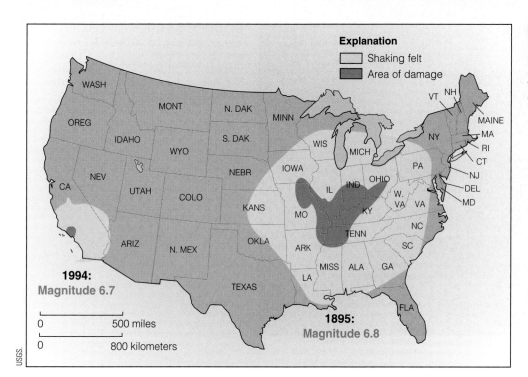

USGS.

Explanation
- Shaking felt
- Area of damage

1994:
Magnitude 6.7

1895:
Magnitude 6.8

| 0 | 500 miles |
| 0 | 800 kilometers |

▶**FIGURE 4-56.** A comparison of similar magnitude earthquakes shows that the damage would be much greater for an earthquake in the Midwest than in the mountainous West.

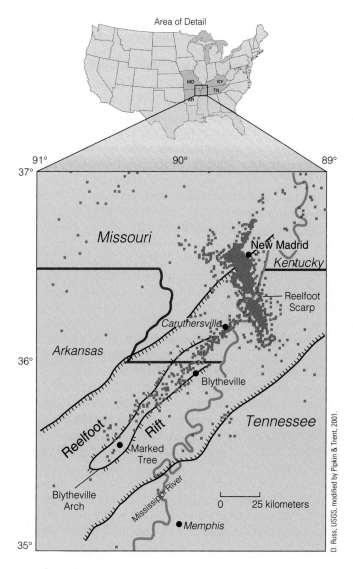

Area of Detail

D. Russ, USGS, modified by Pipkin & Trent, 2001.

▶**FIGURE 4-57.** Recent microearthquake epicenters in the New Madrid region appear to outline three fault zones responsible for the earthquakes of 1811 and 1812. Two lateral-slip faults are offset by a short fault that pushed the southwestern side up over the northeastern side.

significant damage for an earthquake of a given size is greater in the East than in the West (Figure 4-56).

The fault movements of 1811 and 1812 produced too little surface offset to provide useful clues to what happened at depth. Modern studies of **microearthquake** activity reveal three apparent faults (Figure 4-57). Radiocarbon dates on offset sediments exposed in a trench across the Reelfoot Fault scarp in Tennessee suggest that the New Madrid fault system has moved three times within the last 2,400 years. Those data, among others, suggest that the recurrence interval for earthquakes in the New Madrid area may be 500 to 1,000 years, but the uncertainty is large.

Such large events have a low probability of occurrence, but the average life of a building is fifty to 150 years. Retrofitting existing buildings to survive these large but rare events would be extremely expensive; it would be much less expensive to construct new buildings to a higher standard. The cost should be weighed against the large number of lives that could be saved if these dollars were spent on improved medical care or mitigating other hazards. This

▶**FIGURE 4-59.** The structural brick walls of this house at 157 Tradd Street in Charleston collapsed during the 1886 earthquake.

leaves scientists and policy makers in the region unsure of the best strategy for codes for new or retrofitted buildings.

CHARLESTON EARTHQUAKE: SOUTH CAROLINA, 1886 By most standards, the passive continental margin of the Eastern seaboard of the United States is not earthquake country. Nevertheless, because damaging earthquakes have occurred there in the past, it seems reasonable to expect more in the future. Consider the Charleston earthquake of 1886.

An earthquake of Mercalli Intensity X struck from an epicenter 20 kilometers northwest of Charleston, South Carolina, on August 31, 1886. Damage in Charleston corresponded to a Mercalli Intensity of IX. No seismographs existed then, but the magnitude was inferred many years later to have been $M_W = 7.3$, $M_B = 6.8$, $M_S = 7.7$. Some 14,000 chimneys fell, and many buildings were destroyed on both solid and soft ground. One hundred people were killed, most of them in areas where the soil liquefied (▶Figures 4-58 and 4-59).

Thirteen earthquakes shook parts of South Carolina between 1698 and 1886. If we can judge from historic accounts, most had Mercalli intensities below VI. But radiocarbon dates on disrupted sediments exposed in trenches suggest a recurrence interval of several hundred years for earthquakes of this size. Current seismic activity, which may be related to the 1886 earthquake, is in crystalline basement rocks beneath sediments of the coastal plain.

THE EAST COAST FAULT SYSTEM The Charleston earthquake and many others may be the result of movements along segments of the East Coast fault system, a swarm of aligned segments that trend generally northeast near the modern East Coast. They were first recognized between South Carolina and Virginia but may extend much farther. The fault zone is near the buried boundary between continental crust of the Piedmont and Atlantic oceanic crust, with the coastal plain dropping. It may be renewed movement on faults associated with the early stages of opening of the Atlantic Ocean more than 200 million years ago (▶Figure 4-60).

An earthquake of Mercalli Intensity VIII struck Virginia on May 31, 1897. It appears to have been the result of fault movement in the crystalline basement rocks, presumably within the East Coast fault system. On November 18, 1755, an earthquake of magnitude 6, Mercalli Intensity VIII, struck Cape Ann in Massachusetts, collapsing chimneys and walls in Boston. At least six other earthquakes shook the region between 1638 and 1744.

The largest earthquake known in the northeast struck the St. Lawrence River Valley northeast of Quebec City in 1663. Its magnitude is estimated as 7.0. It caused large landslides and fallen chimneys in eastern Massachusetts, 600 kilometers away. An 1870 event had a Mercalli Intensity of IX. An earthquake of estimated magnitude 5.2 from an epicenter near the mouth of the Hudson River in New York was felt from Maryland to New Hampshire in 1884. Occasional mild earthquakes felt between eastern Pennsylvania and Maine suggest that these areas may not be as safe from damaging earthquakes as their residents assume.

The Charleston earthquake of 1886 sent hundreds of people into the streets and toppled several chimneys in Lancaster, Ohio, 800 kilometers away. It shook plaster from the walls on the fourth floor of one building in Chicago, 1,200 kilometers away. Although large earthquakes are more frequent in the West, a few large earthquakes have occurred in eastern North America. These have been much more damaging than those in the West (Figure 4-56) because energy is more efficiently transmitted through the ancient crustal rocks in the East.

Earthquake hazards may be significant in at least some parts of the eastern United States, where past earthquakes left little memory or lasting concern. Good land use planning and building codes for new structures cost little anywhere and may someday save enormous loss of life and property damage in a city that does not now suspect it is living dangerously.

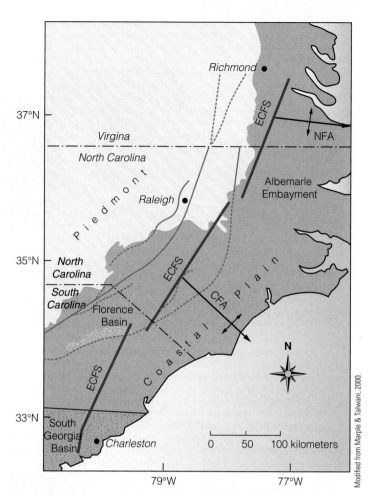

Modified from Marple & Talwani, 2000.

▶**FIGURE 4-60.** This map of the East Coast fault system between South Carolina and Virginia shows how the fault zone lies close to the buried boundary between the continental crust of the Piedmont and the Atlantic oceanic crust. The Piedmont is rising.

The Potential for Still Larger Events

We can look at the problem of potentially much larger earthquakes—that is, events that would be larger than anything ever recorded—from different viewpoints:

- In Chapter 1, we noted that there are many small events, fewer large ones, and only rarely a giant event (see Figure 1-3, page 5). Quantitatively, that translates as a "power law." Plotted on a graph of earthquake frequency versus earthquake magnitude (e.g., Figure 3-24, page 48), the power law can be plotted as a log scale: 10^1 or 10 to the power of 1 is 10; 10^2 or 10 to the second power is $10 \times 10 = 100$; $10^3 = 10 \times 10 \times 10 = 1,000$, and so on. So Figure 3-24, for an example fault, tells us that if we plot all known earthquakes of a certain size against their frequency of occurrence, we get a more or less straight line that we can extrapolate to events larger than those on record. Those giant events will be rare, which is presumably why we have not had any in the historic record. They can occur—not a comforting thought, though the odds are low.

■ Is there a limit to the size of that giant earthquake? In practical terms there is. The magnitude of an earthquake depends, among other things, on the rupture length of the fault (Figure 3-28, page 50). A short fault only a few kilometers long can have many small earthquakes, but its largest cannot be especially large because the whole fault is not long enough; it does not break a large area of rock. The San Andreas Fault, though at 1,000 kilometers long, could conceivably break its whole length in one shot. Certainly, the San Andreas has had some large earthquakes, but even those have broken only half the length of the fault at one time. The potential is still there for a much larger earthquake to break the full length of the fault. For the San Andreas, that would be a truly catastrophic event. However, the continuously creeping section of the fault south of San Francisco seems to be a zone of weak rocks that would be unable to build up a significant stress. Perhaps that means that no more than half of the length of the fault is likely to break suddenly at one time. Slip of a 500-kilometer length of the fault could still generate an earthquake of magnitude 8.

KEY POINTS

✓ Earthquake forecasts, which are now fairly reliable, specify the probability of an earthquake in a magnitude range within a region over a long time period such as a few decades. Earthquake predictions involve statements as to specifically when and where an earthquake is likely to occur; they are generally not reliable. **Review pp. 2, 4–5, 7, 64–73; Figures 1-7, 4-11 to 4-14.**

✓ Earthquake precursors that suggest that an earthquake is imminent include foreshocks, changes in Earth magnetism, radon gas, and groundwater level. **Review pp. 65–67.**

✓ Seismic gaps, where movement has not occurred on a fault segment, are likely areas for a future fault movement. Earthquakes migrating along a fault with time and earthquakes at regular intervals can provide guidance for where and when the next shock may be. **Review pp. 67–68; Figures 4-2 and 4-3.**

✓ Paleoseismology, the study of past earthquake times and sizes, provides guidance as to the expected frequency and sizes of future earthquakes. **Review p. 69; Figures 4-6 and 4-7.**

✓ The San Andreas transform fault, running much of the length of coastal California, is the dominant earthquake fault in North America. Its most dangerous areas are the large population centers of the San Francisco Bay area (**review pp. 73, 75–81; Figures 4-16, 4-17, 4-28 and 4-29**) and the Los Angeles area (**pp. 81–85; Figures 3-22, 4-30 and 4-31**). "Blind thrusts" occur off the main fault. **Review pp. 81–83; Figures 4-30, 4-31 and 4-36.**

✓ The probability of a major earthquake on the San Andreas Fault can be assessed from the relationship between frequency and magnitude or strain accumulated by the overall slip rate on the fault for the San Francisco Bay area (**review pp. 75–76 and 79–80; Figure 4-29**) or the Los Angeles area (**review pp. 84–85**).

✓ Large earthquakes can sometimes progress laterally along a major fault, as shown by the North Anatolian Fault in Turkey. **Review pp. 85–86; Figure 4-39.**

✓ Long subduction zones can generate extremely large earthquakes as shown by the Chile event of 1960. **Review p. 87–88.**

✓ Some large earthquakes over an active subduction zone are not on the subduction zone fault itself but on faults in the overlying slab. **Review p. 88; Figure 4-44.**

✓ Continent–continent collision zones can generate extremely large and disastrous earthquakes, as shown by the 2003 Bam earthquake in Iran and the Bhuj earthquake in India. **Review pp. 88–91; Figures 4-47–4-50.**

✓ Continental spreading zones can generate large earthquakes as in the Basin and Range. **Review pp. 91–93; Figure 4-51.**

✓ Intraplate earthquakes, as in the case of the New Madrid, Missouri, events of 1811–12, though less frequent, can also be quite large. **Review pp. 93–94; Figures 4-56 and 4-57.**

✓ Large earthquakes along the eastern fringe of North America are less frequent but can be significant and highly damaging. **Review pp. 95–96; Figure 4-60.**

IMPORTANT WORDS AND CONCEPTS

Terms

blind thrusts, p. 81
earthquake forecasts, p. 71
forecasts, p. 64
foreshocks, p. 65
microearthquakes, p. 94

migrating earthquakes, p. 68
predictions, p. 64
recurrence interval, p. 71
seismic gap, p. 67

QUESTIONS FOR REVIEW

1. List several of the precursors that have been used to indicate that an earthquake may be coming.

2. What is a seismic gap, and what is its significance?

3. Once a trench has been dug across an active fault, what kinds of information should be collected to determine when the fault last moved, how much it moved, and how frequently?

4. Where in the United States and Canada is the most seismically active zone? Be specific.

5. What is the nature of the fault boundary or boundaries over the extent of that zone?

6. Where are three other significant and active earthquake zones in the continental United States and southern Canada?

7. The North Anatolian Fault in Turkey caused more than 30,000 deaths in 1999. What North American fault is it similar to and in what way?

8. Why does the North Anatolian Fault kill many more people than its North American counterpart?

9. Why does the ground along the coast drop dramatically during a major subduction-zone earthquake? Explain clearly—with a diagram if you wish.

10. What specific evidence is there for major subduction-zone earthquakes along the coast of Washington and Oregon?

FURTHER READING

Assess your understanding of this chapter's topics with additional quizzing and conceptual-based problems at:

 http://earthscience.brookscole.com/hyndman.

CHAPTER 5

In 1946, a large tsunami arrived unexpectedly in Hilo Bay, Hawaii, nearly 4,000 kilometers from the Alaskan earthquake that caused it. Note the people and large truck at left.

K. Fuji collection, Pacific Tsunami Museum.

TSUNAMI
The Great Wave

Harbor Waves

Tsunami, the Japanese name for "harbor waves," are so named because the waves rise highest where they are focused into bays or harbors. They are also called **seismic sea waves** because they are most frequently caused by ocean-floor earthquakes (▶Figures 5-1 and 5-2). Note that the term *tsunami* is both singular and plural. The colloquial use of "tidal wave" is not appropriate because tsunami have nothing to do with tides.

Sumatra Tsunami: December 2004

At 7:59 A.M. on December 26, 2004, a giant earthquake, the fourth largest in the world since 1900, shook the gently northeast-dipping subduction zone just west of northern Sumatra, Indonesia (▶Figure 5-1). The Indian Plate is moving northeast at 6 centimeters per year relative to the Burma Plate. In the ten years preceding this event, there were forty events larger than magnitude 5.5 in the area, but none generated tsunami. In the last 200 or so years, several other earthquakes larger than magnitude 8 have generated moderate-

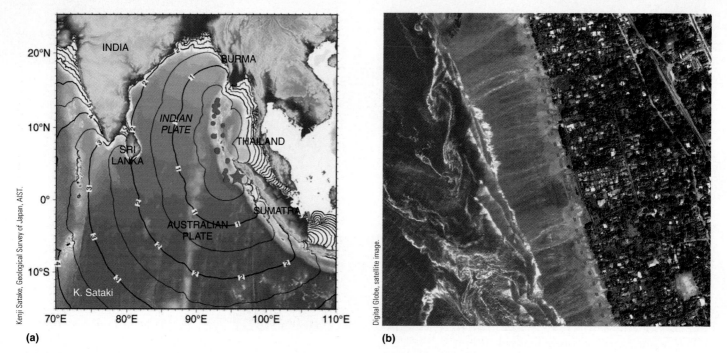

(a)   (b)

▶**FIGURE 5-1.** **(a)** Tsunami wave-front travel times (in hours) are shown emanating from the rupture zone, which spans from the earthquake epicenter (red star) through the area of aftershocks (red dots). **(b)** A broad offshore beach is exposed at Kalutara, Sri Lanka on December 26, 2004, as the first wave of the tsunami drains back to the ocean. The red dotted line is the normal beach edge.

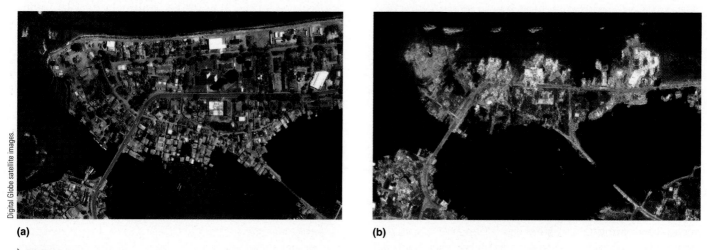

(a)   (b)

▶**FIGURE 5-2.** **(a)** The northern part of Banda Aceh, Sumatra, on June 23, 2004, before the tsunami. **(b)** The same area on December 28, 2004, after the tsunami. Note that virtually all of the buildings were swept off the heavily populated island. The heavy rock riprap along the north coast of the island before the tsunami remains only in scattered patches afterward. A large part of the island south of the riprap has disappeared, as has part of the southern edge of the island between the two bridges, where closely packed buildings were built on piers in the bay.

sized tsunami that have killed as many as a few hundred people. Paleoseismic studies show that giant events occur in the region on an average of once every 230 years.

The subduction boundary had been locked for hundreds of years (see Figure 2-14, page 20), causing the overriding Burma Plate to slowly bulge like a bent stick; it finally slipped to cause a magnitude 9 earthquake. Given the size of the earthquake, offset on the thrust plane was some 15 meters, with the sea floor rising several meters. The subduction zone broke suddenly, extending north over approximately 1,200 kilometers of its length (▶Figure 5-1a), or the entire length of the segment between the India and

Burma Plates; that presumably relieved much of the strain in that zone. The Australia Plate–Burma Plate section of the boundary immediately to the southeast did not break. It may do so in a future event as it did in an estimated magnitude 9.3 quake in 1833.

The consequences of this disaster were tragic. Sumatra, Thailand, and nearby countries are mountainous, with huge populations living near sea level along the coasts. That made them vulnerable to tsunami waves that rose to 10 meters and more above normal sea level. Although the tsunami waves in the open ocean are often only a half meter or so high, they drag on the bottom in shallower water, causing them to slow and rise into much larger waves. Given that most homes and other buildings average only 3 meters high, only buildings such as some coastal tourist hotels remained above the incoming waves. Although tsunami waves can break onshore like normal waves (see the opening chapter photograph), they can flow inland for more than a kilometer in fairly flat regions, becoming a torrent that sweeps up everything in its path: cars, people, buildings, and fragments of their crushed remains. Even strong swimmers have little chance of survival because of impacts from all of the churning debris. As the wave recedes into the trough before the next wave, the onshore water and its debris flowing back offshore are almost as fast and dangerous to those caught in it as the initial tsunami (▶ Figure 5-1b). Because the time between tsunami waves is often more than a half hour, the wave trough is well offshore; people and debris are carried out to sea.

With the retreat of the first wave, the survivors felt that the danger had passed, only to be overwhelmed by the next still larger wave. Hours later, debris and bodies washed up on beaches, in some places accumulating like driftwood. The dead included not only locals but many foreigners vacationing in the region's warm weather. As efforts continued to find survivors, other threats loomed. Bodies began decomposing in the tropical heat. Concern quickly shifted to the danger of contaminated water, cholera, typhoid, hepatitis A, and dysentery. Compounding the nightmare were pools of stagnant water that can foster breeding of mosquitoes that may carry malaria and dengue fever. Often-futile attempts were made to identify the dead before burial, using fingerprints or merely by posting photos. Then there was the problem of burying the dead in ground that was completely saturated with water. Some people who survived the waves died from infected gashes, lacerations, broken bones, and other wounds as both antibiotics and health care workers were in short supply.

The magnitude 9 earthquake shook violently for as much as eight minutes. For people nearby in Sumatra, the back-and-forth distance of shaking, with accelerations greater than that of a falling elevator, made it impossible to stand or run while poorly reinforced buildings collapsed around them. The sudden rise of the ocean floor generated a huge wave that moved outward at speeds of more than 700 kilometers per hour; it reached nearby shores within 15 minutes. Those who were living on low-lying coastal areas had little or no warning of the incoming wave. Most people were preoccupied with the earthquake, and few were aware of even the possibility of tsunami. For some who did not happen to be looking out to sea, the first indication was apparently a roaring sound similar to fast-approaching locomotives. Elsewhere there was no sound as the sea rose.

For the vast majority of the people, there was no official warning. A tsunami warning network around the Pacific Ocean monitors large earthquakes and then transmits warnings to twenty-six participating countries of the possibility of tsunami generation and arrival time. In the Indian Ocean, however, there is no such warning network. The Indian Ocean continental margins do not have active subduction zones, except along the southwest coasts of Sumatra and Java. A warning system would not have been able to save most of the lives in the most devastated region of Sumatra because the time between the earthquake and wave arrival was short; however, it could have saved many lives in more distant locations such as Sri Lanka and India.

Although the massive earthquake was recorded worldwide, people along the affected coasts were not notified of the possibility of major tsunami. The reasons were several. The Pacific Tsunami Warning Center in Hawaii alerted member countries around the Pacific and tried to contact some countries around the Indian Ocean that a tsunami might have been generated. Because tsunami in the Indian Ocean are infrequent, no notification framework was in place to rapidly disseminate the information between or within countries. Compounding the problem was the lack of knowledge, even among officials, that a large earthquake could generate large tsunami. On the other hand, a ten-year-old girl who had recently learned in class about tsunami, saw the sea recede before the first wave and yelled to those around her to run uphill. A dock worker on a remote Indian island had seen a National Geographic special on tsunami, felt the earthquake, and ran to warn nearby communities that giant waves were coming. Together, these two saved more than 1,500 lives. Knowledge of hazard processes can save lives.

An official in West Sumatra recorded the earthquake and spent more than an hour unsuccessfully trying to contact his national center in Jakarta. An official in Jakarta later sent e-mail notices to other agencies but did not call them. A seismologist in Australia sent a warning to the national emergency system and to Australia's embassies overseas but not to foreign governments because of concern for breaking diplomatic rules. Officials in Thailand had up to an hour's notice but apparently failed to disseminate the warning. Among the public, few people had any knowledge of tsunami or that earthquakes can produce them. Ironically, the country's chief meteorologist, now retired, had warned in the summer of 1998 that the country was due for a tsunami. Fearing a disaster for the tourist economy, government officials labeled him crazy and dangerous. He is now considered a local hero.

Although scientists have expressed concern about the lack of a warning system in the Indian Ocean, most officials in Thailand and Malaysia viewed tsunami as a Pacific Ocean problem and the tens of millions of dollars it would cost to set up a network left it a low priority in a region with limited finances. In addition, it was Sunday and the day after Christmas, so few people would have been at work. Compounding the problem was the time delay in determining the size of the earthquake. The location of the earthquake could be determined quickly and automatically from the arrival times of seismic waves from several locations. The magnitude was apparently large, initially estimated by Indonesian authorities as magnitude 6.6, a size that would not generate a significant tsunami. However, because the magnitude of giant earthquakes is determined by the amplitude of the surface waves and such large earthquakes have lower frequency waves, it often takes more than an hour to determine the magnitude. By that time, it would have been too late because waves had already battered Sumatra.

The first reports in northern Sumatra indicated that the earthquake severely damaged bridges and knocked out electric power and telephone service. Buildings were heavily damaged. People ran into the streets in panic. Smaller earthquakes quickly followed farther north along the subduction zone. A short time later, tsunami waves 5 meters high struck northernmost Sumatra, wiping out 25 square kilometers of the provincial capital of Banda Aceh (▶Figure 5-2). Locally, the wave swept inland as far as 8 kilometers; it had a 24-meter-high run up on one hill almost a kilometer inland.

In less than two hours, the first of several tsunami waves crashed into western Thailand, the east coast of Sri Lanka, and shortly thereafter the east coast of India (▶Figure 5-1a); seven hours later, it reached Somalia on the coast of Africa. In Sri Lanka, a coastal train carrying 1,000 passengers was washed off the tracks into a local swamp. More than 800 bodies have been recovered. By 4 P.M. local time, it was apparent that early reports drastically underestimated the level of destruction. Indonesia reported 150 deaths, Thailand 55, Malaysia 8, India 1,000, and Sri Lanka 500. Many villages were completely washed away, leaving no one to identify or bury the bodies. In the following days, the official death count rose rapidly.

By January 13, more than 283,000 people were presumed dead and tens of thousands more remained missing. Even with many hours between the earthquake and the first waves, hundreds of people died in Somalia on the northeast African coast. At least 31,000 died in Sri Lanka, 10,750 in India, and 5,400 in Thailand. In Indonesia, at least 230,000 are dead or missing. In Banda Aceh alone, 30,000 bodies may yet remain in the area in which no buildings were left standing. Relief organizations were overwhelmed by the unprecedented scale of the disaster encompassing eleven countries. Affluent countries around the globe quickly pledged millions of dollars in aid, in the form of food and water, medical and technical help, and relief funds. With almost everyone affected by the disaster, people came together to help one another. Desperation, however, led to some fights over relief food and water. Many people in the region refused to go near the beaches, fearing that the many large aftershocks could generate more tsunami. Five million people in the region lost their homes; hundreds of thousands of survivors huddled in makeshift shelters. This tsunami was the most devastating natural disaster of its kind on record.

Chile Tsunami: May 1960

The largest earthquake in the historic record (magnitude $M_w = 9.5$) was on the subduction zone along the coast of Chile on May 22, 1960. Fifteen minutes after the earthquake, the sea rose rapidly by 4.5 meters (14.7 feet). Fifty-two minutes later, an 8-meter-high second tsunami arrived at 200 kilometers (124 miles) per hour, crushing boats and coastal buildings. A third slower wave was 10.7 meters high. More than 2,000 people died.

In Maullín, Chile, the tsunami washed away houses on low ground or carried them off their foundations. Some of those houses were carried more than a kilometer inland; others were demolished or washed out to sea. Many people wisely ran for higher ground. Some who ran back for valuables were not so lucky. One group survived by climbing to the loft of a barn; several others climbed trees. One person in a tree watched water rise to his waist. One farmer who watched his house on a river floodplain collapse later found 10 centimeters of sand covering his fields (▶Figure 5-3). Forests on low ground dropped abruptly below sea level, permitting saltwater to flow in and kill the trees.

Fifteen hours later, as predicted, the tsunami reached Hawaii. Coastal warning sirens sounded at 8:30 P.M. When the 9 P.M. news from Tahiti reported that waves there were only 1 meter high, many Hawaiians relaxed. Few people in Hawaii realized that well-developed reefs protect the Tahitian islands. Warned hours earlier by radio and sirens, a third of the people in low areas of Hilo evacuated; others did not because some previous warnings involved small tsunami that caused little damage. The first wave just after midnight was little more than 1 meter high (▶Figure 5-4). Many people thought the danger was over and returned to Hilo. At 1:04 A.M., a low rumbling sound like that of a distant train became louder and louder, followed by crashing and crunching as buildings collapsed in the 4-meter high, nearly vertical wall of the largest wave; 282 people were badly injured and sixty-one died, all in Hilo, including sightseers who went to the shore to see the tsunami. The waves destroyed water mains, sewage systems, homes, and busi-

FIGURE 5-3. A tsunami wave swept material from the nearby beach near Maullín, Chile, in 1960. It left a sand layer over a soil horizon on a farmer's field.

CHILE

Grass

Sand deposited by 1960 tsunami

Tidal peat

Topsoil

Brian Atwater, USGS.

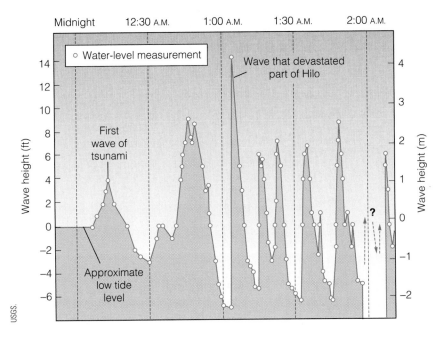

USGS.

FIGURE 5-4. This tide gauge record shows the tsunami waves in Hilo, Hawaii, May 23, 1960, following the Chilean earthquake. In this case, the first wave is relatively low, followed by successively higher waves to more than 4 meters above the low tide that preceded the tsunami. After the first couple of waves, their wavelength and frequency increased.

nesses. Most of the deaths were avoidable; people heard the warnings but misinterpreted the severity of the hazard.

Hilo is the most vulnerable location on the Hawaiian Islands. Although it has a particularly good harbor, Hilo Bay also focuses the damage (▶ Figure 5-5). Even tsunami waves that come from the southeast **refract** in the shallower waters around the island to focus their maximum height and energy in the bay (▶ Figure 5-6). The bay also has the unfortunate form that as a first tsunami wave drains back offshore, it reinforces the incoming second wave that arrives about a half hour later.

Nine hours after the tsunami hit Hawaii, it reached the island of Honshu, Japan. Although its wave height decreased to only 4.5 meters, 185 people died, 122 of them on Honshu. Following a few unusual waves up to 1 meter high, the first sign of the tsunami was retreat of the sea accompanying a rapid 1.5-meter drop in sea level (▶ Figure 5-7). Then the first tsunami wave arrived more than 4 meters above the

USGS photo.

FIGURE 5-5. The 1960 tsunami destroyed the waterfront area at the head of Hilo Bay.

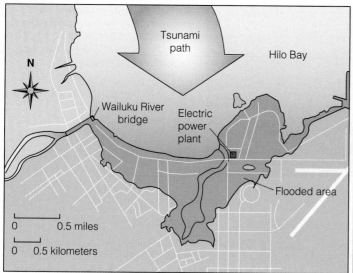

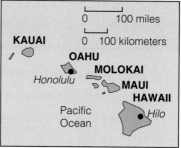

▶**FIGURE 5-6.** This low-lying area of downtown Hilo was destroyed by the Chilean tsunami of 1960.

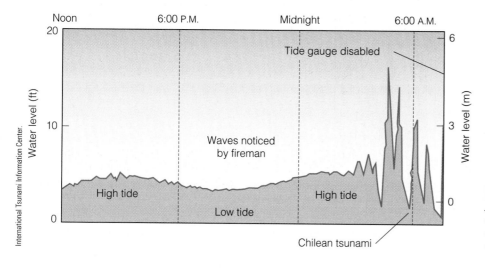

▶**FIGURE 5-7.** The tide gauge at Onagawa, Japan, recorded a dramatic drop in sea level as the May 23–24, 1960, Chilean tsunami arrived. Such a drop provides as much tsunami warning as a rapid rise in sea level.

previous low. It was that withdrawal, followed immediately by rapid rise, that caught people off guard and drowned many of them. Five waves over six hours culminated in a huge wave more than 5 meters high that disabled the tide gauge and further record of the tsunami (right edge of Figure 5-7). Note that the highest wave was far from the first and that the waves can be an hour or more apart.

Tsunami Generation

When the subject of tsunami comes up, most people immediately think of earthquakes. Actually, tsunami are also generated by a variety of other mechanisms that cause sudden displacement of water. These include volcanic eruptions, landslides, rockfalls, and volcano flank collapse. We consider these causes individually.

Earthquake-Generated Tsunami

Most tsunami are generated during shallow-focus underwater earthquakes associated with the sudden rise or fall of the seafloor, which displaces a large volume of water. Earthquake tsunami occur most commonly by displacement of the ocean bottom on a reverse-movement subduction-zone fault and occasionally on a normal fault. Strike-slip earthquakes seldom generate tsunami because they do not displace much water. Waves with short periods (▶Figure 5-8) form with small earthquakes, and tsunami waves with long periods form with larger earthquakes. It is the long-period earthquakes that displace the most water and create the largest waves.

Subduction-zone earthquakes off Japan, Kamchatka, Aleutian Islands and Gulf of Alaska, Mexico, Peru, and Chile are the most frequent culprits. The subduction zone off the

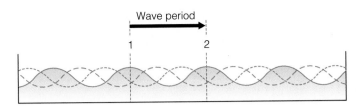

Wave period

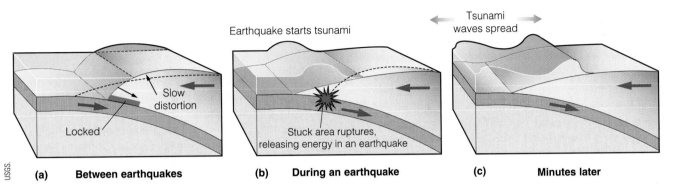

USGS.

(a) Between earthquakes

Earthquake starts tsunami

Slow distortion

Locked

(b) During an earthquake

Stuck area ruptures, releasing energy in an earthquake

Tsunami waves spread

(c) Minutes later

FIGURE 5-9. A subduction-zone earthquake snaps the leading edge of the continent up and forward, displacing a huge volume of water to produce a tsunami.

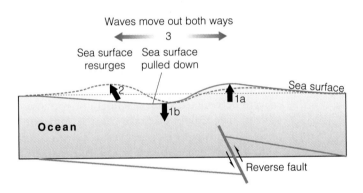

Waves move out both ways

3

Sea surface resurges   Sea surface pulled down

Sea surface

Ocean

Reverse fault

FIGURE 5-10. The sequence of events that create a tsunami that is generated by a subsea reverse or thrust fault in the ocean floor are: (1a) Seafloor snaps up, pushing water up with it; (1b) sea surface drops to form a trough; (2) displaced water resurges to form wave crest; and (3) gravity restores water level to its equilibrium position, sending waves out in both directions.

coast of Washington and Oregon is like a tightly drawn bow waiting to be unleashed. Major tsunami somewhere around the Pacific Ocean occur roughly once a decade. A 30-meter high wave forms somewhere around the Pacific Ocean, on average, once every twenty years. The leading edge of the continent, overlying the descending oceanic crust at a subduction zone, is typically "locked" for many years before breaking loose in a large earthquake (Figure 5-9a). While locked, the overlying continental edge is pulled downward, causing upward flexure of the overlying plate (Figure 5-9b). When the locked zone finally breaks loose in an earthquake, the leading edge of the continent snaps oceanward and up, commonly over a considerable length parallel to the coast. That moves a lot of water and causes a tsunami (Figures 5-9c

and 5-10). See the Cases in Point for Anchorage, Alaska, and Hokkaido, Japan, and the introductory discussion of the Sumatra tsunami.

Volcano-Generated Tsunami

Tsunami are also caused by volcanic processes that displace large volumes of water. Possibilities include collapse of a near-sea-level caldera that pulls down a large volume of water from the surrounding sea. Water is also driven upward by fast-moving ash flows or submarine volcanic explosions into a large body of water. More than one of these mechanisms can occur at an individual volcano. One of the most infamous and catastrophic events involving a volcano-generated tsunami was at Krakatau Volcano in 1883. See "Case in Point: Krakatau," page 111.

Tsunami from Fast-Moving Landslides or Rockfalls

When major fast-moving rockfalls or landslides enter the ocean, they can displace immense amounts of water and generate tsunami. At first thought, the height of the tsunami might be expected to depend primarily on the volume of the mass that displaces water. However, a more important parameter is the height of fall. A striking example was the Lituya Bay, Alaska, tsunami in 1958. It happened when a large cliff fell into a coastal fjord to cause the highest tsunami in historic record (see "Case in Point: Lituya Bay, Alaska," p. 109). That event killed only two people, but the potential may exist for a similar but more catastrophic event as more people move into similar environments (see "Case in Point: Glacier Bay, Alaska," p. 110).

Effects Close to the Epicenter

The giant magnitude 8.6 (M_W = 9.2) Anchorage, Alaska, earthquake on March 27, 1964, showed large vertical effects of slip on the subduction zone. The slab sinking beneath the Aleutian oceanic trench slipped over a length of 1,000 kilometers and an area more than 300 kilometers wide. A strip 500 kilometers long by 150 kilometers wide of extremely shallow seafloor rose 10 meters above sea level and moved 19.5 meters seaward. Another belt onshore from the coast, fully comparable in length and more than 100 kilometers

wide, sank as much as 2.3 meters. Low-lying coastal areas actually dropped below sea level. Twenty-seven years later, the same areas had slowly rebounded to again be above sea level (▶ Figure 5-11).

The sudden change in seafloor elevation displaced the overlying water into giant tsunami waves that washed ashore on the Kenai Peninsula within nineteen minutes and onto Kodiak Island in thirty-four minutes (▶ Figures 5-12 and 5-13). The maximum tsunami wave run-up occurred where it funneled into Valdez Inlet, just west of the Kenai Peninsula, where the earthquake caused a submarine landslide. Of the 131 people killed in the earthquake, 122 drowned in the 61-meter waves that funneled into and devastated waterfront areas in Valdez and Seward. Smaller wave heights extended all the way to Crescent City, northern California, destroying much of the waterfront area.

1964

Effects Far from the Epicenter

Eight minutes after the Anchorage earthquake, an alarm sounded at the Honolulu Observatory. The location and magnitude of the earthquake were determined from seismograms within an hour. The California State Disaster Office received warnings of a possible "tidal wave" two hours after the earthquake. The county sheriff at Crescent City in northern California received the warning one and one-half hours later and notified people in low-lying coastal areas to evacuate. An hour after that, a 1.5-meter high wave reached Crescent City, amplified by the **shallowing water near shore** and **narrowing of the harbor.** According to the Del Norte Historical Society files, the curator of Battery Point Lighthouse, on a small rocky island just offshore, recalled that it was a clear moonlit night and the waves were clearly visible as they pitched into the town. The first wave carried giant logs, trees, and other debris, demolishing buildings and cars. The debris-laden wave receded as quickly as it arrived, leaving battered cars, houses, logs, and boats. The sea receded to a kilometer offshore.

After a second wave came into the harbor, some people returned to clean up. A third and larger wave then washed inland more than 500 meters, drowning five people; it knocked out power and ignited a fire before the sea receded even farther. The fourth and largest

1991

Brian Atwater photos, USGS.

(b)

▶**FIGURE 5-11.** **(a)** Immediately after the Alaska earthquake of 1964, the coastal area flooded when the coastal bulge collapsed. **(b)** This is the same area twenty-seven years after the bulge again began to rise.

USGS.

▶**FIGURE 5-12.** A parking lot full of cars was thrown about at the head of a bay like so many toys following the March 1964 Prince William Sound tsunami in Alaska. Large fishing boats joined them onshore.

National Geopysical Data Center, NOAA.

▶**FIGURE 5-13.** A large fishing boat and crushed fuel truck rest on shore in Resurrection Bay, Seward, Alaska, following the March 1964 Prince William Sound tsunami.

wave was 6.3 meters high; it killed ten people who went back down to check their houses. That wave submerged the damaged Citizen's Dock and lifted a big, loaded lumber barge, setting it down on top of the dock, crushing it. Fuel from ruptured tanks at the Texaco bulk plant spread to the fire and ignited. One after another the tanks exploded. Pieces of everything imaginable drained back offshore with the outgoing wave. The fifth wave was somewhat smaller. In all, the tsunami destroyed fifty-six blocks of the town.

On July 12, 1993, a large magnitude 7.8 earthquake in the Sea of Japan, off the west coast of Hokkaido in northern Japan, generated one of the largest tsunami in Japan's recorded history. A large slab of ocean floor at a depth of 15 kilometers moved up on a thrust fault dipping gently to the east, sending a tsunami onshore within only two to five minutes. Average run-ups were 10 to 20 meters but reached as high as 30.6 meters near one small coastal village. The tsunami inundated the small island of Okushiri, killing almost 200 people and causing $600 million in property loss. The hardest hit community was Aonae, where the first wave from the north-east swept over a massive 4.5-meter breakwater to run up to heights of 3 to 7 meters throughout the town. A second wave ran up to 5 to 10 meters throughout the town. Nearby areas of coast, beyond the protection of the breakwater and a dune field, saw run-up heights of 10 to 20 meters.

The damage in Aonae is apparent in the before-and-after images (▶ Figure 5-14a, b). The residential area and port facilities identified in Figure 5-14a were completely destroyed (▶ Figures 5-14b and 5-15). Fires were fueled by plentiful propane and kerosene used for heating, the flames fanned by strong northeast winds.

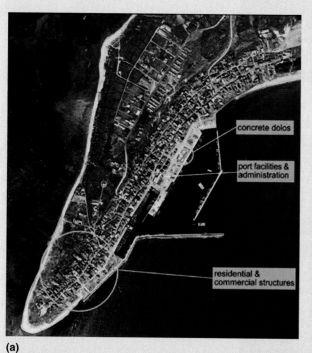

concrete dolos

port facilities & administration

residential & commercial structures

(a)

Aonae, Okushiri Island
July 13, 1993

(b)

▶**FIGURE 5-14.**
(a) This photo of the small village of Aonae on Okushiri Island, west of the southwestern end of Hokkaido Island, Japan, was taken in 1976; compare the areas damaged in Figure 5-14b. **(b)** The small village of Aonae was heavily damaged by the 1993 tsunami. The smoke comes from tsunami-caused fires. Note that strong refraction carried the wave around the end of the island and resulted in major damage on the east side of the island.

▶**FIGURE 5-15** A fishing boat and crushed fire truck lay among the debris on the side of the island away from the incoming 1993 tsunami wave at Aonae, Okushiri Island, Japan.

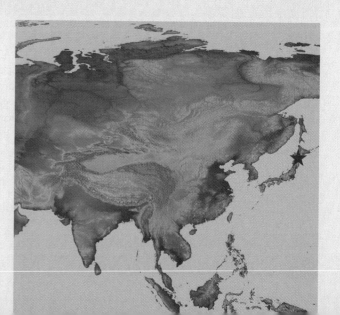

One of the most spectacular tsunami resulting from a land-based rockfall was in Lituya Bay, Alaska, on July 9, 1958. Lituya Bay, a deep fjord west of Juneau, Alaska, and at the west edge of Glacier Bay National Park, was the site of one of the highest tsunami run-ups ever recorded (▶ Figures 5-16 and 5-17). On July 9, 1958, 60 million cubic meters of rock and glacial ice, loosened by a nearby magnitude 7.5 earthquake on the Fairweather Fault, fell into the head of Lituya Bay. The displaced water created a wave 150 meters high, or the height of a fifty-story building. It surged to an incredible 524 meters over a nearby ridge and removed forest cover up

▶ FIGURE 5-18. This view of Lituya Bay shows trimlines from two previous tsunami that were even larger than the 1958 event. Do narrow, steep-sided inlets elsewhere show similar healed trimlines? Could they provide hazard information for future rockfall tsunami?

to an average elevation of 33 meters and up to 152 meters over large areas (▶ Figures 5-16 and 5-17). Older trimlines in the forest and damaged tree rings of various ages document previous similar tsunami in 1936, 1874, and 1853 or 1854 (▶ Figure 5-18). This was a huge wave compared to common tsunami that may be 10 to 15 meters high; it swept through Lituya Bay at between 150 and 210 kilometers per hour.

Although three fishing boats, with crews of two each, were in the bay at the time, only those on one boat died when their boat was swept into a rocky cliff. On another boat on the south side of the island in the center of the bay, Howard Ulrich and his seven-year-old son hung on as their boat was carried high over a submerged peninsula into another part of the bay. They were actually able to motor out of the bay the next day. On the third boat, Mr. and Mrs. William Swanson, anchored on the north side of the bay, were awakened as the breaking wave lifted their boat bow first and snapped the anchor chain. The boat was carried at a height of 25 meters above the tops of the highest trees, over the bay mouth bar, and out to the open sea. Their boat sank, but they were able to climb onto a deserted skiff and were rescued by another fishing boat two hours later.

Examination of the forested shorelines of Lituya Bay, above the **trimline** of the 1958 event, shows two much higher trimlines produced by earlier tsunami. U.S. Geological Survey scientists examined trees at the level of these higher trimlines and found severe damage caused by the earlier events. Counting tree rings that had grown since then, they determined that the earlier tsunami occurred in 1936 and 1874 (▶ Figure 5-18).

▶ FIGURE 5-16. A huge rockfall into the head of Lituya Bay, Alaska, generated a giant tsunami wave that stripped the forest and soil from a ridge. This view to the northeast shows the broad areas of forest that the tsunami swept from the fringes of the bay. The scarp left by the rockfall is visible at the head of the bay (arrow).

▶ FIGURE 5-17. This photo details the tsunami damage at the crest of a ridge 524 meters above the bay.

Potential Future Tsunami?

Glacier Bay, 50 kilometers east of Lituya Bay, is in a similar spectacular environment. It is a deep fjord bounded by precipitous rock cliffs and glaciers. It lies between two major active strike-slip faults, the Fairweather Fault and the Denali Fault, each 50 or 60 kilometers away, and each capable of earthquakes of magnitudes greater than 7. Glacier Bay is a prominent destination for cruise ships touring from Seattle or Vancouver to Alaska, so a tsunami generated by a large landslide into the bay is a concern. Study by USGS geologists suggests that an unstable rockslide mass on the flank of a tributary inlet to Glacier Bay would generate waves with more than 100-meter run-up near the source and tens of meters within the inlet (▶ Figure 5-19a).

In the deepwater channel of the western arm of Glacier Bay, the wave amplitude would decrease with distance out into the bay. Ships near the mouth of the tributary inlet could encounter a 10-meter wave only four minutes after the slide hit the water, then 20-meter high waves after twenty minutes. The waves would likely strike the cruise ships broadside as shown in Figure 5-19b. If the ships kept to this central channel, the largest waves would likely only be approximately 4 meters high. The response of a ship to waves near the mouth of the tributary inlet would depend on the wave height, the wave frequency relative to the ship's rocking frequency, and the height of the lowest open areas on the ship. However, because the cruise operators are now aware of this risk, they can avoid the dangerous near-inlet waters.

(a)

Glacier Bay

~ 4 m

~ 20 m

Gerald F. Wieczorek, USGS.

(b)

▶ **FIGURE 5-19.** **(a)** The tidal inlet landslide mass next to Glacier Bay, Alaska, includes the rock face from the new higher scarp to below water; **(b)** Photo was taken from the apex of the slide with the tidal inlet in the foreground; two cruise ships in Glacier Bay are visible in mid-photo.

Lituya Bay

Glacier Bay

August 27, 1883

People in towns on the west coast of Java, in Indonesia, awoke to the sounds of Krakatau rumbling, 40 kilometers to the west. At 10:02 A.M., the mountain exploded in an enormous eruption. This was the climactic eruption of activity that had been going on for several months. Thirty-five minutes later, a series of waves as high as 30 meters flattened the coastline of the Sunda Strait between Java and Sumatra, including its palm trees and houses. Only a few who happened to be looking out to sea saw the incoming wave in time to race upslope to safety. More than 35,000 people died. Studies of the distribution of pyroclastic flow deposits and seafloor materials in the Sunda Straits between Krakatau and the islands of Java and Sumatra show only pyroclastic flow deposits, suggesting that an enormous flow entered the sea to produce the tsunami.

Computer simulations of three possible causes, however, suggest that two of the reasonable hypotheses fit the data poorly. They were (1) that the waves were caused by **caldera collapse** that pulled down a large mass of water or (2) that pyroclastic flows entering the sea displaced the water. The third hypothesis fits the data well. It suggests that seawater seeping into the volcano interacted with the molten magma to generate huge underwater explosions and upward displacement of a large volume of seawater.

Tsunami from Volcano Flank Collapse

GIANT PREHISTORIC EVENTS IN HAWAII The flanks of many major oceanic volcanoes, including those of the Hawaiian Islands in the Pacific Ocean and the Canary Islands in the Atlantic Ocean, apparently collapse on occasion and slide into the ocean. Hawaii's volcanoes grow from the seafloor for 200,000 to 300,000 years before breaking sea level, then build a reasonably solid lava shield above sea level for a similar time. Mega-landsliding occurs near the end of the shield-building stage when the growth rate is fastest, heavy basalt load on top is greatest, and slopes are steepest and thus least stable. The lower part of each volcano, below sea level, consists largely of loose volcanic rubble formed when the erupting basalt chilled in seawater and broke into fragments. It has little mechanical strength. Above sea level, oceanic islands are built from heavy, reasonably solid basalt lava flows.

The three broad ridges that radiate outward from the top of the volcano spread slightly under their own enormous weight, producing rift zones along their crests. The volcano eventually breaks into three enormous segments that look on a map like a pie cut into three slices of approximately equal size. One or more of the three volcano segments may begin to move slowly seaward (▶ Figures 5-20 and 5-21).

The rifts between the sediments provide easy passage to molten magma rising to the surface. Those rifts that become the sites of most of the eruptions also form weak vertical zones in the volcano. There is a long history of one or more of the volcano segments breaking loose to slide into the ocean, sometimes slowly but sometimes catastrophically. Studies of the ocean floor using side-scanning radar around the Hawaiian Islands reveal sixty-eight **giant debris avalanche** deposits, each more than 20 kilometers long. Some extend as far as 230 kilometers from their source and contain several thousand cubic kilometers of volcanic debris. Such slides must have moved rapidly to reach such a distance.

At least some of those deposits are the remains of debris avalanches that raised giant tsunami waves that washed high onto the shores of the Hawaiian Islands. Boulders of

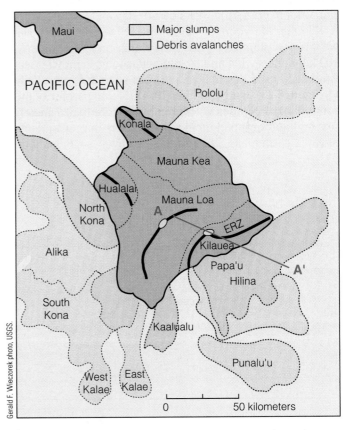

▶ **FIGURE 5-20.** This map of the island of Hawaii shows the major slumps and debris avalanches formed by collapse of the island's flanks. ERZ = East Rift Zone.

coral litter the flanks of some of the islands to elevations of more than 400 meters above sea level and more than 6 kilometers inland; the flanks of some of the islands have lost much or all of their soil to a similar elevation. It now seems clear that both the displaced boulders and the scrubbed slopes are evidence of monstrous waves that washed up the flanks of the islands as one of the enormous pie segments plunged into the ocean. The headscarps of such collapsed segments become gigantic coastal cliffs, some more than 2,000 meters high and among the highest cliffs in the world (▶ Figure 5-22).

The landslides appear to occur during major eruptive cycles and have a recurrence interval of roughly 100,000 years. Headscarps of the slides are the giant **"pali"** or cliffs that mark one or more sides of each of the Hawaiian Islands. Despite the existence of such evidence, the frequency of these horrifying events remains unclear. If we can judge from the age of coral fragments washed onto the flanks of several islands, the most recent slide detached a large part of the island of Hawaii 105,000 years ago. That slide raised tsunami waves to elevations of as much as 326 meters on the island of Lanai.

Mauna Loa, the gigantic volcano on the big island of Hawaii, the youngest and largest of the Hawaiian Islands, has collapsed repeatedly to the west. Two of these collapses were slumps and two were debris avalanches. Most were **submarine collapses,** though the head scarp of the North Kona slump grazes the west coast of Hawaii.

Kilauea, the youngest and most active volcano in Hawaii, is now slumping. The Hilina slump on its south flank, 100 kilometers wide and 80 kilometers long, is moving seaward at 10 to 15 centimeters per year, sometimes suddenly (▶ Figure 5-23). On November 19, 1975, a big sector of the south flank of Kilauea volcano moved more than 7 meters seaward and dropped more than 3 meters during a magnitude 7.2 earthquake. The resulting relatively small tsunami

▶**FIGURE 5-22.** Giant cliffs or *pali* amputate the lower slopes of the big island of Hawaii.

drowned two people nearby, destroyed coastal houses, and sank boats in Hilo Bay on the northeast side of the island. What this portends for further movement is not clear. Will the flank of the volcano continue to drop at unpredictable intervals or could it fail catastrophically?

Hawaiian geologists wondered for years about blocks of coral and other shoreline materials strewn across the lower slopes of the islands. It now seems clear that huge tsunami formed when enormous masses of one of the islands collapsed into the ocean and the waves washed material up the slopes from the beach. Tsunami formed by island flank collapse are documented from tsunami deposits consisting of a cemented mix of fragments of limestone reef and basalt, in many places tens of meters above sea level. On Molokai, they left deposits 70 meters above sea level; on Lanai, they left blocks of coral as much as 326 meters above sea level. An eventual repetition of those events seems inevitable; it would kill much of the population of the Hawaiian Islands.

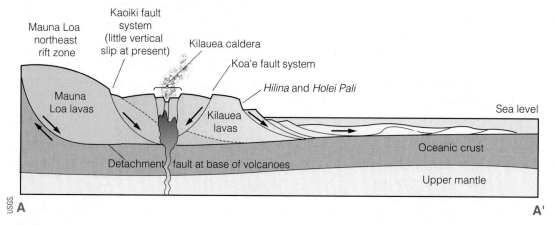

▶**FIGURE 5-21.** This northwest–southeast cross section of Kilauea volcano shows the probable failure surfaces that lead to collapse of the volcano's flanks. See cross-section location as line A–A′ in map Figure 5-20.

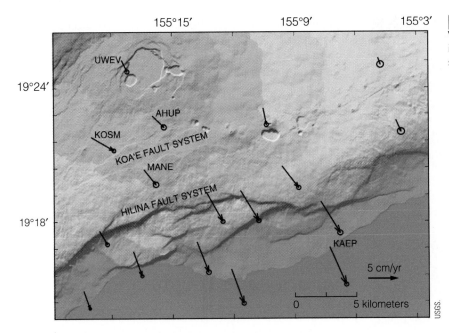

▶**FIGURE 5-23.** The south flank of Kilauea Volcano is slowly slumping seaward. The arrows indicate directions and rates of movement as measured by Global Positioning Systems (GPS).

The potential for a future collapse of the flank of Kilauea volcano on the big island is emphasized in scarps forming in the 80-kilometer-long coastal area southeast of Kilauea. A slab almost 5 kilometers thick is sliding ocean-ward at 10 to 15 centimeters per year. If this huge area collapses into the sea suddenly, perhaps triggered by a large earthquake or major injection of magma, it would likely generate tsunami greater than 100 meters high. Many coastal communities in Hawaii would be obliterated with little warning. However, to put these numbers in perspective, if 100,000 people were killed in such an event every 100,000 years, the average would be one person per year. Although unimaginably catastrophic when it does happen, there are certainly greater dangers, on average, in one person's lifetime.

The danger is not limited to Hawaii. If the flank of Kilauea, now moving seaward, should fail catastrophically, it could generate a tsunami large enough to devastate coastal populations all around the Pacific Ocean. Those in Hawaii would have little warning. The Pacific coast of the Americas would get several hours. It remains to be seen how many people could be warned and how many of those would heed the warning. Certainly major urban centers such as San Francisco and Los Angeles could not be evacuated in time. We hope that the next event will not be any time soon, but we have no way of knowing.

CANARY ISLANDS Like other large basaltic island volcanoes, Tenerife—in the Canary Islands off the northwest coast of Africa—shows evidence of repeated collapse of its volcano flanks. Tenerife reaches an elevation of 3,718 meters, almost as high as Mauna Loa. It is flanked by large-volume submarine debris deposits that left broad valleys on the volcano flanks. Lavas filling these valleys are as much as 590 meters thick and overlie volcanic rubble along

an inferred detachment surface that dips seaward at about 9 degrees.

Collapse may have been initiated by subsidence of the 11- to 14-kilometer-wide summit caldera into its active magma chamber. The most recent caldera and island flank collapse was 170,000 years ago. That event carried a large debris avalanche from the northwest coast of Tenerife onto the ocean floor. It carried 1,000 cubic kilometers of debris, some of which moved 100 kilometers offshore. The much larger El Golfo debris avalanche detached 15,000 years ago from the northwest flank of El Hierro Island. It carried 400 cubic kilometers of debris as much as 600 kilometers offshore.

An average interval of 100,000 years between collapse events on the Canary Islands may be long, but the consequences of such an event would be catastrophic. And the interval is merely an average. The next collapse could come at any time, and the giant tsunami caused by collapse would catastrophically inundate not only heavily populated coastal areas around the north Atlantic Ocean but also reach coastal Portugal in two hours, Great Britain in little more than three hours, and the east coasts of Canada and the United States in six to seven hours (▶Figure 5-24). Because large populations live in low-lying coastal cities and on unprotected barrier islands along the coast, millions would be at risk. Even if warning were to reach endangered areas as much as six hours before arrival of the first wave, we know from experience with hurricanes, that evacuation would likely take much longer that that. Imagine hundreds of thousands of people trying to evacuate without a well-thought-out plan and in traffic that is heavy under normal circumstances. What about congestion on the single two-lane bridges that link most barrier islands to the mainland? How many would ignore the warning,

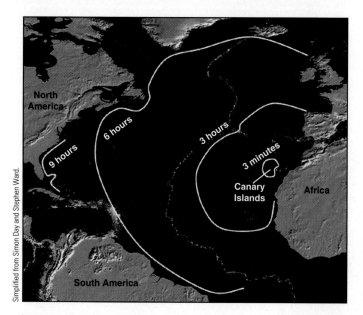

FIGURE 5-24. A large landslide from La Palma, Canary Islands, could generate immense tsunami waves that would fan out into the Atlantic Ocean. Computer simulations suggest that huge waves would reach the east coast of North America in six to seven hours.

Simplified from Simon Day and Stephen Ward.

not realizing the level of danger? The death toll could be staggering.

Similar situations are now known to exist on Reunion Island in the Indian Ocean, Etna in the Mediterranean Sea, and the Marquesas Islands in the Pacific Ocean.

Recently discovered fractures along a 40-kilometer stretch of the continental shelf 100 miles off Virginia and North Carolina suggest the possibility of a future undersea landslide. Such a slide could generate a tsunami like one that occurred 18,000 years ago just south of those fractures. Tsunami-deposited sand layers have been discovered as well at several sites on islands west of Norway. That tsunami 8,150 years ago formed from the immense Storegga submarine slide on the continental slope off the coast of Norway. The tsunami carried sand, pieces of wood, and marine fossils onto peat as much as 20 meters above sea level. A much more recent but smaller subsea slide in 1998 generated a tsunami that killed 2,200 people in Papua New Guinea.

Tsunami from Asteroid Impact

Because anything that suddenly displaces a large volume of water would generate a large wave, the impact of a large asteroid into the ocean would generate large tsunami that would radiate outward from the impact site, much as happens with any other tsunami. (See also the related discussion in Chapter 17 on asteroid impact.) The average frequency of such events is low, but a 1-kilometer asteroid falling into a 5-kilometer-deep ocean might generate a transient 3-kilometer-deep cavity. Collapse of the ocean cavity walls to refill the cavity will reach supersonic speeds to send a

plume high into the atmosphere. Initial kilometer-high waves crest, break, and interfere with one another. Waves with widely varying frequency radiate outward. The behavior of such complex waves is not well understood, but they are thought to decrease fairly rapidly in amplitude away from the impact site. The different wave frequencies would, however, interfere and locally pile up on one another to cause immense run-ups at the shore.

The chance of a 1-kilometer asteroid colliding with Earth are only once every million years, so such a hazard, though significant in scale, is not major in terms of human lifetimes. The chance of catastrophic tsunami from flank collapse of an oceanic island such as Hawaii or the Canary Islands is perhaps 10 times as great.

Velocity and Height

Tsunami wave velocities can be as high as 870 kilometers per hour. Because tsunami wave heights in the open ocean are small, and the average tsunami wavelength is 360 kilometers, slopes on the wave flanks are extremely gentle. The time between waves may be half an hour; thus, it takes that long for a ship to go from the wave trough to its crest and back to the trough. As a result, ships at sea hardly notice them.

Water particles in wind-driven waves travel in a circular motion—the water does not travel with the wave (see Sidebar 5-1). That circular motion fades downward. Because waves touch bottom at depths less than approximately half their wavelength, tsunami waves drag bottom everywhere in the ocean. Their velocity does not depend on wave-

---

Sidebar 5-1

The velocity of tsunami waves depends on the water depth and gravity.

$$C = \sqrt{gD}$$

where

    C = velocity in meters per second

    D = depth in meters

    g = gravitational acceleration (9.8 m/sec.^{2})

Thus,

    $C = 3.13\sqrt{D}$

For example, if D = 4,600 meters (deep ocean):

    $C = 3.13\sqrt{4,600} = 3.13 \times 67.8$ meters per second

        or 763 kilometers per hour

        (the speed of some jet aircraft!)

If D = 100 meters (near shore):

    $C = 3.13\sqrt{100} = 3.13 \times 10 = 31.3$ meters per second

        or 112.7 kilometers per hour

        (the speed of freeway traffic)

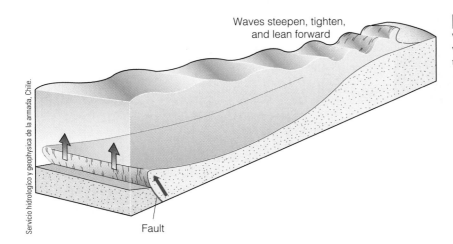

Waves steepen, tighten, and lean forward

Fault

Servicio hidrologico y geophysica de la armada, Chile.

▶**FIGURE 5-25.** Both wind waves and tsunami waves drag on bottom near shore, becoming shorter in wavelength and higher in amplitude before breaking at the shore.

▶**FIGURE 5-26.** The December 2004 tsunami leveled almost all of the homes in Banda Aceh, Sumatra, Indonesia.

Tyler Clements, U.S. Navy photo.

length as do wind-driven **deepwater waves,** but on water depth. As tsunami waves reach shallower water, such as on the continental shelf, they slow down and build in height because their circular motions at depth drag on the ocean bottom and slow down. Still closer to shore, they slow and build even higher before breaking on shore. The waves are forced to slow dramatically in shallower water close to shore, causing the waves to change to a much shorter wavelength with a dramatically greater height, Because the deep part of the wave is slowed the most, the crest of the wave rushes ahead and the wave begins to break near the shore (▶ Figure 5-25).

Equally important is the height of the tsunami wave. A tsunami wave 3 meters high in the open ocean could shorten dramatically in shallow water to roughly one-sixth of its wavelength and rise to 6 times its open ocean height. The same wave in the open ocean could rise in shallow water to

18 meters! On shore its **run-up** height could be even higher (▶ Figure 5-26), causing massive damage.

At a velocity of 760 kilometers per hour and a wavelength of 200 kilometers, a wave would pass any point or arrive onshore every 360 kilometers (720 km/hr), or approximately every half hour. Needless to say, it would not be wise to go down to the beach to see the damage after the first wave had receded. Many tsunami deaths have resulted from people doing just that. What seems like a calm sea or a sea in retreat can be the trough before the next wave.

Run-up heights, the height that a wave reaches onshore, varies depending on distance from the fault rupture and whether the wave strikes the open coast or a bay. For the largest earthquakes such as the subduction event in Alaska and the 1960 earthquake in Chile, run-up heights were generally 5 to 10 meters above normal tide level. Local run-up reached as high as 30 meters in Chile. Water levels can change

H. Helbush, National Geophysical Data Center.

▶**FIGURE 5-27.** This sequence of photos shows the arrival of a tsunami wave onto the beach at Laie Point on Oahu, Hawaii. This tsunami was generated on March 9, 1957, as the result of a magnitude 8.6 (M_w) earthquake that struck the Aleutian Islands of Alaska approximately 3,600 kilometers away.

rapidly, as much as several meters in a few minutes. Run-up is typically about perpendicular to the orientation of the wave crest, but return flow drains downslope as controlled by local topography. The sequence of photos in ▶Figure 5-27 shows a tsunami wave pushing onshore in Hawaii following an earthquake in Alaska in 1957. Driftwood, trees, and the remains of boats, houses, and cars commonly mark the upper limit of tsunami run-up.

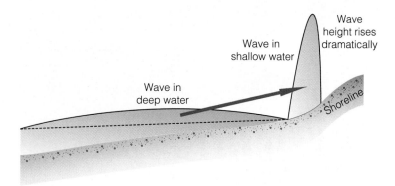

▶**FIGURE 5-28.** A 1-meter-high tsunami wave in the open ocean slows in shallower water near shore, so if the wave volume remains the same, its wavelength shortens and its amplitude rises.

Coastal Effects and Vulnerability

Dragging on the continental shelf, a 1-meter open ocean tsunami wave may slow to 150 to 300 kilometers per hour and dramatically rise in height to 15 meters or more. Water piles up as it slows. The front of the wave slows first and the rear keeps coming. Because the volume of the wave remains the same, its height must rise dramatically (▶Figure 5-28). Similarly, as the first wave slows, the following waves catch up and thus arrive more frequently. In **harbors,** tsunami have wave periods of ten to thirty-five minutes and may last for up to six hours (▶Figure 5-4, p. 103).

Areas most at risk are the low-lying parts of coastal towns, especially near the mouths of rivers and inlets that funnel the waves and dramatically raise their height. If waves arrive at high tide, their height is amplified. Sloshing back and forth from one side of a bay to the other can constructively interfere with one another to raise wave level. Because most coastal towns and seaports are located in bays, the damage resulting from these waves is enhanced. That is also the reason the Japanese call them *tsunami*—that is, harbor waves.

Low-lying Pacific and Caribbean islands would seem likely to be extremely vulnerable to incoming tsunami waves, but some are actually less vulnerable than would be expected. Many are surrounded by offshore coral reefs that drop steeply into deep water. Thus tsunami waves are forced to break on the reef, providing some protection to the islands themselves.

Tsunami warnings have now been perfected for **far-field tsunami,** or those far from the source that generated them. A world network of seismographs locates the epicenter of major earthquakes, and the topography of the Pacific Ocean floor is so well known that the travel time for a tsunami to reach a coastal location can be accurately calculated. In addition, an environmental satellite takes readings from tidal sensors along the coasts, and ocean bottom sensors detect ocean surface height as the waves radiate outward across the Pacific Ocean (▶Figures 5-29 and 5-30). This

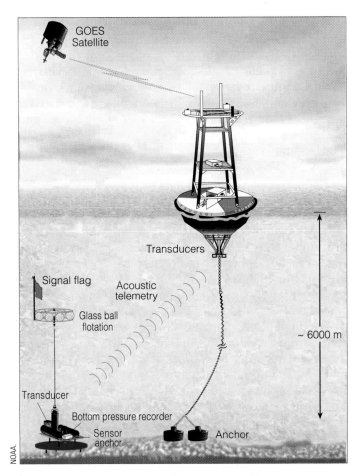

FIGURE 5-29. A pressure sensor on the ocean floor detects changes in wave height because a higher wave puts more water and therefore more pressure above the sensor. The pressure sensor transmits a signal to a buoy floating at the surface and to the warning center via satellite.

information now permits prediction of tsunami arrival times at any coastal location around the Pacific Ocean within five minutes. Pacific tsunami warning centers are located at the National Weather Service Tsunami Warning Center in Honolulu, Hawaii, and the Alaska Tsunami Warning Center in Palmer, Alaska. Some low-lying areas such as parts of Hawaii are equipped with warning sirens mounted on high poles to warn people who are outdoors in coastal areas.

Tsunamis are most likely to appear within a few minutes to several hours after an earthquake with major vertical motion of the seafloor, depending on the distance from the epicenter. A nearby earthquake will be felt but allow people little time to move to higher ground and no time for official warning. Tsunami warning signs in coastal Oregon suggest moving to higher ground if you feel an earthquake (▶Figure 5-31). However, many of those coastal areas have no nearby hills. Quickly moving inland can still help because wave energy, height, and speed dissipate rapidly on land. An earthquake thousands of kilometers away will not be felt but may allow time for official warning. A wave reaching shore may either break on the beach or rush far up onto the beach in a steep front. Large tsunami waves can reach as much as 1.6 kilometers inland.

Tsunami dangers include not only drowning in the incoming wave but also severe abrasion by being dragged along the ground at high speed, being thrown against solid objects, being carried back out to sea in the outgoing wave, and being hit by debris carried by the wave. Such debris can include boards and other fragments of houses, trees, cars, and boulders. Even when the wave slows as it drags on shallow bottom, it moves, for example, at 55 kilometers per hour, much too fast to outrun.

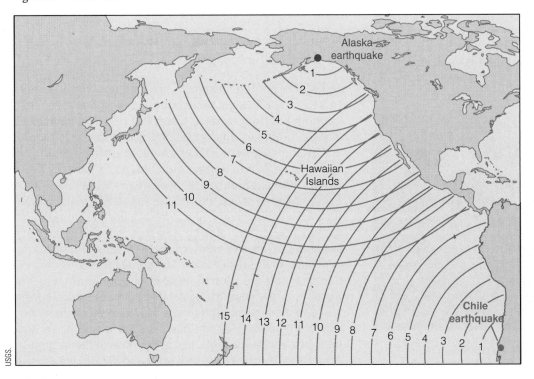

▶**FIGURE 5-30.** Tsunami travel times across the Pacific Ocean from the Chile, 1960, and Alaska, 1964, subduction zone earthquakes. Concentric arcs are travel time estimates in hours after each earthquake. From the Alaska earthquake, for example, the first tsunami wave reached Hawaii in approximately six and one-half hours. It would reach the north island of Japan after nine hours.

▶**FIGURE 5-31.** This sign warns of potential tsunami along the Oregon coast.

Survivors of tsunamis often report an initial withdrawal of the sea with a hissing or roaring noise. In many cases, curious people drown when they explore the shoreline as the sea recedes before the first big wave or before subsequent waves. In Hilo in 1946, assuming the danger had past, people went out to see the wide, exposed beach with stranded boats and sea creatures. There they were caught in the second and larger wave. Adding to the danger is the fact that tsunami waves may continue for several hours and the first wave is often not the highest. In Hawaii in 1960, in spite of several hours of tsunami warning, people went down to the beach to watch the spectacular wave, only to be overwhelmed by it (▶Figure 5-32).

The most vulnerable parts of the United States are Hawaii and the Pacific coasts of Washington, Oregon, California, and Alaska. Locally generated subduction-zone earthquakes, landslides, and volcanic events in the Caribbean can affect Puerto Rico, the U.S. Virgin Islands, and other islands in the Caribbean.

▶**FIGURE 5-32.** People run from a huge tsunami in Hilo, Hawaii, in 1946. The wave is visible in the center of the image behind the people.

Hawaii, hit by disastrous tsunami in 1946 and 1960, saw tsunami run-up elevations from 1.5 to 6 meters above sea level. In Hilo, where the worst damage occurred, making the waterfront area at the head of the bay into a park has minimized future damage. Even though the source of the 1960 tsunami was in Chile, far to the southeast, and Hilo Bay faces northeast, refraction of the waves around the island left the head of the bay vulnerable to waves as much as 4 meters above sea level.

Residents on an island coast opposite the direction of an incoming tsunami should not be comforted. On December 12, 1992, a magnitude M_s 7.5 earthquake in Indonesia generated a tsunami in the Flores Sea. The southern coast of the small island of Babi, opposite the direction from which the waves came, was hit by 26-meter tsunami waves, twice as high as the northern coast. In this case, the waves reaching the northern coast split and refracted around the circular island, constructively interfering with one another on the opposite coast. More than 1,000 people died.

The mound of water suddenly appearing at the sea surface, in response to a major event, generates a series of waves that may cross the whole Pacific Ocean. Because the initial mound of water oscillates up and down a few times before fading away, it generates a series of waves just like a stone thrown into a pond. The magnitude of a tsunami wave depends on the magnitude of the shallow-focus earthquake, area of the rupture zone, rate and volume displaced, sense of motion of the ocean floor, and depth of water above the rupture. The height of the tsunami wave is initially more or less equal to the vertical displacement of the ocean floor. Because the maximum fault offset is typically 15 meters in a giant earthquake, the maximum earthquake tsunami height in the open ocean is roughly 15 meters.

Tsunami from Great Earthquakes in the Pacific Northwest

Slabs of oceanic lithosphere sinking through an oceanic trench at subduction zone boundaries typically generate earthquakes from as deep as several hundred kilometers. Such a boundary undoubtedly exists offshore along the 1,200 kilometers between Cape Mendocino in northern California and southern British Columbia (B.C.) (▶Figure 5-33), so the apparent absence of those deep earthquakes in the Pacific Northwest has worried geologists for years. Several lines of evidence now show that major earthquakes do indeed happen but at such long intervals that none have struck within the period of recorded Northwest history.

Finally, in the 1980s, Brian Atwater of the U.S. Geological Survey found the geologic record of giant earthquakes in marshes at the heads of coastal inlets. It consists of a consistent and distinctive sequence of sedimentary layers. A bed of peat, consisting of partially decayed marsh plants that grew just above sea level, lies at the base of the sequence. Above the peat, lies a layer of sand notably lacking the sort of internal layering contained in most sand deposits. Above

the sand is a layer of mud that contains the remains of sea-water plants (▶Figure 5-34). That sequence tells a simple story that begins with peat accumulating in a salt marsh barely above sea level. It appears that a large earthquake caused huge tsunami that rushed up on shore and into tidal inlets, carrying sand swept in from the continental shelf. The sand covered the old peat soils in low-lying ground inland from the bays as the salt marsh suddenly dropped as much as 2 meters below sea level. Then the mud, with fossil seaweed, accumulated on the sand (▶Figure 5-35). The sequence of peat, sand, and mud is repeated over and over. In some cases, forests were drowned by the invading salt water or were snapped off by a huge wave (▶Figure 5-36). Huge tsunami-flattened forests in low-lying **coastal inlets** are found all down the Pacific coast from British Columbia to southern Oregon. These stumps are now at and below sea level because the **coastal bulge** dropped during the earthquake.

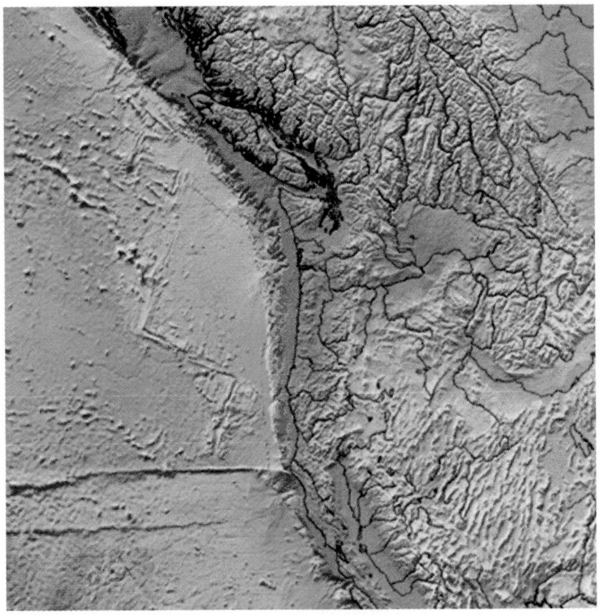

(a)

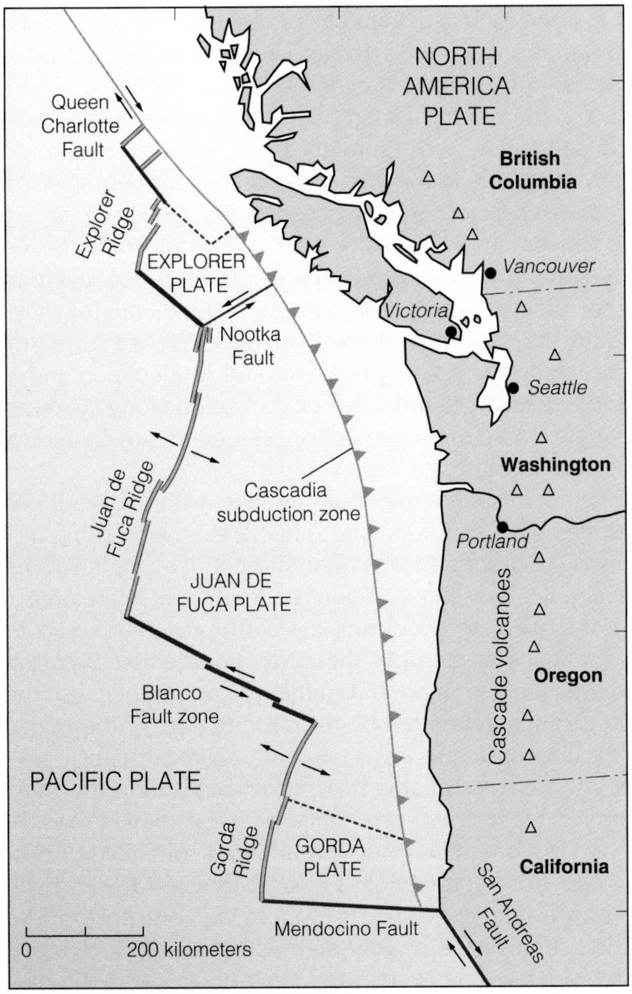

130°   126°   122°W

NORTH AMERICA PLATE

Queen Charlotte Fault

Explorer Ridge

EXPLORER PLATE

Nootka Fault

British Columbia

Vancouver

Victoria

52°N

Juan de Fuca Ridge

Cascadia subduction zone

JUAN DE FUCA PLATE

Seattle

Washington

Portland

48°

Blanco Fault zone

PACIFIC PLATE

Cascade volcanoes

Oregon

44°

Gorda Ridge

GORDA PLATE

Mendocino Fault

California

San Andreas Fault

40°

0   200 kilometers

(b)

▶**FIGURE 5-33.** The Cascadia oceanic trench to the north and the San Andreas transform fault to the south dominate the Pacific continental margin of the United States. **(a)** Seafloor topography. **(b)** Map of plate boundaries.

John Clague photo.

Tsunami sand

Peat

▶**FIGURE 5-34.** Tsunami sand from a megathrust earthquake deposited in 1700 over dark brown peat in a British Columbia coastal marsh. The scale is in tenths of 1 meter.

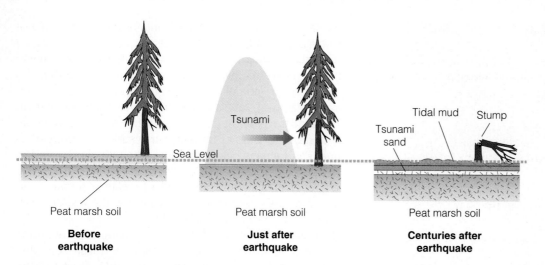

▶**FIGURE 5-35.** Simplified sketch showing tsunami sand deposited immediately after a subduction earthquake when a tidal marsh suddenly drops below sea level.

Tsunami

Sea Level

Peat marsh soil

Before earthquake

Peat marsh soil

Just after earthquake

Tidal mud   Stump

Tsunami sand

Peat marsh soil

Centuries after earthquake

Donald Hyndman photo.

▶**FIGURE 5-36.** This ancient Sitka spruce forest in the bay at Neskowin, Oregon, was felled by a giant tsunami following the huge subduction zone earthquake of January 1700. Stumps of the giant trees punctuate low tide at this beach some 25 kilometers north of Lincoln City. The forest with trees as old as 2,000 years, was suddenly dropped into the surf during a megathrust earthquake and then felled by the huge tsunami that followed.

Radiocarbon dating of leaves, twigs, and other organic matter in the buried soils at Willapa Bay, Washington, indicates seven of those giant events in the past 3,500 years, an average of one per 500 years. Elsewhere along the coast, the records show that twelve have occurred in the last 7,000 years since the eruption of Mount Mazama in Oregon deposited an ash layer on the seafloor at an average interval of 580 years. The intervals between them range from 300 to 900 years. The last one was some 300 years ago, so the next could come at any time.

Those analyses indicate similar dates at most, though not all, sites all along the coast between Cape Mendocino and southern British Columbia. That probably means that the fault generally broke simultaneously along this entire 1,200-kilometer length of coast, an extremely long rupture that would likely correspond to an earthquake of about magnitude 9. Such an enormous earthquake offshore would surely start a wave large enough to cross the Pacific Ocean.

The **sand sheets** were deposited at elevations to 18 meters above sea level. Tsunami of this size expose coastal communities to extreme danger. The larger cities of Seattle, Portland, and perhaps Vancouver would not be at significant tsunami risk from such a subduction zone earthquake because they are well up inlets or rivers; the waves would largely dissipate before reaching them. Communities on the open coast or smaller coastal bays, however, are in real danger. An earthquake near the coast could generate a tsunami wave that would reach the shore in less than twenty minutes, which would leave too little time for warning and evacuation of those in danger. Feeling an earthquake along the coast, people should immediately move inland to higher ground. The first indication along the coast of Oregon, Washington, or British Columbia may be an unexpected rise or fall of sea level.

Shaking in such a major earthquake, with accelerations of at least 1 g, would make it difficult to stand. Strong motion would continue for several minutes, leaving little time to evacuate. Thus, the first defense for people in the area is to protect themselves during the earthquake—take cover from falling objects until the earthquake ends. Then immediately move inland and to higher ground, because the large tsunami generated by the sudden shift of the ocean floor will arrive at the west coast within fifteen to thirty minutes. If hills are available nearby, evacuation on foot may be preferable because of traffic jams and damaged roads. If trapped in a broad, flat area, a reinforced concrete building may offer some protection, but only as a last resort. Even climbing a sturdy tree has saved more than one person. The sudden drop of the coastal area will raise sea level compared with the land even more. Thus, the tsunami will rush ashore to higher levels than would otherwise be expected. Calculations suggest that a 7- to 8-meter tsunami will invade some coastal bays. The record of the last event indicates that waves were as high as 20 meters where they funneled into some inlets.

Knowledge of when the last event happened would provide some indication of when to expect the next one. Radiocarbon dating of the peat and buried trees places the last of those events within a decade or two of the year 1700. In a separate analysis, careful counting of tree rings from killed and damaged trees indicates that the commotion happened shortly after the growing season of 1699.

In a clever piece of sleuthing, geologists of the Geological Survey of Japan found old records with an account of a great wave 2 meters high that washed onto the coast of Japan at midnight on January 27, 1700. No historic record tells of an earthquake at about that time on other Pacific-margin subduction zones, Japan, Kamchatka, Alaska, or South America. That leaves the Northwest coast as the only plausible source. Correcting for the day change at the international date line and the time for a wave to cross the Pacific Ocean, the earthquake would have occurred on January 26, 1700, at approximately 9 P.M.

Coastal Indians have oral traditions that tell of giant waves that swept away villages on a cold winter night. Archeologists have now found flooded and buried Indian villages strewn with debris. These many lines of data help confirm the timing of the last giant earthquake on the coast of the Pacific Northwest.

It seems likely that the oceanic plate sinking through the trench off the Northwest coast is now stuck against the overriding continental plate. If so, the continental plate should bulge up; precise surveys confirm that expectation (▶Figures 5-37 and 5-38). The locked zone is 50 to 100 kilometers off the coasts of Oregon, Washington, and southern British

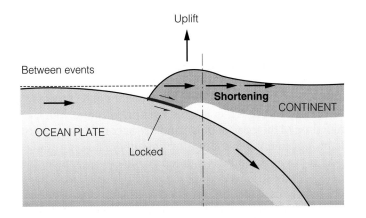

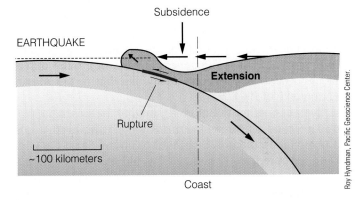

▶**FIGURE 5-37.** Denser oceanic plate sinks in a subduction zone. As strain accumulates, a bulge rises above the sinking plate while an area landward sinks. Those displacements reverse when the fault slips to cause an earthquake.

Roy Hyndman, Pacific Geoscience Center.

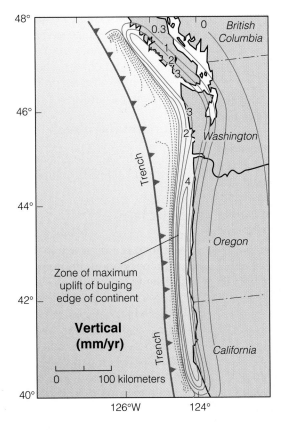

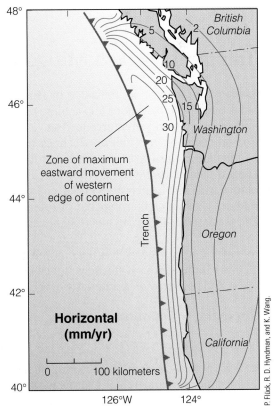

▶**FIGURE 5-38.** The subduction zone is locked between the oceanic trench at the landward edge of the Juan de Fuca Plate and halfway to the coast. Convergence of the plates causes bulging of the edge of the North American Plate. Uplift rates are as high as 4 millimeters per year, and eastward transport is as high as 30 millimeters per year. Rates shown on both maps are in millimeters per year (mm/yr).

P. Flück, R. D. Hyndman, and K. Wang.

Columbia. Just inland, the margin is now rising at a rate between 1 and 4 millimeters per year and shortening horizontally by as much as 3 centimeters per year.

In these convergence zones, accumulating stress eventually breaks the bond between the sinking slab and the continental margin. Then the raised continental crust snaps back down along the coast and rises just offshore. The sudden drop of the coastal bulge and rise accompanying thrust fault movement generates a giant earthquake and a huge ocean wave. Computer numerical models estimate the heights of those waves as approximately 10 meters offshore. These heights would be amplified by a factor of two to three in some bays and inlets. Port Alberni, at the head of a long inlet on the west coast of Vancouver Island, B.C., in the 1964 Alaska earthquake, for example, had a run-up amplified by a factor of three compared with the open ocean. Although approximate, similar numbers are obtained from studies of onshore damage. The general pattern of ground movements described above is exactly like that in the Alaska earthquake of 1964.

Even in Southern California, south of the Cascadia subduction zone, a near-field vertical-motion earthquake poses a potential problem. An earthquake on the Santa Catalina Fault offshore from Los Angeles would reach the community of Marina Del Rey, just north of the Los Angeles Airport, in only eight minutes. Given the large population and near sea-level terrain, the results could be tragic.

Not all tsunami are in the Pacific Ocean. One in 1929 killed fifty-one people on the south coast of Newfoundland. On November 1, 1755, a series of large earthquakes in the Atlantic Ocean southwest of Lisbon, Portugal, wrecked the city and killed tens of thousands of people. The associated 10-meter-high tsunami waves washed ashore, killing still more.

Tsunami Hazard Mitigation

Tsunami hazards can be mitigated by land use zoning that limits building to elevations above those potentially flooded and by engineering structures to resist erosion and scour. Coastal developments that orient streets and buildings perpendicular to the waves survive better that those that are aligned parallel to the shore. They limit debris impact and permit waves to penetrate without building higher. Landscaping with vegetation capable of resisting wave erosion and scour can help, as can trees that permit water to flow between them but slow the wave. But the trees need to be well rooted or they can themselves become missiles. A large ditch placed in front of houses can help reduce the level of the first wave, and may provide a little extra evacuation time.

Surviving a Tsunami

In summary, most tsunami are caused by earthquakes.

■ For a nearby subduction zone earthquake, you do not have much time before the first wave arrives, possibly fifteen to thirty minutes. You need to get to high ground or well inland immediately. A road heading directly inland is an escape route, but blocked roads and traffic jams are likely. Climb a nearby slope as far as possible, certainly higher than 30 meters.

■ Do not return to the shore after the first wave. Although the sea may pull back offshore for a kilometer or more following that first wave, other even higher waves often arrive for several hours. Wait until officials provide an all clear signal before you return.

■ Never go to the shore to watch a tsunami. Tsunami move extremely fast, and traffic jams in both directions are likely to require abandoning your vehicle where you least want to do so.

■ Even without warning, an unexpected rise or fall of sea level may signal an approaching tsunami. Move quickly to high ground.

■ Stay tuned to your radio or television.

Tsunami waves appear much like ordinary breaking waves at the coast, except that their velocities are much greater and they are much larger. Some come in as high breaking waves, a high wall of water that destroys everything in its path. Others advance as a rapid rise of sea level, a swiftly flowing and rising "river" without much of a wave. Even those are extremely dangerous because they advance much faster than a person can run. Loose debris picked up as the waves advance act as battering rams that impact both structures and people. Even a strong swimmer caught in the swift current as the wave retreats will be swept out to sea.

The Pacific Tsunami Warning System has two levels: a **tsunami watch** and a **tsunami warning.** A watch is issued when an earthquake of magnitude 7 or greater is detected somewhere around the Pacific Ocean. If a significant tsunami is identified, the watch is upgraded to a warning and civil defense officials order evacuation of low-lying areas that are in jeopardy.

Tsunami Examples

Some of the largest tsunami events on record include those shown in Table 5-1 (compiled from many sources).

Seiches

A **seiche** is a big wave in a large lake or enclosed bay that sways back and forth from one end of a basin to the other. The same thing happens in a bowl or bathtub if you move much water toward one end of the tub. Seiches form in larger bodies when water is disturbed by a large earthquake or landslide, a change in atmospheric pressure, or a storm surge.

Seiches were first studied in Lake Geneva, Switzerland, in the 1700s, when people noticed that the water level at each end of the lake rose and fell almost a meter once an hour or so after a period of strong wind along the length of the

Table 5-1 Examples of Large Tsunami

Time and Place	Cause	Tsunami Arrival Site	Height (m)	Deaths
B.C.1620 Santorini, Greece	Caldera collapse of ancestral Santorini volcano	Eastern Mediterranean	6	Destroyed Minoan culture on Crete
July 21, 365 A.D.	Earthquake in eastern Mediterranean	Greece, Egypt, Sicily		50,000 in Alexandria
Jan. 26, 1700	Subduction earthquake (magnitude 8–9), coastal Washington and Oregon See pp. 118–122	West coast of Washington, Oregon (near field; would be 30–40 minute delay) Japan (far field)	18	Felled forests at heads of bays. Probably many deaths along coast.
Nov. 1, 1755 Lisbon, Portugal	Earthquake	Lisbon	10	30,000 + 20,000 in resulting fire
1837 Chile		Hilo, Hawaii		14
Aug. 27, 1883 Krakatau, Indonesia	Volcano collapse See p. 111	Sumatra and Java	6–36	>35,000
June 15, 1896 Japan	Earthquake		29	27,000
Mar. 2, 1933 Japan	Earthquake		20	3,000
April 1, 1946 Unimak Island, Alaska	Subduction earthquake	Aleutian Islands, Alaska; Hilo, Hawaii (far field) Waves 15 minutes apart	30 15 11 (Oahu)	159 (96 in Hilo)
July 9, 1958 Lituya Bay, Alaska	Rockfall See p. 109	Lituya Bay, Alaska	33 to 524	2
May 22, 1960 Chile	Subduction earthquake	Coast of Chile; Hilo, Hawaii (far field) Honshu, Japan (far field)	10.7 5.3 4.5	>2,000 61 in Hilo, Hawaii 122 in Honshu, Japan
March 27, 1964 Prince William Sound, Alaska	Subduction earthquake See pp. 106–107	Anchorage and Seward, Alaska (near field; 30-minute delay) Port Alberni, British Columbia	6 30	125
Dec. 12, 1992 Indonesia	Magnitude 7.5 earthquake	Flores Island, Indonesia (near field)	26	>1,000
Sept. 1, 1992 Nicaragua	Earthquake	Masachapa (near field)	10	150
July 12, 1993 Hokkaido, Japan	Earthquake See p. 108	Okushiri, Japan (near field; 5-minute delay)	11	200
July 17, 1998 Papua New Guinea	Undersea landslide triggered by earthquake	Villages, north coast Papua, New Guinea (5–10 minute delay)	12	Officially 2,134 (possibly 3,000)
Dec. 26, 2004 Sumatra	Subduction earthquake off northwest Sumatra See pp. 99–102	Sumatra, Thailand, Sri Lanka, India, Somalia	>10	>283,000

lake. During a seiche, the displaced water moves the length of the lake, causing rise at the far end. That higher water sinks again to cause rise at the other end of the lake, and so on (▶ Figure 5-39). Because it takes time for movement of a large amount of water over a significant distance, larger basins have longer periods of oscillation. The same longer period, or lower frequency, is true of a longer pendulum, a child's swing, or a larger earthquake. Their large wavelength and frequency of moving back and forth depend heavily on the basin size.

Although tsunami have periods of approximately eight to eighty minutes in the open ocean, seiches in lakes or other more confined bodies of water typically have periods of less than ten minutes in small water bodies to several hours in larger ones.

Wind-driven seiche effects are common in the Great Lakes. Strong westerly winds in mid-November 2003 caused a seiche with more than 4 meters of difference in water level from Toledo on the western end of Lake Erie to Buffalo on the eastern end of the lake (▶ Figure 5-40). Lake Erie

▶**FIGURE 5-39.** A seiche in an enclosed basin oscillates back and forth at a frequency controlled by the size of the basin.

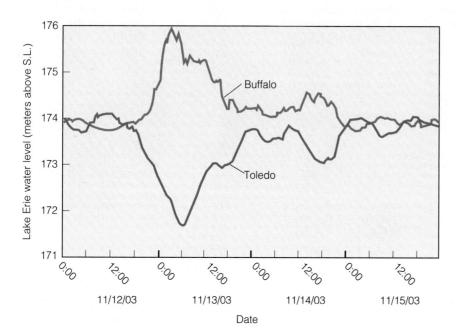

▶**FIGURE 5-40.** Seiche levels in Lake Erie from Buffalo, New York, to Toledo, Ohio, November 12–15, 2003.

commonly experiences seiche effects because the winds tend to blow from west to east along the primary axis of the lake. The impact of such seiche events on coastal erosion is amplified because of large wind-driven waves. The winds that created the November 2003 seiche on Lake Erie gusted to more than 90 kilometers per hour, causing 3- to 5-meter waves on top of the seiche.

Even hurricanes can cause pile up water. As the eye of Hurricane Frances passed close to Lake Okeechobee, Florida, in September 2004, it caused the south end of the lake to rise 3.6 meters higher than the north end. As the winds shifted later in the day, the north end of the lake rose to that height as the south end fell.

Nearly 1,000 years ago, a large earthquake on the Seattle Fault (see Chapter 4, "Earthquake Prediction and Tectonic Environments") deposited tsunami sand deposits along the shores of Puget Sound just north of Seattle. That earthquake also caused a landslide that submerged three areas of forest, drowned trees, and generated a seiche in Lake Washington, a narrow, 12-kilometer-long lake in the eastern part of Seattle. If a major earthquake were to occur on the Seattle Fault today, such a seiche would probably cause many deaths and severe damage to expensive housing around the lakeshore.

The Potential for Giant Tsunami

Tsunami in the historic record have been dramatic and sometimes catastrophic. Are even larger tsunami possible? Because tsunami waves form by sudden displacement of a large mass of water, they are generated by earthquakes, volcanic eruptions, and landslides underwater or into water. Earthquake-generated tsunami are most frequent, but their size in the open ocean is limited to the maximum displacement on an earthquake fault. Horizontal displacement underwater would not displace significant water. The magnitude-versus-displacement graph of Figure 3-27 (page 50) is based on limited data but must be approximately correct. Extrapolation of its trend suggests that an earthquake of moment magnitude 8 from vertical displacement on a normal fault, the most likely type to displace significant water, could have a vertical offset of 15 meters. A thrust-fault movement might have greater offset, but its gentler dip would likely cause a lesser vertical displacement of water. Because the tsunami wave height approximates the vertical displacement on a fault, the maximum wave height from an earthquake is a few tens of meters. As noted above, wave heights are amplified when waves are pushed into shallow water and bays.

Tsunami generated by volcanic eruptions are occasionally catastrophic, but they are poorly understood; their maximum size is unknown. We do not know enough about the mechanism of **water displacement** from an underwater eruption to do much more than wildly speculate.

Tsunami generated by landslides into water can be truly gigantic. The Lituya Bay tsunami of 1958 described earlier in this chapter created a 150-meter-high wall of water that surged more than 500 meters over a nearby ridge. The event began with a large earthquake on a nearby fault. The shock waves dislodged a large slab of rock and ice that fell into the bay to cause the tsunami. In this case, the earthquake was the initial culprit but not the direct cause of water displacement. The large size of a rock mass and the height from which it fell led to the tsunami's great height.

Gigantic **submarine landslides** in Hawaii and other oceanic volcanoes can be even more impressive. Occasionally, a huge slice of an island tens of kilometers wide collapses into the ocean, suddenly displacing thousands of cubic kilometers of water. The resultant tsunami, reviewed above, can be hundreds of meters high. None have happened in historic time, but it is only a matter of time. When it does happen, the low-elevation populations of Hawaii are in deep trouble. Presumably such a giant tsunami would also cross the Pacific to obliterate coastal communities of western North America, Japan, and elsewhere. Collapse of a flank of the Canary Islands, off the northwest coast of Africa, could generate a giant tsunami that would cross the Atlantic Ocean to obliterate coastal cities on the eastern coast of North America and perhaps those in western Europe. In these cases, we could have a catastrophe without needing coincidental overlapping events. Collapse on Reunion Island, a similar volcano in the Indian Ocean, could cause tsunami inundation and destruction of many coastal areas, including the dense sea-level populations of Bangladesh.

Thus, it seems likely that a catastrophic tsunami, many times larger than any in historic time, is likely to come from the flank collapse of an oceanic volcano. Our geologic record of such events is clear enough to indicate that they have happened and will again. Limited evidence for collapses in Hawaii suggests an approximate recurrence interval of 100,000 years, but that is only a crude average. As noted throughout this chapter, however, the result could be truly cataclysmic.

KEY POINTS

✓ Tsunami have such long wavelengths that they always drag on bottom. Their velocity depends on water depth. **Review p. 115; Figure 5-25.**

✓ Tsunami are caused by any large, rapid displacement of water, including earthquake offsets or volcanic eruptions underwater, landslides, and asteroid impacts into water. **Review pp. 104–105, 111–114; Figures 5-9 and 5-10.**

✓ Tsunami, sometimes misnamed "tidal waves," have nothing to do with tides. **Review p. 99.**

✓ Tsunami come as a series of waves, often tens of minutes apart. The largest waves are often the third or later to arrive. **Review pp. 103–104, 118; Figures 5-4 and 5-7.**

✓ Tsunami can reach the coast within a few minutes from a nearby earthquake or many hours later from a distant quake. **Review pp. 102–105, 116–118.**

✓ A subduction-zone earthquake can suddenly drop a low-lying coastal zone below sea level. **Review p. 105; Figure 5-16.**

✓ A volcano flank collapse that suddenly moves an enormous amount of water can generate giant tsunami that would be catastrophic for much of the East Coast of North America, especially low-lying coastal communities. **Review pp. 113–114; Figure 5-24.**

✓ The impact of a large asteroid into the ocean would displace a huge amount of water and generate a massive tsunami. **Review p. 114.**

✓ Tsunami waves in the open ocean are low and far apart but move at velocities of several hundreds of kilometers per hour. They slow and build much higher in shallow water near the coast, especially in coastal bays. **Review pp. 114–116; Figures 5-25 and 5-28.**

✓ Dangers from tsunami waves include drowning, impact from tsunami-carried debris, and severe abrasion from being dragged across the ground. **Review p. 117.**

✓ Tsunami from a Pacific coast subduction earthquake come every few hundred years and would come onshore within twenty minutes of the earthquake to destroy coastal communities, particularly those in bays and inlets. The safest areas are more than a kilometer inland and several tens of meters above sea level. **Review pp. 118–122.**

✓ The record of subduction-zone tsunami is based on sand sheets over felled forests and marsh vegetation in coastal bays. **Review pp. 118–121; Figures 5-34 and 5-35.**

✓ In between earthquakes, the leading edge of the continental plate slowly bulges upward before suddenly dropping during the earthquake. **Review pp. 121–122; Figures 5-37 and 5-38.**

✓ Danger signals for tsunami include a large earthquake and a rapid rise or fall of sea level. You can survive a tsunami by running upslope or driving directly inland immediately upon feeling an earthquake. **Review p. 122.**

✓ Seiches are back-and-forth swaying motions of water in an enclosed basin. **Review pp. 122–124.**

✓ Seiche surges in lakes are caused by strong winds that push up mounds of water as much as a few meters. **Review pp. 123–124; Figures 5-39 and 5-40.**

IMPORTANT WORDS AND CONCEPTS

Terms

caldera collapse, p. 111
coastal bulge, p. 119
coastal inlets, p. 119
deepwater waves, p. 115
far-field tsunami, p. 116
giant debris avalanche, p.111
harbors, p. 116
narrowing of a harbor, p. 106
pali, p. 112
refract (waves), p. 103
run-up, p. 115

sand sheets, p. 120
seiche, p. 122
seismic sea wave, p. 99
shallowing water near shore, p. 106
submarine collapse, p. 112
submarine landslides, p. 125
trimline, p. 109
tsunami, p. 99
tsunami warning, p. 122
tsunami watch, p. 122
water displacement, p. 125

QUESTIONS FOR REVIEW

1. What are three of the main causes of tsunami?

2. Of the three main types of fault movements—strike-slip faults, normal faults, and thrust faults—which can and which cannot cause tsunami? Why?

3. About how high are the largest earthquake-caused tsunami waves in the open ocean?

4. How does the height of a tsunami wave change as it enters a bay? Why?

5. How many tsunami waves are generated by one earthquake?

6. How fast do tsunami waves tend to move in the deep ocean?

7. Do tsunami speed up or slow down at the coast? Why?

8. Why is even the side of an island away from the source earthquake not safe from a tsunami?

9. For a subduction-zone earthquake off the coast of Oregon or Washington, how long would it take for a tsunami wave to first reach the coast?

10. Because the Atlantic coast experiences fewer large earthquakes, what specific other event could generate a large tsunami wave that would strike the Atlantic coast of North America?

11. What specific evidence is there for multiple tsunami events having struck coastal bays of Washington and Oregon?

12. What is a seiche? Explain what happens and what causes it.

FURTHER READING

Assess your understanding of this chapter's topics with additional quizzing and conceptual-based problems at:

 http://earthscience.brookscole.com/hyndman.

CHAPTER 6

Mount St. Helens erupted violently on May 18, 1980.

Austin Post photo, USGS.

VOLCANOES
Materials, Hazards, and Eruptive Mechanisms

In 1980, Mount St. Helens was well known to be the most active volcano in the Cascade range and the most likely to erupt. Its smooth, symmetrical shape showed that it must have erupted recently to fill deep valleys that mountain glaciers had carved in its flanks during the last ice age, perhaps 10,000 years ago. The U.S. Geological Survey had recently mapped and determined the age of the young eruptive deposits and concluded that indeed St. Helens could erupt again at any time. Then, in March 1980, it showed renewed signs of life—swarms of small earthquakes and blasts of steam and ash. Detailed studies of the mountain and its activity began in earnest, culminating with the climactic eruption in mid-May.

At 8:32 A.M. on May 18, Dr. David Johnston, the lone USGS volcanologist stationed high on the volcano at the time, radioed back to the Cascades Volcano Observatory in Vancouver, Washington, "Vancouver, Vancouver—this is it!" He died in the eruption. The night before he had reluctantly agreed to replace someone else at that post. Ironically, he had expressed the most concern about the hazards of an impending eruption; he had repeatedly referred to St. Helens as "a dynamite keg with the fuse lit."

May 18, 1980, was a brilliant late spring day throughout the Pacific Northwest. Early that morning, a few people were camping or logging on forest land in the restricted zone north and west of Mount St. Helens in southwestern Washington. Some had sneaked around official safety barriers erected in response to two months of minor eruptions. Some of the interlopers wanted to watch St. Helens from nearby. They hoped to see a major eruption. A bulge high on the north flank of the volcano had been growing for weeks, showing that a large mass of magma was rising within the cone. At 8:32 A.M., they got more than enough excitement. A few people survived to tell what they experienced.

Keith and Dorothy Stoffel were ready to drive home after having attended a geoscience conference in Yakima, southeast of St. Helens, when they impulsively decided to hire a pilot to fly them around for a quick look at the mountain. As the plane rounded the north flank of the volcano, they saw that the snow had melted off the bulge overnight. Then they watched the bulge detach in a great landslide as an enormous cloud of steam black with ash spouted behind the slide and then blossomed to fill the sky (▶ Figure 6-1). The pilot dove out of range and headed for Spokane.

▶ **FIGURE 6-1.** In the initial eruption of Mount St. Helens at 8:32 A.M. on May 18, 1980, the growing bulge above the magma chamber began to collapse in a landslide. The collapse released pressure on the magma, permitting its gases to expand explosively.

Keith and Dorothy Stoffel photo.

P. and C. Hickson were 17 kilometers northeast of the crater when they saw the north side of the volcano suddenly look "fuzzy." Then it began to slide, the lower part faster. Meanwhile, a densely black cloud blossomed from the summit and the north flank seemed to explode (▶ Figure 6-2). Other people nearby saw the horizontal blast and a shock wave racing ahead of the cloud. It looked like those they had seen in photos of nuclear explosions.

J. Downing, who was on the flank of Mount Rainier 75 kilometers to the north, saw two distinct "flows" 300 to 600 meters (1,000 to 2,000 feet) thick that hugged the ground, disappeared into valleys, and then "hopped" over ridges. Those were ash flows, masses of steam dark with suspended ash that made them so dense they hugged the ground.

C. McNerney was 13 kilometers northwest of the crater, within the area doomed to imminent devastation, when he watched the north side of the volcano collapse. The leading wall of the black cloud climbed over a ridge and a hot wind began to blow from the volcano. Two minutes after the eruption began, he started driving west at 120 kilometers per hour, but the black cloud was gaining on him, so he sped up to 140 kilometers per hour. The base of the black ash flow cloud advanced "like avalanches of black chalk dust," one after another, like waves lapping onto a beach.

G. and K. Baker were 17 kilometers northwest of the crater, also within the area destined for immediate devastation. They saw a "big, black, inky waterfall" a few miles up the valley and began driving west on Highway 504 at 160 kilometers per hour. Even so, the black cloud almost reached them within three to four minutes. The cloud looked like it might be boiling oil with bubbles 2 meters in diameter. This, of course, was the same ash flow that others northwest of the volcano were fleeing.

B. Cole was logging 20 kilometers northwest of the crater with three companions when the ash flow reached them. A "horrible crashing, crunching, grinding sound" came from the east. The air around them became totally dark and intensely hot. Cole and his companions gasped to breathe. The insides of their mouths and throats burned, and they were knocked down along with all the trees. Everything was covered with a foot of gray ash. The heat burned large parts of their bodies but not their clothing. Cole's three companions later died.

D. and L. Davis and A. Brooks were 19 kilometers north of the crater. They watched a black, boiling cloud bear down on them, another view of the ash flow. Its leading edge snatched trees out of the ground and tossed them into the air. Then the black cloud swallowed them and everything was pitch black and burning hot. A physician later said their burns were similar to those caused by a microwave.

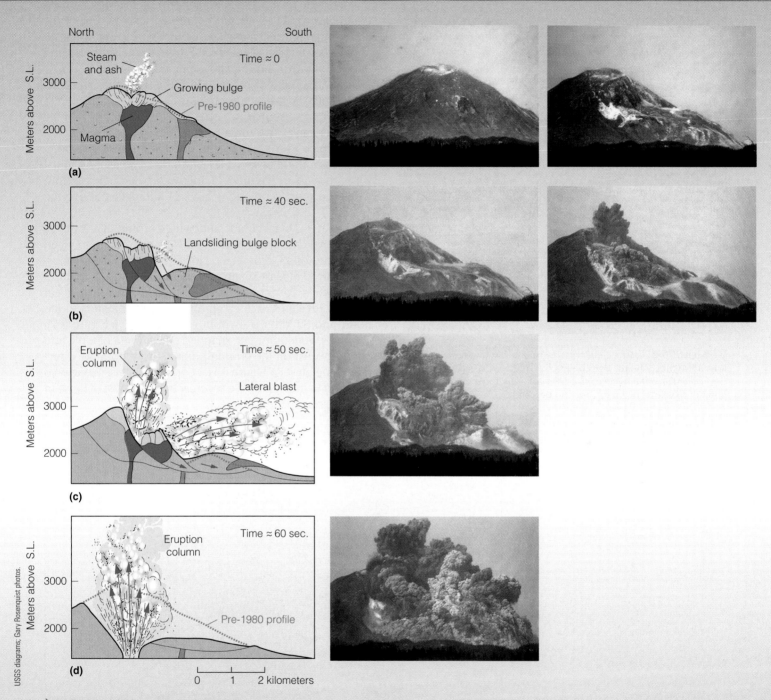

▶**FIGURE 6-2.** Mount St. Helens eruptive sequence, May 18, 1980: **(a)** Magma approaches the surface and weakens the dome. **(b)** The dome collapses in a giant landslide. **(c)** Pressure release on magma permits gases to bubble out rapidly. The lateral blast blows out to north. **(d)** A full Plinian-type eruption ensues.

Mount St. Helens Eruption, May 18, 1980

The morning of May 18, an immense vertical column of volcanic gases black with ash continued to blow out of the crater for much of the day (▶ Figure 6-3). The event was far larger than the U.S. Geological Survey geologists who had been monitoring the volcano expected.

An earthquake of magnitude 5 triggered the landslide, which relieved the steam pressure in the magma beneath. The effect was similar to popping the cork out of a bottle of champagne. The mass of steaming magma expanded into pumice as it felt the relief of pressure, then exploded into a dark cloud of steam heavily laden with ash. An initial surge shot north (Figure 6-1), followed by a huge ash flow, while the eruption cloud soared 24 kilometers above the volcano.

The lateral blast cloud, somewhere around 350°C, must have moved north at 140 to 150 meters per second to reach a distance of 13 kilometers. Hot ash flows (>700°C) must have moved at least 36 meters per second to reach as far as 8 kilometers north of the crater. The lateral blast of steam and ash leveled the forest in an area 20 kilometers to the north and 30 kilometers from east to west (▶ Figure 6-4). The trees were uprooted, snapped off, stripped of bark, and charred. Most on the ground pointed away from the direction of the blast (▶ Figure 6-5). People in the blast zone who were out in the open were severely burned; most of those in vehicles showed few ill effects.

▶ **FIGURE 6-3.** In the climactic eruption of Mount St. Helens in May 1980, hot ash-laden gas boiled out of the vent, expanding into the cooler surrounding air. The darker ash-rich upper and outer parts of the cloud begin to cool and descend.

D. Swanson photo, USGS.

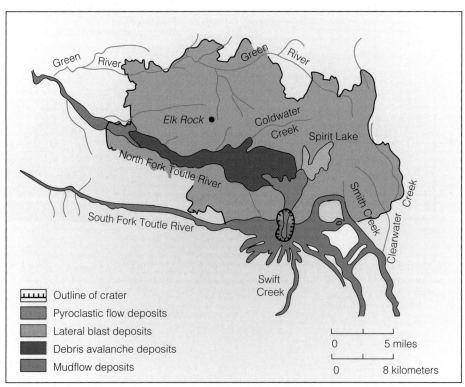

▶ **FIGURE 6-4.** A map of Mount St. Helens after the 1980 eruption shows the crater open to the north, with pyroclastic flow, lateral blast, and mudflow deposits.

Green River
Green River
Elk Rock •
Coldwater Creek
Spirit Lake
North Fork Toutle River
Smith Creek
Clearwater Creek
South Fork Toutle River
Swift Creek

⊓⊔⊓⊔ Outline of crater
▨ Pyroclastic flow deposits
▨ Lateral blast deposits
▨ Debris avalanche deposits
▨ Mudflow deposits

0   5 miles
0   8 kilometers

USGS

The bulge on the north flank of the volcano collapsed to become a huge debris avalanche and mudflow with a volume of 2.8 cubic kilometers. Much of the debris followed the Toutle River all the way downstream to the Columbia River, where it blocked navigation until a deep channel was dredged weeks later.

Within less than ten minutes and for the next nine hours, steam and carbon dioxide blowing from the crater supported a column of ash more than 20 kilometers high, a **Plinian-style eruption.** The vigorously erupting ash partially collapsed at times to drop ash flows down the north flank of the volcano. Rapidly melting snow and ice sent catastrophic floods and mudflows down nearby valleys.

The collapsing bulge displaced the water from Spirit Lake at the base of the volcano and carried part of it over the ridge to the north. Most of that debris avalanche raced down the north fork of the Toutle River at speeds of 110 to 240 kilometers per hour. It continued 22 kilometers down the valley, leaving a hummocky deposit averaging 45 meters thick.

The new crater that lopped 400 meters off the top of the original peak was 1.5 by 3 kilometers across, oval, and open

▶**FIGURE 6-6.** Mount St. Helens before the cataclysmic May 18, 1980, eruption that destroyed the peak.

▶**FIGURE 6-7.** A huge steaming crater replaced the peak of Mount St. Helens after the May 18 eruption.

▶**FIGURE 6-5.** These trees were flattened by the lateral blast from Mount St. Helens. Ash from the eruption coats the trees.

to the north (▶ Figures 6-4, 6-6, and 6-7). The total volume of newly erupted magma (before it spread out as ash) was approximately equivalent to 0.2 cubic kilometers of solid rock. Although the eruption of St. Helens in 1980 was widely regarded as a major catastrophe, this eruption was in fact a modest effort by volcanic standards. By contrast, Mount Mazama (the remains of which now surround Crater Lake in Oregon) erupted 35 times that volume (see the section in Chapter 7 on Mount Mazama).

Preamble to the May 18 Eruption

When Mount St. Helens erupted on May 18, the eruption was certainly the most closely observed and exhaustively studied of any volcanic event to that time. St. Helens had been quiet since it produced a dome of pale andesite in 1843 and darker andesite **lava** flows in 1857. Ash flows from earlier eruptions reached at least 20 kilometers down the valleys around the volcano, mudflows at least 75 kilometers.

Radiocarbon dates on wood charred and buried during previous eruptions were available before 1980 to show

▶**FIGURE 6-8.** The huge bulge on the flank of Mount St. Helens grew a few weeks before the climactic eruption in May 1980. The umbrella shades the surveying instrument.

that most of the modern volcanic cone grew during the last 2,500 years. That pace of activity in the geologically recent past had already persuaded most geologists to consider St. Helens the Cascades volcano most likely to erupt. The U.S. Geological Survey issued hazard forecasts more than two years before St. Helens erupted in 1980. The forecasts proved generally accurate, although they understated the probable size and violence of the upcoming eruption.

Preliminary activity started suddenly in the afternoon of March 23, 1980, when seismographs began to detect swarms of small earthquakes, many with magnitudes greater than 4 on March 24. Several research organizations rapidly installed arrays of portable seismographs around the volcano. Then, just after noon on March 27, St. Helens produced a large cloud of white steam to the tune of a loud boom. The steam rose 600 meters above the volcano while a new crater began to open in its summit. Observers watched a fissure 1,500 meters long open high across the north flank of the volcano, while lesser fractures opened and closed.

Those first clouds of steam were neither hot enough nor dark enough with suspended ash to convince geologists that they were the first phase of a genuine eruption. Many volcanoes blow off great clouds of steam in late spring and early summer as melting snow sends water percolating down into the hot rocks within. St. Helens again blew off clouds of steam about once per hour for seventeen hours starting in the early hours of March 28. Those blasts opened a second crater at the west edge of the first one. Flames began to light the craters at night on March 29; their pale blue color suggested burning methane. The ejected clouds of steam became hotter and much darker with ash. Intense study of the seismic records showed a longer-period wavelike pattern that came to be known as **harmonic tremors.** These began on April 1 and continued intermittently and with growing strength for twelve days. Although the frequency of steam eruptions decreased to one per day between April 12 and April 22, they were becoming richer in ash. The situation was beginning to seem serious.

By the end of April, the two craters had combined into a single oval crater 300 to 500 meters across. Seismographs recorded more than thirty earthquakes with magnitudes greater than 3 every day. The arrays of seismographs revealed that the movements were originating below the north flank of the volcano at depths that became shallower day by day. Microscopic study of the newly erupted ash showed that the eruptions were still producing only fragments of old volcanic rocks that steam had altered within the volcano. There was no new erupted magma.

A **bulge** slowly grew on the north flank of the volcano in early May; it eventually expanded to 106 meters outward from the original slope and about 1,067 meters across (▶Figure 6-8). A large mass of magma was clearly rising into the north flank of the volcano. Meanwhile, the clouds of steam began to drop ash that was clearly derived from new magma rising from below. It had the color and composition of pale andesite, or, strictly speaking, dacite.

At this point, it seemed abundantly clear that St. Helens was ready to erupt, but the type of eruption remained open to vigorous debate. Many geologists expected a dome eruption in which the extremely viscous magma would rise quietly through the north flank of the volcano for months or a few years. They pointed to the bulge that swelled in the flank of the volcano. Other geologists reminded everyone that dacite magma is perfectly capable of absorbing water at depth and then erupting violently in great clouds of steam and ash. They pointed to the blankets of ash that cover much of the surrounding countryside. The debate hinged on the question of water and whether the magma was dry or heavily charged with steam. No one could tell.

Those who expected a dome eruption contended that if the mass of magma had sufficient steam to drive an explosive eruption, then it should be venting enough steam to melt the snow off the swelling dome—yet the bulge was still white. And many agreed that its growth, at a rate of 1 meter per day, would soon make it steep enough to pose an imminent landslide threat.

The U.S. Geological Survey informed the public of the dangers that existed and tried to dispel imaginary dangers. The agency did not attempt to predict specific eruption times, though many people believed it could and should have done so. It provided information but did not dictate which areas should be closed because such public policy decisions were beyond its authority.

News of the eruption spread slowly. Many people in eastern Washington had no inkling that anything had happened until the **eruption cloud** suddenly appeared overhead and began to dump ash on their Sunday afternoon picnics. Meanwhile, the eruption cloud moved directly east, dropping significant amounts of ash as far as 800 kilometers downwind in western Montana, with some continuing farther southeast to Colorado. The **ash fall** hampered transportation, utility systems, and outdoor activities until a heavy rain cleared the air four days later (▶Figure 6-9).

Few people in eastern Washington and the northern Rocky Mountains had experienced a volcanic eruption.

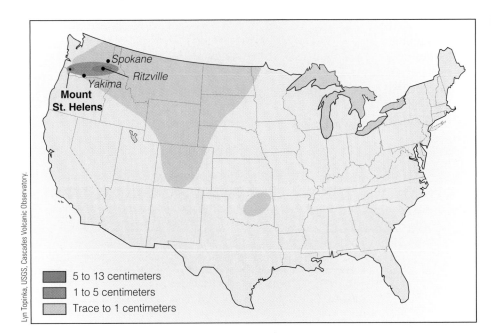

Lyn Topinka. USGS. Cascades Volcanic Observatory.

▶**FIGURE 6-9.** This map shows the ash fall distribution and thickness after the May 1980 eruption.

■ 5 to 13 centimeters
■ 1 to 5 centimeters
□ Trace to 1 centimeters

Many found the ash fall frightening. As far east as Missoula, Montana, the late afternoon western sky turned a ghastly greenish black before fine white ash began falling like persistent snow. Under an otherwise blue sky with no wind, the finer particles hung suspended in the air for several days. People stayed indoors with the windows closed, and many wore masks for brief excursions to work or for groceries.

Smaller eruptions on May 25, June 12, July 22, August 7, and October 16 to 18 produced airborne ash from eruption columns 10 to 14 kilometers high and ash flows that again poured down the volcano flanks. Dacite domes that appeared after each eruption were partially blown out by the next. Observers heard loud blasts of escaping carbon dioxide, sulfur dioxide, and hydrogen sulfide between the eruptions.

Water from melting snow, Spirit Lake, and the Toutle River created mudflows on May 18 that poured down the river's north fork at 16 to 40 kilometers per hour (▶Figures 6-10 and 6-11). They flushed thousands of cut logs downstream,

▶**FIGURE 6-10.** Huge mounds were left by the May 18 debris avalanche and mudflow that raced down the Toutle River. The flow moved from left to right.

Donald Hyndman photo.

Lyn Topinka photo, USGS, Cascades Volcano Observatory.

▶**FIGURE 6-11.** Lines from the Toutle River mudflow were left high on trees after the flow continued to drain down valley.

destroyed twenty-seven of thirty-one bridges, and deposited sediment in navigation channels including the Columbia River shipping channel near Portland, Oregon. Mudflows diverted streams and raised valley floors as much as 3 meters and channel beds as much as 5 meters. Airborne ash blocked Interstate 90 and other highways. Airports more than 400 kilometers downwind were closed for a week while crews cleared 6 to 10 centimeters of ash.

Types of Volcanic Hazards and Products

Lava Flows

Where fluid basalt lava flows dominate the volcanic eruptions, they are called **Hawaiian type lava,** after one of the places that best characterizes them (see Table 6-1 for an overview of volcanic materials). Some basalt flows have smooth tops with a smooth ropy or billowy surface generally called **pahoehoe,** a Hawaiian term (▶Figures 6-12 and 6-13). Pahoehoe surfaces develop on lavas rich in steam and other volcanic gases. The ropy form develops when fluid lava drags a thin crust into small wrinkles or folds. These flows commonly develop open passages as molten lava flows out from under a solid crust. Those are important because they make naturally insulated pipes in which lava can flow for distances of at least several kilometers without much cooling.

Basalt lavas charged with less steam and other gases and with a greater degree of crystallization develop an extremely rubbly and clinkery surface called **aa,** another Hawaiian term. An aa surface is easily capable of ruining a nice pair of boots during a short walk over the sharp and rough rocks (▶Figures 6-12 and 6-14). The rubble tumbles down the slowly advancing front of the flow, which runs over it the way a bulldozer lays down and runs over its tread (▶Figure 6-13).

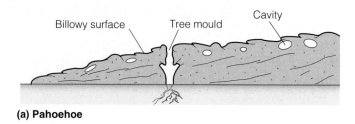

(a) Pahoehoe

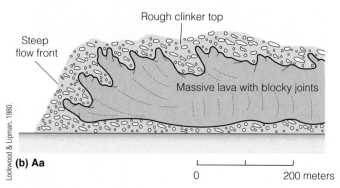

Lockwood & Lipman, 1980.

(b) Aa

0   200 meters

▶**FIGURE 6-12.** These cross sections show **(a)** pahoehoe and **(b)** aa lava flows.

J. D. Griggs photos, USGS.

▶**FIGURE 6-13.** Yes, it might be advisable to stop here! Creeping pahoehoe lava flows over a road. The black, clinkery basalt on the surface of the flow covers the red-hot molten basalt.

Table 6.1	**Generalized Products of Volcanoes**		
		Volcanic Products	
	Lava	**Pyroclastic Material (Ash and Ash Flow Products)**	**Mudflows**
Definition	Molten magma that flows out and onto the Earth's surface.	Fragments of solidified magma blown out of a volcano. May be deposited by a pyroclastic flow or ash flow (hot flowing and billowing ash) or by air fall (cool, falling out of ash cloud).	Volcanic ash and fragments transported downslope with the aid of water.
General characteristics	Molten magma that solidified as coherent sheets or broken jumbles of volcanic rock.	Fragments range from less than 2 mm ash to tens of cm. Larger pieces may be broken from older volcanic rocks on the sides of the vent.	Angular to rounded; unsorted particles of all sizes.

Basalt flows commonly advance at speeds from 1 meter per second near the vent to less than one-tenth that as the lava cools. Most travel up to 25 kilometers from the vent but may reach to more than 50. Andesite flows are more viscous and travel shorter distances. Because of their slow speeds and limited coverage, basalt or andesite flows generally pose little threat to human life. The clearest exception is an alkali-rich lava flow from Vesuvius that killed 3,000 people in 1631. Average flow speeds of the lavas in that eruption were 8 kilometers per hour. Lava that erupted from Vesuvius in 1805 traveled from the crater to the base of the mountain in four minutes, twice the speed of an Olympic sprinter.

Lava flows rarely kill people, but they take a heavy toll on houses, parked cars, and other human-made structures. Although basalt typically forms lava flows, it may occasionally contain enough gas to produce a violently streaming Plinian eruption. Mount Etna in southern Italy sometimes erupts that way (see the section in Chapter 7 on basaltic volcanoes and Mount Etna). Even well down slope from the vent, where the lava is no longer red on the outside, it is still hot enough to ignite wood structures such as houses (▶Figure 6-15). Green woody vegetation may be dehydrated first but eventually burns. Even if an object does not burn, it is still overwhelmed and often buried by the flow (▶Figures 6-16 and 6-17). Where basalt lava flows surround a green tree, the heat boils moisture in the tree trunk and chills the lava next to the tree. The woody tissue either chars or later rots away leaving a cast of the tree trunk. Excellent

Donald Hyndman photo.

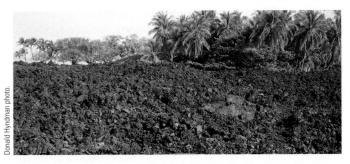

▶**FIGURE 6-14.** This ragged, clinkery-looking surface is aa lava from Mauna Loa Volcano, Waikoloa, Hawaii.

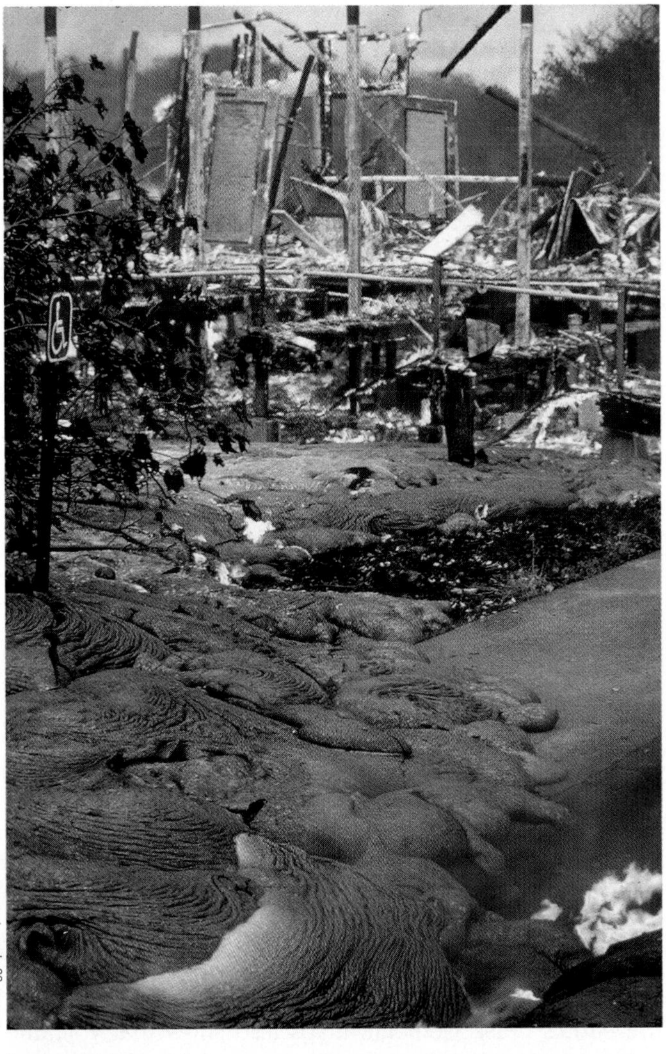

J. D. Griggs photo, USGS.

▶**FIGURE 6-15.** Dark but still hot basalt lava from Kilauea Volcano torches the Wahaula Visitor Center.

USGS.

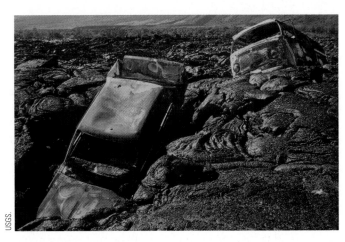

▶**FIGURE 6-16.** Parked cars succumbed to pahoehoe basalt lava in Hawaii.

Donald Hyndman photo.

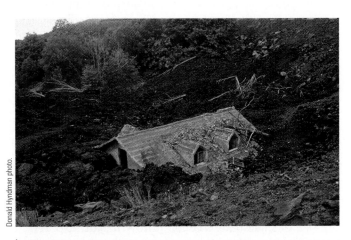

▶**FIGURE 6-17.** This is all that remains of a house that was in the path of a 1983 Mount Etna lava flow north of Nicolosi, Sicily.

examples of such casts are on Kilauea Volcano and at Lava Cast Forest, south of Bend, Oregon.

Attempts to slow or divert lava flows have brought only partial success. Perhaps the most effective approach is to cool and solidify the front of the flow with copious amounts of water delivered from fire hoses. A large basalt flow advancing on the town and harbor of Heimaey, Iceland, was cooled, slowed, and partly diverted that way in 1973. Flows erupting from Mauna Loa were bombed in 1935 and 1942 to break the solid levee of cooling lava along the edge of the flow, thus diverting it into another path. The attempts were not successful enough, however, to inspire a continuing program of bombing lava flows. Emergency diversion barriers were erected in nine days in 1960 to divert a flow erupting from Mauna Loa near Kapoho in westernmost Hawaii. These were partly successful in diverting the flow away from beach property and a lighthouse. Bulldozing levees of broken basalt lava in the path of a basalt lava flow proved unsuccessful presumably because the bulldozed levee was less dense than the flowing lava.

Oceanic Ridges

Iceland is the only place where an oceanic ridge stands above sea level. Its crest makes a broad valley that follows a dogleg course across the island, its opposite walls moving a few centimeters farther away from each other every year. Every few hundred years, a long **fissure** opens in the floor of the valley and erupts a large basalt flow. The most recent such occasion was in 1783 when the Laki fissure erupted 12 cubic kilometers of basalt lava that spread 88 kilometers down a gentle slope. If the events of the last 1,000 years of written Icelandic record provide any sort of volcanic calendar, another such fissure eruption may be about due.

In Iceland as everywhere along the oceanic ridge system, the hot dark rock of the Earth's mantle, peridotite, rises at depth to fill the gap between lithospheric plates separating at an oceanic ridge. The decrease in pressure on the rising mantle rock permits it to partly melt to make basalt magma, which erupts in the ridge. The considerable fluidity of the erupting basalt lava permits it to spread into a thin flow. Similar lava flows erupt in the crest of the oceanic ridge system along its entire length. They build the basalt lava flows that cover the entire ocean floor.

Watch wax dribbling down a candle and think about molten basalt lava erupting on the ocean floor. The dribble of wax soon acquires a thin skin of cooler wax. Then the molten wax within bursts through that skin and dribbles down a new path. As that happens again and again, the original dribble of wax proliferates into a network of dribbles. That is similar to how basalt erupting into water behaves. A chilled skin of solid basalt forms on the outside of the flow; then the molten basalt within bursts out and pours off in a new direction. The result is a pile of basalt cylinders about the size of small barrels called **pillow basalt.** Exposed in vertical section in a cliff or road cut, they look like a pile of oversized pillows in dull shades of greenish black.

Continental Flood Basalts

Flood basalt flows are simply giant basalt flows, typically with volumes more than 100 times those of ordinary basalt flows: several hundred cubic kilometers. Individual flows generally cover areas of tens of thousands of square kilometers to depths of 30 to 40 meters within a period of a week or two. Some in the Pacific Northwest cover areas almost the size of the state of Maine. Flood basalts erupt to build a stack of flows at least several hundred kilometers across.

Flood basalt provinces typically start erupting suddenly, remain vigorously active for a million or so years, then go out of business in another million or so years. The Pacific Northwest flood basalts, for example, began to erupt 17 million years ago and were essentially finished 15.5 million years ago. The few eruptions after that time added a negligible volume to the province.

M. R. Rampino and R. B. Stothers suggested in 1988 that a strong correlation exists between the times when flood basalts erupt and the times when global mass extinctions happened. The correlation is strong enough to suggest a cause-and-effect relationship. Plant fossils in sediments sandwiched between lava flows in the Pacific Northwest flood basalt province are the remains of hardwood forests similar to those that now thrive in Florida. They leave no doubt that a wet and tropically warm climate then prevailed in a region that is now mostly dry and cold. At the same time, an extremely dry climate prevailed in the southeastern United States.

The reasons for these dramatic climatic events are not completely clear. But it is clear that if carbon dioxide erupts in the same proportion to volume of magma in flood basalt eruptions as in ordinary basalt eruptions, then a single flood basalt flow could produce enough carbon dioxide to cause a global greenhouse effect (discussed in Chapter 10, "Climate and Weather-Related Hazards"). The dozens or perhaps hundreds of giant basalt flows that erupt in a flood basalt province could produce an extreme greenhouse effect on a global scale (see discussion in Chapter 10). Perhaps that is at least part of the cause of global mass extinctions. In any case, it seems clear that a single flood basalt eruption would be a catastrophe for the entire human race.

It is fortunate, then, that continental flood basalt provinces do not erupt often. The most recent flood basalt eruptions happened 15 million years ago in the Pacific Northwest. At least thirteen continental flood basalt provinces have erupted in the last 250 million years, an average of one every 20 million years or so.

None of those flood basalt provinces have apparently erupted anywhere near an existing plate boundary, but most seem to have started new plate boundaries. Perusal of geologic maps reveals that oceanic ridges start from flood basalt provinces at the time the flood basalts were erupting. Evidently, whatever started the flood basalts also cracked the earth's crust. Geologic maps also show that volcanic hotspot tracks also begin at the same time and place as flood basalt provinces.

▶**FIGURE 6-18.** A gigantic eruption cloud rose from Redoubt Volcano, southwest of Anchorage, Alaska, on April 21, 1990. The ash cloud spread laterally at the top, where the density of the hot ash was no longer lighter than the atmosphere.

Many geologists attribute the origin of flood basalt provinces to masses of magma that rise through the Earth's mantle, fracture the crust where they encroach upon it from below, then erupt the flood basalts. They explain the volcanic hotspot as the eruption of the residual magma left after the flood basalts have erupted. Other geologists propose that flood basalt flows erupt from a lava lake of molten basalt that floods an enormous crater that opened where an asteroid struck the Earth. They argue that the impact of the asteroid cracks the Earth's crust to start the oceanic ridge. After the sides of the original crater collapse into it, the remaining crater still relieves the pressure on the rocks in the upper mantle enough to cause them to partially melt. The melt has the composition of basalt.

Ash Falls

Volcanic ash is composed of bits of pumice less than 2 millimeters across, light enough to drift some distance on the wind (▶ Figures 6-9, 6-18, 6-19, and 6-20). Ash erupts

▶**FIGURE 6-19.** Air-fall ash from Mount Pinatubo in the Philippines collapsed many roofs.

▶**FIGURE 6-20.** Heavy air-fall ash from Rabaul Caldera, in September 1994, collapsed the roof on the left and thickly coated the roof on the right.

suspended in a cloud of steam that condenses into water droplets as it expands and cools. Much of the water coats the particles of ash, which fall like snow downwind of the vent. Some ash may linger for several years in the upper atmosphere, where it blocks radiation from the sun. In one notorious case, a giant eruption of Tambora Volcano in Indonesia in April 1815 blew an immense amount of ash into the upper atmosphere. It hung there, above the altitudes of weather, for several years. Sulfur dioxide from the eruption caused white coatings on the fine ash that effectively reflected the incoming solar energy. The following year was known as "the year without a summer." Crops failed in New England, Britain, and elsewhere, and tens of thousands of people died of starvation.

Heavy ash falls are a threat to people who do not move out from under a dense plume of ash by heading at right angles to the wind direction. Roofs collapsing under the weight of fallen ash kill many people during some eruptions (▶Figures 6-19 and 6-20). The hazard can be especially severe in regions where people flock to their places of worship when an eruption begins. Twenty centimeters of ash is enough to collapse most roofs; less than that is required in warm regions where roofs are not designed to bear a load of snow. Wet ash is much heavier than dry ash. Rain falling during an eruption is common because ash eruptions commonly generate their own **volcanic weather.** The rising hot ash heats and lifts the surrounding air, which expands and cools, causing its dissolved water vapor to condense and fall as rain.

A heavy ash fall can quickly drop visibility to zero, turning day to night and making evacuation difficult or impossible. Even in a vehicle, headlights cannot penetrate the falling ash; roads and familiar landmarks disappear. Ash in the air or stirred up from the ground can clog a radiator and be drawn into an engine, causing it to seize up.

Many ash columns rise more than 6 kilometers and some more than 20 kilometers. The largest particles fall closer to the vent where they form dangerous projectiles. Finer particles are carried downwind; especially fine ash is carried high into the jet stream and carried around the world.

Airplanes blundered into fairly dense clouds of volcanic ash twenty-three times between 1976 and 1990. In June 1982, a Boeing 747 with 263 passengers and crew on a flight from Malaysia to Australia flew into an ash cloud erupting from a volcano in Java. All four engines failed. The plane fell from 11,470 meters to 4,030 meters before the crew was able to restart the engines. In December 1989, a Boeing 747 heading from the Netherlands to Japan with 245 passengers and crew flew into an ash cloud, lost power in all engines, and dropped from 7,500 meters to 3,500 meters in twelve minutes before the pilot was able to restart the engines. The plane managed to reach Alaska but with millions of dollars of damage. Ash entered the jet intakes, melted, filled the fuel injectors, and coated the turbine vanes.

Monitoring systems installed near many active volcanoes now warn of potential danger. Even so, airplanes still fly into clouds of volcanic ash. Pilots flying unexpectedly into a volcanic ash plume are instructed to slow the engines to lower their operating temperatures below the melting point of the ash and to fly back out of the plume.

Ash Flows, Glowing Avalanches, and Surges

Ash flows are mixtures of hot volcanic ash and steam that pour downslope because they are too dense to rise (▶Figure 6-21). Geologists variously call them **pyroclastic flows,** "nuee ardente," glowing avalanches, or ignimbrites. In this book, we simply call them ash flows.

Both steam and ash in the hottest ash flows are at a dull red heat, between 800° and 850°C. They glow in the dark. The ash particles in those flows are still extremely hot and "plastic," still hot enough to fuse into a solid mass as they come to rest. They become hard rocks, sheets of **welded ash** that may cover hundreds of square kilometers, all laid down during a single eruption, in a matter of hours.

M. Yount photo, USGS.

▶**FIGURE 6-21.** An ash flow races down the flank of Augustine Volcano in the Aleutians in April 1986. The ash flow proper hugged the ground near the bottom of the photo while loose ash billowed above.

▶**FIGURE 6-22.** Lenses of black obsidian mark densely welded parts of an ash flow in northern California west of Alturas.

Thick ash flow sheets can weld to become lenses (▶Figure 6-22) or even sheets of black obsidian sandwiched between sheets of much softer unwelded ash. The unwelded parts (▶Figure 6-23) cool enough in contact with the ground and the air that they do not weld.

Glowing hot ash flows can race down the flank of a volcano from 50 to more than 200 kilometers per hour, incinerating everything flammable in their path, including forests and people. Many flows develop when rapidly erupting steam carries a large volume of ash in a column that rises high above the volcano. When the rush of steam slows with greater height, part of the column collapses, and the cloud of ash pours down the flank of the volcano. The main flow hugs the ground, its less dense part billowing above as loose ash is stirred into the turbulent air above. Flows tend to hug valley bottoms, but their high velocity can carry them over intervening hills and ridges (▶Figure 6-24). In other cases, ash flows blow directly from the flank of the volcano in a lateral blast of steam and ash.

▶**FIGURE 6-23.** These layers of Minoan ash 10 meters thick (with surge cross beds in the lower part are in a quarry south of Thera, Santorini, Greece.

North   Flow   South

▶**FIGURE 6-24.** The directed blast of the May 1980 eruption of Mount St. Helens stripped and flattened trees on the windward side of a hill and snapped off their tops on the leeward side.

▶**FIGURE 6-25.** This forest was flattened by the lateral blast from the 1980 Mount St. Helens eruption. People circled in the lower right show the size of these large trees.

▶**FIGURE 6-26.** This car was singed, abraded, and crumpled by ash flow from Mount St. Helens 11 kilometers north of the crater.

In some cases, a high speed ash-rich shock wave called a **surge** may race across the ground ahead of an ash flow. Surges commonly pick rocks off the ground and carry them along, leaving in their wake a deposit of ash mixed with rocks. Surges deposit dunes of ash with cross beds much like those in sand dunes (▶Figure 6-23). Surges are generally less dense than standard ash flows because they contain a larger proportion of steam. They commonly originate as lateral blasts of ash and steam in the first stage of an ash-flow eruption. Some surges hug the ground at speeds up to 600 kilometers per hour and may cover hundreds of square kilometers. They flatten forests and kill almost every living thing they meet, by heat, abrasion, and impact. People engulfed in an ash flow face certain death unless they are near its outer fringes, preferably in a building or vehicle. Locations many kilometers from the base of the volcano may not be safe. Ash flows and surges move fast enough to flow over the tops of large hills to impact and incinerate anything flammable on the far side (▶Figures 6-24 and 6-28).

It is somewhat surprising to find that ash flows can cross rivers and bays or lakes. The lower part of an ash flow is the main flowing mass. Much less dense ash billows into the air above the descending flow (▶Figures 6-21 and 6-27). Although ash flows are too dense to rise into the air, only the main flow is dense enough to sink into water. The less dense ash cloud above it may skim across a body of water, even some that are tens of kilometers across, leaving a trail of floating pumice chunks bobbing in its wake. During the eruption of Vesuvius in 79 A.D., a glowing ash flow crossed some 30 kilometers of the Bay of Naples. Similarly, 2,000 people died in hot ash flows that crossed more than 40 kilometers of open sea in southeastern Sumatra during the eruption of Krakatau in 1883. In 1902, ash flows raced down the slopes of Mount Pelée in Martinique and continued offshore to capsize and burn boats and ships anchored in the harbor. It seems that a body of water is not an effective safety barrier.

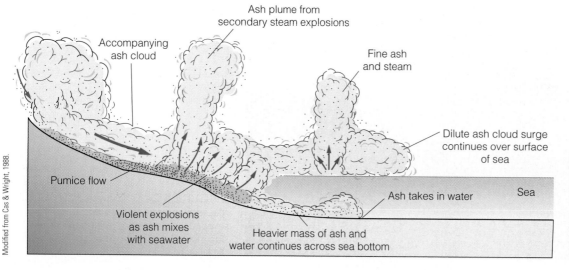

▶**FIGURE 6-27.** One explanation for the movement of hot ash flows over water is that the dense part of a pumice flow sinks as the lighter part continues over the water surface.

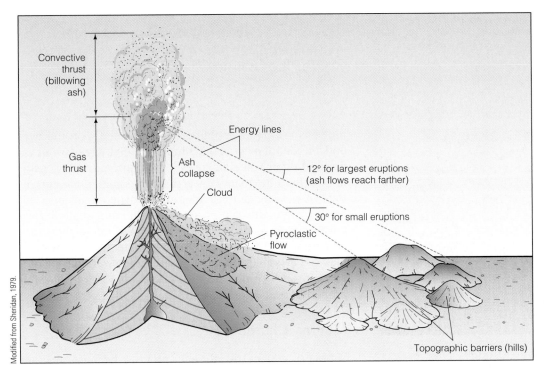

Modified from Sheridan, 1979.

Convective thrust (billowing ash)

Gas thrust

Energy lines

Ash collapse

Cloud

Pyroclastic flow

12° for largest eruptions (ash flows reach farther)

30° for small eruptions

Topographic barriers (hills)

▶ **FIGURE 6-28.** An energy line sloping from the top of the heavy "gas-thrust zone" of an eruption can estimate the height of hills that might be surmounted by an ash flow from a stratovolcano eruption. The energy line has a lower slope from larger, more ash-rich eruptions, so their ash flows reach farther from the vent.

Because ash flows pour downslope, the most dangerous place to be is in a valley bottom. Given their high velocity, inertia typically carries them over hilltops. Thus an intervening hill may not offer much protection. You can estimate the travel distance of an ash flow and which hills it might cross by imagining an **energy line** sloping downward from the top of the gas-thrust zone of the eruption (▶ Figure 6-28). For a small ash eruption, the slope is approximately 30 degrees. For the largest Plinian eruptions, the slope may be as low as 12 degrees, so the danger zone extends much farther from the volcano. The gas-thrust zone is the ash-rich part of the eruption column carried upward by violently expanding steam in the magma. The convective-thrust zone is

the billowy cloud of ash carried to still higher levels by air drawn in and heated by the hot ash. The best place to be when an ash flow forms is somewhere far away.

Ash after deposition forms a rock called **tuff.** Ash-fall material becomes ash-fall tuff; ash-flow material becomes ash-flow tuff. Either may contain lumps of **pumice,** the frothy rock that is typically light enough to float on water. In assessing the potential hazards from a volcano, it is important to be able to distinguish deposits of ash-fall tuff from ash-flow tuff. Ash flows are among the most hazardous products of an eruption. The most obvious distinction is that **ash-fall tuff** is generally distinctly layered, whereas **ash-flow tuff** is unlayered (▶ Figures 6-29 and 6-30; see also Figures 6-31

Peter Francis photo, USGS.

▶ **FIGURE 6-29.** This thick layer of air-fall pumice lies under thin layers of air-fall ash in Arequipa, Peru.

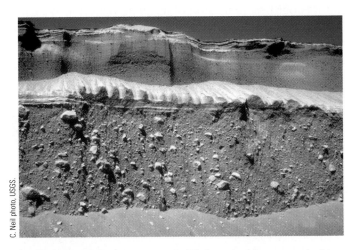

C. Neil photo, USGS.

▶ **FIGURE 6-30.** These new ash-fall deposits (the gray and white upper third of the photo) from Redoubt Volcano lay over the top of an ash flow deposit containing big lumps of pale pumice. The shovel at the bottom provides scale.

▶FIGURE 6-31. Ash falls, surge deposits, and ash flows can be distinguished by their different distributions over hills.

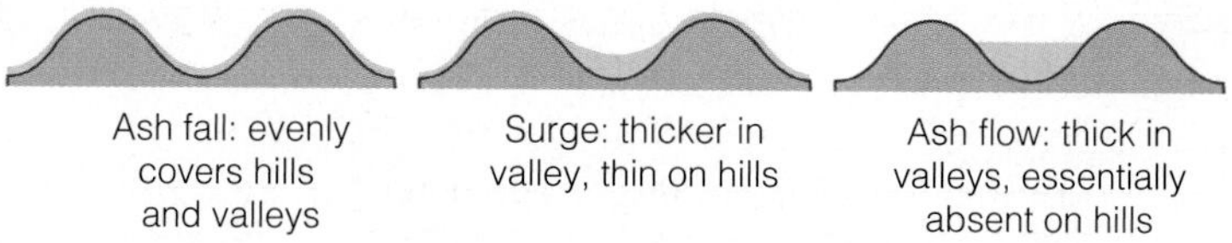

Ash fall: evenly covers hills and valleys

Surge: thicker in valley, thin on hills

Ash flow: thick in valleys, essentially absent on hills

William E. Scott photo, USGS, 1991.

▶FIGURE 6-32. Ash-flow deposit from the 1991 Mount Pinatubo eruption filled the valley bottom of Marella River. The foreground vegetation was stripped by hot ash in the upper part of the same ash flow. Ash coating surfaces in the foreground either settled from the billowy upper part of the ash flow or settled later from air-fall ash.

and 6-32). Other distinguishing characteristics are listed in Table 6-2.

Magma Chambers and the Driving Force behind Eruptions

The past behavior of a volcano can be extremely variable, yet it is our best indication of its future behavior. We can learn from eyewitness accounts as well as from interpretation of the erupted materials. No two volcanoes are quite alike; nor are any two eruptions at the same volcano. What

happens during an eruption depends mainly on the type and amount of magma that erupts, the quantity of water vapor and other volcanic gases it contains, and its viscosity, or how fluid the magma is. It is not possible in the current state of knowledge to predict exactly when a volcano will erupt, or in many cases whether it will erupt at all. But we have learned much about the history of many volcanoes and monitored the activity of many others. That knowledge makes it possible to assess the likely future behavior of a volcano and to mitigate the hazards it poses to life and property. Great strides have been made in forecasting.

Magma Chambers

Masses of magma rise though the crust because they are less dense than the surrounding rocks. Magmas may even rise into rocks of lower density, much as the ice of an iceberg rises into the air because ice has lower density than the water surrounding it (see also Figure 2-11, page 19). As long as the column of magma in the Earth is less dense than the surrounding rock, the magma will float in the rock. Magma may rise through cracks, sometimes breaking off and incorporating pieces of the adjacent rocks. **Magma chambers** are large masses of molten magma in the Earth's crust. They may reveal their presence in the behavior of earthquake waves, minute irregularities in the Earth's gravity field, and a slight rise or tilt of the ground surface. Erosion eventually exposes and dissects old magma bodies that are now crystallized into masses of solid igneous rock.

Magma bodies may be of almost any shape. Idealistic, unrealistic drawings often show them as spheres or large masses shaped like inverted droplets tapering downward. Some, particularly large masses of granite or rhyolite, may be broad, thinner masses. Others, particularly magmas of

Table 6.2 Distinguishing Features of Rhyolitic Ash-Fall and Ash-Flow Deposits	
Rhyolite Ash-Fall Tuff	**Rhyolite Ash-Flow Tuff**
Thin layers on a scale of centimeters (Figures 6-29 and 6-30)	Unlayered but may be layered farther from vent (Figure 6-30) Surge part is commonly cross-bedded (Figure 6-23)
Coarser particles toward base	Pumice fragments generally coarsen upward in layer
Ash evenly coats hills and valleys (Figure 6-31)	Thicker in valleys, thin or absent on hilltops (Figures 6-31 and 6-32)
Not normally welded	Thick flows may be welded (fused to layers or streaks of black obsidian) just above base of the flow, mainly within a few tens of kilometers of the base of the volcano (Figure 6-22)
Deposits generally thicker near volcano but extend far downwind	

large basalt volcanoes, may be intersecting sheets of magma filling fractures having a variety of orientations, some parallel to old lava flows, some at other angles.

It is common to find that a large mass of magma has crystallized into rocks of quite different compositions, a group of complex phenomena known as **magmatic differentiation.** Geologists have proposed many ways in which magma may differentiate to form different compositions. Most likely many quite different processes play a role. Differentiation may explain why some stratovolcanoes produce different kinds of lava during a single eruption.

One well-documented process involves the crystallization of high-temperature, relatively insoluble minerals early in the magma's crystallization. In most cases, those crystals are also denser than the remaining melt so they can settle to the bottom of the magma chamber, leaving the remaining melt composition changed. The differentiation process most relevant to volcanic hazards involves volatiles such as water, carbon dioxide, and sulfur oxides. These can seep upward to concentrate in the upper parts of a magma chamber. Here they may remain as a time bomb, ready to explode violently in the next volcanic eruption. Volatiles remain dissolved in the magma at high pressures but can separate as gases as the magma rises toward the Earth's surface.

Explosive Eruptions

The more viscous the magma and the greater its gas content, the more likely it is to explode. Steam bubbles separating from an expanding mass of rhyolite or pale andesite magma expand to foam it into a frothy mass of glassy bubbles, pumice. The bubbles expand as the magma continues to rise until they burst into ash, which consists mostly of curving shards of glass, formerly the walls of bubbles. When the pumice bursts into ash, the steam escapes, and the whole mass explodes into a cloud of volcanic gases and suspended ash (▶Figures 6-1, 6-18 6-33, and 6-34). (Note that the terms **tephra** and **pyroclastic** mean substantially the same thing. *Pyro* is derived from "fire" in Greek, and *clastic* means "pieces"—thus, pieces of hot rock.) Most basalt magmas do not explode because they are fluid and contain modest amounts of gas, mainly water and carbon dioxide. It bubbles harmlessly out of the fluid melt.

Volcanic Mudflows

Mudflows pour down the flanks of stratovolcanoes at the full range of highway speeds and with densities resembling those of wet concrete. They are extremely dangerous. **Stratovolcanoes** are notorious for being covered with ash and broken rocks. In many cases, the amount of snow and ice on the volcano becomes critically important, because the hot ash can rapidly melt or mobilize the ice to create a slurry of water and ash that becomes a mudflow (▶Figure 6-35). An erupting stratovolcano not only coats the volcano with large amounts of hot ash but also creates its own

▶**FIGURE 6-33.** This explosive eruption of St. Helens on July 22, 1980, pushed an ash cloud to a height of 14 kilometers, dwarfing the mountain. However, it released only 1 percent of the energy released in the climactic eruption of May 18.

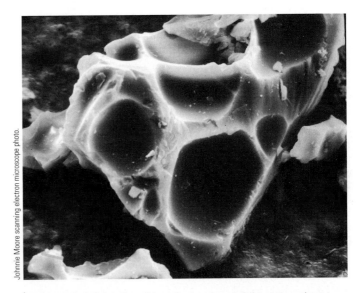

▶**FIGURE 6-34.** These bits of volcanic ash fell in Missoula, Montana, 800 kilometers downwind from Mount St. Helens in 1980. Magnified 3,000 times.

▶**FIGURE 6-35.** In this March 21, 1982, view from the north, a dark mudflow can be seen running from the crater of Mount St. Helens.

▶**FIGURE 6-37.** This village was buried in a mudflow after the eruption of Mount Pinatubo in 1991. Air-fall ash still mantles the roof on the right.

▶**FIGURE 6-36.** Mudlines high on tree trunks show the depth of the Toutle River mudflow on May 18, 1980. The person in yellow on the right is almost six feet tall.

▶**FIGURE 6-38.** Mudflows from Mount Pinatubo buried these local stores up to their rooftops.

▶**FIGURE 6-39.** Mudflows from Mount Unzen in Japan completely buried the first story of many houses.

weather because of the rapidly rising heat plume. Even on an otherwise clear day, the eruption cloud may create thunderstorms and heavy rain. Rain on loose ash washes it down slope as a heavy slurry—a mudflow.

Volcanic mudflows typically start as water mixes with volcanic ash to make a thick slurry with enough buoyancy to help lift large rocks to the surface of the flow. As the mud moves downslope, it gathers rocks of all sizes and carries them along, accelerating as it goes. Many mudflows carry rocks the size of cars. The only apparent limit to the size of rocks they carry is the size of available rocks. Mudflows are truly ugly. Racing down valleys at the velocities of deep floodwaters, they spread over the lower slopes where people live. They killed roughly 80 percent of the more than 28,000 people who died in volcanic eruptions during the 1980s. And they caused untold millions in property damage (▶ Figures 6-36 to 6-39).

Water pouring down the slope of a stratovolcano may develop into a mudflow whether or not the volcano is erupting. That explains why volcanic mudflows come in the

complete range of temperatures from ice water to boiling. Technically, a hot mudflow is called a **lahar,** though many volcanologists use the term for any mudflow. Big blocks of rhyolite magma buried within a hot mudflow may blow steam for months before they finally cool. Steam within a hot mudflow alters the ash and rocks within the mudflow, sometimes for months, converting them to clay and in many cases staining them red with iron oxide.

The best way to avoid volcanic mudflows is to stay out of the floors of stream valleys that drain away from volcanoes. Trying to outrun a mudflow is almost certainly suicidal. Climbing well up the valley side is a much better idea. Broad floodplains are especially dangerous because the valley sides may be too far away. Indonesia, a country with many volcanoes, has built safety hillocks several meters high where people can hopefully climb above the mudflows.

Warning systems high in valleys on the flanks of active volcanoes use temperature sensors to detect the passage of hot mudflows and pass the word downslope. New Zealand, for example, has used them for many years on ski slopes on the flank of Ruapehu, an active volcano. Most towns develop in valley bottoms; these are the very places that mudflows follow as they move down slope. The U.S. Geological Survey installed temperature sensors on the flank of Rainier to provide warning for the people of Orting, who live in a broad valley below its northwest flank (Figure 1-2, page 3).

Mudflows are covered in more detail in Chapters 8 and 9 on landslides.

Poisonous Gases

Carbon dioxide is a familiar gas in the air we breathe but deadly to people, animals, and trees in high concentrations. The high concentration expelled from some volcanoes can pour downslope where it concentrates in depressions. Because it is colorless and odorless, it can suffocate people and other animals without warning.

A tragic case in 1986 in Cameroon, West Africa, involved carbon dioxide that bubbled out of Lake Nyos, a volcanic crater lake some 200 meters deep. The carbon dioxide had seeped into and dissolved in the deeper waters of the lake over any years. On the evening of August 21, 1986, the carbon dioxide–saturated deep waters of the lake rumbled loudly and belched out an immense volume of the gas. Heavier than air, it swept rapidly down through several small villages as a stream 50 meters thick and 16 kilometers long. More than 1,700 people, some 3,000 cattle, and uncounted numbers of small animals died of asphyxiation. Ten percent carbon dioxide in the air can kill people; this was nearly 100 percent of the air. A rainy season landslide into the lake, pushing some of the deep water upward, may have prompted overturn of the lake waters. Drop in pressure on the dissolved carbon dioxide would have rapidly released it from solution.

In another example, carbon dioxide seeping from within the Long Valley Caldera north of Bishop, California, has killed trees in several areas around Mammoth Mountain on the rim of the caldera (▶ Figures 6-40 and 6-41). On March 11, 1990, Fred Richter, a forest ranger found refuge from a blizzard in an old cabin surrounded by snowdrifts. Entering the

K. McGee photo, USGS.

▶**FIGURE 6-40.** Carbon dioxide gas killed these trees on the south side of Mammoth Mountain, California, in September 1996.

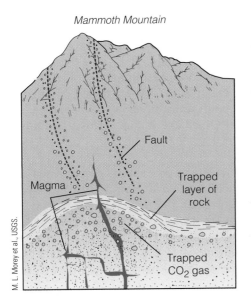

M. L. Morey et al., USGS.

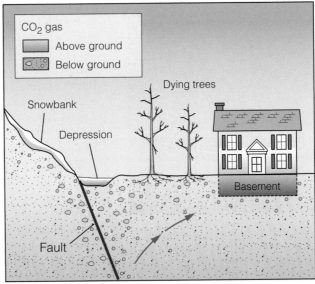

▶**FIGURE 6-41.** Carbon dioxide gas rises along faults from molten magma at depth.

cabin through a hatch, he almost suffocated before fighting his way back up the ladder into fresh air. The denser-than-air carbon dioxide had collected and concentrated in the nearly sealed cabin. Although people walking through an area are probably not in danger, someone camping in a depression out of the wind or entering a confined space below ground could be asphyxiated. The trees, located near caldera faults, often succumb to the carbon dioxide concentrating around their roots.

Some eruptions produce large quantities of sulfur dioxide, a noxious gas with a sharply acrid smell and choking effect (▶Figures 6-42 and 6-43). Sulfur dioxide is harmful to animals and extremely poisonous to plants, even in small concentrations. It reacts with oxygen in the atmosphere to make sulfur trioxide, which reacts with water vapor to make minute droplets of sulfuric acid. They hang suspended in the atmosphere for months, partially blocking the incoming sunlight and thus cooling the climate.

▶**FIGURE 6-43.** Volcanologists sample gases on Mount St. Helens.

▶**FIGURE 6-42.** Volcanic gases in the crater of Ngaruhoe Volcano, New Zealand, deposit bright yellow coatings of sulfur (view from above). The crater floor is roughly 5 meters wide.

Some volcanoes also erupt poisonous hydrogen sulfide, various chlorine compounds, and small quantities of many other gases. Most have little to do with events during an eruption, though fluorine breaks bonds and can make magmas less viscous.

The tremendous pressures at depths of more than a few kilometers confine gases dissolved in the magma. The pressure drops as the magma rises, and that permits its gases to **exsolve,** or come out of solution. Thus, an increasing volume of escaping gases commonly precedes and may warn of an impending eruption.

Volcanic Behavior: Relationships to Viscosity, Volatiles, and Volume

VISCOSITY The nature of a volcanic eruption depends heavily on the composition and therefore the viscosity of the magma and its content of water and other gases. The spectrum ranges from low-viscosity dark basaltic magmas that are low in water to high-viscosity pale rhyolitic magmas that have more water. See also Appendix 2, Rocks.

Basalt is black and fluid, by the sluggish standards of lavas, about as fluid as cold molasses. It is fluid enough to let the small amounts of water and carbon dioxide that it may contain escape easily without causing much commotion. Basalt erupts at temperatures of 1,100° to 1,200°C, a bright red heat. It generally pours down stream valleys or spreads across flat ground like pancake batter across a griddle, making relatively thin flows that commonly cover dozens of square kilometers.

Rhyolite comes in white or pale shades of gray, yellow, pink, green, and lavender. Rhyolite lava is extremely viscous, causing it to flow stiffly if at all. Obsidian flows are rare; the few that exist are so viscous that thick flows move and solidify to form flow fronts 50 to 100 meters high and 45-degree slopes (▶Figure 6-44). Rhyolite may contain almost no water or as much as 10 percent water by weight

▶**FIGURE 6-44.** The steep front of the Big Obsidian Flow at Newberry Volcano south of Bend, Oregon, is visible in the center of the photo. The steepness emphasizes the viscosity of the magma.

and many times that by volume. Rhyolite erupts at temperatures between 800° and 900°C, a dull red heat. If rhyolite lava reaches the surface nearly devoid of water, it erupts quietly as a bulging dome of rhyolite that may reach the size of a small mountain. If it arrives at the surface with a heavy charge of water, rhyolite lava explodes into horrible clouds of steam full of foamy pumice and white rhyolite ash.

Andesite, as commonly used, covers most of the range of compositions between basalt and rhyolite, rocks involving a complete spectrum of grays between those extremes. The viscosities of andesite lavas range almost from the low viscosity of basalt to the high viscosity of rhyolite. The gas content of the dark andesite lavas is generally quite low. Dark andesite lavas generally erupt in sluggish lava flows considerably thicker than basalt flows or they blow out chunks of cinders. Pale andesites tend to erupt more vigorously to make fields of broken rubble and clouds of ash (Table 6-3).

The **viscosity** of a magma depends on its chemical composition and therefore its internal arrangement of atoms and molecules. By far the most abundant atoms are oxygen and, to a lesser extent, silicon and aluminum. Almost

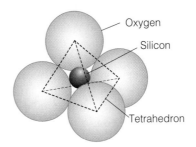

▶**FIGURE 6-45.** Four oxygen atoms surround each silicon atom to make a silica tetrahedron.

every silicon atom in its natural state is surrounded by four oxygen atoms arranged in the shape of a tetrahedron (▶Figure 6-45). All silicate rocks and minerals, including the common volcanic rocks, consist of an array of **silica tetrahedra** that generally are linked to atoms of aluminum, iron, magnesium, calcium, potassium, sodium, and other elements. The chemical bonds between silicon and oxygen atoms are too strong to bend or break easily. Silicate structures are rigid, like an assemblage of Tinkertoy parts.

None of the common magmas contain enough silicon atoms to consist only of silica tetrahedra. The dark magmas

Table 6.3   Colors, Compositions, Viscosities, and Water Contents of the Common Lavas*

Basalt to	Andesite to Dacite to	Rhyolite
Black	Dark shades of gray and other intermediate colors	Pale colors, including white, pink, yellow
Magnesium, iron, and calcium-rich and silica-poor		Potassium and silica-rich
Fluid (low viscosity)		Extremely viscous
Mostly lava flows	Lava flows, ash, broken fragments	Mostly ash
Little water content unless magma encounters groundwater. May contain significant carbon dioxide.		Generally lots of water

*All of these rock types are gradational.

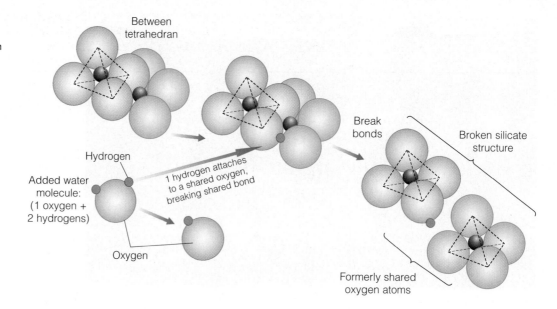

FIGURE 6-46. A water molecule reacts with a silicate structure and its shared oxygen atoms.

Between tetrahedran

Hydrogen

Added water molecule: (1 oxygen + 2 hydrogens)

1 hydrogen attaches to a shared oxygen, breaking shared bond

Oxygen

Break bonds

Broken silicate structure

Formerly shared oxygen atoms

and rocks fall especially short. Their fewer silica tetrahedra are either not linked directly to others or more commonly share two oxygen atoms to form long chains of tetrahedra. The tetrahedra or chains link to others with weaker bonds provided by charged atoms of calcium, magnesium, and iron. The chains of tetrahedra can wriggle about in the magma like worms in a can. Their freedom of motion explains why the dark magmas are so fluid.

Pale magmas (Table 6-3) contain more silicon atoms to attach to oxygen atoms. Most of the silica tetrahedra share oxygen atoms to build frameworks of tetrahedra linked in all directions. The numerous strong silicon–oxygen bonds make the framework exceptionally rigid. The other atoms, including aluminum, sodium and potassium, do not provide much freedom of movement so pale magmas are extremely viscous.

Geologists speak of water dissolving in magmas, but things are not quite that simple. Some of the water that enters molten magmas reacts chemically at the position of one of the shared oxygens. The oxygen comes off the water molecule (H_2O) with one of the hydrogen atoms still attached, while the remaining hydrogen atom attaches itself to one of the shared oxygen atoms (▶Figure 6-46). That breaks the silicate structure at the point of the reaction, thus reducing its rigidity and decreasing the viscosity of the magma. Thus magmas rich in water are less viscous than dry magmas. When the magma cools and crystallizes, that reaction reverses, and the water again separates.

Rhyolite magmas are extremely sensitive to water, which greatly decreases the viscosity of the magma as it breaks bonds between oxygen and silicon atoms. Addition of 2 percent water, by volume, decreases the viscosity of rhyolite more than 100 times. Water has much less effect upon the viscosity of basalt magma because basalt contains many fewer shared oxygens.

VOLATILES Many large stratovolcanoes emit a plume of white steam that quietly drifts out of the summit crater like smoke from a factory chimney. In most such cases, the steam is simply meltwater from snow on the high slopes that boils as it sinks into the hot rocks below. Although it probably does not portend an eruption, it does tell us that the interior of the volcano is extremely hot. The rocks in there are stewing in the hot water and steam that fills the fractures. Fresh andesite comes in all shades of gray from almost black like basalt to almost white like rhyolite. But most ancient andesite comes in medium shades of brown, red, and even green. When steam alters the original minerals to clays and related minerals, the new colors form by oxidation of iron in the rocks.

Water, the most abundant gas in many magmas, is mainly what inspires volcanoes to violence. Magmas that contain little water erupt quietly as lavas. Water at the high temperature of a magma expands as steam at low pressures near the Earth's surface. Those that contain large amounts of dissolved water are likely to explode unless they are fluid enough to let the steam fizz quietly away. Steam cannot easily escape extremely viscous magmas (Table 6-3).

Among **volcanic gases,** water is the most important in governing what happens when the magma erupts. Under volcanic conditions, water in magmas exists mostly as steam—not familiar teakettle steam, but as steam so hot that it glows and instantly ignites any flammable thing it may touch.

The amount of water a magma can contain decreases dramatically as it rises into levels where the rock pressure is lower. A rhyolite magma with 2 percent water at a depth of 20 kilometers could contain only half that much at a depth of 2 or 3 kilometers. Carbon dioxide separates from magmas at more than twice the depth at which water separates. By the time the magma reaches the low pressure of the Earth's

Table 6.4 General Characteristics of Common Volcanoes

Volcano Type	Viscosity	Slope	Volatiles	Volume
Shield volcano (basalt)	Low	Gentle	Low	Large to giant
Cinder cone (basalt)	Low	Steep	Moderate	Small
Stratovolcano (andesite)	Moderate	Moderate	Moderate to high	Moderate
Lava dome (rhyolite)	High	Steep	Low to moderate	Small
Continental caldera* (rhyolite)	High	Very gentle	Moderate	Giant

*A volcanic caldera is formed by collapse or subsidence of the overlying rock, generally into an underlying magma chamber as magma is withdrawn during an eruption. A volcanic crater, in contrast, is formed by blasting material out of the volcanic vent. Craters are generally smaller—less than 2 kilometers across.

surface, it can dissolve virtually no water or other gases; any gases that were dissolved at depth must separate from the magma and bubble out. It is those separating gases that drive explosive volcanic eruptions.

Even if magma stops rising, steam can separate and build up pressure to drive an eruption. Many of the new crystals growing in the magma contain little or no water. As they grow, they displace the water or steam in the magma into an ever-smaller volume under a steadily rising pressure that may eventually drive an eruption.

Carbon dioxide is normally second to water in abundance among the volcanic gases, but it has much less influence on the explosive nature of an eruption. It is relatively more abundant in the dark magmas than in those closer to rhyolite in composition.

VOLUME OF MAGMA The volume of magma expelled in a single eruption has a significant bearing on the degree of hazard. It affects the size of a lava flow, the volume and time span of an ash eruption, and the size of the volcano produced. Recognition of various types of volcanoes from their size and the slopes of their flanks permits us to interpret their behavior, even from a distance. Table 6-4 provides an overview of these characteristics.

KEY POINTS

✓ Landslides sometimes trigger explosive volcanic eruptions by removing load from above the magma and releasing pressure on its dissolved gases. **Review pp. 128–130; Figure 6-1.**

✓ The lateral blast of the initial stages of an eruption can move outward at as much as 250 kilometers per hour and level an adjacent forest. **Review pp. 130–131.**

✓ Precursor signs for the Mount St. Helens eruption included swarms of small earthquakes and harmonic tremors, steam blasts, opening fractures, burning methane, growth of a bulge, and small ash eruptions. **Review pp. 127, 130–133.**

✓ Basalt lava flows can be *pahoehoe*, with smooth, sometimes ropy-looking tops; or *aa*, with ragged, clinkery tops. **Review pp. 134–136; Figures 6-12 to 6-16.**

✓ Volcanic ash can deposit as air-fall ash drifts down like snow from eruptive columns as much as 6 kilometers high. **Review pp. 137–138; Figures 6-29 and 6-30.**

✓ Ash flows, glowing avalanches, and surges form when ash-laden air collapses onto the flank of a volcano to race downslope at 50 to 200 kilometers per hour, hugging valley bottoms but able to climb over ridges and even cross bays or lakes. **Review pp. 138–142; Figures 6-21, 6-24, and 6-27.**

✓ An ash-rich surge can blast out ahead of the ash flow at velocities as high as 600 kilometers per hour to flatten forests. **Review p. 140; Figures 6-24 and 6-25.**

✓ An imaginary energy line sloping down from the gas-driven part of the eruption cloud gives an idea of how far an ash flow might extend and what hills it might cross en route. **Review p. 141; Figure 6-28.**

✓ Ash-fall deposits are distinguished by thin layers, coarser particles toward the base of a layer, and even thickness over hills and valleys. Ash-flow deposits show thick or no layers, coarser particles toward the top of a layer, and greatest thickness in valley bottoms. **Review pp. 141–142; Figures 6-29 to 6-31; Table 6-2.**

✓ Magma chambers can be segregated into different compositions by processes of magmatic differentiation, including settling of heavier mineral grains or venting of dissolved magmatic gases. **Review p. 143.**

✓ Mudflows form when volcanic ash becomes saturated with water and pours down nearby valleys.

These kill more people than any other volcanic activity. **Review pp. 143–145; Figures 6-35 to 6-39.**

✓ Poisonous gases venting from magmas include carbon dioxide, sulfur dioxide, and related gases. **Review pp. 145–146; Figures 6-40 to 6-43.**

✓ The viscosity, or fluidity, of magma depends on its silica content. **Review pp. 146–148; Figures 6-44, 6-45, and 6-46; Table 6-3.**

✓ Water and other volatiles, along with the viscosity of a magma, affect its explosivity. Because magma can dissolve a lower percentage of volatiles at lower pressure, volatiles separate from the magma as it approaches the surface. It is those separated gases that drive explosive eruptions. **Review pp. 148–149.**

✓ The volume of magma erupted during a single eruption, along with the viscosity of the magma and its volatile content, affects the size and slope of a volcano. **Review p. 149; Table 6-4.**

IMPORTANT WORDS AND CONCEPTS

Terms

aa, p. 134
andesite, p. 147
ash fall, p. 132
ash-fall tuff, p. 136
ash flows, p. 138
basalt, p. 146
bulge, p. 132
carbon dioxide, p. 145
energy line, p. 141
eruption cloud, p. 132
exsolve, p. 146
fissure, p. 136
flood basalt, p. 136
harmonic tremors, p. 132
Hawaiian type lava, p. 134
lahar, p. 144
lava, p. 131
magma chamber, p. 142
magmatic differentiation, p. 143

mudflows, p. 143
pahoehoe, p. 134
pillow basalt, p. 136
Plinian-style eruption, p. 131
pumice, p. 141
pyroclastic, p. 143
pyroclastic flows, p. 138
rhyolite, p. 146
silica tetrahedra, p. 147
stratovolcanoes, p. 143
surge, p. 140
tephra, p. 143
tuff, p. 141
viscosity, p. 147
volcanic ash, p. 137
volcanic gases, p. 148
volcanic weather, p. 138
welded ash, p. 138

QUESTIONS FOR REVIEW

1. What factors control the violence or style of an eruption?

2. What properties of basalt magma control its eruptive behavior?

3. What properties of rhyolite magma control its eruptive behavior?

4. What drives an explosive eruption?

5. How does pahoehoe lava differ from aa lava?

6. How fast do basalt flows typically move?

7. How much ash on a roof would generally cause it to collapse?

8. Which is more dangerous to a person—an ash fall or an ash flow? Why?

9. Why is erupting volcanic ash dangerous to jet aircraft flying at an altitude of 8 to 10 kilometers (30,000 feet)?

10. What causes a big bulge to slowly grow on the flank of an active Cascades volcano?

11. If you visit Mount St. Helens, Washington, you will see thousands of trees lying on the ground, all parallel to one another. Explain how they got that way.

12. If an ash flow approaches you from across a kilometer-wide lake, are you likely to be safe or not? Explain why.

13. What characteristics of an old ash-fall tuff will permit you to distinguish it from an old ash-flow tuff?

14. Which of the hazards of erupting volcanoes kill more people than anything else? Why are they so dangerous?

15. What can happen to heat, pressure, and water content to melt rock and create magma?

FURTHER READING

Assess your understanding of this chapter's topics with additional quizzing and conceptual-based problems at:

 http://earthscience.brookscole.com/hyndman.

Rainier is heavily cloaked with snow and ice. Debris avalanches and mudflows in downstream valleys are the greatest threat. The community of Orting, pictured here, is built on one such old mudflow (see Figure 7-20).

VOLCANOES
Types, Behavior, and Risks

Volcanoes can commonly be classified at a distance from their appearance, primarily their size and the slopes of their flanks (▶see Figure 7-1). Knowledge of volcano type can be used to infer the magma composition that produced it and its volatile content; both affect the types of associated hazards and risks. In this chapter, we describe the main differences between volcano types, using examples to illustrate the range of behaviors and hazards.

Basaltic Volcanoes

Shield Volcanoes

Many basalt eruptions within a small area eventually build a pile of thin basalt flows. More than a century ago, geologists decided to call those piles **shield volcanoes** because of a fancied resemblance to the shape of an ancient Roman shield (▶see Figures 7-2 and 7-3). The flows are characterized by their low viscosity, their low volatile content, and their large to giant size.

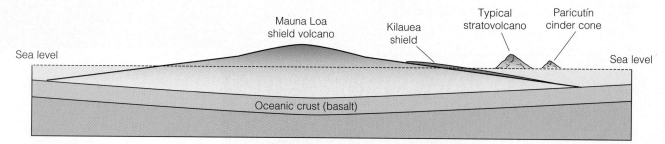

CHAPTER 7

▶**FIGURE 7-1.** The different types of volcanoes have dramatically different sizes or volumes. Mauna Loa, the giant shield volcano in Hawaii, is roughly 220 kilometers in diameter and rises some 9,450 meters above the sea floor. A typical stratovolcano is 2,000 to 3,000 meters high on a base near sea level.

Robert I. Tilling photo, USGS.

▶**FIGURE 7-2.** The gentle slopes of snow-capped Mauna Loa are typical of giant shield volcanoes. Its 4,169 meters above sea level is less than half its total height.

Shield volcanoes are broad and have gently sloping sides. Most of the lava of shield volcanoes erupts not from the summit but from the crests of three broad, equally spaced ridges that radiate outward from the volcano summit. Gravity pulling down on each ridge causes it to spread slightly, which causes rifting of the ridge crest. Lava rising within a rift spreads out and flows down the flanks of the ridge (▶Figures 7-3, 7-4, and 7-5). The shield volcano may have a **collapse caldera** in the summit where the surface rocks subside into the magma chamber below. Many have cinder cones or lava cones along the crests of the three main **eruptive rifts** that radiate outward from the summit, sometimes flowing downslope to populated areas (▶Figures 7-3 and 7-7). The largest volcano on Earth, Mauna Loa in Hawaii, rises 9,450 meters (31,000 feet) above its base on the floor of the Pacific Ocean, 4,270 meters (14,000 feet) above sea level. Somewhat smaller basaltic shield volcanoes include Newberry Volcano in central Oregon and Medicine Lake Volcano in northernmost California, both just east of the Cascades. Small shield volcanoes may cover as little as 10 to 15 square kilometers and rise as little as 100 meters above the surrounding countryside.

NASA astronaut photo.

▶**FIGURE 7-3.** The big island of Hawaii has three major volcanoes: Mauna Loa, Mauna Kea, and Kilauea (see Figure 7-4 for locations). Lava flows erupted from Mauna Loa's southwest rift zone are visible as dark fingers in the lower left part of the photo.

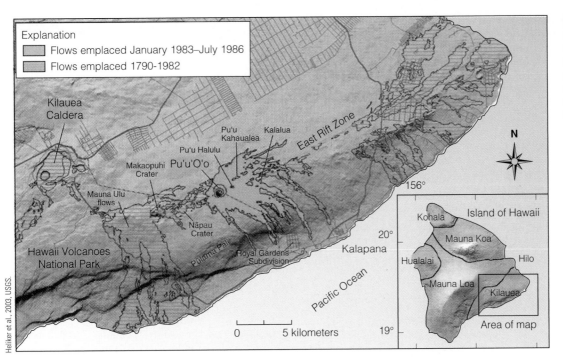

▶**FIGURE 7-4.** This map of Kilauea shows the summit caldera, southwest and east rift zones that erupt most of the lava flows, and major fault zones on which the south side is sliding toward the ocean.

Explanation
Flows emplaced January 1983–July 1986
Flows emplaced 1790-1982

Kilauea Caldera

Pu'u Kahaualea
Kalalua
East Rift Zone

Pu'u Halulu
Makaopuhi Crater
Pu'u'O'o

Mauna Ulu flows
Nāpau Crater

Hawaii Volcanoes National Park
Pulama Pali
Royal Gardens Subdivision

Kalapana
Pacific Ocean

0   5 kilometers

156°
20°
19°

Island of Hawaii
Kohala
Mauna Koa
Hualalai
Hilo
Mauna Loa
Kilauea
Area of map

Heliker et al., 2003, USGS.

National Park Service photo.

▶**FIGURE 7-5.** Lava erupts along the southwest rift of Kilauea in September 1971.

J. D. Griggs photo, USGS.

▶**FIGURE 7-6.** Eruptions of Pu'u O'o crater, Kilauea volcano, Hawaii, send basalt lava flows downslope.

D. Weisel, USGS.

▶**FIGURE 7-7.** The May 2, 1990, lava flow from Pu'u O'o cone on Kilauea Volcano invaded a road and park.

Table 7-1 Volcanic Explosivity Index (VEI) for Different Types of Individual Eruptions*

VEI	Volume of Ejecta (m³)	Eruption Column Height (Km)	Eruption Style	Duration of Continuous Blast (Hrs.)	Eruption Frequency (Approximate)	Example Eruption
0	<10⁴	<0.1	Hawaiian	<1		Kilauea, Hawaii
1	10⁴–10⁶	0.1–1	Hawaiian Strombolian	<1	100 per year	Kilauea, Hawaii Stromboli, 1996
2	10⁶–10⁷	1–5	Strombolian Vulcanian	<1	15 per year	Unzen, Japan, 1994
3	10⁷–10⁸	3–15	Vulcanian	1–6	2–3 per year	Nevado del Ruiz, Columbia, 1985
4	10⁸–10⁹ (0.1–1 km³)	10–25		6–12	1/2 year	El Chichon, 1982 Papua New Guinea, 1994
5	10⁹–10¹⁰ (1–10 km³)	>25	Plinian	>12	1/10 years	Mount St. Helens, 1980
6	10¹⁰–10¹¹ (10–100 km³)	>25	Plinian	>12	1/ 40 years	Krakatau,1883, Pinatubo, 1991 Thera (Santorini), 1600 B.C.
7	10¹¹–10¹² (100–1,000 km³)	>25	Plinian	>12	1/200 years	Tambora, Indonesia, 1815
8	>10¹² (1000–10,000 km³)	>25	Yellowstone	(off scale)	1/2,000 to 1/1,000,000 years	Yellowstone, 600,000 B.C. Long Valley, California, 730,000 B.C. Taupo, New Zealand, 186 B.C.

*Modified from Newhall & Self, 1982.

MAUNA LOA AND KILAUEA: BASALT GIANTS OVER AN OCEANIC HOTSPOT

Mauna Loa. Mauna Loa, Hawaii's largest volcano, has erupted almost forty times since 1832, fifteen times since 1900 (refer to Figures 7-3 and 7-4, and Table 7-1). Recent eruptions include those in 1926, 1940–1942, 1949–1950, and 1975. Seven of its flows, including one in 1984, came within 6.5 kilometers of Hilo. Eruptions are often classified according to a Volcanic Explosivity Index (VEI) that attempts to quantify eruption size based on volume of ash erupted, height of the eruption column, eruption duration, and eruption frequency (see Table 7-1; see also pages 171–173). Details are provided below. Mauna Loa and Kilauea exhibit the mildest of eruptions with VEIs of 0 to 1.

Oceanic hotspot volcanoes go through three stages of activity. The first is a long series of eruptions below sea level that build the broad base of the volcano, a great heap of volcanic rubble with little mechanical strength. In the second, or main stage, eruptions produce basalt lava flows that build the main visible mass of the volcano. The late stage of activity comes as the volcano moves off the hotspot. Eruptions become smaller and less frequent. Mauna Loa is in this late stage.

Kilauea. Kilauea is the successor to Mauna Loa, the new volcano at the active end of the hotspot track. It is much younger than Mauna Loa, still just a small fraction of its size, and it produces much smaller eruptions. Kilauea looks like a big ledge on Mauna Loa's southeastern flank.

Kilauea is now in its main growth stage. Approximately 95 percent of the part of Kilauea above sea level has grown within the last 1,500 years. It has erupted more than sixty times since 1832. Some of its more recent eruptions happened in 1955, 1961–1974, 1977, and almost continuously from 1983 to 2004.

Even though Kilauea is still young, its edifice has already split into three pie-slice segments that have begun to spread. Most of its eruptions are along the **rift zones** that radiate from the summit and separate the segments. The south flank of Kilauea broke along a scarp more than 500 meters high, Hilina Pali, and has begun slowing sliding toward the ocean.

Like many volcanoes, Kilauea announces an impending eruption with swarms of small earthquakes that originate at shallow depths, along with harmonic tremors that record magma movement. Sensitive tiltmeters may detect an inflation of the volcano summit as magma pressure increases. In some cases, these symptoms raise a false alarm.

Kilauea's eruptions typically produce only basalt. Although lava flows occasionally torch buildings (▶Figure 6-15, page 135; and Figure 7-7), then bulldoze and enter the ruins, they rarely injure or kill anyone. But hot gases, presumably water vapor and carbon dioxide, did drive a surge in 1790 that killed eighty warriors from King Keoua's army as they crossed a high flank of Kilauea.

MOUNT ETNA, SICILY, ITALY

The huge mass of Mount Etna broods over Catania and other cities in eastern Sicily. It is 3,315 meters high, the largest continental volcano on Earth, the most active volcano in Europe, and the highest point on Sicily. It typically erupts basalt lava flows and cinders (▶Figure 7-8). Etna stands on an unstable base of soft muds deposited from seawater onto oceanic crust.

International Space Station image, NASA Earth Observatory.

▶**FIGURE 7-8.** Mount Etna erupted a prominent plume of dark ash on October 30, 2002, while fires were ignited by lava pouring down the north flank. Light-colored plumes are gas emissions from a line of vents along a rift north of the summit. Roads and towns on the flanks of the volcano are visible in the upper right and across the bottom of photo.

Flank eruptions from rifts that divide the volcano into three large pie slices have built three radial ridges similar to those of the Hawaiian volcanoes. Escaping gases blow cinders out of vents in the radial ridges to build basalt cinder cones that quietly produce lava flows that burst from the bases of the cones (▶Figure 7-9). People easily avoid the flows, but their houses cannot. Ruined houses litter the flanks of the volcano (▶Figure 6-17, page 135).

Etna erupts almost continuously. It has also staged at least twenty-four sub-Plinian eruptions in the past 13,000 years. Plinian eruptions of basalt volcanoes are rare but hazardous, especially with towns creeping ever higher on Etna's flanks. A Plinian eruption in 122 B.C. brought havoc to the Roman town of Catania 25 kilometers south of Etna. The main Plinian deposit was 20 centimeters of coarse ash that started fires and collapsed roofs. As the eruption rate dropped, magma apparently boiled groundwater into steam that carried clouds of ash into the air.

The low viscosity and water content of basalts normally prevent Plinian eruptions. Etna's may develop when the magma rises so rapidly that gases cannot escape as they normally do. Instead, bubbles of steam explode within the rapidly decompressing basalt magma.

Etna's frequently erratic eruptive style includes occasional violent episodes (▶Figure 7-8). On July 22, 1998, one of the craters produced an eruption column 10 kilometers high that dumped ash over a wide area to the east. On September 4, 1999, basalt lava fountains rose as high as 1.5 to 2 kilometers. They rained ash and cinders on towns on the east flank of the volcano, causing significant damage. On July 9, 2001, strong earthquakes followed a cinder cone eruption. Then, on July 17, a new fissure erupted a lava flow that spread south and southeast. The next day, the flow blocked the main ac-

NASA.

▶**FIGURE 7-9.** This satellite view of a cinder cone in Lassen National Park, California, shows the basalt lava flow that spilled from the base of the cone after subsurface water dried up.

cess road to the south flank of Etna and slowly approached within 4 kilometers of the town of Nicolosi. Light ash fell on Catania to the south, temporarily closing the airport.

Cinder Cones

Cinder cones are characterized by their small size, low viscosity, steep sides, and moderate volatile content. They erupt where rising basalt magma encounters groundwater, and escaping steam coughs cinders of bubbly molten lava out of the vent. The smallest shreds drift downwind in a black cloud and fall as basalt ash. The larger cinders are too heavy to drift on the wind. They fall around the vent to build a loose pile around a small crater in the summit, where cinders are blasted out (▶see Figures 7-10 and 7-11).

Fluid basaltic magma feeds a cinder cone by rising along a fracture. On shield volcanoes such as Mauna Loa and Kilauea in Hawaii (▶Figures 7-3 and 7-5), the main fractures are along the crests of the three main ridges that radiate out from the top of the volcano. Typically, the basalt pours out in fluid lava flows. Where basaltic magma encounters water in the ground, the water bubbles up through the basalt magma; the expanding steam blows the molten magma into bubbly fragments that fall around the vent as cooling and solidifying cinders. Cinder cones can also form on the flanks of the darker, more basaltic varieties of stratovolcanoes. Along extensional faults such as the Great Rift

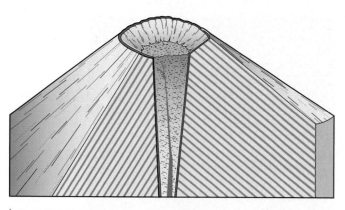

FIGURE 7-10. The internal structure of a cinder cone shows the layers that build up by falling around the explosion crater. The magma, rich in expanding steam bubbles, rises in the feeder conduit to blast out basalt cinders.

Donald Hyndman photo.

FIGURE 7-12. Vesicular scoria in a cinder cone 1 kilometer west of Prineville Junction, Oregon, contain volcanic bombs twisted and tapered as the molten magma flew through the air. The red color forms later as volcanic gases percolate up through the porous pile. The pocket knife in the lower left is 8 centimeters long.

Cinder cones:

Donald Hyndman photo.

FIGURE 7-11. In this view of the cinder cones in Haleakala Caldera, Maui, Hawaii, the inside of a cinder cone crater can be seen in the foreground; the bottom of the crater is to the lower left.

R. E. Wilcox photo, USGS.

FIGURE 7-13. Paricutín Volcano in Mexico is a large cinder cone. Red-hot cinders blasted out of the vent fall and tumble down the cone.

at Craters of the Moon in southern Idaho, the basalt rises to feed lava flows, but where it encounters water, the water again flashes into steam to build cinder cones.

Most cinder cones erupt over only one short period, a few months or a few years. They typically build a pile of cinders 100 to 200 meters high. When the supply of steam is exhausted, a cinder cone usually erupts a basalt lava flow from its base (see Figure 7-9). The lava flow emerges from the base in the same way that water poured on top of a pile of gravel emerges from the bottom of the pile. The next eruption in the area will probably build a new cinder cone where water is present rather than reactivate an old one. Many volcanic areas such as Haleakala on Maui in Hawaii, Lassen in northern California, and Newberry in Oregon are liberally peppered with cinder cones, many visible within a single view (Figure 7-11).

Cinder cones provide an exciting nighttime fireworks display as glowing cinders arc through the air and then roll, still glowing, down the flanks of the cone (e.g., see Figures 7-12 and 7-13). The upper parts of lava flows look like glowing red streams flowing downslope; the lower parts darken as they cool.

Most people are sensible enough to stay out of the rain of glowing cinders on the growing cinder cone and agile enough to avoid cremation in its lava flow. These eruptions endanger only property. The usual death toll in cinder cone eruptions is zero.

The dark basalt ash that drifts downwind from erupting cinder cones weathers into extraordinarily fertile soil that supports abundant and nutritious crops if water is available. Such areas generally support large numbers of people.

R. McGimsey, USGS.

Andesitic Volcanoes: Stratovolcanoes

Andesitic volcanoes are characterized by their moderate volume and size, moderate viscosity and slope, and moderate to high volatile content. Big **stratovolcanoes** such as those of the High Cascades are dominated by andesitic compositions. Darker varieties range to basaltic andesite; lighter varieties range to dacite and locally rhyolite. The name *stratovolcano* conveys the fact that it typically consists of layers of lava flows, fragmental debris, and ash. The slopes of the flanks of stratovolcanoes depend on two factors. The magma has moderate viscosity so lavas are not especially fluid; they flow only on moderate slopes before they cool and solidify. Escape of their dissolved volatiles through the viscous magma typically causes formation of large amounts of ash that concentrates near the vent and build the cone higher near the vent (▶ see Figures 7-14 and 7-15).

When most people imagine a volcano, they envision a stratovolcano—a large, steep-sided cone such as that typically found in the High Cascades, including Mount St. Helens (see "Case in Point: The High Cascades" and the introduction to Chapter 6), as well as Mount Fuji in Japan and Mount Vesuvius in Italy (see "Case in Point: Vesuvius and Its Neighbors," pages 180–184). Volcanoes of the Pacific Northwest, in fact, are good examples. Most stratovolcanoes grow in long chains above a slab of ocean floor that is subducting into the interior of the Earth. The volcanic chain trends parallel to the oceanic trench that is swallowing the ocean floor and is somewhere between 75 and 200 kilometers inland.

Stratovolcanoes mostly consist of andesite. They are also known as *composite* volcanoes, a name that refers to their internal structure with a mixture of rubble and ash layered

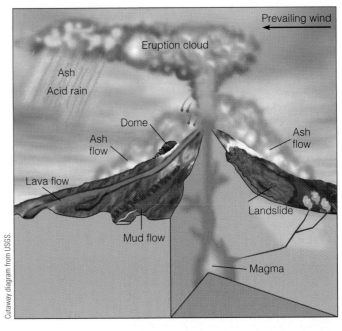

Cutaway diagram from USGS.

▶**FIGURE 7-15.** Most stratovolcanoes produce a broad range of hazards.

together with a few relatively solid lava flows. The lava flows give the volcanic cone enough mechanical strength to hold an internal column of magma. That enables the volcano to erupt repeatedly from a summit crater and permits it to grow into a tall cone.

Most stratovolcanoes erupt different sorts of andesite, along with some basaltic andesite and perhaps some rhyolite. Many of the large ones begin their careers by building a large basalt shield, then erecting a tall andesite cone on that base. Finally, they begin to erupt rhyolite and may end

This typical chain of stratovolcanoes sitting over an active subduction zone is active in northern California, Oregon, Washington, and southwestern British Columbia. They erupt along a nearly continuous line inland from where the Pacific Ocean floor descends through the oceanic trench between Cape Mendocino in northern California and southwestern British Columbia. The oceanic trench with a parallel volcanic chain is the universal signature of a subduction plate boundary (▶Figure 7-16).

Until the 1970s, most of the Cascades volcanoes were thought to be extinct or at least dormant. The rule of thumb was that if a volcano had significant glacial erosion that had not been erased by later eruptions, it probably had not erupted since the last ice age some 10,000 years ago. The inference was that because it had not erupted in such a long time it was unlikely to erupt again. It seems that our human time frame colored our view of what a volcanic timeframe should be. Mount Lassen in northern California staged a series of eruptions that started in 1914 and continued with declining vigor until 1921, but most geologists regarded it as an unusual event. This eruption did not persuade them that the High Cascades are an active volcanic chain. Even more recently, the apparent lack of a deep oceanic trench offshore remained enigmatic; such trenches are parallel to most active volcanic chains. That view changed with recognition of the role of plate tectonics in the 1960s and geophysical studies that showed that there was a trench off the West Coast, although it was filled with sediment.

In the late 1960s, the U.S. Geological Survey embarked on a vigorous campaign of research in the High Cascades in an effort to assess volcanic hazards. Those investigations revealed that the chain is far more active than most geologists had supposed; future large eruptions in the High Cascades should not surprise anyone.

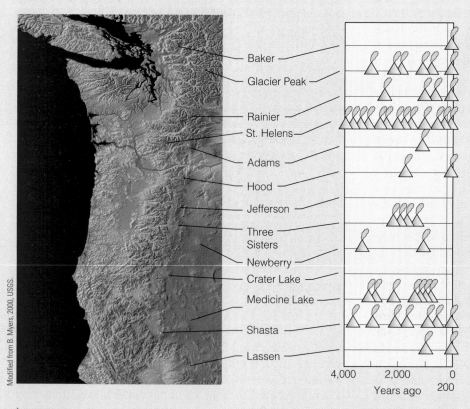

Modified from B. Myers, 2000, USGS.

▶**FIGURE 7-16.** This diagram shows the major Cascade Range volcanoes and their eruptions through time.

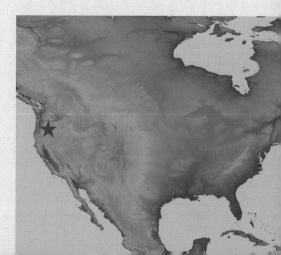

their careers in a massive debacle of rhyolite that destroys the andesite cone. In some of those cases, renewed activity builds a new andesite cone in the ruins of the old one. The eruptive behavior and intervals between eruptions varies widely. Mount St. Helens has erupted at least fourteen times in the last 4,000 years. Mount Lassen in northern California remained dormant for 27,000 years before again erupting in 1914.

Mount St. Helens

Mount St. Helens is discussed in detail in Chapter 6; pages 127–134. See also Table 7-1.

Mount Hood

Mount Hood, the spectacular volcano 75 kilometers east of Portland, Oregon, has not erupted since the 1790s. That is not a significant hiatus for a large andesite volcano (Figure 7-17). Radiocarbon dates on charcoal preserved in volcanic deposits reveal two major eruptions during the last 2,000 years, the last one only some 200 years ago. Volcanic domes near Crater Rock have repeatedly collapsed over the lifetime of the volcano to shed numerous hot ash flows down the southwestern flank. Some reached as far as 11 kilometers from the peak.

There is little doubt that Mount Hood will produce more eruptions, probably like those of the past 2,000 years, but it is not currently possible to predict when. Close study of the deposits that record the most recent eruptions provides the best available guide to what might happen during the next eruption.

Another eruption would probably send ash flows as far as 10 kilometers down the south and west flanks of the cone. Mount Hood does not generally erupt large amounts of ash in clouds that drift downwind. Water from melting snow and ice would lift ash and rubble from the surface of the volcano to generate mudflows that would race down the Sandy River and its tributaries toward the eastern edge of Portland and probably north toward the town of Hood River (see Figure 7-18).

In all likelihood, Mount Hood will provide ample warning of its next eruption with swarms of small earthquakes and clouds of steam dark with ash. The slopes of the volcano are almost uninhabited, so property damage there would be limited. However, mudflows pouring down river valleys toward Portland or Hood River would certainly inflict immense property damage. If warning did not come a few days before an eruption, mudflows would probably kill large numbers of people.

Mount Rainier

Mount Rainier is the largest, highest, and most spectacular volcano in the High Cascades. It rises from sea level to 4,393 meters, and dominates the skyline of a large part of western Washington (see Figure 1-2, page 3; and Figure 7-19).

Mount Rainier is arguably the most dangerous of the High Cascades volcanoes simply because enormous numbers of people live close to its base. Rainier's status as a national park prevents most kinds of development that might attract even more people to crowd in too close.

Although Rainier has not erupted in historic time, nothing suggests that it is dead. A substantial eruption could dump heavy ash falls on cities built on old ash falls downwind, most likely to the east and southeast. Several smaller towns lie within range of large rockfalls or ash flows. Rainier's greatest threat probably lies in the mudflows likely to pour down the broad valleys to the north and west. They could develop whether or not the volcano erupts.

Large stratovolcanoes are inherently unstable piles of lava, rubble, and ash. That is especially true of Rainier be-

David E. Wieprecht photo, USGS.

▶**FIGURE 7-17.** Mount Hood watches over Portland, Oregon. Although Portland is to its west, occasional storms that bring winds from the east could possibly dump large amounts of ash on the city.

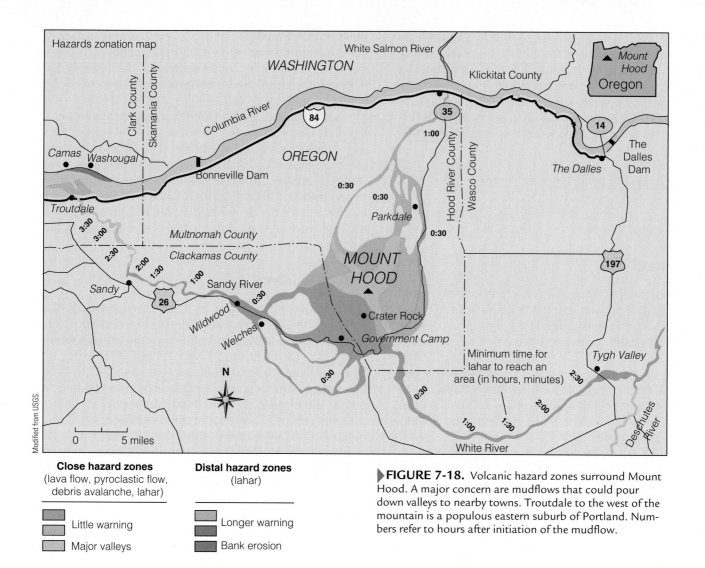

Hazards zonation map

WASHINGTON

White Salmon River

Klickitat County

Clark County

Skamania County

Columbia River

84

35

1:00

OREGON

Mount Hood

Oregon

The Dalles

The Dalles Dam

14

197

Camas   Washougal

Bonneville Dam

0:30

0:30

Parkdale

Hood River County

Wasco County

Troutdale

3:30

3:00

2:30

Multnomah County

2:00

1:30

Clackamas County

0:30

MOUNT HOOD

Sandy

26

1:00

Sandy River

Wildwood

0:30

Welches

Crater Rock

Government Camp

0:30

N

0:30

Minimum time for lahar to reach an area (in hours, minutes)

Tygh Valley

2:30

2:00

0:30

1:30

1:00

White River

Deschutes River

Modified from USGS.

0   5 miles

Close hazard zones
(lava flow, pyroclastic flow, debris avalanche, lahar)

Little warning

Major valleys

Distal hazard zones
(lahar)

Longer warning

Bank erosion

▶**FIGURE 7-18.** Volcanic hazard zones surround Mount Hood. A major concern are mudflows that could pour down valleys to nearby towns. Troutdale to the west of the mountain is a populous eastern suburb of Portland. Numbers refer to hours after initiation of the mudflow.

▶**FIGURE 7-19.** Mount Rainier looms over Tacoma, Washington. Mudflows from its flanks have reached all the way to Puget Sound, including the areas of Seattle and Tacoma.

Lyn Topinka photo, USGS/Cascades Volcano Observatory.

cause it consists of weaker material than most. Hot volcanic gases and groundwater have degraded much of its andesite ash and rubble into soft clay that becomes even softer when wet. The enormous snow and ice deposits on the higher slopes of the mountain make it even less stable. A season of unusually rapid snowmelt or shaking by a major subduction-zone earthquake could easily mobilize immense volumes of ash and rubble into enormous mudflows. Every 500 to 1,000 years, large mudflows reach more than 100 kilometers from Rainier to cover large parts of the Puget Sound lowland. The mountain is a subtle but enduring threat.

Meanwhile, the flat valley floors that drain from Mount Rainier are attracting large housing developments to accommodate the rapidly growing population of the Seattle–Tacoma area. Many thousands of people now live on the surfaces of geologically recent mudflows. The Electron mudflow ran 48 kilometers down the Puyallup River valley 500 years ago, then spread onto the Puget Sound lowland, including the area where Orting now stands (see Figure 1-2, page 3; and Figure 7-20). The Osceola mudflow poured 73 kilometers down the White River 5,000 years ago, burying the broad valley floor beneath as much as 20 meters of mud. The towns of Enumclaw and Buckley, home to tens of thousands of people, stand on that deposit. The Paradise mudflow filled the upper Nisqually River valley south and west of the mountain between 5,800 and 6,600 years ago. Future years will certainly see more mudflows follow those old routes, covering the old mudflows and everything on

them. Every new development along those old mudflow paths increases the hazard posed by Mount Rainier.

Mount Lassen

Visitors to Mount Lassen today see a quiet mountain that provides little indication of a new eruption any time soon. Lassen is a grossly oversized **lava dome** of pale andesite that descends from a long and distinguished heritage of volcanic violence. It rose within the wreckage of Tehama, a large andesite volcano that sank into a caldera during a cataclysmic eruption 350,000 years ago. Brokeoff Volcano is a remnant of one of its flanks. Tehama Volcano had earlier risen in the caldera of an even larger volcano that similarly destroyed itself 450,000 years ago. Perhaps we should not be surprised to see more eruptions in that vicinity. Age dates show that six lava domes called Chaos Crags rose near Lassen 1,100 years ago. That was some 27,000 years after its previous large eruption.

Most of the people who lived in northern California in 1914 thought Mount Lassen was thoroughly dead. Then on May 30, with no apparent warning, it suddenly produced an enormous cloud of steam to the tune of loud booming noises. Newspaper accounts tell of raw panic in the streets of Redding and Sacramento and places between. After that dramatic preamble, several lesser eruptions overflowed the summit crater with extremely viscous pale dacite magma that broke off as blocks that tumbled downslope. On

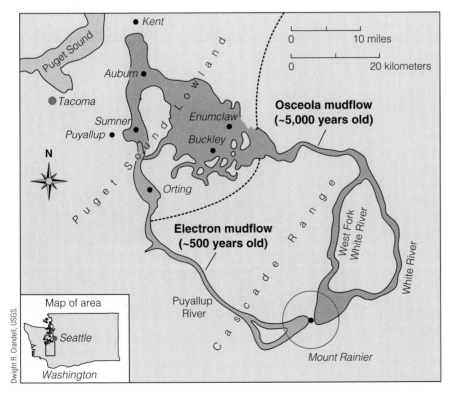

▶**FIGURE 7-20.** The Osceola mudflow inundated areas now occupied by hundreds of thousands of people. The Electron mudflow filled the floor of the valley that Orting now occupies.

R. E. Stinson Photo, National Park Service.

▶**FIGURE 7-21.** On May 22, 1915, Mount Lassen blasted ash and steam high into the air, as seen here from the town of Red Bluff, 64 kilometers west of the volcano.

May 19, 1915, the north slope collapsed, sending a mass of hot blocks racing down over a 10-square-kilometer area at the base of the mountain. Eruptions climaxed on May 22 with a dark mushroom cloud of steam and ash that reached a height of 8 kilometers (▶Figure 7-21). Melting snow sent an enormous mass of water and mud down the east slope as a hot mudflow that filled 50 kilometers of the Hat Creek valley overnight. Some of the blocks of hot lava that were buried in the mud spouted steam for months. After that climax, Lassen continued to produce sporadic clouds of steam until 1921.

The area near Lassen is thinly populated, so any future eruptions will probably cause minimal loss of life or property damage. Redding, Red Bluff, and other towns to the west are 50 to 75 kilometers downslope and against the prevailing wind for the most likely hazards of falling ash. Mudflows could possibly reach that far. Susanville, a similar distance to the east, could be heavily impacted by a major ash eruption.

Mount Mazama (Crater Lake)

Mount Mazama was an enormous andesite volcano that hovered over a large area of southwestern Oregon until around 7,700 years ago when it destroyed itself in a gigantic rhyolite eruption. Crater Lake now floods the caldera that opened where Mazama once stood (▶Figure 7-22). Study of

the geologic record of the Mazama eruption provided much of our present understanding of such events.

The rim of Crater Lake is the stump of Mazama. It is easy to reconstruct an approximation of Mazama by projecting the outer slopes of the rim upward at the usual angle for large andesite volcanoes, then lopping something off the top for a crater. The result is a volcano fully on the scale of Mount Shasta. Enormous amounts of rhyolite pumice and ash surround Crater Lake to a distance of tens of kilometers. Rhyolite ash that spread over much of the Pacific Northwest and northern Rocky Mountains provides evidence of an extremely large and violent eruption. Radiocarbon dates on the remains of plants charred in that eruption indicate that it happened 7,700 years ago.

By most estimates, the Mazama eruption produced at least thirty-five times as much magma as the 1980 St. Helens eruption. It dumped at least 40 centimeters of volcanic ash across large areas of the northern Rocky Mountains. If an eruption of that size were to happen today, the news media would portray it as an unqualified natural disaster, one without precedent.

Deep deposits of the culminating ash flow from the Mazama eruption are pale but grade upward into much darker rock (▶Figure 7-23). Most geologists interpret that gradation in color as evidence that the eruption tapped a mass of magma that had differentiated into pale magma above and much darker magma below. The pale rock erupted first to make the lower part of the ash flow, and the darker magma erupted last as the deeper part of the magma chamber emptied.

NASA - Astronaut Photo.

▶**FIGURE 7-22.** Crater Lake is the caldera depression formed when Mount Mazama erupted 7,700 years ago. The event was many times the size of the 1980 eruption of Mount St. Helens. Wizard Island is in the upper left part of the lake.

Willie Scott, USGS photo./Cascades Volcano Observatory.

▶**FIGURE 7-23.** A thick off-white ash flow from Mazama grades up to a darker cap, evidence that the eruption tapped deep into a differentiated magma chamber.

For many years, geologists generally agreed that Mazama had blown itself to smithereens in an enormous steam explosion that opened a gaping crater 10 kilometers across. Then groundwater filled the caldera to make Crater Lake. But more detailed study revealed an absence of old and altered volcanic rock in the debris blanket around Crater Lake. All the debris is freshly erupted volcanic ash and pumice (▶Figure 7-24). That discovery led geologists to conclude that Mazama simply sank into the emptying magma chamber beneath the volcano. The basin that holds Crater Lake is a collapse caldera, not an **explosion crater**.

One thousand years ago, a new volcano grew in the floor of the caldera until it barely emerged above the surface of Crater Lake to become Wizard Island (▶Figure 7-22). Evidently, the eruption that destroyed the volcanic edifice of Mazama was not the end of volcanic activity on the site. A new version of Mazama may yet rise among the ruins of the old one.

Crater Lake has become a metaphor for peace and serenity. It probably does not present any great hazard because the region supports few people and has enough roads to enable them to evacuate easily. However, we too easily forget that it is the site of an extraordinarily large and violent volcanic eruption.

Three Sisters

Among the lesser known of the Cascade volcanoes are the Three Sisters, a spectacular trio of volcanic cones just west of Bend, Oregon. Although any of the Cascade volcanoes could reawaken, the Three Sisters were not high on most people's list of those most likely to erupt anytime soon. All that changed in March 2001 when U.S. Geological Survey scientists using satellite radar interferometry discovered a low but broad bulge that had been rising since early 1997. By spring 2002, the 16-kilometer-wide bulge had risen a total of about 16 centimeters— 3.3 centimeters per year. The crest of the bulge is 5 kilometers west of South Sister (▶Figure 7-25). Its shape suggests a magma chamber 5 or 6 kilometers below. A large eruption on the east flank of South Sister could be catastrophic for Bend, with a population of 60,000 people. To date, however, there have not been other indications of impending volcanic activity such as earthquakes or increased emission of carbon dioxide.

Donald Hyndman photo.

▶**FIGURE 7-24.** Pumice exposed in this quarry east of Crater Lake was deposited from the eruption of Mazama 7,700 years ago. Exposure is approximately 2 meters high.

▶**FIGURE 7-25.** This view of South Sister is from west of Bend, Oregon.

Donald Hyndman photo.

Mount Shasta

French naval Captain Jean-Francois de la Perouse and his crew cruised north along the coast of California in 1786 shortly after failing to see the Golden Gate and the great bay behind it, probably because of coastal fog. But they did report seeing a volcano in eruption some distance inland at the latitude of Mount Shasta. That was indeed Shasta in its most recent eruption, the last of eleven in the past 3,400 years.

Shasta, at 4,318 meters, is the second highest of the High Cascades volcanoes (▶Figure 7-26). It looms over a large area of northern California, a constant hovering presence in the lives of hundreds of thousands of people.

Some geologists note that Shasta is approximately the same size as Mount Mazama was before it destroyed itself to leave us Crater Lake. And it may be significant that Shasta has produced small quantities of rhyolite during its most recent eruptions instead of the standard medium andesite that makes most of the volcano. Finally, seismograph records of earthquake waves that pass beneath Shasta show weak S waves or none at all. The only way to explain that is to conclude that the earthquake waves passed through a liquid, presumably molten magma in a shallow chamber, on their way to the seismographs. Some geologists interpret those observations as evidence that Shasta is getting ready to erupt a monstrous amount of rhyolite ash or lava. Of course, no one can tell when or even whether that eruption may happen.

Nevertheless, Shasta is a major hazard just as it stands. It is essentially an enormous pile of loose andesite ash and rubble stacked at a steep angle to a height of 3,000 meters with only a few lava flows holding it together. Shasta is likely to shed large masses of rubble with little or no provocation.

Donald Hyndman photo.

▶**FIGURE 7-26.** Mount Shasta, as seen from the north, punctuates the skyline of northern California. Shastina is the younger cone on its right flank.

Ash flow

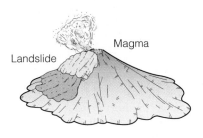

Landslide   Magma

Collapse of dome with or without gas explosion (e.g., Mt. Pelee, 1902; Unzen volcano, Japan, 1993)

Landslide of bulge releases pressure on magma, initiates eruption (e.g., Mount St. Helens, 1980)

Continuous eruption with continuous or intermittent column collapse (e.g., Mount St. Helens, 1980, after initial blast)

Magma rises into vent with resulting collapse.

Modified from Williams, 1932; Cas & Wright, 1988

Shasta did drop much of its north side in a giant rock avalanche 300,000 years ago (▶ Figure 8-70; page 225). The debris covers an area of at least 350 square kilometers in the floor of Shasta Valley, which reaches 51 kilometers northwest from the peak. Later eruptions repaired the scar on the volcano. The towns of Mount Shasta City and Weed stand on older rock avalanche deposits directly downslope from valleys on Shasta. They also lie within easy range of large ash flows.

Rhyolitic Volcanoes

Lava Domes

Rhyolitic volcanoes are characterized by their small to moderate size, high magma viscosity, steep flanks, and low to moderate volatile content. Rhyolite or pale andesite magmas sometimes erupt with little steam. They emerge slowly and quietly expand over months or years like a spring mushroom to make a hill as large as a small mountain. As the eruption proceeds, the lava on the outside of the dome solidifies while the still molten but extremely viscous lava within continues to rise. The solid rock on the outside cracks off in pieces and tumbles down the side of the growing dome to make steep talus slopes of angular rubble. The typical result is a steep mountain so cloaked in sliding rubble that solid rock is exposed only on the summit.

Like cinder cones, a single lava dome typically erupts only once, though it may be replaced by another dome as magma below continues to rise. The most dangerous lava domes form a single bulge on the flank of a volcano (▶ Figure 6-8, page 132). Collapse of a big dome growing on the flank of Mount St. Helens in 1980 released pressure on the underlying magma. That was the final trigger for its catastrophic eruption. Sometimes magma that solidifies in the throat of a stratovolcano is slowly extruded as magma below continues to rise; that magma may boil forth as an ash flow that pours down the volcano flanks. A lava dome rarely poses much danger to either lives or property—that is, until it erupts. When a dome collapses to form an ash flow, it is extremely dangerous to anything in its path.

Many ash flows develop in the collapse of an expanding volcanic dome (▶ see Figure 7-27 and "Case in Point: Mount Pelée, Martinique, West Indies"). Ash flows, from whatever source, commonly travel distances of 20 kilometers or more—in a few cases, much farther. Continuing expansion and expulsion of steam, and heating and expansion of air trapped beneath, maintain the internal turbulence and high speeds of ash flows.

St. Pierre was a beautiful small city of 28,000 people near the north end of the island of Martinique, one of the larger islands in the Lesser Antilles, north of South America. It was picturesquely sited along the curving head of a small bay with the symmetrical cone of Mount Pelée looming 10 kilometers to the north. St. Pierre made and shipped large amounts of sugar and rum. It also served as a port for a fishing fleet and for large numbers of tourists from Europe and North America.

Pelée is a typical andesite volcano approximately 1,400 meters high. Historical records contain brief accounts of minor eruptions in 1762 and 1851 but no hint of major or dangerous activity. The people of St Pierre in 1902 were quite aware that Pelée was a volcano but did not let that worry them. Surprising violence was about to follow a long period of mild activity.

Swarms of small earthquakes began shaking Pelée in early spring 1902. As magma rose into its summit, Pelée began to pour occasional glowing andesite ash flows down its southwest flank in the general direction of St. Pierre, but without reaching that far (▶ see Figure 7-28). The increasing level of volcanic activity soon convinced many people who lived on the flanks of Pelée that they were in grave danger. St. Pierre was still nearly unscathed, so many refugees stopped there; many others fled farther south along the coast to Fort de France, the capital of Martinique.

As concern mounted, the French colonial governor convened an advisory committee of five prominent citizens who were knowledgeable in various aspects of public affairs and science. Only two of them considered the situation even mildly dangerous.

At noon on May 5, a hot mudflow boiled out of the shallow lake that flooded the crater and down the Riviere Blanche to destroy and bury the sugar factory at its mouth at the northern edge of St. Pierre. It killed forty people and left an expanse of black mud studded with boulders that was more than 50 meters deep and 1,500 meters across the mouth of the river. People in St. Pierre were not familiar with mudflows.

Thursday, May 8, was Ascension Day, a holiday in Martinique. The sky was blue except for the plume of ash and gases drifting west from the summit crater of Pelée and before the sky darkened to a dirty gray as the plume of ash spread across it. People swarmed the streets of St. Pierre. At 8 A.M., men on the deck of a ship 16 kilometers offshore and coming in to St. Pierre saw a dark cloud gush from the crater of Pelée and glow internally. They heard an accompaniment of sharp booming noises that sounded like heavy artillery. As the eruption cloud rushed down the mountain, its less dense part rose into the air as a plume of steam dark with ash. The denser part poured down the southwest flank of the volcano directly toward St. Pierre at a speed later estimated at 190 kilometers per hour. It engulfed the town within two minutes and then rushed out across the bay, making great roaring noises. That part of the cloud was too heavy to rise into the air but not dense enough to plunge into the water (▶ Figures 7-29 and 6-27, page 140). The glowing cloud was an ash flow, a cloud of steam and ash hot enough to glow in the dark. As the ship came into port, the people on board saw seventeen or eighteen large ships in the harbor unmanned and in flames, along with a large number of smaller boats. The ship's crew saw people dashing about the dock area just before the fatal cloud swept over them, and the entire city burst into flames.

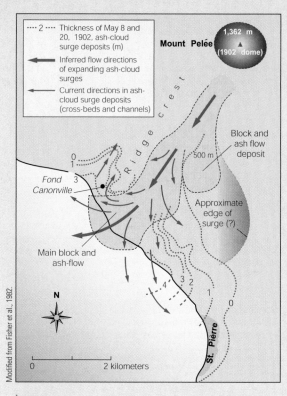

▶**FIGURE 7-28.** This map of ash flows from Mount Pelée in 1902 shows the thickness of the deposits. St. Pierre lies near the southern limit of the deposit.

▶**FIGURE 7-29.** Ash flow from Pelée entered the sea north of St. Pierre, 1902, the first record of an active ash flow.

People who visited St. Pierre soon after the tragedy were amazed at the extremely modest amounts of volcanic ash; little drifts lay here and there, but nothing even remotely resembling a blanket. The glowing steam had been the major agent of death and destruction. It had arrived with enough force to smash walls that stood perpendicular to the flow direction (▶ Figure 7-30).

People who had been on the street when the glowing ash flow arrived were cremated where they fell, leaving little piles of bones and ashes on the sidewalk. People who were inside when the ash flow struck may have survived its blow, but the suffocating gases and intense heat of the glowing steam instantly ignited everything flammable in its path. Thousands of fires soon merged into a monstrous firestorm that flared high as it destroyed what remained of St. Pierre and sucked the oxygen out of the air.

Only two of the people in the direct path of the glowing cloud survived. One of the two survivors was a shoemaker. The other was a prisoner locked in a dungeon awaiting trial. Both were badly burned in the ash and steam. The authorities dropped the charges against the jailed survivor because all of the records of his case had been destroyed in the eruption. He went on to become a minor celebrity on the carnival circuit as he toured Europe and North America displaying his dreadful scars. The exact death toll in St. Pierre is unknown, but the best estimates place the figure close to 30,000. That would include the citizens of St. Pierre who did not evacuate, the refugees from the flanks of the volcano, and those who died on boats in the harbor. Property damage amounted to the total value of the entire town plus that of the boats in the harbor.

Nothing could have been done to mitigate the property damage, but any reasonable attempt to evacuate St. Pierre would have greatly reduced the death toll. That did not happen because the available advice suggested that the danger was not great enough to justify an evacuation order. In 1902, no

▶**FIGURE 7-31.** In 1987, Mount Pelée and the community of St. Pierre offered a serene vista long after the tragic eruption of 1902.

one understood the destructive potential of clouds of volcanic gases loaded with enough ash to keep them on the ground.

In the months after the catastrophic ash flow erupted on May 8, an enormous spine slowly rose from the crater of Pelée. It eventually looked like an enormous obelisk standing on the summit of the volcano, viscous lava extruding through the throat of the volcano. It lasted a few weeks before it crumbled.

A French geologist, Alfred Lacroix, visited St. Pierre a few weeks after the catastrophe to try to reconstruct and explain the events of the morning of May 8. He studied and photographed the ruins of St. Pierre, as well as some of the later but smaller glowing cloud eruptions that continued through the fall of 1902. Lacroix recognized that the lavas likely to explode are those that are too viscous to rid themselves of a large content of steam and other volcanic gases. Instead of bubbling freely from the magma, the water vapor expands within the lava until it explodes, shattering the magma into a cloud of ash.

A few thousand people eventually moved into the ruins, cleared many of the streets, and built modest homes on the foundations of those that burned in 1902 (▶ Figures 7-30 and 7-31).

▶**FIGURE 7-30.** The total destruction of St. Pierre in 1902 can be seen in this Alfred Lacroix photo.

Giant Continental Calderas

Giant rhyolite volcanoes are characterized by their high viscosity and high volatile content but gently sloping flanks because of their predominant ash content that is spread over large areas (▶see Figures 7-32 to 7-37). However, these truly horrible monsters hardly look like volcanoes, certainly not like what most people think of as proper volcanoes. Few of the millions of people who visit Yellowstone National Park realize that they are on one of the world's largest volcanoes. The Yellowstone Volcano is a typical giant rhyolite volcano (see "Case in Point: Yellowstone Volcano").

These giants erupt rhyolite, typically in enormous volume, most of it explosively. Giant rhyolite volcanoes erupt enormous ash flows that cover tens of thousands of square kilometers and sheets of airborne ash that cover millions of square kilometers. The volumes of magma involved are typically in the range of hundreds to more than a thousand cubic kilometers.

As the emptying magma chamber withdraws support from the ground above, the surface collapses to open a basin called a **caldera** (▶Figures 7-33 and 7-36). A caldera is a collapse basin, not an explosion crater. Many large eruptions of pale andesite magmas open calderas from several to 25 kilometers across. Giant rhyolite caldera eruptions commonly produce enough rhyolite to open calderas 50 or more kilometers across. Those eruptions generally fill the sinking caldera with enough rhyolite ash to ensure that only

▶**FIGURE 7-32.** Collapse of the immense column of expanding rhyolite ash in a giant continental caldera produces thick, incinerating sheets of incandescent ash that race outward in all directions.

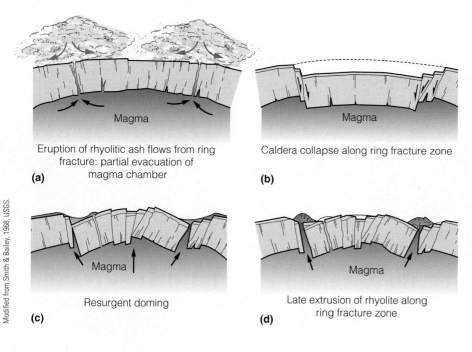

(a) Eruption of rhyolitic ash flows from ring fracture: partial evacuation of magma chamber

(b) Caldera collapse along ring fracture zone

(c) Resurgent doming

(d) Late extrusion of rhyolite along ring fracture zone

▶**FIGURE 7-33.** The surface above an erupting rhyolite magma chamber subsides to make a caldera during the eruption of a giant ash flow; it then domes again as new magma refills the magma chamber.

▶ **FIGURE 7-34.** Charcoal logs are all that remain of the forest that was overwhelmed by the Taupo ash-flow eruption in New Zealand some 2,000 years ago.

Donald Hyndman photo.

Donald Hyndman photo.

▶ **FIGURE 7-35.** A basalt lava flow with columnar joints overlies thick ash-flow tuffs in the Yellowstone River canyon north of Tower Falls, Yellowstone National Park.

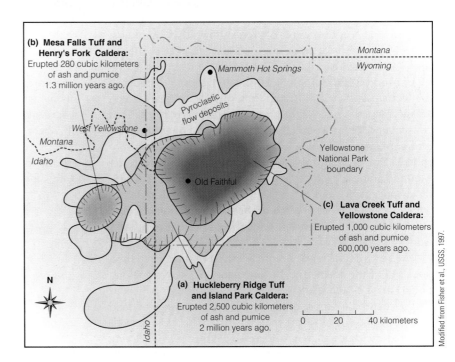

(b) Mesa Falls Tuff and Henry's Fork Caldera:
Erupted 280 cubic kilometers of ash and pumice 1.3 million years ago.

Montana
Wyoming

● *Mammoth Hot Springs*

Pyroclastic flow deposits

● *West Yellowstone*

Montana
Idaho

Yellowstone National Park boundary

● *Old Faithful*

(c) Lava Creek Tuff and Yellowstone Caldera:
Erupted 1,000 cubic kilometers of ash and pumice 600,000 years ago.

N

(a) Huckleberry Ridge Tuff and Island Park Caldera:
Erupted 2,500 cubic kilometers of ash and pumice 2 million years ago.

Idaho

0   20   40 kilometers

Modified from Fisher et al., USGS, 1997.

▶ **FIGURE 7-36.** The Lava Creek ash erupted 600,000 years ago as the immense Yellowstone Caldera collapsed.

A rhyolite giant over a continental hotspot, the Yellowstone Caldera does not resemble the towering andesite stratovolcanoes that grow above active subduction zones. But it does look exactly like a typical large **resurgent caldera** with a nearly flat summit and gently sloping sides. The Yellowstone Volcano we visit today is the relic of three monstrous eruptions that occurred 2 million, 1.3 million, and 600,000 years ago, so the recurrence interval is, crudely, 700,000 years. Each of the great eruptions produced more than 1,000 times the volume of lava erupted in the Mount St. Helens eruption of 1980 (▶Figures 7-35 and 7-36; see also Table 7-1, page 152). Each left collapse calderas approximately 50 kilometers across. Because the caldera collapsed into the magma chamber below, that magma body must have been at least 50 kilometers across. Much of the ash erupted in gigantic rhyolite ash flows that reached more than 100 kilometers down adjacent valleys, filling most of them to the brim. They were hot enough when they finally settled to weld themselves into solid sheets of rock tens of meters thick. Airborne ash drifted east and south on the wind and settled on the High Plains (▶Figure 7-37). Ash from the last eruption covered the High Plains as far east as Kansas, at least as far south as the Mexican border. A large proportion of the North American wheat crop grows on soil developed in Yellowstone rhyolite ash.

Resurgent caldera eruptions are by far the largest and presumably most destructive of all types of volcanic eruptions. No eruption even remotely comparable to the most recent eruption of the Yellowstone resurgent caldera has happened in historic time, perhaps not since the appearance of modern human beings.

If such an eruption were to happen now, it would almost certainly be the most cataclysmic natural disaster in the entirety of human experience. The destruction or disruption of transportation, communication, and energy systems in the western and central United States would be nearly total. Those impacts would hardly matter relative to the serious consequences that would almost certainly follow from enormous volumes of rhyolite ash injected into the upper atmosphere. They would probably resemble what happened after Mount Tambora in Indonesia erupted in 1815 (see Chapter 6, page 138), but on a vastly larger scale. All that ash would probably block enough of the sun's radiation to cause much colder climates within only a few weeks, leading to a worldwide agricultural disaster and probably famine.

Will Yellowstone erupt again? The picture does seem ominous. All of the steam and boiling water in Yellowstone National Park is clear evidence of tremendous amounts of heat at shallow depth. It is more ominous that a broad fringe of dead trees surrounds most of the thermal areas. This can mean either that the thermal areas are growing larger, overwhelming trees that were thriving just a few years ago, or that there is greater circulation of carbon dioxide gas up to the roots of

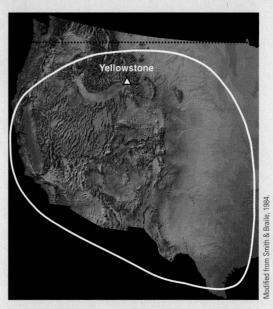

▶**FIGURE 7-37.** The Lava Creek ash from the Yellowstone Caldera, at the northeast end of the Snake River Plain, covered most of the central plains of the United States. The hotspot track is marked by the white arrow.

Modified from Smith & Braile, 1984.

the trees. Temperature measurements taken since the 1870s show that the thermal areas are also getting hotter.

Detailed studies of seismograph records show that the shear waves of earthquakes that pass beneath the Yellowstone resurgent caldera arrive at the seismograph later and weaker than they would in other areas. They clearly show that molten magma exists at a depth of 6 kilometers or so under the caldera. The swarms of small earthquakes that frequently rattle small areas in the resurgent caldera include harmonic tremors, which probably means that the magma is moving. Two resurgent bulges in the floor of the caldera are rising, presumably because magma is rising below them. All of these manifestations are typical symptoms of a volcano getting ready to erupt. No one knows whether they also portend eruptions of such supervolcanoes.

Water-saturated magma (rare)

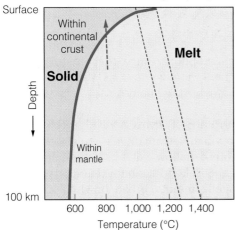

(a) Melting curve moves to lower temperature (at higher pressure) if more water is available

Common subduction-zone magma (e.g., 2% water by weight)

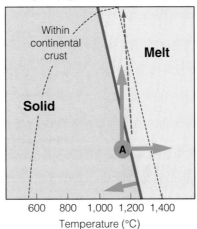

(b) Green arrows indicate trends that lead to melting of a high-temperature solid at A.

Dry magma (most basalts)

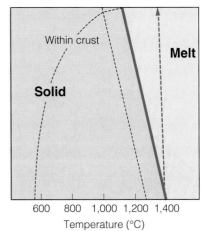

(c) Melting curve moves to higher temperature (at higher pressure)

▶**FIGURE 7-38.** Basalt magma (almost dry) is formed at high temperature deep in the mantle above the subduction zone. It rises into the continental crust that it heats and partly melts with what little water is there to form granite (rhyolite) magma. Both magmas rise to erupt in strato-volcanoes like those of the Cascades. See text for further discussion.

a subdued depression remains in the landscape. It is hard to see an old caldera as an element of the scenery except on an air photo or satellite image.

As magma continues to rise beneath the filled caldera, it raises a **resurgent bulge** in its surface. The bulge may become as large as a small mountain, and it may or may not develop into a new eruption, perhaps after hundreds of thousands of years. Some ancient rhyolite calderas still display an obvious resurgent bulge.

Many giant rhyolite calderas erupt several times at intervals of hundreds of thousands of years. It is fortunate that their eruptions are so infrequent because their ash flows would likely incinerate everything in valleys for 100 or more kilometers from the caldera (compare also Figure 7-34). They would also almost certainly inject enough ash into the upper atmosphere to cause drastic climate change over a large region for years afterward.

Volcanic Eruptions and Products

Volcanoes do not erupt just anywhere. They exist in a limited number of geologic settings, most of which are reasonably well understood. Volcanoes form wherever rock melts at depth and the magma can rise to erupt at the surface. To understand this process, we need to know what causes

rock to melt. The obvious answer is high temperature, and that is certainly required. However, the **melting temperature** depends on the depth in the Earth and the amount of available water. Consider ▶ Figure 7-38. Note that a hot rock deep in the Earth (e.g., at "A" on the middle diagram) may melt by (1) increasing its temperature, (2) decreasing its pressure, or (3) adding water to shift the melting curve to lower temperature.

Magmas rising toward the surface may not always reach it. The magma may crystallize by cooling or loss of water, or a water-rich magma may rise to a depth at which it crosses the crystallization curve and therefore must solidify (see dashed arrow in Figure 7-38a).

Because basalt magmas ascend from deep in the Earth's mantle, they may rise into a water-rich rhyolite magma within the continental crust. This heats the rhyolite, causing boiling of its water, rapid expulsion of steam near the surface, and explosive eruption of the rhyolite magma. Dark inclusions of basalt in pale rhyolite are common; this process may be a frequent trigger for eruptions (▶ Figures 7-39 and 7-40). The size of an ash eruption depends on many factors including the amount of magma, the magma viscosity, and its water-vapor content. A **Volcanic Explosivity Index** (VEI) crudely quantifies that eruption size (▶ see Figure 7-41 and Table 7-1).

Steam explosions eject particles of magma at speeds that depend upon the amount and pressure of the steam. Those factors dictate the rate at which the steam can expand. Gas

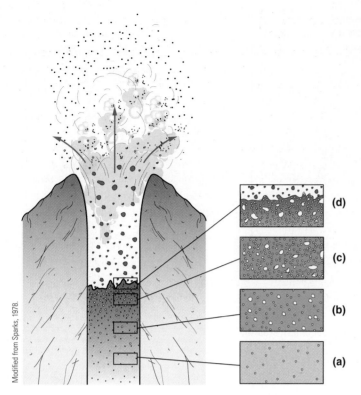

Modified from Sparks, 1978.

▶FIGURE 7-39. (a, b) Steam separating from a magma opens bubbles. **(c, d)** As the bubbles grow, the magma begins to froth. That decreases pressure on the magma below, launching a chain reaction of increasingly rapid bubble formation that leads to an explosive eruption.

slows as it expands and cools. Erupting steam commonly moves at rates in the range between 400 and 700 kilometers per hour but may exceed the speed of sound in air—1,200 kilometers per hour. Shreds of magma smaller than a few centimeters in diameter, ash or cinders, slow rapidly as atmospheric drag overcomes their small inertia. Steam

explosions may throw large blocks and blobs of magma—bombs—as far as 5 to 10 kilometers.

Explosive Eruption Styles

Styles of explosive eruptions range from frequent, mild eruptions to infrequent, violent eruptions. Many are typified by volcanoes that provide their names.

PHREATIC AND PHREATOMAGMATIC ERUPTIONS
Phreatic refers to violent steam-driven explosions generated by vaporization of shallow water in the ground, a nonvolcanic lake, a crater lake in a stratovolcano, or shallow sea. The result is often a **maar,** a broad, bowl-shaped crater surrounded by a low rim. If steam dominates, the process is called phreatic; if magma is more abundant, the process is **phreatomagmatic.** In either case, such water–magma eruptions can be especially dangerous locally; the erupting column of steam and ash can collapse on itself to form a base surge that sweeps rapidly outward. The surge carries hot sand and rock fragments that can sandblast and uproot trees and overwhelm and kill people.

STROMBOLIAN ERUPTIONS
Stromboli, off the west coast of Italy, is fed by fluid basalt or andesite magma that interacts with groundwater or seawater. Rapidly expanding steam bubbles in the magma blow it into cinders, scoria, and some bomb-size blocks that fall around the vent and tumble down steep slopes to form a cinder cone, a style called a **Strombolian eruption** (▶Figures 7-11 and 7-13). The magma is extremely fluid, so the eruptions are generally mild. Quiet lava eruptions may follow.

VULCANIAN ERUPTIONS
Vulcano, an island off the north coast of Sicily, is fed by highly viscous, andesitic magmas that are rich in gas. Dark eruption clouds blow out blocks of volcanic rock, along with ash, pea-size "lapilli," and bombs. With **Vulcanian eruptions,** ash-fall eruptions

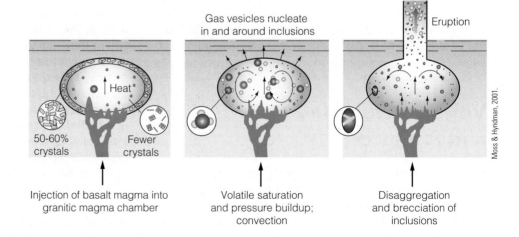

▶FIGURE 7-40. Injection of basalt magma into a rhyolite magma chamber can cause boiling and eruption of the rhyolite magma.

Moss & Hyndman, 2001.

50-60% crystals | Fewer crystals | Heat

Gas vesicles nucleate in and around inclusions

Eruption

Injection of basalt magma into granitic magma chamber

Volatile saturation and pressure buildup; convection

Disaggregation and brecciation of inclusions

172 CHAPTER 7

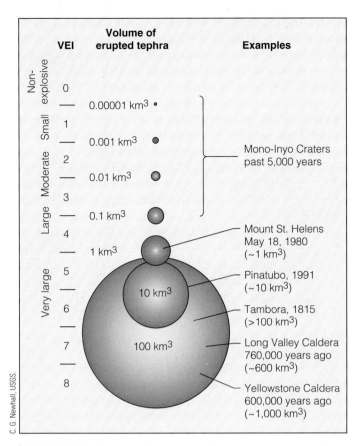

FIGURE 7-41. The Volcanic Explosivity Index (VEI) shows the relative magnitudes of eruptions. A few examples are provided.

dominate, but ash flows and base-surge eruptions can develop with or follow the ash-fall eruptions.

PELÉAN ERUPTIONS Mount Pelée in Martinique, in the West Indies, erupted in 1902 to obliterate the town of St. Pierre and its 28,000 inhabitants (See "Case in Point: Mount Pelée, Martinique, West Indies," pages 166–167). Its rhyolite, dacite, or andesite eruptions can be violent, especially early in the eruption when high columns of ash may be ejected. Some of those columns collapsed to form incandescent ash flows. Most distinctive, however, was the growth of Pelée's huge, steep-sided dome that ultimately reached a height of 900 meters. At intervals, the sides of the expanding dome collapsed to form searing hot block-and-ash flows. At times, the magma below slowly pushed viscous lava spines outward from the surface of the dome, some as much as 300 meters high.

PLINIAN ERUPTIONS The great eruptions of Vesuvius in 79 A.D. (described in detail by Pliny the Younger) and Mount St. Helens in 1980 produced powerful continuous blasts of gas that carried huge volumes of pumice to high elevations. Even larger than **Peléan eruptions, Plinian eruptions** can be truly catastrophic for any nearby population. Silica-

rich ash falls accompany ash and pumice flows. Ejection of a large volume of magma often causes collapse to form a caldera, as in the case of Mount Mazama (Crater Lake), Oregon, 7,700 years ago, Krakatau in Indonesia in 1883, and Santorini, Greece, in approximately 1620 B.C. (see "Case in Point: Santorini, Greece").

Small eruptions typically form craters at the **eruptive vent**; a depression or crater forms where the vent walls are blown out by the violent eruption. Large eruptions from shallow magma chambers often develop much larger depressions where the overlying rock collapses into the magma chamber. The distinction between crater and caldera is not so much size but the mechanism of formation of the depression. A cinder cone would blow out vent material to form a crater. A giant Plinian eruption such as the one that formed Crater Lake, Oregon, 7,700 years ago, would collapse into the emptying magma chamber to form a caldera. Crater Lake should have been called "Caldera Lake."

Assessment of Hazard and Risk of Volcanoes

Ancient Eruptions

Is that volcano active? Such a simple question begs for a simple yes or no answer. Recall that several decades ago, the answer for a Cascades volcano was that it was active if eruptions had covered evidence of glacial activity from the last ice age 10,000 to 12,000 years ago. We now know that some volcanoes lay dormant for much longer periods before erupting again. Mount Lassen in northern California paused for some 27,000 years before again erupting in 1915. In the eastern Mediterranean Sea, Santorini rests for an average of 30,000 years between major eruptions, though the intervals are becoming shorter. And Yellowstone, the gigantic volcano in northwestern Wyoming, last erupted 600,000 years ago. Not only is that about the average time between its massive eruptions but also seismic studies show that molten magma lies just a few kilometers beneath the surface. Careful surveying within the Yellowstone Caldera shows resurgent domes that periodically rise and fall over a period of decades or centuries, apparently with magma movements underground. In 2002, geysers that had not erupted for a long time became quite active. Clearly, there is no simple answer for whether a volcano is active.

If asked to anticipate future events, scientists commonly look to see what has happened in the recent past and then forecast more of the same. The past behavior of a volcano may help guide us to what it will do in the future, but in many cases no written record of past eruptions exists. The only archives are commonly in the rocks.

The number of eruptions in the last 100 or even 1,000 years may be documented in populated areas with long historic records, such as in Italy or Japan. Elsewhere, such as in

Santorini is an island volcano in the eastern Mediterranean Sea, south of mainland Greece. It now appears as a ring of islands, high places along the rim of an otherwise submerged caldera 6 kilometers across. Santorini is one of the most spectacular caldera volcanoes on Earth, similar in origin and size to Crater Lake in Oregon (▶ compare Figures 7-22, 7-42, and 7-43).

This stratovolcano has had long recurrence intervals; it also led to the destruction of a civilization. Santorini has staged twelve major explosive eruptions during the last 360,000 years, or an average of one every 30,000 years. Caldera eruptions happened 180,000, 70,000, 21,000, and 3,600 years ago. The intervals between eruptions become progressively shorter with time. After each caldera collapse, a new andesite volcano grew within the old caldera until it sank into a new caldera during the next catastrophic eruption.

In approximately 1620 B.C., a series of catastrophic Plinian eruptions of rhyolite ash and pumice evacuated the huge magma chamber and culminated in the most recent caldera collapse (▶ Figures 7-42 and 7-43). The main islands of Santorini are high points along the rim of that caldera. The remnant flanks of the volcano slope gently outward from the much steeper cliffs that face into the caldera. Thera, the main town, is on the steep cliffs. Many of the homes and tourist hotels have rooms excavated from the caldera wall into the deep ash that fell in 1620 B.C.

Events during the initial stages of collapse included eruption of enough white rhyolite pumice to make a layer as much as 4 meters thick. Its white lower part grades upward to dark gray andesite ash, presumably because the eruption was tapping progressively deeper levels of a differentiated magma chamber. A similarly graded ash flow exists around Crater Lake, Oregon, presumably for the same reason (see Figure 7-23, page 163). Seawater pouring into the collapsing caldera probably caused tremendous steam explosions. Some estimates suggest that the eruption raised a plume of ash 36 kilometers into the atmosphere. It may have lasted for weeks.

▶FIGURE 7-42. Santorini's main islands surround the collapse caldera that opened during the great eruption of 1620 B.C. The eruption ended the Minoan civilization.

NOAA image from Terra Satellite. Caldera outlines from McCoy & Heiken, 2000.

Donald Hyndman photo.

▶**FIGURE 7-43.** The inside wall of Santorini caldera exposes white Minoan pumice at the top right. White houses in Fira, along the lower caldera rim to the left, are mostly built into the same pumice erupted in 1620 B.C.

Many of the rhyolite ash deposits contain inclusions of chilled basalt magma and compositionally banded pumice, indicating that basaltic magma rapidly injected the rhyolite magma chamber from below. That superhot magma injected into cooler, water-bearing rhyolite magma would cause rapid boiling to drive a major Plinian eruption.

The eruption of 1620 B.C. buried and preserved the remains of Akrotiri, a wealthy town on the lower south flank of the volcano, to a depth of some 70 meters. Akrotiri's Minoan civilization was one of the crowning glories of the Bronze Age. The town had paved streets, underground sewers lined with stone, beautiful wall paintings, decorated ceramics, and attractive jewelry. The people raised sheep and pigs, farmed using surprisingly modern techniques, made barley bread and wine, gathered honey, and imported olives and nuts.

Many archaeologists believe that the account of the sudden demise of the Minoan empire was handed down in the oral histories of Egyptian priests. Twelve centuries later, Plato told of the disappearance "in a single day and night" of the island empire of Atlantis. Plato's descriptions include earthquakes, an account of the island disappearing in the caldera collapse, along with devastating floods, presumably giant waves. Similar giant tsunami waves appeared as Krakatau collapsed into its caldera on the sea floor during the great eruption of 1883.

Most of the population of Akrotiri must have been sufficiently frightened to evacuate with their possessions before the main eruptions. Those who remained were buried under several meters of deposits dropped from the Plinian column, and another 56 meters of pumice and ash erupted during caldera collapse. Earthquakes, falls of pumice and ash, mudflows, and debris flows completed the destruction of buildings.

Renewed volcanic activity began in 197 B.C. and continued sporadically until 1950. Resurgent domes of andesite and dacite rose in the floor of the caldera. Earthquakes took a further toll in 1570 and 1672. In 1956, an earthquake of magnitude 7.8 wrecked Thera and raised a giant wave that rose along the shores to heights between 25 and 40 meters. Hazard mitigation for Santorini since then involves restrictions on building on steep slopes of loose pumice in the caldera wall and also includes monitoring minor earthquakes and volcanic gases.

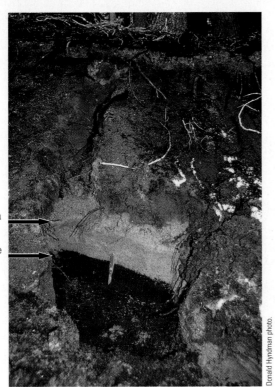

Mt. Mazama ash

Cinder-cone scoria

Donald Hyndman photo.

▶**FIGURE 7-44.** Pale-colored Mount Mazama (Crater Lake) ash covers black scoria from a nearby cinder cone at Mount Bachelor ski area west of Bend, Oregon (exposed in a trench dug for paleovolcanology information). The small pocketknife at the contact provides scale.

the western United States, the historic record may cover only the past century or so. That means little if the volcano erupts once or twice in 10,000 years. In such **paleovolcanology** cases, information is provided by interpreting deposits from prehistoric eruptions and reconstructing a record using age dates on plant material charred in past eruptions or dates on the volcanic rocks themselves (▶Figures 7-43 and 7-44).

The record in the rocks is open to judgment. Does each ash fall, ash flow, or lava flow record a separate eruption? Or do they record episodes of a single eruptive sequence that may have lasted a matter of a few days or weeks? Such questions, if not resolved, make a shaky platform for statistical assessment of long-term recurrence intervals.

Eruption Warnings: Volcanic Precursors

Forecasting volcano behavior for the long term is one thing. It is quite a different matter to predict what a volcano may do in the next few days or weeks. The scientific challenges of volcanoes are likely to prove less daunting than the political problems. The record of politicians and some government agencies does not inspire confidence.

The U.S. Geological Survey was reasonably successful in **predicting** a major **eruption** of Mount St. Helens in 1980. But many people deeply resented the effects of the prediction upon their personal freedoms. The local loggers and timber companies fought closure of the nearby forests because they feared loss of income. The civil authorities relented and permitted their continued access. Loggers died and logging equipment was destroyed in the ash flows and mudflows that accompanied the eruption of May 18, which also destroyed millions of trees (▶Figure 7-45). It is unfortunate that those civil authorities allowed them into the area, and that the loggers who pressed for such access were not aware of the extent of the geological hazards and the consequences of ignoring them. Local sightseers were equally ignorant of the real dangers, and some paid a heavy price. (See introductory material on Mount St. Helens survivors at the beginning of this chapter.)

U.S. Geological Survey scientists also predicted an eruption of Nevado del Ruiz, Colombia, in 1985. When the eruption began, the local authorities announced that the eruption was minor, which it was, and unfortunately urged people not to worry. When the eruption melted a thick pack of snow and ice on the volcano, a mudflow entombed 23,000 people.

Seismograph records of volcanic earthquakes have been used to infer magma movement underground and to project the likelihood that the volcano might erupt. **Harmonic earthquakes**, the rolling earthquake waves, precede many eruptions. U.S. Geological Survey volcanologists recorded a series of minor earthquakes originating beneath Kilauea Volcano in Hawaii in 1959. Over a period of two months, the earthquakes became more frequent as they rose from a depth of 60 kilometers, finally reaching the surface as the volcano erupted. Similar earthquakes have been recorded below many volcanoes, and in some cases a rapid increase in their frequency and magnitude preceded an eruption. That happened in the New Hebrides Islands during the 1950s and 1960s. But the frequency and magnitude of earthquakes did not change much at St. Helens in 1980 during the two months between the first activity and the climactic eruption of May 18.

In March 1991, Philippine and American volcanologists watched Pinatubo volcano getting ready to erupt. They saw the frequency of earthquakes increase and their origins migrate from deep below the north side of the volcano to shallow levels near the summit. Small explosions of steam and ash began on June 3 and dome growth began on June 7. The volcanologists predicted a major eruption for June 14. Some 60,000 people were evacuated from the area within 30 kilometers of the volcano. The big eruption, with a volume of approximately 4 cubic kilometers, came on June 15 when a mushroom cloud of steam and ash rose to an elevation of 34 kilometers. This was one of the few successful (short-term) predictions of an eruption. See "Case in Point: Mount Pinatubo."

Telescopes fitted with a thermometer instead of an ordinary optical eyepiece can measure temperatures of dis-

Phil Carpenter photo, USGS.

▶**FIGURE 7-45.** Logging trucks (yellow) and logs lie in a chaotic heap after the mudflows that poured down the Toutle River during and after the climactic eruption of May 18, 1980.

tant objects with amazing accuracy. They can be used to measure surface temperatures of volcanoes and the steam they erupt. But their usefulness in predicting eruptions is limited because it is sometimes hard to know whether the measured temperature changes are the result of volcanic causes or such extraneous factors as rainfall cooling the rocks.

Small changes in summit elevations and slope steepness associated with eruptions have been observed at some Japanese volcanoes and at Kilauea Volcano in Hawaii. The **tiltmeters** first installed at Kilauea in the 1920s were simple water tubes 25 meters long used as levels. They could detect tilts as gentle as 1 millimeter in a kilometer. Far more precise modern instruments use lasers to measure changes in both elevation and distance. They show that the volcano summit swells as lava rises into it and then deflates as it erupts (▶Figures 6-8, page 132; and Figure 7-46).

Measurement of the magnetic field over an active volcano provides an indication of how high the magma has risen in the volcano. Most volcanic rocks are weakly magnetic. A rise in temperature weakens the strength of all magnets, and they lose their magnetic properties entirely at high temperatures. Thus, it is not surprising to find that irregularities in the Earth's magnetic field near a volcano may occur as lava rises into it, perhaps months before it erupts.

As magma rises toward the surface, the magma releases steam and other gases. At some volcanoes, magma rise is marked by abrupt increases in sulfur or the ratio of sulfur to chlorine. Geologists collect and analyze fumarole gases to watch for this ominous sign (▶Figure 6-43, page 146).

Accurate eruption predictions are especially critical in areas where a large population lives close to the base of a stratovolcano. Mount Vesuvius, near Naples, Italy, is such a case (see "Case in Point: Vesuvius and Its Neighbors").

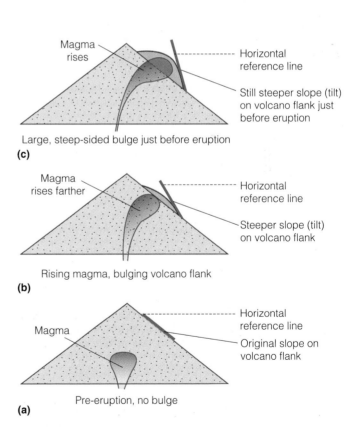

▶**FIGURE 7-46.** When rising magma gets close enough to the surface, it often pushes the overlying rocks ahead of it to create a bulge in the flank of the volcano. A tiltmeter measures the changing increase in slope on the volcano flank as the bulge grows.

Mount Pinatubo is an andesite volcano 1,745 meters high on the Philippine island of Luzon, some 90 kilometers northwest of Manila. On April 2, 1991, steam explosions suddenly piled ash on Pinatubo's upper slopes. This surprised and frightened the people who lived on the flanks of the mountain, the rice farmers on the plains below, and the 300,000 people who lived 25 kilometers from the crater in Angeles City. Pinatubo had not erupted in more than 400 years, and few people were aware that it was an active volcano.

Philippine volcanologists, with the help of USGS scientists, installed portable seismographs on and around the mountain to observe its earthquake activity. Geologic mapping soon showed that an eruption 600 years ago had spread hot ash flows across densely populated areas south and east of the summit and over the site of Clark Air Force Base. Those ash flows reached 20 kilometers east of the crater. Deep deposits of ash and widespread mudflows reached much farther.

The early blasts of steam and ash seemed fairly harmless to the volcanologists at the site because they contained only fragments of old rock, no freshly solidified ash from new magma. By June 5, the numbers of small earthquakes and volumes of sulfur dioxide emissions had increased dramatically. Occasional ash flows swept down valleys. The volcanologists thought a major eruption could happen within two weeks. That led them to recommend evacuation of the area within 10 kilometers of the summit. The volcanologists worked closely with public officials, carefully explaining the looming dangers. Then the officials went to great lengths to educate the public.

On June 7, a viscous lava dome began growing. Many small earthquakes and harmonic tremors suggested that magma was moving at depth. At that point, the volcanologists thought a major eruption might start within twenty-four hours. A large eruption did happen on June 9, but it was not the climactic event. Continuous ash eruptions accompanied expansion of a thick lava dome in the crater. On June 12, authorities evacuated people to a radius of 30 kilometers from the crater.

The climactic event, a classic Plinian eruption (VEI = 6), finally began early on June 12 with a lateral blast and a huge plume of steam and ash. The eruption became continuous by early afternoon. It climaxed in late afternoon when the eruption cloud towered to a height of 35 to 40 kilometers (▶ Figure 7-47). Ash flows reached 16 kilometers from the old summit.

Ash was as much as 30 centimeters thick at a distance of 40 kilometers from the volcano. Then, in an unfortunate coincidence, the climactic Plinian eruption coincided with passage of Typhoon Yunya, which brought intense rains. Heavy loads of wet ash collapsed many roofs, and mudflows rushed down regional valleys. By June 16, another 200,000 people

▶ **FIGURE 7-47.** Mount Pinatubo's climactic eruption was on June 12, 1991.

D. Harlow photo, USGS.

fled the mudflows. The rains continued to trigger mudflows every few days. By December 1991, almost every bridge within 30 kilometers had been destroyed (▶ Figures 7-48 and 7-49).

After Pinatubo erupted 4 to 5 cubic kilometers of lava, a crater 2 kilometers wide yawned at its summit. In spite of a major program to educate everyone and to evacuate 58,000 people, some 350 people died, mostly when heavy wet ash collapsed buildings. Another 932 died later from disease. Although the death toll seems large, there is no doubt that the timely warnings and broad evacuations saved tens of thousands of lives. But little could be done to reduce property damage. The American government soon abandoned its historic Clark Air Force Base.

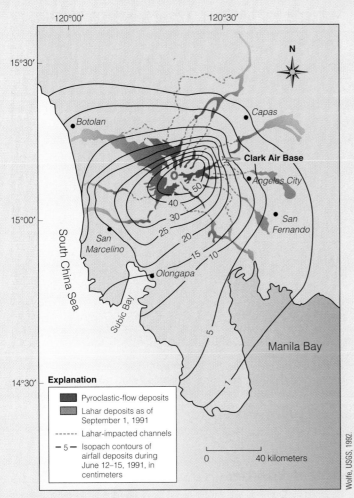

Wolfe, USGS, 1992.

▶**FIGURE 7-48.** This map of Mount Pinatubo shows the depths of ash laid down between June 12 and 15, 1991, and the distribution of mudflow deposits two months after the eruption.

T. J. Casadevall photo, USGS.

▶**FIGURE 7-49.** An air view over the Abacan River shows a bridge collapsed on August 12 by mudflows in Angeles City, Philippines, near Clark Air Force Base. In the lower left, people cross the river on temporary footbridges.

Twenty million tons of sulfur dioxide gas had combined with water in the atmosphere to make minute droplets of sulfuric acid. They hung in the air, reflecting 2 to 4 percent of the incoming ultraviolet radiation. Mean temperatures dropped as much as 1°C in parts of the northern hemisphere, approximately 0.5°C globally. Spectacular sunsets with broad streaks of green continued for another two years.

In general, the efforts to predict and mitigate the hazards of a large eruption were an outstanding success. The volcano provided ample warning, the geologists correctly anticipated most of the major volcanic events, the local officials efficiently managed evacuations, and the local people cooperated. It is hard to imagine a better outcome given the current state of knowledge.

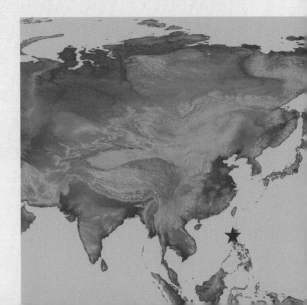

The northeastern part of the African Plate descends under southern Europe along a short collision boundary that raises the Alps and drives a chain of volcanoes in Italy and the eastern Mediterranean region (▶Figures 7-50 and 7-51). These volcanoes have devastated populations and caused enormous property damage for thousands of years. Some have even completely destroyed ancient civilizations. This is one of the most dangerous volcanic environments on Earth.

The long history of interaction between volcanoes and large populations of educated people laid the foundations for much of our understanding of volcanic processes and hazards. Italy, in particular, is the birthplace of modern volcanology.

Large numbers of people living in close proximity to extremely active volcanoes in Italy provide glimpses into the increasing volcanic hazards in North America. People do not differ much from one place to another. Their behavior during the eruption of Vesuvius in 79 A.D. was probably similar to that of people now living in towns near an erupting volcano.

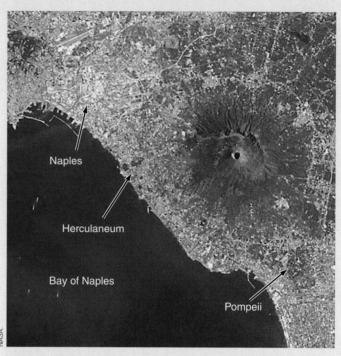

▶**FIGURE 7-51.** This NASA Landsat image shows Mount Vesuvius in the right center. Campi Flegrei, a Yellowstone-type caldera volcano, extends from the west edge of Naples (white area in upper left) to west of this view. The white areas are mostly towns and highways.

Mount Vesuvius

Naples, with 3 million people, nestles between Vesuvius, just 13 kilometers to the east, and Campi Flegrei, another large volcano centered 13 kilometers to the west. Mount Vesuvius is a volcano best known for destroying Pompeii in 79 A.D. (▶Figure 7-52). Its populace is living on borrowed time.

Seven hundred thousand people live within the immediate danger zone of the next major eruption of Vesuvius. Major eruptions occurred in approximately 5960, 3580, 1740, and 600 B.C. In 62 A.D., a major earthquake heavily damaged Pompeii and nearby towns. Then between August 24 and 25 of 79 A.D., during the reign of Nero, Vesuvius erupted catastrophically, burying Pompeii and Herculaneum and killing 4,000 of the 20,000 people who lived there. Details of the sequence of events in 79 A.D. were reconstructed from eyewitness descriptions in the letters of Pliny the Younger and modern studies of the volcanic deposits.

Local earthquakes began a few months before the main event, along with reports of rising ground levels. On the morning of August 24, 79 A.D., a series of steam explosions dropped a few centimeters of coarse ash close to the volcano.

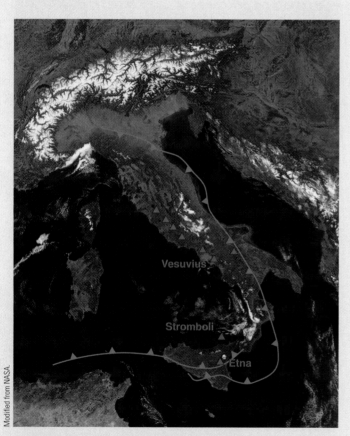

▶**FIGURE 7-50.** A chain of active volcanoes, including Vesuvius and Etna, runs down the western side of Italy. Subduction zones are shown in green lines with pointers sloping down dip.

That was a surprise because people had forgotten during its 700 years of quiet that Vesuvius was capable of erupting. Pliny the Younger sat across the bay, watching the eruption unfold. Meanwhile, his uncle, Pliny the Elder, who was commander of the Roman fleet in the Bay of Naples, dashed about in a ship equipped with oars, trying to establish some sort of order. Vesuvius went into full eruption shortly after noon as ash and pumice rose in a scalding column of steam. High altitude winds carried the ash plume directly over Pompeii, a town of 20,000 situated 8 kilometers south and east of the summit. Approximately 12 to 15 centimeters of white pumice fell per hour until evening, a total of 1.3 meters. Roofs must have begun to collapse under the weight of the ash after the first few hours.

The volatile content of the magma decreased, or the vent widened, or both, after twelve hours of continuous eruption. Either way, the proportion of ash in the erupting column of steam increased to reach a height of 33 kilometers. The increasing content of ash finally made the rising column of steam so dense that it collapsed onto the flanks of the volcano. It became a series of ash flows that killed everyone in their paths (▶ Figures 7-53 and 7-54), including many who had escaped west to the shore of the Bay of Naples. Volcanologists call volcanic events of that type *Plinian eruptions* in gratitude for the eyewitness account of Pliny the Younger.

▶**FIGURE 7-52.** Vesuvius looms over the excavated ruins of Pompeii.

▶**FIGURE 7-53.** These are casts of the bodies as they were found in Pompeii on top of the air-fall pumice.

▶**FIGURE 7-54.** Bodies in an active excavation in Pompeii are found in the top layer of the air-fall pumice. Skulls and bones are preserved. Cross-beds of the surge deposit that killed the victims are exposed at the top of the photo.

▶**FIGURE 7-55.** Boat chambers at Herculaneum can be seen in the foreground. The current town of Ercolano in the middle ground rests on top of the ash that buried Herculaneum. Vesuvius looms in the background.

Burial ash

Donald Hyndman photo.

Volcanologists have determined that the outward distribution of particle sizes falling from an ash cloud depends on both the height of the eruption cloud and the prevailing wind velocity. Detailed measurements of particles in ancient deposits can therefore be used to infer column heights and wind velocities. Lack of wind during an eruption, for example, leaves the particles evenly distributed in all directions around the vent; a strong wind spreads them in a long, narrow plume downwind. Large particles greater than 10 centimeters in diameter distribute themselves based on their exit velocity from the vent. Smaller particles are carried to higher altitudes that depend on total vent discharge rate and their size and settling velocity. Such calculations correlate well with actual measurements made during eruptions elsewhere.

In the early hours of August 25, a series of ash flows overwhelmed Herculaneum. They covered the 6 kilometers from the summit of Vesuvius in four minutes. Archeologists found only six skeletons in the town of 5,000, hundreds of others in the boathouses along the bay where they had sought shelter. They died instantly as they inhaled the hot ash and steam.

The ash-flow deposits are 20 meters deep at Herculaneum (▶Figure 7-55). They toppled walls of some buildings and buried nearly all of the others. The people who built the current town of Ercolano on top of these flows did not realize that its ancestor was entombed below. A farmer digging a well discovered part of Herculaneum in the early 1700s.

By 8 A.M. or so on August 25, 2.4 meters of pumice had accumulated on Pompeii, caving in most roofs and burying much of the town. The eruption also produced six ash flows. Many of the survivors walked around on the pumice during a pause in the activity. Then another big surge of scalding steam and ash swept over them and buried what remained of Pompeii, killing another 2,000 people.

Archeologists found many of those last victims inside their houses, preserved as hollow casts in the ash. They were lying on the earlier deposits of ash and pumice, some holding cloths over their faces. Perhaps they had not left their homes, because it surely seemed safer there than the frightening scene outside. Ultimately, Pompeii and 4,000 of its people were buried under 2.5 meters of ash and pumice.

At breakfast time on August 25, a culminating sixth surge of ash and pumice swept across Stabiae, 15 kilometers south of Vesuvius, where it dumped 2 centimeters of ash. This surge killed Pliny the Elder, who had gone ashore to provide aid and to see the eruption at closer hand.

Pliny the Younger wrote about the scene in Misenum, across the Bay of Naples, 28 kilometers west of the crater, where he waited in vain for his uncle (▶Figures 7-51 and 7-56). He fled with others as the dense black cloud of hot ash and steam raced toward them across the surface of the water. Complete darkness fell as they ran for high ground. The dark cloud followed them, hugging the ground—behind it, "fire." His must have been a narrow escape.

Rapid expansion of the eruption column quickly cooled the hot steam, condensing its water vapor on particles of ash, which fell as muddy rain onto the loose debris on the flanks of Vesuvius. Much of it poured rapidly down slope as mudflows that killed more people. As usual with Plinian eruptions, no lava flows appeared. The total volume of magma erupted was 3.6 cubic kilometers within less than twenty-four hours,

Donald Hyndman photo.

▶ **FIGURE 7-56.** Naples and Naples Bay lie in the shadow of Mount Vesuvius (seen in the background).

Donald Hyndman photo.

▶ **FIGURE 7-57.** A modern house has been built right at the toe of a 1944 lava flow from Mount Vesuvius. A repeat of this event would overwhelm many houses at the base of the mountain.

approximately eighteen times the volume of magma that erupted from St. Helens in 1980.

Vesuvius erupted at intervals of seven to 180 years in the millennium after 79 A.D. Then activity almost stopped for nearly 500 years, during which time people came to believe the volcano was defunct; they again encroached onto its flanks. A major eruption in 1631 produced heavy ash flows and ash falls that killed 4,000 people. Ash falls or lava flows partly destroyed villages near the volcano in 1737, 1779, 1794, and 1872. Some of those eruptions killed hundreds of people. A major eruption in April 1906 killed more than 500 people. Steam explosions during that eruption enlarged the crater enough to remove 170 meters from the top of the volcano. Lava flows partly destroyed the communities of San Sebastiano and Massa in March 1944. The lack of significant activity since 1944 is again leading to complacency (▶ Figure 7-57).

Seismic and tiltmeter studies suggest that molten magma now exists 5 to 10 kilometers below the crater of Vesuvius. Tiltmeters detected a half-meter of expansion from 1982 to 1984 with an increase in numbers of minor earthquakes. However, in approximately half of such cases no eruption follows. This large uncertainty leaves volcanologists wondering how to advise the civil authorities.

So far as anyone knows, Vesuvius could erupt anytime. If past experience is any guide, some 0.5 cubic kilometer of magma may have accumulated beneath Vesuvius after more than sixty years of inactivity. Some volcanologists think the next eruption of Vesuvius will probably resemble those of 1906 and 1944 with lava flows and heavy ash falls collapsing roofs. If it erupts ash, it could collapse as many as 20 percent of the roofs in Naples and stop all traffic and activity.

Vesuvius experts suggest that the long pause since the last major eruption suggests that the next may be the largest since 1631. Ash flows racing down slope at more than 100 kilometers per hour and a temperature of 1,000°C would reach populated areas between five and seven minutes after the eruption column collapsed. Of course, a large Plinian eruption, as in 79 A.D., is also possible with much more severe hazards over much larger areas. Modern Pompeii and a dozen other towns, each with thousands of people, ring the base of Vesuvius hardly more than 6 kilometers from the crater (▶ Figures 7-51 and 7-57). Instead of several thousand deaths, the toll of a new eruption like that of 79 A.D. would likely kill several hundred thousand to more than a million people and cause property damage beyond reckoning.

The hazard area around Vesuvius is a major economic and cultural center. In the current state of knowledge, disaster prevention depends on early sensing of volcanic warnings such as swarms of small earthquakes. With a lot of luck, that could make it possible to issue warnings early enough to allow a timely evacuation of the people in greatest danger. But avoiding Vesuvius presents almost insurmountable problems, both physical and political.

Without a lot of luck, a warning of an eruption could become a false alarm that would destroy the credibility of all involved. The social and economic consequences of a needless evacuation of hundreds of thousands of people would be devastating, if it were even possible. How could 700,000 people leave on roads and rail lines that are taxed beyond their capacity on normal days? And where could so many people find convenient refuge on short notice?

A special commission formulated an eruption contingency plan in 1996, assuming that a warning could be issued

Donald Hyndman photo.

▶**FIGURE 7-58.** In the Roman market Temple of Jupiter Serapis in Pozzuoli, gray pitted parts of columns (left of the bracket) show borings from marine mollusks when part of the columns were underwater.

Pitted columns

at least two weeks before an eruption and 600,000 people evacuated within a week. Those who have driven in Naples, or used trains and buses almost anywhere in Italy, might consider such an evacuation plan hopelessly optimistic. Vocal critics of this plan argue that it is impossible to predict an eruption more than a few hours or days in advance. They suggest that even with the best available monitoring, an evacuation alarm would almost certainly come too late. The alternative of major urban planning in a region that already contains hundreds of thousands of people would require truly ruthless resettlement on a scale almost beyond imagining! If you were in charge, what would you do?

Campi Flegrei

Campi Flegrei is an unquestionably active resurgent caldera within the western suburbs of Naples (▶Figure 7-58). Two million people live on the floor of the caldera, with 400,000 residing within the most active part. They make the Campi Flegrei one of the most hazardous volcanic areas in the world.

Major rhyolite eruptions opened a collapse caldera 16 kilometers in diameter 37,000 years ago, another 21 kilometers in diameter, at the same site, 12,000 years ago. Intense explosive activity along the faults that define the margin of the caldera has happened repeatedly during the past 12,000 years. Eruptions came at intervals of approximately fifty to seventy years from 12,000 to 9,500 years ago; 8,600 to 8,200; and 4,800 to 3,800 years ago. That is an extremely high level of activity for any volcano, especially for a giant rhyolite caldera volcano.

Most of those eruptions dumped ash to a depth of a meter or more over much of Naples. If such an eruption were to happen now, it would kill many thousands of people, perhaps

hundreds of thousands. Recall that 1 meter of volcanic ash is heavy enough to collapse almost any roof, especially if rain falls on it—and volcanic eruptions typically generate their own weather.

Pozzuoli, a city of 100,000 people on the western outskirts of Naples, stands near the center of Campi Flegrei (▶Figure 7-58). The lower parts of the columns of the Temple of Jupiter Serapis, a Roman marketplace within Pozzuoli, show borings of mussels that live in seawater. They leave no doubt that the market sank about 12 meters after its construction in the second century B.C. to be submerged in sea water. Magma rising under the caldera then pushed it back above sea level. The market generally subsided since 1538, but from 1969 to 1972 and from 1982 to 1984 it rose a total of 3.5 meters.

The thought of the potential death toll and property damage from a resurgent caldera eruption in Campi Flegrei simply boggles the mind. It is hard to imagine any workable evacuation plan. This is almost like a situation where a school bus is parked squarely on the railroad tracks and no one seems able to find the keys.

Violent Eruptions and Active Subduction Zones

What controls the most violent eruption over an active subduction zone? We can consider the possibility from several viewpoints.

- For andesite to rhyolite gas-driven, explosive eruptions, the volume of pyroclastic material erupted depends on the volume of magma reaching the Earth's surface, the viscosity of the magma, and the proportion of dissolved gas it contains. Because the dissolved gases tend to migrate toward the top of the magma chamber and the expanding gases drive the eruption, the upper gas-rich part of the magma (the upper third or so of the magma chamber) erupts explosively. The remaining magma stays underground or a little may erupt as a lava flow. Because these magmas are highly viscous, it may take decades to centuries for more gas to collect in the upper part of the magma chamber to again drive an eruption.

- The total volume of magma rising under a subduction-zone volcano depends on the subduction rate, the temperature and water content of the descending slab, the composition, temperature, water content (and thus degree of melting) of the overlying crustal rocks, the ease with which the magma can rise though the crust, and many intangible factors.

- Magmas that erupt are dry enough to approach the surface and wet enough to drive an explosive eruption. For an individual volcano, most of these factors are unlikely to change much over time. Thus, behavior of an individual volcano is likely to be more-or-less predictable. However, melting of crustal rocks under a volcano may change the composition of the remaining crust, mak-

ing further melting more difficult. The complexities are almost unfathomable.

A Look Ahead

Despite our knowledge of volcanic hazards and our ability to monitor volcanic activity, the opportunities for volcanic catastrophe to people and property are greater today than ever before. Growing populations and the great fertility of volcanic soils encourage people to nestle close to active volcanoes. The chance of dying in an eruption is small enough that most people ignore the hazard.

But the dangers are real. Many times in the past, mudflows and ash flows rushed down the valleys that drain the flanks of Cascade volcanoes and onto the broader expanses beyond. A 1987 analysis of specific volcanic hazards and their distribution around the High Cascades volcanoes showed them to be far more dangerous than previously expected.

Is that volcano alive or dead? Decades ago, any volcano that had erupted during historic time was officially classified as active. Those that had not were considered dormant if they looked fairly fresh, extinct if they were considerably eroded. The abundance of age dates now available shows that some volcanoes may remain dormant for many thousands of years, far longer than anyone had supposed.

Most large volcanic eruptions are both dangerous and destructive. Some are genuine catastrophes that kill tens of thousands of people and destroy billions of dollars worth of property. And we still have not seen the worst. No really big volcanic catastrophe such as a resurgent caldera or flood basalt eruption has happened during historic time. Nevertheless, the aggregate volcanic toll of death and devastation pales in comparison to those of earthquakes, landslides, and floods.

KEY POINTS

✓ By observing the size of a volcano and the slope of its flanks, you can infer volcanic type, magma composition, and volatile content. Basalt shield volcanoes, which mostly erupt fluid lava flows with low volatile content, are especially large with gently sloping flanks. Examples are Mauna Loa and Kilauea in Hawaii. **Review pp. 152–157; Figure 7-2.**

✓ Basalt cinder cones, which explosively erupt bubbly cinders, are small and steep-sided. They generally erupt a single basalt lava flow once the water in the ground dries up. **Review pp. 155–156.**

✓ Andesitic stratovolcanoes, which consist of a combination of explosively erupted ash and quiet lava flows, are steep-sided and intermediate in size.

The lavas have intermediate viscosity and moderate to high volatile content. **Review pp. 157–164.**

✓ The Cascade volcanoes are representative of a chain of continental-margin arc stratovolcanoes. **Review pp. 158–165.**

✓ Rhyolitic lava domes, which frequently produce dangerous ash flows, are small to moderate in size, steep-flanked, and have low to moderate volatile content. **Review pp. 165–167.**

✓ Giant rhyolite collapse caldera volcanoes, which erupt mostly ash from high viscosity magmas with abundant volatiles, are gigantic and have gently sloping flanks. The top of such a volcano is marked by a huge collapse caldera that sank into the underlying magma chamber, such as Crater Lake. **Review pp. 168–171; Figures 7-22 and 7-33.**

✓ The temperature required to melt rocks at depth depends on the composition of the rocks and the available water. High-silica and high-water contents melt at lower temperatures. **Review p. 171.**

✓ An eruption tends to be more explosive when the magma has higher silica and higher water contents. **Review pp. 171–173.**

✓ A range of explosive eruption violence and volumes is given names based on volcanoes that exhibit those styles. **Review pp. 172–173; Figure 7-41; Table 7-1.**

✓ Volcano risk can be assessed using paleovolcanology, the study of the deposits of past eruptions. **Review pp. 173, 176.**

✓ Eruption warnings or precursors commonly include frequent small earthquakes, harmonic earthquakes, a rise in heat and tilt of the slopes, and an increase in the sulfur and chlorine contents of the venting gases. **Review pp. 176–177.**

✓ Mount Vesuvius, Italy, exemplifies a highly active volcano with large numbers of people living in the hazard zone; hazards include heavy air-fall ash, ash flows, and fast-moving lava flows. **Review pp. 180–184.**

IMPORTANT WORDS AND CONCEPTS

Terms

caldera, p. 168
cinder cone, p. 155
collapse caldera, p. 152
eruptive rift, p. 152
eruptive vent, p. 173
explosion crater, p. 161
harmonic earthquakes, p. 176
lava dome, p. 161
melting temperature, p. 171
maar, p. 172

paleovolcanology, p. 176
Peléan eruption, p. 173
phreatic, p. 172
phreatomagmatic, p. 172
Plinian eruption, p. 173
predicting eruptions, p. 176
resurgent bulge, p. 171
resurgent caldera, p. 170
rift zone, p. 154
shield volcano, p. 151
stratovolcano, p. 157

strombolian eruption, p. 172
tiltmeter, p. 177

Volcanic Explosivity Index, p. 171
Vulcanian eruption, p. 172

QUESTIONS FOR REVIEW

1. What products of a volcano can kill large numbers of people long after the eruption has ceased?

2. If a mudflow is heading down valley toward you, what is the best way to survive?

3. What is the driving force behind the explosive activity of a cinder cone? Where does it come from?

4. On a huge shield volcano such as Mauna Loa, what is the main type of eruptive site? Where on the volcano is (or are) such a site (or sites)?

5. Yellowstone Park has two huge calderas, each more than 20 kilometers across. How do such calderas form?

6. How can you tell whether a plume rising from a stirring volcano contains new magma that may soon erupt?

7. If a gigantic flow of flood basalt magma were to erupt, why might that cause a sudden increase in global warming?

8. How does rhyolite magma form in the line of arc volcanoes such as the Cascades?

9. Why do the Hawaiian Islands form a chain of volcanoes?

10. Scientists once believed that if a volcano had not erupted in the last 10,000 to 12,000 years, it was extinct. What evidence is there that this is not correct? Give at least one example (be specific).

11. What evidence do scientists use to decide whether a volcano may be getting ready to erupt?

FURTHER READING

Assess your understanding of this chapter's topics with additional quizzing and conceptual-based problems at:

 http://earthscience.brookscole.com/hyndman.

CHAPTER 8

The 1995 La Conchita landslide destroyed several houses. Note the pronounced tilt of window frames in the house on the right, indicating that the upper floor pushed to the right.

Donald Hyndman photo, March 1998.

LANDSLIDES AND OTHER DOWNSLOPE MOVEMENTS
Falling Mountains

Landslides can be extremely costly. Few insurance policies cover them or any other type of ground movement. Landslides in the United States cost more than $2 billion* and twenty-five to fifty deaths per year. Worldwide, landslides caused an average of 1,550 deaths per year for the period 1969–1993. As with other hazards, major landslide disasters increase with the growth of world populations as people settle in less suitable areas.

Factors Controlling Downslope Movement of Landslides

The ability of a slope to resist sliding depends on the total driving force pulling it down versus the friction holding it in place. Rounded dry sand grains on a pile will stand no

---

* In 2002 dollars.

The small community of La Conchita sits below a 100-meter-high terrace on the narrow coastal plain along the California coast southeast of Santa Barbara. On March 4, 1995, following a period of especially heavy rainfall, some residents of La Conchita noticed cracking of house walls and movement in the steep slope above. Within a short time, the slide began to move rapidly but not catastrophically. Several residents whose homes were threatened watched the slide move until its increasing rate made them feel so uneasy that they ran out of its path. The 200,000-cubic-meter slide buried several houses and wrecked others on its lower fringes (▶ Figure 8-1).

Although the face of the high terrace stands almost vertically, it is made of soft, weak, and porous sediments that are not cemented into rock. An irrigated avocado orchard at the top of the terrace soaked it with water. A primitive road excavated across the face of the steep slope may have permitted the infiltration of even more water, and it certainly made the slope above weaker, perhaps enough to promote sliding. The presence of an old landslide indicated that the slope was

unstable. It probably contained preexisting slip surfaces that could easily be reactivated. A professional examination of the area before development would have recognized the hazard.

What Went Wrong?

Soft, weak sediments on especially steep slopes, heavy irrigation of an orchard above, and possible oversteepening by road construction.

A tragic footnote: As this book went to press, a week of heavy rains in southern California saturated the 1995 landslide deposit and adjacent slopes. On January 10, 2005, a huge mass of additional material flowed rapidly downslope to bury fifteen more houses under 10 meters of landslide. Ten people died, several were injured, and six remained missing and presumed dead after several days of frantic digging.

(a)

Robert L. Schuster photos, USGS.

(b)

▶**FIGURE 8-1.**
(a) The community of La Conchita, California, huddles at the base of a soft marine terrace that gave way when soaked with too much water. **(b)** The irrigated avocado field is visible at the top of the terrace in this photograph, which was taken the day after the landslide in 1995.

steeper than approximately 30 degrees (▶ measured down from the horizontal; see Figure 8-2), which is their **angle of repose.** The same is true of oranges in a grocery store. Adding more sand grains or oranges to the pile will cause some of them to roll down or patches of them to slide until the slope angle flattens to its angle of repose. Different materials stand at different angles of repose, depending mainly on the angularity and size of the grains and the moisture content.

Compared with dry sand, damp sand can stand much

steeper, even vertical. However, with too much water, the sand loses its cohesion, becomes weak and sloppy, and flows almost like a fluid because sand grains in water become buoyant enough that they do not bear so heavily on the grains beneath.

Gravity pulls a rock on a slope vertically downward unless the soil or rock beneath it prevents movement in that direction. However, the rock can move subparallel to the

David Hyndman photo.

▶ **FIGURE 8-2.** The talus slope of basalt blocks at the bottom of this image formed as rocks fell from the cliff and rolled down the slope to rest at their angle of repose (on the Columbia River Plateau near Palouse Falls, southeastern Washington). The material will remain at this slope as new material is continually added.

slope or roll down the hill.

The main features of a common landslide are shown in (▶ Figure 8-3). The **zone of depletion** is the area where a mass is displaced from the hillside and transported downslope. The **zone of accumulation** is the area where material is added to the slope below.

Many factors affect whether or not a slope will fail (Figure 8-3): load or weight of material making up the slope, **slope angle,** strength of the material, friction or **resistance to sliding,** water content, clay content, packing of grains, and internal surfaces and their orientation. Some of these are nearly independent; others are closely related. We address each factor below.

Load

The steeper the slope, the greater the downslope force. A greater proportion of the load that the rock imposes will be directed parallel to the slope (▶ Figure 8-4).

Slope

The load imposed by the rock can be resolved or separated into two components, one pulling parallel to the slope and the other perpendicular to the slope. The steeper the slope, the greater the gravitational pull parallel to the slope and thus tending toward slope failure (Sidebar 8-1 and Figure 8-4).

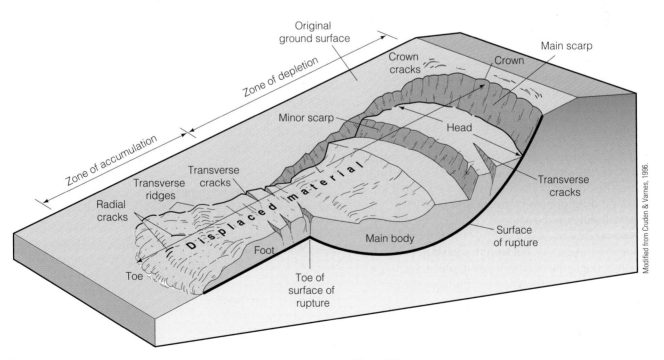

Modified from Cruden & Varnes, 1996.

▶ **FIGURE 8-3.** This diagram shows the main features of a rotational landslide.

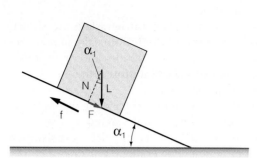

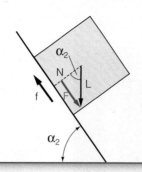

For a moderate slope of 30°:
Force parallel to the slope =
Load × sin 30° = Load × 0.5
(e.g., 100 kg × 0.5 = 50 kg).

For a steep slope of 60°:
Force parallel to the slope =
Load × sin 60° = Load × 0.87
(e.g., 100 kg × 0.87 =87 kg).
Clearly the mass on the steeper slope
is more likly to slide.

α = Slope angle
L = Load or weight
N = Force perpendicular to
    the slope
F = Force parallel to the slope
f = Friction holding the mass
    from sliding

▶**FIGURE 8-4.** These two diagrams show the forces on a mass resting on a slope. A steeper slope has a larger force parallel to the slope (red arrow) and is therefore more likely to slide. Note that for the gentler slope, the friction force (f) is larger than the force pulling parallel to the slope (F), whereas for the steeper slope the opposite is true.

Slope Material: Strength and Friction

"Soil" as used by geotechnical engineers who deal with landslides is any loose material above the bedrock. We follow this usage here.

Loose aggregates such as soil are inherently weak. So are loose sedimentary deposits not yet cemented into solid rocks or soft sedimentary materials such as clay and shale. These materials are the most likely to slide. Most solid rocks such as granite, basalt, and limestone are inherently strong and unlikely to slide. The exceptions are rocks that contain weak zones such as sedimentary layering, parallel micas in metamorphic rocks, or strongly developed zones of fracturing. Such rocks are especially likely to slide if their zones of weakness are nearly parallel to a slope.

Friction

Friction is the resistance to sliding. It depends on the force pressing down on the slope and the "roughness" of the slippage surface. The area of contact between a mass and the underlying slope does not affect the friction coefficient, so a small mass will slide on the same slope as a large mass of the same material. The mass will slide, or the slope will fail, when the force exceeds the **frictional resistance.** If the friction is high enough, the mass will stay in place. Anything that reduces the friction on the slope will increase the likelihood that the rock will move.

The friction resisting movement depends on the slope angle (α) and the load (L) of the body. Movement occurs when the force (F) exceeds the frictional resistance (f) (Figure 8-4). We add an additional factor, **cohesion** (C), a

force holding soil grains together. It is generally provided by the surface tension of water or cement between the grains (Sidebars 8-2 and 8-3).

Thus, several factors will increase the likelihood that the slope will fail: the load, the slope angle, and cohesion:

- increasing the load on the slope (e.g., adding a building or soil fill, or soaking the slope with water);

- increasing the slope angle (e.g., undercutting it or adding fill above); and

- decreasing the strength of the slope by adding too much water.

Loose soils have from 10 to 45 percent pore space. As with loose sand in a child's sand castle, the amount of water between the grains is all-important. A small amount of water provides cohesion, the surface tension force that helps hold grains together (▶Figure 8-5). Too much water fills the pore spaces and pushes the grains apart. Thus, rain saturating a slope will both add considerable load and decrease the strength of the slope.

Sidebar 8-2

Mass will slide if

    Force > frictional resistance + cohesion

    $F > f$ or $F > L \times \sin \alpha + C$
   driving   resisting
   force   force

where

    L = Load

    α = Slope angle

    C = Cohesion

Cohesion results from the static charge attraction between minute clay particles, the surface tension attraction of water between the grains, or even the strong chemical bonds of a cementing material. Most particles have tiny static charges on their outer surfaces. If you walk across a carpet on a cold day, you feel the shock when you touch a doorknob and the negative-charged electrons you picked up from the carpet jump to the doorknob. Small particles have a relatively large surface area for their total mass, so they have a relatively large surface charge. Billions of extremely fine particles of clay (<30 microns or 0.03 millimeter) have sufficient charges to hold tightly together.

For somewhat larger grain sizes such as sand, capillary cohesion or surface tension becomes more important. The thin films of water between sand grains hold them together (Figure 8-5).

Cements precipitated from chemicals dissolved in water can hold grains together to make solid, strong rocks. Natural cements are primarily calcium and magnesium carbonates, silica, and iron oxides.

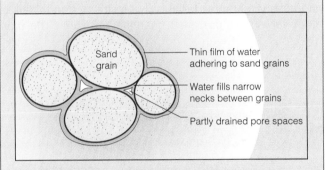

▶**FIGURE 8-5.** Capillary cohesion in the narrow necks between grains pulls the grains together.

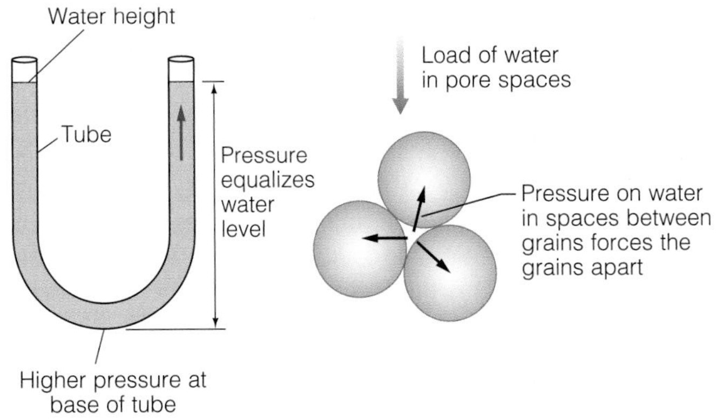

A Little Water

The child playing with sand quickly learns that a little water makes the sand stick together so it can be molded into sand castles. That little bit of water is sufficient to wet the surfaces of sand grains but not to fill the spaces between them (Figure 8-5). Adults also learn that two wet boards are hard to separate, as is a piece of glass on a wet countertop. The surface tension attraction of water in a narrow space pulls the glass to the countertop. That cohesion or **surface tension**, the pulling together of surfaces or particles, holds sand grains together and keeps them from separating (see also Sidebar 8-3).

Too Much Water

Most children who have spent time at a stream or rocky beach understand that rocks are easier to lift under water than in the air. And that is not an illusion. Water buoys up rocks by an amount equal to the mass of the volume of water they displace. Water buoys rocks just as it does boats.

▶**FIGURE 8-6.** Water pressure at depth is equal to the load or weight of the overlying water.

The difference is that the buoyancy of the boat is enough to float it.

If the pore spaces between the mineral grains in the soil are wet enough, surface tension and cohesion disappear. That weakens the soil because each grain is thoroughly wet and does not bear down on those below as heavily as it otherwise would. The pore spaces in a saturated soil are generally connected, so we can imagine them as continuous vertical columns within the soil. The **water pressure** at the base of each column is under the load of the water above it. More water in the slope raises the level of water in the soil and the pressure in the pore spaces at depth (▶Figure 8-6). In some cases, you can tell the level of water in the ground by the fact that it is wet up to a certain height (▶Figure 8-7).

▶**FIGURE 8-7.** Water (dark) seeps out of the soil below the sharply defined saturation level exposed in a road cut in Glacier National Park, Montana.

The water pressure tends to push the mineral grains apart, which further weakens the soil and makes it more likely to slide.

Slopes in wet climates are generally more prone to landsliding, other things being equal. However, if those slopes have been tectonically stable for geologically long periods, slope angles adjust to near-equilibrium values controlled by the local environment; that is, they reflect the slope material, climate, and thus the water content of the soil. Those slope-controlling processes can include ongoing landslides. Changes in slope imposed by external factors—such as undercutting of the slope by a stream or building a road, loading of the upper part of the slope by construction, addition of water by various means, or removal of vegetation—tend to destabilize the equilibrium and promote sliding.

Periods of heavy or prolonged rainfall tend to saturate the soil, increasing the pore water pressure and causing slides. Prolonged watering of lawns or excessive crop irrigation will raise the soil water level, as will too many septic drain fields in a small area. Leaking water or sewer pipes or cracked swimming pools inject water into the soil (▶Figure 8-8). Filling a reservoir behind a dam may raise the el-

evation of soil saturation enough to cause slides around the edges of the reservoir. The rising water fills pore spaces between loose grains to cause rotational slumps into the reservoir (▶Figure 8-9). In some cases, the rising water fills fractures in surrounding sedimentary layers sloping toward the reservoir, causing massive sliding into the reservoir (see "Case in Point: The Vaiont Landslide," pages 210–211).

Because the additional water reduces the shear strength of the slope, it is more prone to slide. Removing water from the soil can increase shear strength, making it less prone to sliding. One of the most effective mechanisms for removal of water is called **evapotranspiration.** Trees and shrubs take up water from the soil in their roots, thereby drying the soil. Water from roots reaches the leaves where it is transpired into the air. In addition, much of the rain falling on the leaves or soil surface evaporates there rather than percolating into the ground.

Some kinds of trees and shrubs take up water from the soil more eagerly than others. In general, the kinds of trees and bushes that grow prolifically on stream floodplains or along lakeshores use great quantities of water. The most notorious of these are cottonwoods, willows, and aspens.

▶**FIGURE 8-8.** After the 1999 Dana Point landslide north of San Diego, California, black plastic was spread on the ground in the lower left to prevent water infiltration. Note that the pale gray concrete pool (right center) was amputated by the headscarp. Did a leaking swimming pool add water to the slope?

Donald Hyndman photos.

Pool

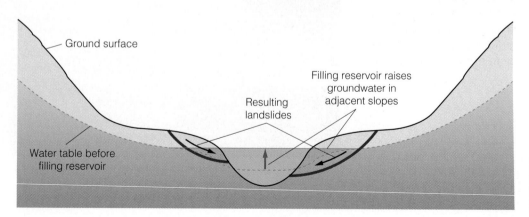

Ground surface

Filling reservoir raises groundwater in adjacent slopes

Resulting landslides

Water table before filling reservoir

▶**FIGURE 8-9.** Filling a reservoir behind a dam raises groundwater in the adjacent slopes, often leading to sliding into the reservoir.

Tamarisk trees may completely drain irrigation ditches in the desert Southwest. Where those same plants grow well above the floodplain, the ground probably contains excessive water and may be in danger of sliding. Planting trees or shrubs that use large amounts of water will help stabilize the slope. People sometimes cover potential slide areas with plastic to prevent water penetration (Figure 8-8), although this also has the unintended consequence of shutting down evapotranspiration.

It is possible to drain and thus stabilize many slopes artificially. One widely used method is to drill holes inclined slightly upward into the slope and insert **perforated pipes.** Water drains into the pipes and trickles out to the surface (▶ Figures 8-10 and 8-11). A more vigorous approach, also widely applied, is to dig deep trenches in the slope with a backhoe, line the trenches with **geotextile fabric** (cloth that permits water but not sediment to flow through), then backfill them with coarse gravel. Water will trickle out through the gravel for years. If the situation is truly desperate, it may help to drill wells into the slope and pump the water out. All these methods reduce the water pressure in the soil, which makes it less prone to slide.

Clays and Clay Behavior

Some soils or rock materials contain clays that absorb water and expand, thereby weakening the rock and even lifting it. Feldspars, the most abundant minerals, are basically aluminum silicates, which also contain calcium, sodium, or potassium in various proportions. Chemical weathering of all minerals consists primarily of their reaction with water. As they weather, feldspars lose most or all of their calcium, sodium, and potassium while their aluminum, silicon, and oxygen reorganize into sheets composed of aluminum or silicon atoms that are each surrounded by four oxygen atoms. Two important clay minerals, kaolinite and smectite, have structures that can lead to landslides (these structures are on the scale of individual molecules and are too small to be seen with a microscope; see more details in Appendix 2).

Kaolinite flakes have no overall charge but have weak positive charges on one side (e.g., on top) and weak negative charges on the other. The weak positive and negative charges attract, holding the layers together. Kaolinite does not absorb water and does not expand when wet. It forms by weathering in warm, wet environments. However, the overall structure is soft and weak, which contributes to landslides.

Smectite flakes have an open structure between their layers, which can be filled with water and cause the clay to dramatically expand; these are **swelling soils.** Because water has virtually no strength, almost any load will cause layers to slide easily over other layers. Adding an artificial load aggravates the problem. Smectite forms readily by weathering of volcanic ash, so soils with old volcanic ash tend to be extremely slippery and prone to landslides when they get wet.

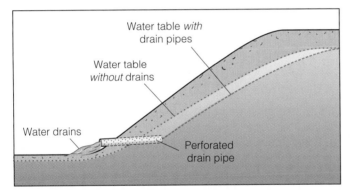

▶**FIGURE 8-10.** Installation of a perforated drainpipe can lower the water table and reduce the chance of sliding.

▶**FIGURE 8-11.** Water drains from perforated groundwater-drainage pipes in shale road cut on U.S. Highway 101 north of Garberville, California. The low permeability of this shale necessitates many drainpipes.

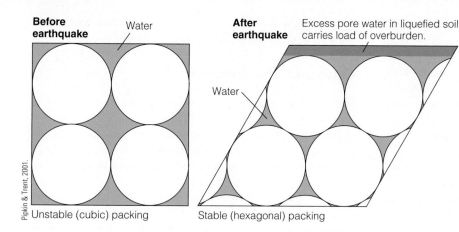

Before earthquake

Water

Unstable (cubic) packing

After earthquake

Excess pore water in liquefied soil carries load of overburden.

Water

Stable (hexagonal) packing

Pipkin & Trent, 2001.

▶**FIGURE 8-12.** Loosely packed grains provide large pore spaces for water. If the grains collapse to a tighter arrangement, much of the water must be squeezed out.

Liquefaction and Earthquakes

Many accounts of earthquakes include reports of sand spouting out of the ground or surfacing in big sand boils (Figure 3-30, page 52). This activity is evidence of **liquefaction.**

When some earthquake waves pass through soil saturated with water, the sudden shock jostles the grains, causing them to settle into a more closely packed arrangement with less pore space. Because that leaves more water than the remaining space can accommodate, water must escape. During this event, the soil grains are largely suspended in water rather than pressing tightly against one another. The soil settles and is free to flow down any available slope.

A loosely packed sand with say 45 percent **porosity** or pore space can collapse during an earthquake by rearrangement of the grains so they fit together more closely (▶Fig-

ure 8-12). The new porosity might be 30 percent, so the excess water between the grains is forced out and the soil behaves nearly like a liquid. You can easily demonstrate the process by filling a glass of water with fine sand. Then repeatedly tap the side of the glass with something hard. The sand will progressively settle as the sand grains rearrange themselves; the displaced water rises to cover the mass of sand grains. Buildings on liquefied soils can tilt and may even fall over when their underpinnings settle or spread (▶Figure 3-31, page 52; and Figures 8-13 and 8-14).

Quick Clays

Water-saturated muds in marine bays, estuaries, and old saline lakebeds are called **quick clays** because they are especially prone to collapse and flow when disturbed. Silt and clay grains are so fine that water cannot move through the tiny pore spaces quickly enough to escape.

▶**FIGURE 8-13.** Three typical types of ground failure occur during liquefaction.

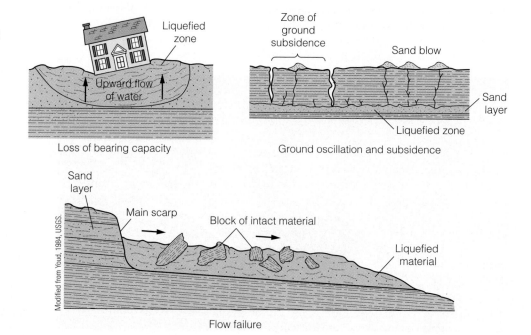

Liquefied zone

Upward flow of water

Loss of bearing capacity

Zone of ground subsidence

Sand blow

Sand layer

Liquefied zone

Ground oscillation and subsidence

Sand layer

Main scarp

Block of intact material

Liquefied material

Flow failure

Modified from Youd, 1984, USGS.

FIGURE 8-14. These houses on San Francisco Bay fill did not collapse during the 1906 earthquake but settled as the fill liquefied.

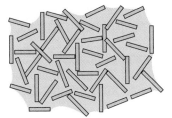

Before collapse, "a house of cards" with water in between

After collapse, much less pore space. The water is displaced.

▶**FIGURE 8-15.** Clay grains deposited in random orientation have especially large pore spaces—a "house of cards" arrangement. After collapse, the compacted clays take up much less space so water in the pore spaces must escape.

▶**FIGURE 8-16.** The Lemieux flow in a horizontal terrace of the Leda Clay of the St. Lawrence River Valley near Ottawa, Ontario, settled and flowed into the adjacent river on June 20, 1993.

When flakes of clay are deposited in salty water, static negative charges on their tiny grains hold them apart; the flakes remain in random orientations so the mass has a total pore space of 50 percent or more. This "house of cards" with water and sea salts in between is unstable. Salt dissolved in water separates into positive-charged sodium ions and negative-charged chlorine ions. Because tiny clay flakes have negative charges, the positive charges in the water hold the combination together. If this loose arrangement is disturbed, the flakes can collapse, permitting the water to escape and "float" the flakes (▶ Figure 8-15). The deposit liquefies and flows almost like water during the minutes while the clay flakes are moving into their new orientation. Then it again becomes solidly stable and will not liquefy again. The puddles of water that appear during the event fill the much larger volume of pore space that existed before the collapse.

When tectonic movements raise a marine or salty lake sediment above the level of the surface water body, fresh water generally washes the salt from between the flakes of clay, leaving the mass even less stable. Water seeping through the deposit eventually removes the positive sodium ions that had attached to the negative-charged clay flakes. Then the deposit becomes a stack of randomly oriented clay flakes with nothing holding them rigidly in position. That is an unstable situation. One day, the deposit liquefies as the clay flakes collapse into their usual parallel orientation like sheets of paper scattered across the floor. That commonly happens without warning, even with little or no triggering event.

Vibrations caused by an earthquake, pile driver, or heavy equipment can cause failure of a quick clay. Even loading of the surface can be enough to send the mass downslope as a muddy liquid. In Rissa, Norway, in 1978, a farmer piled soil at the edge of a lake, thereby adding a small load. It triggered a quick clay slide covering 33 hectares of farmland. The widespread Leda Clay in the St. Lawrence River Valley of Ontario and Quebec is another such sensitive marine clay. On June 20, 1993, a nearly flat 2.8 million cubic meter clay terrace settled and slowly flowed down a gentle slope into the South Nations River near the town of Lemieux (▶ see Figure 8-16).

Quick clays are common along northern coasts, including those of Canada, Alaska, and northern Europe. They make perilous foundations for almost any building.

Earthquakes Trigger Many Landslides

Many eyewitness accounts tell of great clouds of dust rising from hillsides during and after an earthquake. In most cases, they rise from slides that the earthquake has just shaken loose. If a slope is at all unstable, an earthquake is likely to send it on its way. Even without water, sudden shaking may trigger failure.

Earthquakes below about magnitude 4 trigger few landslides. Progressively larger earthquakes trigger more and more landslides, especially closer to the earthquake epicenter (▶Figure 8-17). Larger earthquakes may also start rockfalls. Earthquakes less commonly reactivate old landslides unless the soil is water-saturated and the earthquake pressurizes the pore water.

Of all the kinds of downslope movement that earthquakes may trigger, rock avalanches and rapid soil flows make up less than 1 percent. However, because they move at high speeds on slopes as gentle as a few degrees, they are even more deadly than rockfalls, often killing more people than the earthquake that triggered them. In some cases, they bury towns or villages several kilometers from their starting points. Slow-moving soil and rock slumps and lateral spreads rarely kill many people, but they do collapse buildings.

Rockfalls are the most abundant type of slope failures likely to accompany or follow an earthquake. Stabilization of rock cliffs or slopes can be expensive (▶Figures 8-18 and 8-19). Some are sprayed with a cement mixture called **shotcrete** or gunite to restrict water access. Some are draped with heavy wire mesh to prevent falling rocks from reaching buildings or highways. Some are drilled and anchored in place by **rockbolts.**

Earthquakes also often cause the failure of slopes that are inherently unstable. Among the most susceptible are recently raised marine terraces composed of soft, wet marine clays and associated sediments. A prominent case involved Anchorage, Alaska, in the magnitude 9.2 (M_w) earthquake of 1964. Much of Anchorage is built on a flat to gently sloping terrace as much as 22 meters above sea level. Shaking in Anchorage lasted seventy-two seconds, causing liquefaction of clays in the terrace and collapse (▶Figure 8-20). Wood-frame houses and other buildings survived moderately well except where they happened to straddle a slump scarp.

Internal Surfaces

Soils or weak rocks that are coherent and homogeneous may lack zones of weakness that might become **slip surfaces.** These often fail along a curving slip surface shaped much like the bowl of a spoon. In the interests of simplicity, engineers often calculate the behavior of such a surface by treating it as a segment of a cylinder (see Figure 8-35, page 207).

(a)

(b)

▶**FIGURE 8-17.** **(a)** The Northridge earthquake caused a swarm of landslides across these hill slopes near the epicenter. **(b)** A major earthquake on the Denali Fault in 2003 sent the side of the mountain down on top of the Black Rapids Glacier, Alaska Range.

▶**FIGURE 8-18.** Workers install heavy wire mesh over a dangerous rockfall area above the Trans Canada Highway.

(a)

(b)

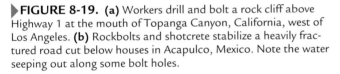

 FIGURE 8-19. **(a)** Workers drill and bolt a rock cliff above Highway 1 at the mouth of Topanga Canyon, California, west of Los Angeles. **(b)** Rockbolts and shotcrete stabilize a heavily fractured road cut below houses in Acapulco, Mexico. Note the water seeping out along some bolt holes.

▶**FIGURE 8-20.** Subsidence and seaward spreading of the marine terrace in Anchorage left most wood-frame buildings intact but severed or tilted some of them. This dropped trough formed at the head of the L Street landslide.

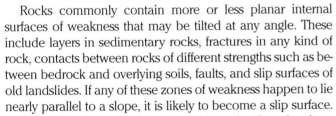

▶**FIGURE 8-21.** Steeply dipping limestone beds slope toward and daylight over a coastal highway near Sorento, Italy.

Rocks commonly contain more or less planar internal surfaces of weakness that may be tilted at any angle. These include layers in sedimentary rocks, fractures in any kind of rock, contacts between rocks of different strengths such as between bedrock and overlying soils, faults, and slip surfaces of old landslides. If any of these zones of weakness happen to lie nearly parallel to a slope, it is likely to become a slip surface.

Internal surfaces that dip at gentler angles than the slope of a hill may intersect the lower slope. These **daylighted surfaces,** which are often exposed at their lower ends by a road cut or stream (▶Figure 8-21), make ideal zones for slippage. Because there is no resisting load holding them back, only the friction between the layers can keep the mass from sliding. The rock above the slip surface does not have to push any rock out of the way to start sliding.

If the geologic structure is simple, the potential slip surfaces may extend a considerable distance. If it begins to move, an enormous amount of material may come down the slope in a translational slide (see page 208). Some of those move fast enough to kill large numbers of people.

Situations such as these explain many peculiar situations in which highways cross the same river again and again. Why did the engineers specify all those expensive bridges back and forth across the same canyon? In many cases, it was to keep the highway out of the paths of potential translational slides.

▶**FIGURE 8-22.** This hummocky landslide terrain is near Gardena, north of Boise, Idaho.

Landslide terrain {

Old Landslides

Old landslides that are not currently moving are more widespread than most people realize. Often the only sign is a hummocky-looking hillside (▶ Figure 8-22); even then, the hummocks may be obscured by vegetation. A stretch of road that becomes cracked or has broad waves in the pavement often requires repaving. That is certainly suggestive of continuously sliding terrain.

Building a road across such hummocky terrain would be unwise. So would construction of any kind of building or removing material from the base of the hummocky slope. Many old landslides are reactivated by removing material from the toe of the slide because that material encroached on a road, railroad, or construction site.

As noted above, old landslides can be reactivated by all of the processes that initiate new landslides—that is, adding water, artificially or naturally steepening the slope, undercutting the toe or removing toe material, loading the upper part of the slope, removing vegetation, or earthquakes. As with earthquakes and volcanoes, major events often occur in response to a combination of these factors.

A combination of any of these influences makes it more likely the slope will fail. As with other catastrophes, self-organized complexity dictates that there will be many small failures for every large failure. A large slide following a long period without new slides does not preclude soon having another large one. In fact, the existence of landslides in an area provides evidence that the circumstances for slides are ripe in that area.

What about reduction in slope following landsliding? Does that make the slope less prone to future sliding? Not necessarily; preexisting slip surfaces and surface fractures that permit further water penetration both contribute to further sliding. Building of roads or structures on an existing landslide merely aids further movement of the slope. Clearly, if the conditions are appropriate for landsliding, preexisting slip surfaces can reactivate.

Types of Downslope Movement

Downslope movements are generally classified on the basis of the type of material, the type of movement, and the rate of movement. Materials include solid bedrock, debris mostly coarser than 2 millimeters, and earth or soil mostly finer than 2 millimeters. Water plays a major role in many of these. Styles of movement include rockfalls, slides, lateral spreads, and flows. Note that any of them can involve rock, debris, or soil (Table 8-1). A continuous range of characteristics exists between most of these types of materials and styles of movement, and one type often transforms into another as it moves downslope.

Rockfalls and Rock Avalanches

The people of New Hampshire fondly regarded the Old Man of the Mountain in Franconia Notch State Park as a symbol of their state. They even put its craggy profile on their state quarter.

The Old Man of the Mountain was an outcrop of granite that projected from the higher part of a high mountain slope. It did indeed look like the profile of an old man if viewed from a good vantage point. Few people were much surprised when the old man broke off the mountain and fell down its lower slopes in spring 2003. For years it shed granite slabs onto the slope below; it looked so precarious that many people had expected it to collapse for more than a century.

Rockfalls develop in steep, mountainous regions marked by cliffs with nearly vertical fractures or other zones of weakness. Large masses of rock separate from a steep slope or cliff, sometimes pried loose by freezing water, to fall, break into smaller fragments, and sweep down the slopes below. They often collect in **talus** slopes, the fans of rock fragments banked up against the base of the cliff (▶ Figure 8-23). Individual rocks, especially large boulders, may bounce or roll

Table 8-1 Classification of Mass Movements

Type	Material: Rock (Often Broken)	Material: Debris (Mostly >2 mm)	Material: Soil (Mostly <2 mm)
Fall	Rockfall	Debris fall	Earth fall
Topple	Rock topple	Debris topple	Earth topple
Slide: rotational	Slump	Debris slide	Earth slide
Slide: translational	Rock slide Block slide	Debris slide	Earth slide
Lateral spreading	Rock spread	Debris spread	Earth spread
Flow	Rock flow	Debris flow	Earth flow

well away from the base of the slope. These can be particularly dangerous because of their size, speed, and distance of travel. There is a growing number of cases in which a giant boulder has completely demolished a house (see "Case in Point: Rockfall Hazards West of Denver," and "Case in Point: Rockville Rockfall, Southwestern Utah").

Robert Webb photo, USGS.

FIGURE 8-23. In Marble Canyon, Arizona, recent rockfall debris forms talus fallen from the visible scar on the cliff above.

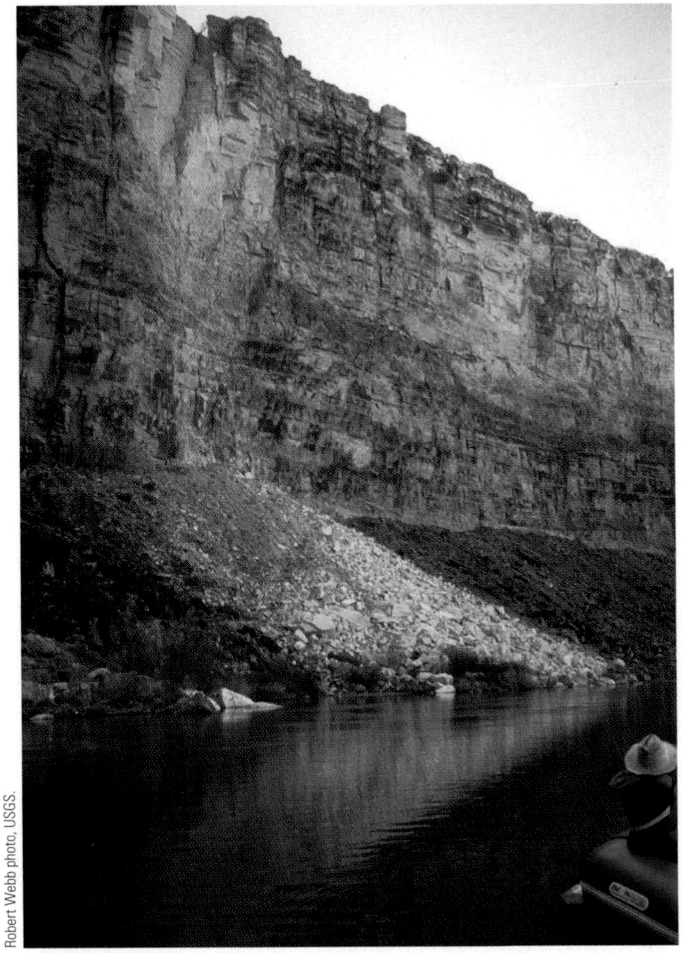

Rockfalls can be hazardous not only to houses and people but also to highways. Road building often follows the base of precipitous cliffs or excavates high road cuts that destabilize the original slope. For example, a huge rockfall occurred along Interstate 70 in Glenwood Canyon, west of Denver, Colorado, on May 9, 2003, and again on November 25, 2004. A relatively small rockfall on November 10, 2003 closed Washington Highway 2 in the northern Cascades for an extended period of time (Figure 8-24).

The largest rocks that fall from a cliff often reach the base of the slope because they roll easily on the smaller material on the slope. Smaller rocks tend to get caught between any larger fragments that may be on the slope.

Even in areas of relatively subdued topography and horizontal sedimentary layers, high road cuts can be prone to rockfalls. Strong layers such as sandstone or thick beds of limestone often break on vertical fractures. When intervening soft layers such as shale weather back to leave an overhang, the strong but fractured layers often break off and fall (Figure 8-25).

Translational slides, as the name implies, involve rapid downslope movements on a nearly planar surface, carrying

Washington State Department of Transportation photo.

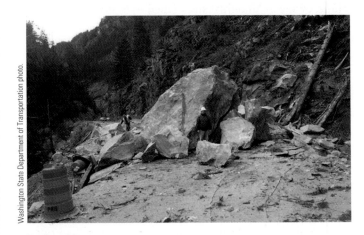

FIGURE 8-24. Rockfalls can be hazardous to highways. Granite has spaced fractures that often create large blocks. This rockslide came down across a state highway in the northern Cascades on November 10, 2003.

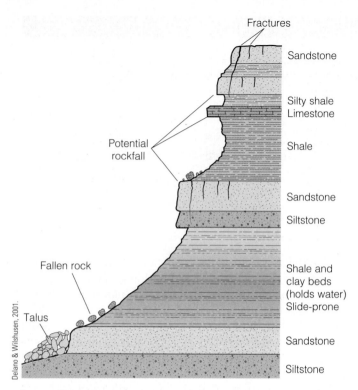

Fractures

Sandstone

Silty shale
Limestone

Potential
rockfall

Shale

Sandstone

Siltstone

Fallen rock

Shale and
clay beds
(holds water)
Slide-prone

Talus

Sandstone

Siltstone

▶**FIGURE 8-25.** Some rockfall problems arise where a strong layer such as sandstone overlies a weak layer such as shale or clay in Pennsylvania.

CASE IN POINT
Rockfall Hazards West of Denver

Denver is on the edge of the High Plains, but communities just to the west are in the foothills where steeply tilted layers of sandstone make ridges and cliffs. Tertiary volcanic rocks cap some bluffs such as North Table Mountain just north of Golden. Big boulders that tumbled from its caprock cliffs litter the steep, grassy slopes below. Houses cover the lower parts of the slopes. Jefferson County zoned undeveloped land immediately north as a no-build area in response to a rockfall hazard map that the U.S. Geological Survey published in 1973.

A developer, determined to build a subdivision, challenged this designation in the 1980s and staged a rockrolling demonstration to prove his point. Unfortunately for him, several boulders rolled into the area of his proposed subdivision, one bouncing high enough to take out a power-line tower at the mountain's base. His application was denied. The base of a steep slope capped by vertical cliffs that have shed big boulders in the past is no place for houses.

a broken mass of angular rock that can be buoyed by air beneath the moving mass or between the fragments. Run out distances are typically greater than the height of the original slide scarp (see "Case in Point: Madison Slide, Montana" and "Case in Point: Frank Slide, Alberta").

Factors that favor rockfalls include cliffs or steep slopes of at least 40 degrees. Rocks that are most likely to cause rockfalls are those that break easily into fragments: granite, metamorphic rocks, and sandstone. Strong earthquakes are likely to trigger rockfalls, as are large explosions, passing trains, or severe undermining of the lower slope.

Debris Avalanches

Rockfalls in which the material breaks into numerous small fragments that flow at high velocity as a coherent stream of fragments are called **debris avalanches.** A classic case at Elm, Switzerland, more than 100 years ago shows how a large rockfall can transform into a fast-moving catastrophic debris avalanche (see "Case in Point: Elm, Switzerland").

Even more catastrophic and deadly debris avalanches struck Yungay, Peru, on January 10, 1962, and again on May 31, 1970. See "Case in Point: Yungay, Peru."

Debris avalanches or debris-fall streams commonly begin as rockfalls but break up, entrain air and water, and often flow downslope at speeds that can reach 100 to 300 kilometers per hour. Many of the largest and most destructive landslides in recorded history started as ordinary rockfalls that developed into debris avalanches. Some contain boulders

as big as houses. The distance is generally related to the height a fragmental rock mass falls—that is, to its potential energy before it accelerates downslope. Large rockfalls can run out horizontally as much as five to twenty times their vertical fall. Thus, a mountainside 1,000 meters high may collapse to run out horizontally as much as 20 kilometers. The same principle guides a ski jumper or a child on a sled. They start higher on the hill to reach farther out on the flat below a hill. The simple formula in Sidebar 8-4 helps to explain the driving force.

Sidebar 8-4

The distance traveled by a debris avalanche depends principally on the height of the mass before it falls and therefore its potential energy. The values are variable, but an approximation is provided by the formula below:

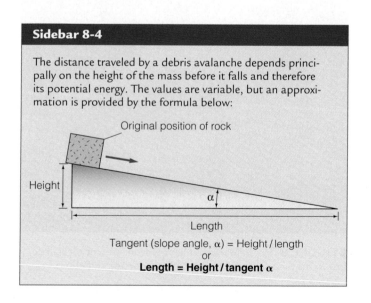

Original position of rock

Height

α

Length

Tangent (slope angle, α) = Height / length
or
Length = Height / tangent α

Jack Burns, an assistant manager for natural resources at Zion National Park, awoke the morning of October 18, 2001, to the sounds of crashing and breaking glass and cringed as a giant boulder mowed through the living room, bathroom, and into the bedroom of his four-year-old home (▶ Figures 8-26 and 8-27). It came within 60 centimeters of his head. The boulder, 3.5 meters by 4.5 meters, was part of a large mass of sandstone and conglomerate that broke loose from a cliff 60 meters above and tumbled down the slope. Weak shales under the massive cliff had eroded back into the slope, leaving the caprocks unsupported. Numerous large boulders litter the lower parts and base of the slope.

▶**FIGURE 8-27.** The source of the boulders was a massive sandstone and conglomerate cliff above the location shown in Figure 8-26. The boulder is partly visible middle right. The man at left provides scale.

▶**FIGURE 8-26.** A 4.5-meter boulder crushed the living room, bathroom, and part of the bedroom of this house at Rockville, Utah, near Zion National Park at 5:38 A.M. on October 18, 2001.

At 11:37 P.M. on August 17, 1959, the magnitude 7.3 Hebgen Lake earthquake triggered a large rockslide near West Yellowstone, Montana (▶Figure 8-28). The bedrock was an obvious rockfall hazard 340 meters above the riverbed. A huge mass of weathered schist in which the layering was nearly parallel to the south canyon wall detached, crossed the valley floor, and moved 200 meters up the opposite slope. It buried twenty-three people in a campground in the valley bottom. A mass of strong dolomite marble buttressed the weak rocks above, at least until the earthquake weakened it.

The toe of the slide in the Madison Valley is 1.5 kilometers wide, twice as wide as the slide scar on the south wall of the canyon. The slide mass spread out in both directions as it reached the valley floor. A blast of air from under the falling slide mass swept away two people and tumbled one automobile.

The massive dolomite marble, at the lower edge of the slide mass, stayed ahead of the schist and came to rest at the toe of the slide, highest on the far valley wall. Many of the largest single blocks of dolomite, 2 meters to 6 meters across, are covered with lichens, indicating that they rode on the slide surface without much internal churning. Nothing suggests that water played a significant role in triggering or moving the slide mass.

What Went Wrong?

A strong mass of dolomite at the base of the mountain held up weak schist and gneiss on a steep mountainside. Foliation in the schist, almost parallel to the steep slopes, formed zones of weakness. A nearby major earthquake with strong vertical motion triggered the collapse.

Donald Murray photo.

(b)

(b) The breakaway scarp of the Madison slide shows layers and parallel orientation of micas in the schist almost parallel to the slope. The soft, crumbly nature of the weathered schist is apparent in the small size of fragments on the slope.

Donald Hyndman photo.

(a)

▶**FIGURE 8-28.** **(a)**The Madison slide collapsed into the Madison River canyon and continued upslope to the north, left to right. It dammed the river to form Earthquake Lake in the foreground. The dark rectangle near the center of the photo is the visitor center building that was built to commemorate the slide.

The small town of Frank mined coal just north of Waterton-Glacier National Park in the Front Ranges of the Canadian Rockies. At 4:10 A.M. on April 29, 1903, 30 million cubic meters of rock fell 762 meters from the steep east face of Turtle Mountain, swept across the town, buried most of it, and killed seventy people. Moving on bedding planes and approximately parallel fractures in crystalline limestone, the event took less than 100 seconds (▶Figures 8-29 and 8-30). It began as a translational slide that developed into a rockfall as it gained speed. The giant pile of rock rubble continued north across the current route of Alberta Highway 3 and moved up the slope to the east to a height of 120 meters above the valley floor. Consensus is that the rubble moved as a "fluid" with compressed air, pulverized rock, and possibly water from the river that crossed its path. The immediate cause of the tragedy was mine cuts that undermined tilted layers of sedimentary rocks. Although seventeen miners were actively working within the mountain, they were below the slide scarp. Rescuers were able to dig down to the shafts and save the miners; for once, a mine was a safer place to be than in the town below the mountain. Local tradition tells of several attempts to sink a shaft through the slide mass to recover a payroll in the vault of the buried bank.

What Went Wrong?

Bedding surfaces and other zones of weakness paralleled the steep face of the mountainside. Underground coal mining weakened support of the steep mountain face.

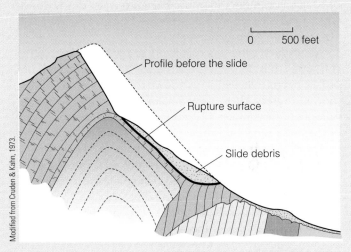

Modified from Cruden & Kahn, 1973.

▶**FIGURE 8-30.** This cross section shows the slope before and after the Frank slide. Fracture sets in some rocks and approximately parallel bedding surfaces in other rocks provided zones of weakness.

David Hyndman photo.

▶**FIGURE 8-29.** The Frank slide peeled off the whole east face of Turtle Mountain and spread a spectacular bouldery deposit across the valley and up the slope to the west. Huge limestone boulders are part of the Frank slide deposit.

On September 11, 1881, a 360-meter-high mountain face at Elm, Switzerland, collapsed into a rockfall that quickly transformed into a debris avalanche. The problem began when amateurs with no mining experience were digging slate for use as chalkboards in classrooms. The quarry opened a 65-meter-deep notch at the base of a high cliff. When the excavation reached a depth of more than 50 meters, a large crack developed above and the cliff above began to creep slowly downward. Quarrying stopped only because small rocks were falling and injuring the workers.

Surviving eyewitnesses recalled that everyone expected the cliff to fall, but no one expected it to shatter into a flood of broken debris that would rush down to the main Sernf Valley, turn 60 degrees, then continue along that nearly horizontal valley floor for another 1.5 kilometers. Before it turned, part of the mass surged up the far slope of the main valley to a height of 100 meters, overwhelming those who were running uphill to escape it. Ironically, dogs and even cattle instinctively ran to the side and survived. The whole event took forty seconds, no time to run far. Sixty-five people died.

An eyewitness watched the mass breaking up as it began to fall. It hit the floor of the slate quarry, completely disinte-grated, and shot out horizontally (▶Figure 8-31). It did not flow along the ground but launched over the slope below and cleared a creek in the valley bottom. Witnesses saw houses, trees, fleeing people, and cattle under the flying debris. The underside of the rockfall was sharply bounded, but the upper side was a cloud of rocks and dust. One witness described its movement along the main valley as a torrential flood with a bulging head that tapered to the rear. It roiled as if boiling.

The flood of rubble certainly did not ride on a cushion of air, at least not everywhere along its path. If it had, it could not have carved parallel grooves in the flat valley floor; in one place it unearthed a water pipe buried to a depth of 1 meter and carried it 1 kilometer downstream. Some houses near the leading edge of the slide were moved off their foundations. One was filled with rocks and broken boards. An old man standing inside was buried up to his neck but otherwise uninjured. Apparently, the material flowed not as a dense mass but swirled almost like a viscous fluid.

The deposit is shaped like that of a glacier or a debris flow with well-defined ridges at the sides and at the toe. Internal waves on the surface of the flow were convex downstream. No rubble sprayed beyond the toe.

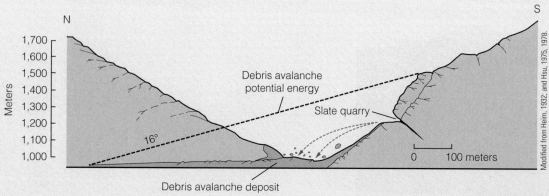

FIGURE 8-31. This cross section shows the Elm, Switzerland, debris avalanche and the location of the slate quarry that set it off.

The 1970 mudflow happened at the end of the wet season with maximum snow and glacier-ice depths on Mount Nevados Huascarán, the highest mountain in Peru. An earthquake of magnitude 7.7 at 3:23 P.M. in the subduction-zone offshore and 130 kilometers away triggered the slide. It began with a loud boom and a cloud of dust as 50 million to 100 million cubic meters of granite, glacial debris, and ice fell 400 meters to 900 meters from a vertical cliff to the surface of a glacier and raced down the valley. It disintegrated and then picked up water from ice, streams, irrigation ditches, and soil (▶ Figure 8-32). It traveled downslope at speeds of 280 kilometers per hour.

When the mudflow reached the ridge of a glacial moraine halfway down the slope, much of the debris, including huge boulders, launched into the air to rain down on houses, people, and animals. Boulders weighing several tons flew as much as 4 kilometers. Trees blew down for at least 0.5 kilometers beyond the area of boulder impact. Mud splattered more than 1 kilometer farther in a blast of wind that was strong enough to knock people off their feet and to shred bare skin. Survivors recalled that that strong wind arrived first, followed by flying rocks, then a huge wave of wet debris with a "rolling confused motion." The 1970 avalanche buried the entire city of Yungay, near the end of the flow, and several smaller towns, killing 18,000 people (Figure 8-32). The 1962 mudflow had killed 4,000 people in the same area.

Only the walls of the main cathedral, some rooftops, and the tops of a group of palm trees in the central plaza protrude above 30 meters of mud that covered the city. Water drained from most of the fluid mud in Yungay for three to four days, leaving behind the larger rocks. The flow continued another 1 to 2 kilometers down valley from Yungay to the fertile Rio Santa Valley and another 50 kilometers down that valley. One lobe temporarily dammed the Rio Santa for thirty minutes, but it took eight days for the river to cut down through the tens of meters of debris in the dam.

Eyewitness accounts indicate that the total time from earthquake to arrival of the flow at Yungay, 14 kilometers downslope, was approximately three minutes, and to the Rio Santa in three and one-half minutes! The average velocity must have been 270 kilometers per hour, but higher on the mountain the initial velocity must have been more than 750 kilometers per hour to fling rocks as far as 4 kilometers. Such high speeds would have resulted from the initial steep drop amplified at constrictions where lateral moraines caused a funnel effect. The victims could not have seen this one coming. They never had a chance.

Equally precipitous cliffs of the avalanche scar now mark the peak, and broad fresh cracks parallel the cliffs in ice on the peak. The hazard of further collapse remains.

What Went Wrong?

A major earthquake coincided with the end of the wet season when snowpack on the mountain and glacier were thickest. It shook loose rock and glacial ice on a high cliff and accelerated down valley as a large, fast-moving debris flow. Its high velocity was driven by the initial drop, steep slopes, water from ice and steams, and jetting between morainal ridges. People should avoid building on fans that show evidence of previous debris deposits, especially after a recent large event such as the 1962 flow.

Some important lessons:

■ Major rockfalls or mudflows on volcanoes with heavy ice and snow can be triggered by relatively small eruptions or by large earthquakes.

■ Mudflow volumes generally increase downslope by addition of water and sediment in the downstream channel.

■ Mudflows confined within valley walls maintain high velocities and can continue for 100 kilometers downstream.

▶ **FIGURE 8-32.** The Mount Nevados Huascarán debris avalanche in 1970 fell from the peak at the top of the photo and raced down the valley to bury the town of Yungay that occupied the lower half of the photo.

George Plafker photo, USGS.

▶**FIGURE 8-33.** This section of terrain, the Mclure slide south of Aspen, Colorado, is notorious for continuing to slide. The car plunged off the severed road in 1994 in the middle of the night, but no one was injured.

For small rockfalls of 500,000 cubic meters (i.e., roughly 80 meters on a side) or less, the angle α is approximately 33 degrees. Thus, a mass falling 100 meters would run out 166 meters. For larger rockfalls such as Elm, Switzerland, in 1887, with a volume of 10,000,000 cubic meters, α is 17 degrees. The mass falling 100 meters would run out 325 meters. For an especially large rockfall with a volume of 1,000,000,000 cubic meters (i.e., 1 kilometer on a side), α is 5 degrees. The mass falling 100 meters would run out 1,110 meters. The relationship is similar to the energy line for volcanic ash flows (see Figure 6-28, page 141; Sidebar 8-4, page 200).

A mass of rock that falls but does not disintegrate will not run far. The mechanism that permits high speeds and long run outs of debris avalanches has been a matter of considerable debate that has still not been resolved. Some authorities argue that rockfalls ride on a cushion of compressed air trapped beneath them. Air entrained between the rock fragments lubricates the mass. But major rock avalanches such as the one at Elm in Switzerland scoured deep furrows and excavated the ground beneath them, so they could not have ridden on a cushion of air.

Many of the characteristics of rockfalls and descriptions of witnesses suggest that they flow as a fluid composed of rock fragments suspended in air. The mechanism is called **fluidization.** If the mechanism works, it is because air cannot readily escape the small spaces between rock fragments during the extremely brief period of movement. With tiny spaces between small particles, rockfalls composed of small grains should be more subject to fluidization than those composed of coarser particles. The air briefly supports the fragments and lubricates their flow.

Rotational Slides and Slumps

For homogeneous and cohesive materials, those that lack a planar surface that guides the shear, landslides commonly move on a curving slip surface concave to the sky (▶Figure 8-33). The curvature of the slip surface rotates the slide mass as it moves, so the upper surface of the slide block tilts backward into the original slope as it moves (▶Figures 8-34 and 8-35). The lower part of the mass moves outward from the slope. Additional **slump** surfaces may also develop within the rotating block. Below the rotating original surface, the excess material above the original slope may collapse as an incoherent flowing mass at the toe of the landslide (Figures 8-3 and 8-35). Examples of **rotational slides** include many of those in coastal southern California such as the 1995 La Conchita landslide reviewed earlier in this chapter and slides in the Malibu area of southern California.

▶**FIGURE 8-34.** Rotation of a slump block on an arcuate surface also rotates everything on the block. These trees growing on an originally horizontal upper terrace (the Blackfoot landslide in Montana) now tilt back toward the headscarp as a result of rotation. Note the person standing at left. The headscarp is on the right.

Headscarp (= slip)

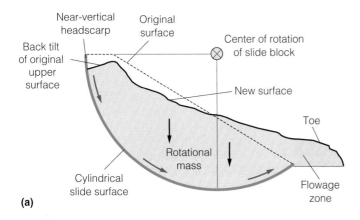

(a)

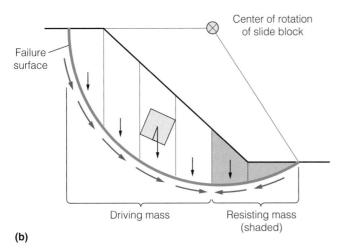

(b)

▶**FIGURE 8-35.** **(a)** This cross section shows a rotational slide. **(b)** Engineers calculate for a rotational slump sliding on a curved surface by considering slices for which they can more easily calculate the forces. They work out the stresses for each slice in the driving mass and compare the results with the stresses for the resisting mass (compare Figure 8-4). Note that the failure surface for the driving mass slopes downward and that the failure surface for the resisting mass slopes back into the slope.

Rotational slides are common in homogeneous and cohesive materials such as deep soils rich in clay or soft sedimentary deposits, including some glacial materials (▶see Figure 8-36). The lower part of the slip surface commonly dips back into the slope. That provides some resistance to further movement.

Stresses on the Shear Surface: Method of Slices

Engineers commonly estimate whether a rotational slump will move by what they call the "method of slices" (Figure 8-35). How can they figure out the shape of the slide surface underground and how deep the slip surface is? Generally, they drill holes in the landslide to find the slip surface at depth. The slip surface may appear as a thin zone of smeared out soil or a thin zone of thoroughly broken rock (▶e.g., Figure 8-37).

▶**FIGURE 8-36.** A house begins to break up in a series of rotational slumps in Colorado River bluffs, Grand Junction, Colorado.

Having determined the shape of the moving mass, engineers calculate the total of the forces pulling the slide downslope (the driving mass) (see Figures 8-4 and 8-35) and compare that with the total forces holding it back (the resisting mass). If the driving mass is large enough to overcome both the force from the resisting mass and the friction along the potential slip surface, then the slide will move. To stop it from moving, highway engineers sometimes pile heavy boulders on the toe area to increase the "resisting mass" (▶Figure 8-38). They may also increase the frictional resistance by draining water from the slide mass (Figure 8-11).

Although tree roots do not have significant resisting effect in deep rotational landslides, because they do not penetrate to the slip surface, they can help prevent initiation of some slides because they remove water from the slope. They can also increase effective friction in thin slides over shallow bedrock.

As the upper surface of the slide rotates to a gentler slope, the driving mass decreases and the resisting mass increases until it slows or stops (Figure 8-35). The new "balance" can be upset by adding material or load at the top or removing material at the toe. We mentioned above some ways in which people upset that balance. Load can also be added naturally by rain or snow. In some cases, backward rotation of the block leaves closed depressions at the top that may collect even more water or snow that can soak into the slide. In addition to load, the water may soak in to increase pore pressure and facilitate further movement. We can, of course, add load to the lower part of the slide to resist movement (Figure 8-38).

On some particularly steep slopes, such as the face of a terrace steepened by wave action, the slope may provide little or no resisting mass. In some such cases, a slump may move rapidly, with tragic results. In one such case, part of

Donald Hyndman photo.

▶**FIGURE 8-37.** Crushed rock is scattered along the failure surface of a landslide at Newport, Oregon (see arrow).

the face of a steep tree and brush-covered bluff collapsed suddenly, crushing a home in which a teacher and his young family were sleeping, instantly killing them (▶Figure 8-39). The home was part of a single row of houses on the narrow strip of beach on Puget Sound near Seattle. The teacher's parents had expressed concern about the home location but had received assurance that the home had been there for seventy years and there had never been a problem.

Sackung

The German term ***sackung*** translates as "slope sagging." Reverse scarps on steep mountain slopes are typically pull-apart features at the head of deep-seated gravity-driven movement zones. An opposing hypothesis proposes that they form by recent movement on earthquake faults. How-

ever, the scarps are restricted to the higher parts of steep slopes and have downslope slickensides (slippage striations), and some follow contour lines. Although some scarps follow ancient inactive fault zones, careful surveys indicate that active movement is creep-like in a downslope direction. Most appear likely to continue to move slowly, but a few have failed catastrophically as rockslides or rock avalanches so their presence is indicative of potential hazard (▶Figure 8-40).

Translational Slides

Translational slides move on preexisting weak surfaces that lie more or less parallel to the slope. These may be planes between sedimentary layers, inherently weak layers such as shale, old fault or slide surfaces, or fractures. Some involve soil sliding off the underlying bedrock. Translational

Donald Hyndman photo.

▶**FIGURE 8-38.** Heavy boulders are often piled on the lower part of a slide to resist movement. This road, cut through a landslide on U.S. Highway 101 near Garberville, in northern California, has been stabilized by loading its toe area.

USGS photo.

▶**FIGURE 8-39.** A section of coastal cliff at the edge of Puget Sound, near Seattle, collapsed, crushing and burying a home on the narrow strip of beach.

slides are especially dangerous because they commonly move faster and farther than rotational slides. The range of internal behavior is large. Some move as coherent masses, others break up in transit to become debris slides. Still others, lacking internal coherence, flow and spread down slope (▶ Figures 8-41 and 8-42). The Thistle slide of Utah (Figure 8-41), the Slumgullion earthflow of southwestern Colorado (▶ Figure 8-43), the destructive Aldercrest slide of southwestern Washington state (▶ Figure 8-44), and many of the small slides around Pittsburgh, Pennsylvania, are good examples of incoherent translational slides. The Vaiont slide of northern Italy is a good example of a coherent translational slide (see "Case in Point: The Vaiont Landslide").

▶**FIGURE 8-42.** In October 1985, a landslide destroyed 120 houses in Mamayes, Puerto Rico, and killed at least 129 people. The catastrophic block slide was triggered by a tropical storm with extremely heavy rainfall. Contributing factors in this densely populated area may have included sewage directly discharged into the ground and a leaking water pipe at the top of the landslide.

▶**FIGURE 8-40.** A series of sackung features in the high mountains of British Columbia approximately follow contours of the slope.

▶**FIGURE 8-43.** The Slumgullion earthflow of southwestern Colorado snakes downslope from the volcanic rock peaks in the background that are heavily altered to slippery clays, spreading out into Cristobal Lake in the lower right. It has been moving slowly for hundreds of years.

▶**FIGURE 8-41.** The 1983 Thistle landslide at Thistle, southeast of Salt Lake City, Utah, began flowing down valley in response to rising groundwater levels from heavy spring rains during the melting of a deep snowpack. Within a few weeks, the slide dammed the Spanish Fork River and took out U.S. Highway 6 and a major railroad line. Water behind the slide dam submerged the town of Thistle. With costs of more than $400 million, it was the most expensive single landslide event in U.S. history.

▶**FIGURE 8-44.** The Aldercrest landslide in southwestern Washington spread slowly downslope soon after forest was cleared and homes were built in a new subdivision.

In a classic case in the southern Alps of northeastern Italy north of Venice, filling a reservoir behind the newly completed Vaiont Dam caused catastrophic mountainside collapse (▶ Figures 8-45 and 8-46). Engineers completed the modern, 264-meter-high, thin-arch concrete dam in 1960 across a narrow, rock-bound gorge. It was designed to provide both flood control and hydroelectric power. When the dam was ready for use, the engineers filled the reservoir behind it to 23 meters below the spillway.

Heavy summer rains in 1963 filled the reservoir to only 12 meters below the spillway of the dam. The mountainside on the south side of the reservoir consisted of limestone and shale layers parallel to the 35- to 40-degree slope that was known to be unstable. An ancient slide plane that was partly exposed was not recognized or perhaps not acknowledged.*

Engineers had monitored the slope just upstream from the dam for three years, and small landslides were expected. The slope had been creeping at 1 to 30 centimeters per week. By September 1963, the rate had increased to 25 centimeters per day; and by October 8, 100 centimeters per day. At that point, engineers finally realized the size of the mass that was moving. They quickly began lowering the reservoir, but it was too late. Continued rain slowed the reservoir draining and saturated the mountainsides. As often happens, grazing animals sensed danger a week before final failure and moved off that part of the slope.

At 10:41 P.M. on October 9, 238,000,000 cubic meters of rock and debris collapsed catastrophically from the mountain face just upstream from the dam. It moved down at 90 kilometers per hour, filled a large part of the reservoir, and moved 260 meters up the far mountainside. The slide displaced all of the water in the downstream half of the reservoir. A wave 125 meters high swept over the dam and downstream, destroying Longarone and four smaller villages only two minutes after the slide began. People had no chance. The towns lay near the confluence with the Piave River, just 2 kilometers downstream from the dam. Another huge wave swept into a town on the upstream end of the reservoir. The two waves killed 2,533 people.

*Sometimes if people want something badly enough, we ignore significant negative aspects. The deep narrow rock gorge seemed an ideal place to build a dam.

Donald Hyndman photo.

Slide mass

▶ **FIGURE 8-45.** The disastrous 1963 Vaiont slide in northeastern Italy moved catastrophically on weak layers of shale within limestone beds parallel to the mountain face. The slip surface is in the upper right; the slide mass fills the center of the view. The former canyon lies under the landslide mass. The highway looping around the toe of the slide mass provides scale.

The resulting slide mass filled 2 kilometers of the length of the reservoir. Amazingly, the well-built dam remained almost undamaged. The slide generated strong earthquakes and a violent blast of air that shattered windows and blew the roof off a house well above the final level of the slide mass.

Engineering failures contributed significantly to the disaster. Exploratory drilling before dam construction intersected zones in which little or no drill core was recovered, a condition suggesting broken rock with much pore space. A tunnel excavated during dam building crossed a strongly sheared zone, but work continued without thorough examination of the implications. A study of the surrounding area would have shown that heavy surface runoff from higher slopes disappeared into innumerable solution fractures upslope of the slide plane. That water would dramatically increase the internal pore water pressure in the rocks above the reservoir.

An important lesson from the disaster is that slopes that seem to be moving slowly may at some point fail catastrophically and that catastrophic landslides sometimes show precursory movement.

In addition to the lives lost, the cost of the landslide was considerable. Loss of dam and reservoir cost $490 million; other property damage downstream came to a similar amount, and civil lawsuits for personal injury and loss of life cost even more.

▶**FIGURE 8-46.** This north–south cross section shows the Vaiont Reservoir area, preslide mass, and landslide.

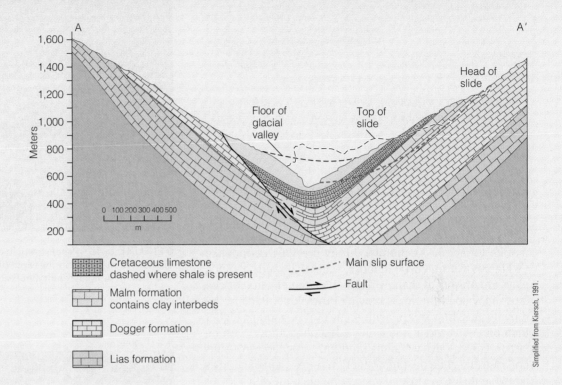

Cretaceous limestone
dashed where shale is present

Malm formation
contains clay interbeds

Dogger formation

Lias formation

Main slip surface

Fault

Simplified from Kiersch, 1991.

What Went Wrong?

■ Soft shale layers, porous limestone dipping parallel to a steep mountain slope, high pore pressure from prolonged heavy rains, and a raised groundwater level from filling a dammed reservoir. Too many overlapping circumstances combined to cause failure.

■ Rain that seeped into the ground likely built up on top of shale. Such material has low **permeability**—that is, water or other fluids cannot move easily through it. The combination of the resulting reduction in friction along the slip surface with the added water at the base of the slope was enough to cause the slope to slide.

■ The Vaiont Dam disaster remains a constant reminder to engineering geologists of the consequences of inadequate planning and geological mapping before designing and building a dam. The engineers either did not recognize an ancient slide plane or underestimated its importance to the slope instability.

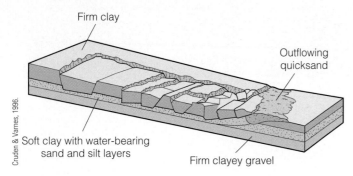

Firm clay

Outflowing quicksand

Soft clay with water-bearing
sand and silt layers

Firm clayey gravel

Cruden & Varnes, 1996.

▶**FIGURE 8-47.** A stronger layer over weak clays may fail by lateral spreading. Parts of the spreading mass may drop as it spreads.

Lateral-Spreading Slides

A variant of the translational slide type is sometimes called a *lateral-spreading* slide. One such slide occurred on the marine terrace in Anchorage during the 1964 earthquake (▶Figures 8-20 and 8-47). If loose, water-rich sands or quick clays are present at shallow depth, then liquefaction or collapse may send the mass moving downslope. Parts of the moving mass may sink, leaving some blocks standing higher than others. Those in quick clay (described above) can be fast and destructive when the randomly oriented clay flakes collapse and glide on thin, water-rich zones. Liquefaction as grains settle into a closer packing arrangement with a lower porosity (also described above) expels a fluid mix of sand and water, a **quicksand.** Liquefaction can occur on a flat surface where it does not cause a slide but it can still destabilize the ground under a building and cause collapse.

Debris Flows

Debris flows are common and extremely dangerous. They are widespread, begin without warning, move quickly, and have tragic consequences for both structures and people. Hundreds of thousands of people live on gravelly alluvial fans in Los Angeles, Phoenix, Palm Springs (▶Figure 8-48), Salt Lake City (▶Figure 8-49), Denver, and elsewhere. Most of these alluvial fans were built up from fast-moving slurries of sand, gravel, and boulders—debris flows. Even arid regions can have periods of intense or prolonged rainfall. People who live on such fans are at considerable risk from debris flows.

Debris flows differ from stream flows in the amount of solid grains suspended in the flow. Because rocks are approximately 2.7 times as dense as water, a slurry of rocks with water in the pore spaces is commonly twice the density of water in a flooding stream. That high density permits them to pick up and carry huge boulders, some as large as a car or even a school bus. Rates of movement can be rapid to slow. Although not intuitively obvious, the high density of debris

Donald Hyndman photo.

▶**FIGURE 8-48.** People living in the modern housing subdivisions in the Palm Springs area of California find the boulders make for great landscaping but seem unaware of how the boulders got there.

Mark Milligan, Utah Geological Survey.

▶**FIGURE 8-49.** A series of debris flows from the huge fault scarp of the Wasatch Front behind Salt Lake City, Utah, in 1983, inundated homes built on alluvial fans at the mouths of steep canyons, including Rudd Creek (pictured here).

flows in steep terrain can propel them at higher velocity than clear water. The tendency of some debris flows to move at impressive speeds shows that they do not owe their ability to move large rocks solely to their high viscosity.

Debris flows are characterized by internal shearing with some parts moving faster than others. Much of the movement of the flow is by the whole mass sliding on the stream bottom, with a lesser amount of jostling between fragments.

Except for all of the boulders and internal shear, movement is something like wet concrete coming down the chute of a cement truck. Slippage at the base of the flow permits it to scour material from its channel to entrain more material in the flow.

Debris flows tend to move in surges. With mixes of particle size, jostling moves the largest pieces to the edges and front of the flow. The same thing happens, because of its size, to a human body caught in a snow avalanche. Boulders buoy up to bob along the surface of the flow (Figures 8-50 and 8-51); they are pushed to the sides and front, typically forming prominent bouldery natural levees (see

▶**FIGURE 8-50.** Bouldery natural levees formed along a major 1996 debris-flow channel, a tributary to the Columbia River, near Dodson, east of Portland, Oregon. Note the large boulders concentrated at the top of the deposit.

▶**FIGURE 8-51.** This field of huge boulders is the upper surface of a debris flow that killed one person, partly buried this house, and severely damaged four others at Slide Mountain, Nevada, southwest of Reno in May of 1983. The fast-moving flow was triggered by rapid snowmelt.

▶**FIGURE 8-52.** This debris flow in the Andes of northwestern Argentina was rich in rocks and boulders until it began losing water and slowing down. That permitted the water-rich finer-grained material (covering boulders in middle left of photo) to catch up and in turn flow over the surface of the head of the flow. A natural gas pipeline is buried beneath the channel, visible as a terraced area in the top third of the photo on the left. Debris flows come down this channel every year, and in one case the flow eroded down to the pipeline, causing a massive explosion.

Figure 8-50). Although water drains easily from large spaces between boulders, movement depends on having lots of water in the pore spaces.

Movement slows when water drains from between the fragments, especially at the toe and flanks of the flow. Fine material has small interspaces that drain water much more slowly so movement continues in the fine-grained rear parts of a flow (Figures 8-52 and 8-53).

Many years of occasional rains tend to wash mud and assorted debris down slopes and into canyon floors. After decades or centuries, canyons contain enough accumulated debris to provide the raw material for a large mud or debris

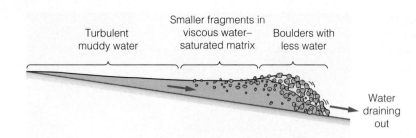

Turbulent muddy water

Smaller fragments in viscous water–saturated matrix

Boulders with less water

Water draining out

flow. That will happen as soon as one of the occasional heavy desert rains flushes the canyon floor and spreads its burden of debris out over the alluvial fan downslope. Then another long period must pass before the canyon can again accumulate enough debris to repeat the performance. It must recharge with loose debris from the sides of the canyon. Many geologists familiar with desert canyons can tell just by looking whether a canyon is sufficiently charged to produce a new debris flow.

Debris flows commonly begin with heavy rainfall or rapid snowmelt that fills pore spaces above less permeable bedrock. This increases the pore pressure that sets the mass in motion, especially if it is disturbed by an earthquake or even a strong gust of wind on a large area of slope. Often the initial movement is a landslide that begins in a swale where soils tend to become thicker and the groundwater level is high. Flows tend to keep moving as long as the positive pore-water pressure is maintained. The lower ends of swales are especially risky places for homes. In coarse colluvium (loose, broken material over bedrock), landslides that develop into debris flows begin on slopes from 33 to 45 degrees, measured down from horizontal. Where root strength is a factor, the friction angle, or slope required for failure, may increase to 40 degrees.

Flows tend to move in surges or waves, each surge with a steep front of especially coarse boulders. The initial main surge is often followed by a series of smaller surges traveling faster than the overall flow. Individual surges may slow and stop, only to be remobilized by a subsequent surge. Although most debris flows are not witnessed, their velocities and peak discharges can be estimated from peak-flow mud lines left at the sides of the channel and higher-elevation banking angle at sharp bends.

Debris flows are most common in steep mountainous deserts such as those in the American Southwest. They are especially common along major active faults where fault movements actively build mountain belts such as the Basin and Range of Nevada, the Wasatch Front at Salt Lake City, and the Andes Mountains of South America (see "Case in Point: A Tropical Climate Debris-Flow and Flood Disaster"). The steep slopes maintain a continuing supply of broken debris. Most debris flows empty onto alluvial fans. As they spread across the fan, they block old channels with dumps of gravel and boulders. Subsequent debris flows then overflow to erode new channels. The details of the channels

change, but the general style of channels on the fan remains the same. Flows continue moving on a broad alluvial fan at slopes less than 10 to 15 degrees but rarely at slopes less than 5 degrees or so. Because debris flows thin as they spread out on a fan, their energy dissipates, and their largest boulders drop near the head of the fan; progressively smaller ones drop downslope. Having dropped most of their bouldery load, water floods can continue to lower slopes.

Even relatively humid areas with higher rainfall, gentler slopes, and deeper soils can be subjected to debris flows. In August 1969, Hurricane Camille, one of only three Category 5 hurricanes to hit the United States in the twentieth century, came onshore in Mississippi where it did severe damage. Before petering out, it stalled over the mountains of central Virginia causing still more damage, primarily in debris avalanches. It dumped an amazing 71.1 centimeters of rain in eight hours, saturating the ground and causing shallow slides that quickly developed into debris avalanches.

At least 1,100 slopes slid, generally where they were steeper than a grade of 17 percent (a 10-degree slope). Most began near the inflexion point between convex rounded hilltops and the concave segments above main river channels where groundwater is closer to the surface. Where slopes were covered with coarsely granular soil, they were generally stripped down to bedrock. Formations that were most vulnerable to sliding contained nonresistant, well-layered, and mica-rich rocks that dipped parallel to the downslope direction. The problem was an unfortunate combination of torrential rainfall and mica-rich rocks with strong foliation that dipped parallel to the slope.

Twenty-six years later, on June 27, 1995, an intense storm stalled over the Blue Ridge Mountains in Madison County, Virginia, where it dumped more than 76 centimeters of rain in sixteen hours. This rain triggered more than 1,000 debris flows; one named the Kinsey run flow northwest of Graves Mill began as a landslide that developed downslope into a 2.5-kilometer-long debris flow. It raced down the slope in surges at 8 to 20 meters per second, eroding and incorporating ground material to depths up to 0.6 meters. Velocities were calculated from the tilt of the flow surface where it poured around bends in the channel. Its deposits included coarse rocks with boulders up to 7 meters by 7 meters by 1.8 meters thick in a matrix of clay to sand.

Although this flow did not impact people, others living elsewhere in the Blue Ridge Mountains are vulnerable. Debris flows are common in the Blue Ridge, and events of this magnitude have recurrence intervals of tens of years.

What triggered the flows? Intense, prolonged summer storms on thick, loose soil.

MINIMIZING THE DANGER OF DEBRIS FLOWS Debris flows can be highly destructive, even on slopes less than 30 degrees that have thick brush or forest. A broad alluvial fan spreading from the mouth of a desert canyon with picturesque boulders littering its surface is not the safest setting for residential development (Figures 8-48 and 8-57).

The obvious solution to this problem is to avoid building on those clearly hazardous alluvial fans. However, the time for that is already past in many areas. Houses, many of them expensive, already cover many of those fans, and the boulders have been artfully integrated into the landscaping (Figure 8-48). And some debris flows do bring new boulders. These events are most likely in watersheds that have lost their cover of brush to a fire within the previous few years. Bare ground does not absorb rain as well as it did when it was covered with plants, so the hills shed water like a roof, and the heavy surface runoff flushes the canyons (see Chapter 16, "Wildfires").

DEBRIS-FLOW HAZARDS As noted above, debris flows are among the most dangerous of downslope movements because of their sudden onsets and high velocities. Building in the paths of debris flows, especially on the broad, lower slopes of alluvial fans fed by debris flows, is dangerous (▶Figures 8-48, 8-54, 8-55, and 8-56). The dangers include not only burial in heavy debris but also huge impact forces from fast-moving boulders.

Detection devices can help warn people of an already moving debris flow. The best are acoustic flow monitors

▶**FIGURE 8-55.** It may not be obvious, but the car this geologist is standing on is on the edge of the roof of a house buried in a bouldery debris flow that flowed across its rooftop (note the roof vent pipe above the car's roof).

▶**FIGURE 8-56.** The small community of Orting, Washington, is built on huge mudflows that poured down a valley from Mount Rainier 500 years ago. Earlier mudflows filled the same valley; undoubtedly, mudflows will do so again.

▶**FIGURE 8-54.** In May 1998, a muddy debris flow from the steep hillside to the right of this house in Siano, Italy, east of Mount Vesuvius, had sufficient momentum to blow right through the walls of the house and out the other side. Debris on both balconies (arrows) provides an indication of the height of flow.

that detect the distinctive rumble frequency of ground shaking caused by debris flows. The sensed motion is telemetered automatically to downstream sirens. People should immediately run to higher ground off to the sides of a debris flow path. Such sensors are in use at Mount Rainier (▶Figure 8-56) and in Alaska, Ecuador, and the Philippines. Where

The rainy season in coastal Venezuela is normally from May through October, so a storm in the first two weeks of December 1999 was unusual. A moist southwesterly flow from the Pacific Ocean collided with a cold front to cause moderately heavy rains of 29.3 centimeters. Then on December 14 to 16, when soils were already soggy, torrential rains arrived. Rainfall near sea level at Maiquetia International Airport amounted to 91.1 centimeters in fifty-two hours! Imagine almost a meter of water on the landscape in a little more than two days, all headed downslope. Higher elevations toward the crest of the mountain range received twice as much rainfall as areas along the coast. This was the area's greatest storm in more than half a century.

Flash floods and debris flows inundated the coastal towns of Maiquetia and La Guaira, 56 kilometers north of Caracas. Most homes and buildings in this area crowd large alluvial fans and narrow valley bottoms at the base of incredibly steep, unstable mountainsides (▶Figures 8-57 and 8-58). Floods began after 8 P.M. local time on December 15. Eyewitnesses who fled back from the river and watched the flood from nearby rooftops reported crashing rocks and debris flows around 8:30 P.M. Massive mudslides and floods killed an estimated 30,000 people; more than 400,000 were home-

less. Exact numbers are difficult to determine because muddy slides buried many people or swept them out to sea in the floods and debris flows. Mud and debris either swept away or buried shantytowns of tin-and-cinderblock shacks that cover extremely steep mountainsides of the Cordillera de la Costa, an area that is especially vulnerable to mudslides during heavy rains. Although the potential costs for property damage is low in these barrios, the potential for loss of life is high.

Losses totaled $1.94 billion (in 2002 dollars), with heavy damage to homes, apartment buildings, roads, telephone and electric lines, and water and sewage systems. Because the only gently sloping land along this coastal part of Venezuela is on alluvial fans (▶Figure 8-57), the large fan at Caraballeda was intensively developed with large multistory houses and many high-rise apartment buildings. At the upper end of the fan, the flood flow was above channel capacity; on reaching the fan, the flood separated into several streams to spread debris throughout the city, in places up to 6 meters thick. Outside the main channel, flows destroyed many two-story houses. They impacted the second stories of several apartment buildings, leaving boulders more than 1 meter in diameter there (▶Figure 8-57b). The ends of several buildings collapsed. Residents described several high stream flows and debris flows that be-

(a)

(b)

▶**FIGURE 8-57.** **(a)** The debris flow and flood in Venezuela destroyed most of the homes on the low-lying Caraballeda fan at the mouth of the canyon. **(b)** The end of this apartment building on the Caraballeda fan collapsed when debris-flow boulders crushed key support columns. The largest boulder is more than 2 meters high.

Matt Larsen photo, USGS.

▶**FIGURE 8-58.** Homes on and at the base of steep hillsides in Venezuela were crushed as the slopes gave way in the saturating rain.

gan the night of December 15 and continued until the next afternoon. With little warning, people caught in their homes were buried by flood debris or carried out to sea.

This was not the first disastrous debris flow in the region. Eight similar events are found in the historic records between 1798 and 1951. Examination of older debris-flow deposits shows that some flows were larger—thicker and with larger boulders—than the 1999 flow.

Food, water, clothing, and antibiotics were in short supply. Evacuation and disaster recovery was especially difficult because many areas were difficult to reach. The single highway along the coast was extensively blocked and cut by debris flows and flood channels. Additional roads and helicopter landing pads would facilitate evacuation and recovery in future disasters.

Donald Hyndman photo.

▶**FIGURE 8-59.** A modern debris-flow collection basin was built in Rubio Canyon, north of Pasadena, California. Downstream is to the upper left.

likely sources of debris flows are a short distance upstream and the channel gradient is high, the warning time may be too short for evacuation. A tripwire installed in the canyon 3 kilometers upstream from heavily populated areas on the Caraballeda fan in Venezuela after the December 1999 disaster would have provided only five minutes' warning for the large population there, almost certainly not enough for most people to move out of danger. Inexpensive tripwire sensors are in use in many areas, but false tripwire warnings can be triggered by falling trees, animals, or vandalism.

The best solution in minimizing the impact of debris flows is to not build in a vulnerable area. Where buildings predate recognition of the hazard, education and zoning can minimize the problem. Warnings of increased hazard such as during prolonged or heavy rainfall can help, but the steep terrain in which most flows occur leaves little time for evacuation once a debris flow has begun moving. This is especially true at night in a heavy rainstorm. The sound of a rainstorm can drown out that of an approaching debris flow.

What else can be done?

- Zone especially hazardous areas as open space, including parks, golf courses, or agriculture.
- Prohibit building on the debris fan and in some cases buy out and remove existing development to open pathways for future flows. Where development is necessary, orient buildings and streets with their lengths parallel to the downslope direction of flow to limit building exposure to flows.
- Build walls to deflect large-volume debris flows to a part of a fan with restricted development.
- Channel the debris flows into a debris basin large enough to contain all of the loose debris in the channel upstream. These must be cleaned out after each flow (▶Figures 8-59 to 8-61).

Michael J. Bolander photo.

▶**FIGURE 8-60.** A 1978 storm filled the basin behind this debris-flow dam and overtopped the dam at La Crescenta, California. The large, yellow caterpillar tractor and the men provide scale.

Donald Hyndman photo.

▶**FIGURE 8-61.** Houses immediately below a debris-flow dam, north of Pasadena, apparently trust the dam to block all debris flows and floods that may come down the canyon.

- Slow the debris flow by constructing check dams in its path. These and other dams must be able to withstand the inertial impact force of a rapidly moving debris flow.

- Trap the debris in a flow by building permeable structures across the channel above the alluvial fan.

- Monitor conditions to provide warning when values reach known thresholds for triggering an event. For example, the U.S. Geological Survey in the San Francisco Bay area now has a two-stage system for landslides based on past experience: (1) a seasonal accumulation of 11 inches (27.5 centimeters) of rainfall (usually by November) and (2) a single storm that provides more than 30 percent of local mean-average precipitation.

TRAPPING DEBRIS FLOWS Structures to trap debris flows in canyons upstream from alluvial fans include permeable dams that stop boulders but permit water to drain, that is, grid dams consisting of cross-linked steel pipes, horizontal beams, vertical steel pipes, or reinforced columns. Widely used in Canada, Europe, Japan, China, Indonesia, and in the United States, they abruptly slow the progress of the debris flow by draining the water. The grid is generally spaced to permit people, animals, and fish to easily travel through the structure.

Danger Signs: Evidence for Former Debris Flows

Where a debris flow has occurred in the past it will occur again in the future. If there is no evidence of their previous occurrence, then debris flows are not likely to develop as long as the conditions upstream remain essentially unchanged. Disturbances such as fire, logging, housing developments, road building, and volcanic activity can change a basin to make it more prone to forming debris flows. Because the maximum amount of material carried by a debris flow is the amount of loose material available in the catchment basin channels, the maximum size of a future debris flow is the volume of that loose debris. A large debris flow that removes most of the loose material in the basin will lessen the maximum size of future flows for years until the loose material again builds up. Some small streams that transport sediment in normal floods can occasionally develop more destructive debris flows in a larger storm.

Evidence of former debris flows include the following:

- A valley floor strewn with boulders that seem much too large for the modern stream to move. In exposed cross sections, huge boulders are perched on top of finer-grained deposits that are massive, unlayered, and have large, angular boulders in a finer matrix (Figures 8-50 and 8-52);

- levees of coarse, angular material next to the stream (Figure 8-50);

- deep, narrow channels cut in those deposits (Figure 8-50);

- fan-shaped deposits forming rounded lobes with coarser material at the outer edges;

- rocks lodged against trees, deposited in tree branches, or embedded in bark (▶ Figures 8-62 and 8-63);

- bark scars high on trees on their upstream sides (Figures 8-62 and 8-63);

▶**FIGURE 8-62.** Huge boulders are lodged against the upstream sides of trees and the trees are debarked along Tumalt Creek, east of Portland, Oregon. Another bouldery natural levee is across the stream in the upper right.

▶**FIGURE 8-63.** Rocks (above and just to the right of the red pocketknife in the center of this photo) are lodged in the branches of a tree, which was debarked to about that height by a debris flow along Sleeping Child Creek, south of Hamilton, Montana, in July 2001.

- lobes of even-age vegetation younger than the surrounding vegetation;
- a drainage basin with large, actively eroding areas; and
- active faulting, which helps supply broken rock to a slope.

The age of a mud flow or debris flow is rarely obvious and generally hard to determine. No lichens grow on freshly deposited mud or debris flows. As the years pass, one kind of lichen after another colonizes the boulders. Most kinds grow slowly, and each kind grows at its own pace, so the crust of lichens changes character with age. Observation of the size and species of lichens growing on dated headstones in nearby cemeteries can provide insight into how long ago the rocks in a mudflow were deposited.

Although insurance cannot be purchased for landslides or other ground movement, debris flows and mudflows may provide some exceptions. The distinction for insurance purposes is generally that if the flow is too watery to be shoveled, then damages can be claimed under flood insurance. Thus, some fast-moving, watery debris flows may be covered.

Mudflows and Earthflows

If mud or clay dominates the solids choking a flowing mass, it is a **mudflow;** if the solids are coarser and highly variable in size, it is a debris flow. Volcanic mudflows are reviewed in more detail in Chapter 6.

Active volcanoes are notorious for spawning mudflows, especially during eruptions because volcanic ash supplies abundant mud-size material. However, mudflows often contain rocks and boulders as well. The collapse of heavily altered ash on the flank of a volcano during heavy rains or rapid melting of snow or ice forms a dense slurry that collects loose volcanic debris as it races downslope (▶Figure 8-64). During volcanic eruptions, hot ash may mix with rain, lake, or stream water downslope to pour down nearby valleys. The mixing of hot rocks and ash with ice and snow causes much more rapid melting than situations in which these hot materials simply fall on the snow surface. Even if the falling ash is too cool to melt snow or ice, large volcanic eruptions often generate their own weather. The heat of the eruption pulls in and lifts outside air, causing it to expand and cool. That causes moisture in the air to condense; locally heavy rains fall on the newly deposited ash, flushing it downslope.

Even long after major ash-rich eruptions, the amount of ash on the volcano flanks provides ample material for catastrophic mudflows. Such volcanic mudflows, or *lahars*, often have especially long run outs, especially where they are confined to valleys. They can continue to move on slopes as low as 1 percent. Prominent examples of recent volcanic mudflows are at Mount St. Helens, Washington, in 1980; Mount Pinatubo, the Philippines, in 1991; and Nevado del Ruiz, Colombia, in 1985.

Like other landslides, mudflows are mobilized by abundant water in the slope and are often triggered by heavy or prolonged rainfall. The tiny pore spaces and low permeability of mud slurries retain water and keep the mud mobile. Boulders are rafted along near the surface of flows that have high density and resemble wet concrete. Earthquakes jar others loose. The tragic landslide in Colonia Las Colinas, El Salvador, which was triggered by a magnitude 7.6 (M_W) earthquake on January 13, 2001, killed 585 people (▶Figure 8-65). It was mostly dry ash that collapsed from the top of a steep

USGS.

▶**FIGURE 8-64.** This is the flow front of one of the many fast-moving mudflows racing right to left down the valley from Mount Pinatubo in the Philippines. Note that the surface of the flow is covered with rocks and pebbles, especially in the nose of the cresting flow.

E. L. Harp photo, USGS.

▶**FIGURE 8-65.** A large earthquake on January 13, 2001, triggered this devastating mudslide in a middle-class subdivision in Colonia Las Colinas, El Salvador. It killed approximately 585 people.

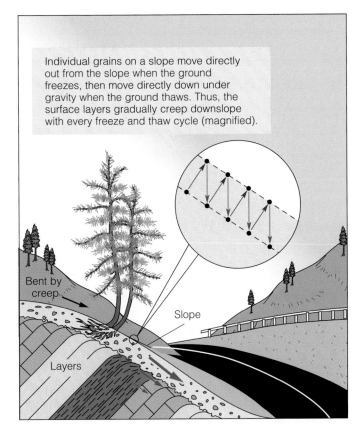

Individual grains on a slope move directly out from the slope when the ground freezes, then move directly down under gravity when the ground thaws. Thus, the surface layers gradually creep downslope with every freeze and thaw cycle (magnified).

Bent by creep

Slope

Layers

▶ **FIGURE 8-66.** Soil creep produces a slow downslope movement of the upper layers of soil or soft rocks.

Donald Hyndman photo.

▶ **FIGURE 8-67.** These trees along California's Highway 1 west of Leggett have been bent by soil creep. They originally grew upright but were tilted outward by creep. Continued upward growth produces the bending.

escarpment 166 meters high; it ran out to 735 meters. The slide was channeled by streets that were parallel to the downslope direction and demolished hundreds of houses.

Soil Creep

Soil creep, the slow downslope movement of colluvium and weak rock, involves near-surface movement and is not especially dangerous. It tilts fences, power poles, and walls. The rate of movement decreases at greater depth because most driving processes operate close to the surface. Alternating expansion and shrinking of the soil from several processes, including wetting and drying or freezing and thawing, cause creep to accelerate. When the soil expands, it moves out perpendicular to the slope; when it shrinks, it moves more nearly straight down under the pull of gravity. The net change is slight movement downslope. When burrowing animals tunnel into the soil, their cavities eventually collapse. Plant roots expand the soil and then later rot, their cavities collapsing. Even trampling by animals and people is significant over a long period of time (▶ Figure 8-66). Trees that stand straight but with their bases curving back into the slope, so-called pistol-butt trees, have trunks that initially grow upward but become tilted downslope with movement by creep (▶ Figure 8-67). Even layers in the bedrock can bend downslope (▶ Figure 8-68).

NOAA photo.

▶ **FIGURE 8-68.** Creep of sedimentary layers near the ground surface bends their upturned ends downslope (to the right).

Table 8-2 Typical Velocities for Various Types of Downslope Movement*

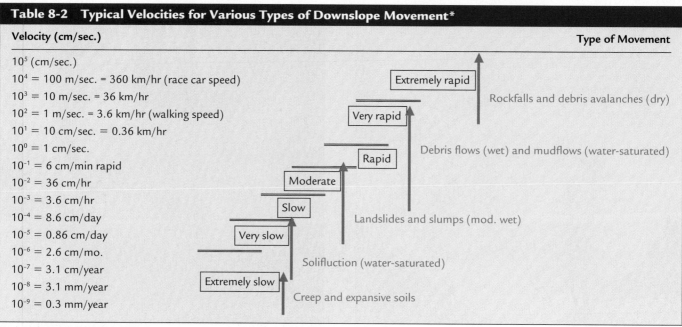

From Finlayson & Statham, 1980.

Solifluction is another type of near-surface downslope movement that occurs in extremely cold Arctic or alpine areas where water-saturated ground freezes to great depth. It is common downslope from snowdrifts on slopes. When near-surface layers thaw, the water cannot drain downward because the soil below is still frozen. The soggy near-surface layers slowly ooze downslope.

Rates of Downslope Movements

Rates of downslope movements are highly variable, even for individual mechanisms of movement. Rates depend on many factors, including, slope steepness, grain size, water content, thickness of the moving mass, clay mineral type and amount, and many other factors. Table 8-2 provides approximate movement rates.

Submarine Slides

Volcano Flank Collapse

The flanks of many major oceanic volcanoes, including those of the Hawaiian Islands and Canary Islands off the northwest coast of Africa, are likely to collapse and slide into the ocean at some point in the future. Large oceanic volcanoes start to grow on the deep ocean floor, approximately 4,000 meters below sea level. These submerged lower parts consist largely of loose volcanic rubble formed when the erupting basalt chilled in seawater and broke into fragments. This rubble has little mechanical strength. Above sea level, oceanic islands are built from reasonably solid basalt lava flows.

The three broad ridges that radiate outward from the top of the volcano spread slightly under their own enormous weight, producing rift zones along their crests. The volcano eventually breaks into three enormous segments that look on a map like a pie cut into three large slices of about the same size. One or more of the three volcano segments may begin to move slowly seaward. The rifts between the sediments provide easy passage to molten magma rising to the surface and become the sites of most of the eruptions. They also form weak vertical zones in the volcano. One or more of the volcano segments often picks up speed and slides into the ocean, sometimes slowly but sometimes catastrophically. Submarine radar topographic surveys of the immense masses of landslide debris on the ocean floor show them extending as much as 70 kilometers from the volcano. They must have moved rapidly.

Each great dump of debris on the ocean floor starts seaward of an enormous coastal cliff on the volcano. Some of those on the Hawaiian Islands are more than 2,000 meters high, the highest coastal cliffs in the world. They appear to be the head scarps of giant detachments, each of which dumped a large part of a volcano into the ocean.

Hawaiian geologists wondered for years about blocks of coral and other shoreline materials strewn across the lower slopes of the islands to an elevation of almost 300 meters above sea level. It now seems clear that huge waves, tsunami, formed when enormous masses of one of the islands collapsed into the ocean and the waves washed material up the slopes from the beach. A future repetition of those events seems inevitable. When that does happen, it will kill much of the population of the Hawaiian Islands. (See further discussion of the hazards associated with volcanic flank collapse in Chapters 5, 6, and 7 on tsunamis and volcanoes.)

Table 8-3 **Prominent Landslide Dams Showing the Type of Landslide and Whether the Dam Failed**

Landslide Dam	Type (and Material)	Year	River Blocked	Dam Height (m)	Dam Length (m)	Dam Width (m)	Lake Length (km)	Dam Failed?	See Figure
Madison, Montana	Rock slide (broken rock)	1959	Madison	60–70	500	1,600	10	No	8-28
Mayunmarca, Peru	Rock slide (broken rock)	1974	Mantaro	170	1,000	3,800	31	Yes	
Gros Ventre, Wyoming	Slide (soil)	1925	Gros Ventre	70	900	2,400	6.5	Yes	
Slumgullion, Colorado	Earth flow (altered volcanic rock, smectite clay, soil)	1300	Gunnison	40	500	1,700	3	No	8-43
Polallie Creek, Oregon	Debris flow (gravel, boulders)	1980	Hood, East Fork	11	—	230	—	Yes	
Thistle, Utah	Earth slide	1983	Spanish Fork	60	200	600	5	No	8-41
Usoy, Tadzhiistan	Landslide (broken rock)	1911	Murgab	500–700	1,000	1,000	60	Partial, may yet fail	

Costa and Schuster, 1988, USGS.

Failure of Landslide Dams

Any moderately fast-moving landslide can block a river or stream to create a dam. Many back up water to form a temporary lake before they eventually fail. Of those dams that failed, roughly a quarter eroded through in less than a day, half failed within ten days, and some lasted for a long time (Table 8-3). What controls this behavior and what are the downstream dangers? The time before dam failure and the size of the resulting floods depend on:

- the size, height, and geometry of the dam;
- the material making up the dam;
- the rate of stream flow and how fast the lake rises; and
- the use of engineering controls such as the excavation of artificial breaches or the construction of artificial spillways or tunnels.

Mudflows, debris flows, and earth flows create many natural dams that block rivers quickly, are not high, are composed of noncohesive material, and overtop and breach soon after formation. Other kinds of flows block rivers much more slowly and are often long-lived. Most landslide dams fail because the water behind them overflows and erodes a spillway that drains the lake. That may not happen if the dam is so permeable that the lake drains by seepage instead of through an overflow spillway. And some landslide dams consist of blocks too large for the overflow stream to move. A dam is much less likely to fail if it consists of large particles because the stream cannot move such large particles, even on a fairly steep gradient. A small landslide dam is likely to fail soon after formation if it blocks a large stream. Permeable, easily eroded sediment is vulnerable to piping, seepage, and undermining that can lead to dam failure.

Examples of landslide dams include those shown in Table 8-3.

To minimize the chance of a catastrophic flood downstream if the dam fails, officials commonly try to stabilize landslide dams by constructing a channelized spillway across or around the dam. The U.S. Army Corps of Engineers handled the Madison rockslide this way. At the Thistle slide, the Corps used pipe and tunnel outlets.

The height and volume of the impounded water, and therefore its potential energy, control the maximum height of a flood from the failure of a landslide dam (Sidebar 8-5).

Sidebar 8-5

Specifically, estimates for the peak discharge for a landslide dam failure are (Costa and Schuster, 1988):

$$Q_{max} = 6.3H^{1.59}$$
$$Q_{max} = 672V^{0.56}$$
$$Q_{max} = 181(HV)^{0.43}$$

where

Q = peak discharge in m³/sec.

V = reservoir volume in millions of m³

H = height in meters

Most dam-failure floods are less than this, and most flood flows decrease rapidly downstream. However, if the flood incorporates significant easily eroded sediment into the flow, then it can bulk up to form a debris flow many times larger than the flow at the dam itself.

Different kinds of landslide dams fail in different ways and at different rates. Constructed dams, especially high-arch concrete and concrete-gravity dams are most susceptible to rapid failure. Such failures often occur by erodible material piping below the dam, high pore pressure in weak porous rock under the dam, or sliding on fault surfaces under the dam. The calculated maximum flows can be used to understand the level of plausible danger downstream from dams of all kinds. Broader valleys will spread out the flood and decrease flood levels much more quickly downstream than will narrow canyons.

Why not construct a useful dam on a landslide dam? Actually, in the twentieth century, at least 167 dams were constructed that way. Most of these were built on top of rockfalls or rock slides because those are most stable, only a few on softer materials. One outstanding example of a landslide-dammed reservoir that produces hydropower is Lake Waikaremoana, the largest landslide-dammed lake in New Zealand.

Only one of these modified landslide dams failed catastrophically. In 1928, the St. Francis high-arch concrete dam failed while the reservoir was being filled because it was built on the toe of a large Pleistocene landslide in schist bedrock. Four hundred fifty people drowned. One seldom addressed concern about constructing a dam atop a landslide is whether other slopes around the new reservoir are also prone to landsliding. If so, filling the reservoir would raise the water table in surrounding slopes, possibly leading to major slope failure that would drive a surge wave of water over the dam.

Landslide Influences and Hazard Maps

Geographic Information Systems (GIS) can be used to build debris-flow and landslide-hazard maps. Such GIS maps can be used to prescribe restrictions in land use such as road building, timber harvesting, or even housing subdivisions. A model commonly used for shallow landslides uses factors that affect landslide development.

Debris flows and landslides are more likely in the following environments:

- steep slopes (Figure 8-4) and clearly mountainous areas;
- when the local slope exceeds the local angle of repose (friction angle or stable angle of repose of a hillside is 33 to 45 degrees);
- areas with abundant loose debris on a slope;
- slopes with low permeability, as in fine-grained soils;

- times when rainfall or snowmelt amounts entering the ground are high;
- when shallow slides commonly develop at the interface between bedrock and the loose colluvium that covers it; and
- locations of previous shallow landslides of any size (these suggest an unstable slope).

Shallow slides are more likely to develop on slopes with sparse vegetation and a lack of significant tree roots to hold surficial material in place. This factor is relevant only for debris flows and small translational slides that are shallower than the depths of root penetration.

In the GIS approach, the area of concern is mapped—for example, on a scale of 1:20,000—and the area is divided into a set of polygons. Each polygon is chosen as having consistent internal attributes such as slope, concave-upward curvature, soil texture and depth, ease of slope drainage, slope-facing direction, type of vegetation, bedrock type, length of roads within the polygon, and presence of slope failures. Polygons with slopes of less than 15 degrees, for example, may be excluded from study because they typically lack evidence of landsliding and are less likely to slide (▶ Figure 8-69).

Why Do People Build in Landslide-Prone Areas?

Aside from convenience of location, the places considered most desirable for building homes are often the most scenic. People build at the tops of steep slopes to take in magnificent views of lakes, rivers, or valleys. Unfortunately, those slopes are steep for a reason. They are there because they are being oversteepened by processes that remove material at the base of a cliff. Waves undercut cliffs at the beach, rivers undercut cliffs on the outsides of meander bends, and roads undercut the base of slopes. People build on the slopes themselves, again for views. Some of those slopes are soft materials that were able to stand for a long time until people dug into them to provide a flat area for the house or its yard. In doing so, they undercut the slope above and overloaded the slope below, making the slope more prone to sliding.

People build at the base of cliffs or top of steep slopes because their homes seem nestled in scenic environments (Figures 8-1, 8-8, 8-27, 8-36, and 8-39). In some cases, they build among large boulders that provide convenient highlights for landscaping (Figure 8-48). They tend to not think much about where the boulders came from. If they did, they would quickly conclude that they came off the cliff above. These are truly dangerous places to live.

Could there be a larger landslide in a particular area, larger than any in historic time?

To summarize, given already weak rocks in an area, the most significant factors in landslide formation include the following:

USGS.

Explanation

Landslide Incidence

- Low (less than 1.5% of area involved)
- Moderate (1.5% –15% of area involved)
- High (greater than 15% of area involved)

Landslide Susceptibility/Incidence

- Moderate susceptibility/low incidence
- High susceptibility/low incidence
- High susceptibility/moderate incidence

- adding a large amount of water to the slope;

- increasing the slope angle, such as by undercutting the toe of the slope or adding fill to the top;

- increasing the load on the slope, such as by adding water, fill, or a large building high on the slope;

- a major earthquake shaking the ground and increasing the pore water pressure to "float" the overlying grains; and

- removal of vegetation by logging, forest fire, overgrazing, or clearing for development.

Most of these factors are independent of one another. The weather does not depend on whether vegetation has been removed, or the slope angle has changed, or load has been added. Nor is it related to the occurrence of an earthquake. Examination of our "chaos net" (Figure 1-5, page 6) shows many of the major influences on landslides. Is it possible that several of these could overlap in time? In one such unfortunate coincidence, the devastating mudflows that accompanied the 1991 eruption of Mount Pinatubo in the Philippines developed in large part because a major tropical typhoon hit the area just in time for the eruption.

Mount Rainier, the highest and largest of the Cascade volcanoes, has not erupted in 2,500 years but is still thought to be active. Its snow-clad flanks are steep but weak; they are heavily fractured and altered to clays. A megathrust earthquake on the subduction zone off Washington and Oregon last happened in January 1700. Consider the following scenario, unlikely though it may seem at first sight. Winters are wet in the Pacific Northwest, with especially heavy snow packs at high elevations. If the next giant megathrust earthquake were also to happen in winter, strong shaking lasting three minutes or longer could cause collapse of a large part of the flank of the mountain, possibly similar to the gigantic flank collapse of Mount Shasta 300,000 years ago (▶ see Figure 8-70). If the collapse were toward the

S. Brantley photo, USGS.

▶ **FIGURE 8-70.** The hummocks that cover the foreground of this photo were deposited as a massive landslide from Mount Shasta, visible in the upper left. This slide occurred between 300,000 and 380,000 years ago.

communities to the northwest, the consequences would be tragic. What about the possibility that water-bearing magma could be resting or rising slowly within the volcano at that time? Sudden decrease in pressure by flank collapse could trigger rapid expansion of magmatic gases and eruption. None of these possibilities is implausible. Perhaps the likelihood of such coincidences is low, but they do happen.

Where one major event could trigger another, the likelihood of simultaneous occurrence is much greater. Eventually, given enough time, several of these could conceivably overlap at the same time, regardless of whether any of the events are cyclic (see Figure 1-4, page 5). The results could be truly disastrous.

KEY POINTS

✓ Factors that influence downslope movements include the slope and the natural angle of repose of a material. **Review pp. 187–188.**

✓ A landslide has many features that reflect the movements that formed it, including the zones of depletion and accumulation, a headscarp, transverse cracks, and ridges. **Review pp. 188–189; Figure 8-3.**

✓ Factors that dictate whether the slope will move include the load, slope angle, material strength and frictional resistance, and water content. **Review pp. 189–190; Figure 8-4.**

✓ Water is the most important factor that affects slope stability. A little water coating the grains can make a slope less prone to sliding, but water filling the pore spaces, and especially a greater load of water above, makes it more prone to sliding. **Review pp. 191–192; Figures 8-5 and 8-6.**

✓ Water can be removed from a slope by inserting perforated drains or by planting vegetation to increase evapotranspiration. Vegetation takes up water in roots and transpires it through the leaves; rain falling on leaves partly evaporates and does not reach the ground. **Review pp. 192–193; Figures 8-10 and 8-11.**

✓ Clays, especially those containing soils that swell when wet, become extremely slippery and prone to sliding. **Review p. 193.**

✓ Loosely packed sandy soil saturated with water can "liquefy," settle into a smaller volume, and spread when shaken with an earthquake. **Review pp. 194–195; Figures 8-12 to 8-14.**

✓ Quick clays, consisting of clay flakes laid down in saltwater, can stand on edge like a house of cards. If the clays are raised, drained, and rinsed with fresh water, they are susceptible to collapse. **Review pp. 194–195; Figures 8-15 and 8-16.**

✓ Earthquakes trigger many landslides, especially rockfalls. **Review pp. 195–196.**

✓ Internal surfaces sloping in the same direction of land surface provide weak zones that facilitate slip, especially if the surfaces daylight—that is, they come back to the surface at their lower end—or if they are old landslide slip surfaces. **Review pp. 196–197; Figure 8-21.**

✓ Rockfalls are facilitated by cliffs with nearly vertical fractures and moisture that can freeze at night. People living below such cliffs are in grave danger. **Review pp. 198–201.**

✓ Debris avalanches are similar to rockfalls except that the fragments disintegrate to tiny pieces that entrain air or water and flow at high velocity much like a dense liquid. Their travel distance depends primarily on their height of fall. **Review pp. 200, 204, and 206; Sidebar 8-4.**

✓ Rotational slides and slumps are common in weak, homogeneous material. The sliding mass rotates on a curving surface. It can be slowed or stopped by piling a heavy load on the toe of the slide or removing load from the top. **Review pp. 206–207; Figures 8-35 to 8-38.**

✓ Translational slides are facilitated by slip surfaces inclined nearly parallel to the ground surface. **Review pp. 208–209.**

✓ Debris flows begin as loose material in a steep canyon that entrains water to form a fast-moving slurry that generally spreads out on an alluvial fan. Larger rocks rise to the top, front, and lateral edges of the flow. People building on fans at the mouth of debris-flow channels place themselves in real danger. To decrease the hazard, much of the solid material can be trapped in debris-flow basins at the head of an alluvial fan. **Review pp. 212–220; Figures 8-48 to 8-55 and 8-59.**

✓ Previous debris-flow deposits can often be recognized as unlayered deposits with a concentration of the larger rocks at the top and along its edges, levees of coarse and angular material next to the stream channel, and rocks lodged against the upslope sides of trees or in their branches. **Review pp. 219–220.**

✓ Mudflows are similar to debris flows except that mud dominates and chokes the flowing wet mass. They are especially common on the lower flanks of volcanoes coated with abundant loose ash. **Review p. 220; Figure 8-64.**

✓ Soil creep involves slow downslope movement of soil or bedrock, especially near the surface, which is caused by alternating heating and cooling, freezing and thawing, and burrowing and trampling by animals. **Review p. 221; Figures 8-66 to 8-68.**

✓ Flanks of major oceanic volcanoes with weak fragmental submarine deposits can collapse, sometimes suddenly, to cause major tsunami. **Review p. 222.**

✓ Landslide dams can fail, depending on the size of the dam, its material, and the rate of rise of the lake behind the dam. **Review pp. 223–224.**

IMPORTANT WORDS AND CONCEPTS

Terms

angle of repose, p. 188
cohesion, p. 190
daylighted surface, p. 197
debris avalanche, p. 200
debris flow, p. 212
evapotranspiration, p. 192
fluidization, p. 206
frictional resistance, p. 196
geotextile fabric, p. 193
liquefaction, p. 194
mudflow, p. 220
perforated pipes, p. 193
permeability, p. 211
porosity, p. 194
quick clay, p. 194
quicksand, p. 212
resistance to sliding, p. 169
rockbolts, p. 196

rockfalls, p. 196
rotational slide, p. 206
sackung, p. 208
shotcrete, p. 196
slip surface, p. 196
slope angle, p. 189
slump, p. 206
smectite, p. 193
soil creep, p. 221
surface tension, p. 191
swelling soils, p. 193
talus, p. 198
translational slide, p. 208
water pressure, p. 191
zone of accumulation, p. 189
zone of depletion, p. 188

QUESTIONS FOR REVIEW

1. The maximum slope at which loose grains will stand depends on many factors. What is the approximate maximum slope for loose, rounded, dry sand grains?

2. List the main factors that affect whether a slope will fail in a landslide.

3. Why is the friction force on a gently sloping slip surface greater than that on a steeply sloping slip surface?

4. Why does raising the groundwater level above a point in the ground often lead to slope failure?

5. List several distinctly different ways in which water can be removed from a wet slope that has begun to slide.

6. What causes liquefaction of sediments? Briefly explain the process.

7. What circumstances lead to formation of quick clays? List the sequence of events.

8. List several ways in which old landslides are commonly reactivated.

9. Why does the top of a rotational slide tilt back into the slope? Be specific.

10. What can be done to slow or stop movement of a rotational slide? Be specific.

11. Why is it dangerous to live on an alluvial fan? Be specific.

12. How can you tell by looking at a canyon that has produced debris flows in the past whether it will produce another within the next few years?

13. If you see a debris flow or mudflow heading down a valley toward you, what should you do? Be specific.

14. What parts of the United States are most susceptible to landslides? Be specific as to geographic location.

15. Why do people build in landslide-prone areas?

FURTHER READING

Assess your understanding of this chapter's topics with additional quizzing and conceptual-based problems at:

 http://earthscience.brookscole.com/hyndman.

CHAPTER 9

Devin Galloway photo, USGS.

A large sinkhole developed May 1981 in Winter Park, Florida. It swallowed a house, a couple of Porsches, and half of a municipal pool. See Figure 9-4 for a ground view of this same sinkhole.

SINKHOLES, LAND SUBSIDENCE, AND SWELLING SOILS

Ground Collapse and Sinkholes

A variety of ground movements cause far more monetary damage in North America than the more dramatic hazards discussed in this book. **Sinkholes** develop when the overlying ground collapses into underground cavities in limestone. Land subsidence often occurs when large amounts of groundwater or petroleum are pumped out, or when an earthquake causes closer repacking of sediment grains. **Swelling soils** typically form by alteration of volcanic ash to clays that swell when wet. All of these ground movements can deform and effectively destroy roads, utilities, homes, and other structures built on them.

Formation of Cavities in Limestone

Some common sedimentary rocks are **soluble** in water, allowing them to dissolve, which poses hazards to those living above. Salt and gypsum are highly soluble; limestone and other carbonate rocks are slowly soluble in acidic rainwater but the effect is distributed over vast regions. In fact, slowly dissolving carbonate rocks underlie more than 40 percent of the humid areas of the United States east of Oklahoma. Such areas can exhibit caves, springs, streams that sink into the ground, and sinkholes (▶ Figures 9-1 and 9-2). **Caverns** form as carbonate rocks near the water table dissolve in groundwater. When the water table drops below the top of the cavern, water percolating through fractures above can precipitate calcium carbonate as stalactites hanging from the roof or stalagmites growing from the floor (Figure 9-2). The roof of a cavern that was formerly supported by water pressure is then susceptible to collapse and the potential formation of a sinkhole.

Where the soluble rocks are close to the surface, the solution cavities and caverns can grow large enough for the

Sidebar 9-1

H_2O (rain water) + CO_2 (carbon dioxide) = H_2CO_3 (carbonic acid in water)

Carbonic acid reacts with limestone to form calcium bicarbonate:

$$H_2CO_3 + CaCO_3 = Ca(HCO_3)_2$$

roof rocks to collapse. Because the reaction between acidic water and limestone acts more rapidly under warm, moist conditions, caverns and sinkholes are most common in tropical climates.

Water droplets in the atmosphere dissolve carbon dioxide to form weak carbonic acid. The slightly acidic rain that falls percolates through the soil and sediment and down to the bedrock. Where the bedrock is limestone (calcium carbonate), the acidic water slowly reacts with the limestone along fractures, dissolving it to widen cracks and leave cavities (see Sidebar 9-1). The rate of solution is slow with mildly acidic rain, of the order of millimeters per 1,000 years. Dissolution is amplified by rain that is more acidic because of air pollution. In both cases, water flows through fractures and cavities in the limestone and carries away the soluble calcium bicarbonate.

Limestone can dissolve above, at, and below the water table. Where the limestone is above the water table, acidic water running down through fractures slowly widens them. Just below the water table, cavities can rapidly widen if horizontal bedding surfaces are open enough to conduct large amounts of water. In that case, the large flow past limestone surfaces brings fresh acidic water to these reactive surfaces and carries away the calcium bicarbonate products. This is the environment in which large caverns generally form and where stalactites and stalagmites develop as the water locally evaporates and loses its carbon dioxide (▶ Figure 9-3). Where such caverns are found high on hillsides above the current water table, the water table has generally dropped as a nearby stream eroded its valley deeper. Fractures can

▶**FIGURE 9-1.** A limestone cavern exposed along Interstate Highway 44 near Springfield, Missouri, shows sagging and fracturing of the rocks above it.

American Institute of Professional Geologists, Wilgus B. Creath photo.

Donald Hyndman photo.

▶**FIGURE 9-2.** The Lewis and Clark Caverns in Montana show stalactites hanging from the ceiling of the cavern that connect to stalagmites that grow from the bottom of the cavern. The final result is shown in the continuous column on the left side of this photo.

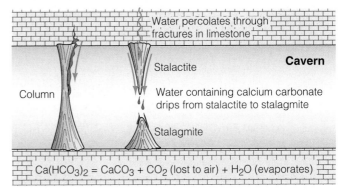

▶**FIGURE 9-3.** Water containing dissolved calcium carbonate seeps down through limestone into a cavern. There the evaporating water loses carbon dioxide and again precipitates calcium carbonate as stalactites and stalagmites.

also widen below the water table, especially where steeply inclined bedding surfaces circulate water to deeper levels before it rises again.

Collapsing Cavities

In some places, especially in limestone terrains of the eastern United States, the ground may suddenly collapse leaving deep holes, sinkholes, that are tens and sometimes hundreds of meters across (▶Figure 9-4).

Limestone bedrock near the water table dissolves along fractures to create an uneven and potholed upper surface. Later erosion of the soil cover may expose that surface as a limestone landscape called **karst** (▶Figure 9-5). In extreme cases, deep solution of the limestone along dominantly vertical fractures can form an extremely ragged, toothy-looking, and often picturesque landscape. Perhaps the best known

Donald Hyndman photo.

▶**FIGURE 9-6.** Extreme karst weathering can be seen in the Stone Forest, south of Kunming, China. The width across the middle of the photo is roughly 300 meters.

S. Navy photo, USGS.

▶**FIGURE 9-4.** A view toward the upper part of the chapter-opening photo shows cars and parts of buildings that dropped into the sinkhole.

J. E. Baker photo, Pennsylvania Geological Survey.

▶**FIGURE 9-5.** In this view of an exposed limestone karst landscape in Pennsylvania, the lumpy appearance of the limestone emphasizes solution as the mechanism of erosion.

of these are in the Guilin and Kunming areas of southwestern China (▶Figure 9-6).

Three types or aspects of sinkhole formation are most common, with gradations between them: **dissolution, cover subsidence,** and **cover collapse.**

1. **Dissolution:** Where the soil cover is thin and highly permeable, acidic groundwater seeps through the soil and dissolves the underlying limestone along fractures. The upper parts of fractures widen to form a lumpy or jagged karst surface (▶Figure 9-5). The overlying soil can slowly percolate or "ravel" down into the fractures to create a surface depression. Where the groundwater level is high or the fracture plumbing becomes clogged with sediment, the depression may fill to form a pond. The depressions are shallow and not generally dangerous.

2. **Cover Subsidence:** Where as much as 60 meters of sandy and permeable sediment exists on top of the limestone bedrock, numerous sinkholes can form as the soil slowly fills expanding fractures and cavities in the limestone. Depressions generally form gradually.

3. **Cover Collapse:** Where a significant amount of clay is present in the overlying sediments (in the **overburden**), this cover will be more cohesive and less permeable. As a result, it does not easily ravel into the underlying cavities in the limestone. This can allow a cavity in the limestone to grow large and unstable, leading to the sudden collapse of the thinning roof. The

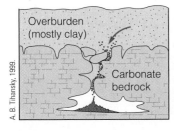
Sediments spall into a solution cavity in limestone.

New cavity forms in cohesive sediments above limestone as sediments trickle into limestone cavity below.

Soil cavity enlarges by progressive roof collapse.

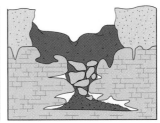
Ground surface suddenly collapses into enlarging soil cavity creating a sinkhole.

Overburden (mostly clay)

Carbonate bedrock

A. B. Tihansky, 1999.

▶**FIGURE 9-7.** This sequence of events can lead to a cover-collapse sinkhole.

Florida Geological Survey photo.

▶**FIGURE 9-8.** A truck-and-drill rig that was drilling a new water well near Tampa, Florida, overloaded the roof over a limestone cavern, causing it to collapse. The truck and equipment eventually sank out of sight in a crater 100 meters deep and 100 meters wide.

Thomas Scott photo, Florida Geological Survey.

▶**FIGURE 9-9.** A large sinkhole opened under this house in Sebring, central Florida, leaving it buckled in the middle and full of water.

lack of warning makes these steep-sided sinkholes destructive and dangerous (▶ Figure 9-7). Cover-collapse sinkholes open with little or no warning, taking with them roads, parking lots, cars, and occasionally even houses and other buildings (▶ Figures 9-4, 9-8, and 9-9).

A few clues can help determine whether an area is likely to develop large sinkholes. Areas with the greatest potential are those where surface water tends to sink into the ground to recharge the aquifer. This potential is greatest where the water table lies below the top of the limestone cavern, allowing for unsupported open space above the water level in cavities. Areas with the least potential for sinkholes are those where water is being discharged to the surface. There the cavities are likely to be filled with water that helps support the roof of the cavity.

For similar reasons, more sinkholes tend to form during dry seasons or when excessive pumping drops groundwater levels. One circumstance that often leads to unusual groundwater use and lowering of the aquifer is pumping during a prolonged spell of freezing weather. Strawberry and citrus farmers spray warm groundwater (at 23°C or 73°F) on the plants to form an insulating coating of ice.

Construction activities can also lead to sinkhole development by increasing the load on the ground surface, dewatering foundations, and well drilling. Loads can be imposed by heavy construction equipment or by the weight of the new structure (▶ Figure 9-8).

Sediments such as salt and gypsum also produce karst landscapes and underlie large areas of the United States. Their relative ease of solution in water can lead to cavity formation in just days. They are widespread in western New York, Michigan, Iowa, western Kansas and Oklahoma, eastern New Mexico, southern Utah and Nevada, eastern Wyoming and Montana, and northwestern North Dakota. In some cases, underground mining of salt and gypsum creates artificial cavities that can be enlarged by solution in water.

Sinkholes in Different Regions

Central Florida Sinkholes

For millions of years, the Florida peninsula was a shallow, warm, submarine platform, an ideal place for the growth of limestone reefs like those in the present-day Bahamas. As

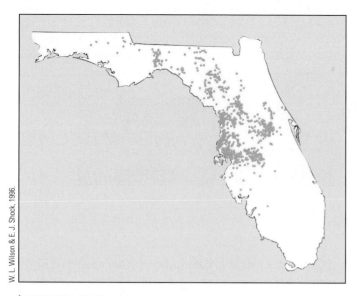

W. L. Wilson & E. J. Shock, 1996.

▶**FIGURE 9-10.** This map shows sinkholes reported in Florida from 1960 to 1991.

William Kochanov photo, Pennsylvania Geological Survey.

▶**FIGURE 9-11.** A sinkhole formed under this house in Palmyra Borough, Lebanon County, Pennsylvania.

William Kochanov photo, Pennsylvania Geological Survey.

▶**FIGURE 9-12.** A large sinkhole formed under the Corporate Plaza building in Allentown, Pennsylvania, in February 1994. Note the pillar missing in bottom center, the buckling of the building, and severe damage over a second sinkhole in the back right corner.

recently as 20,000 years ago, during the last ice age, when much ocean water was tied up in continental ice sheets, the whole carbonate platform of Florida was above sea level. The peninsula was then roughly twice as wide as at present. The groundwater level was much lower, so many caverns were above the water table. When the ice sheets melted, sea level rose and the water table rose to fill most of the caverns.

In Florida, sand and clay up to 60 meters thick cover most of the cavernous limestone that is weathered to a karst surface. Sinkholes are prominent in central Florida (▶Figure 9-10), and many of the lakes and ponds in west-central Florida are water-filled sinkholes. Most are merely a nuisance, but some cause considerable damage. Anytime the water table drops, either because of a lengthy period of dry weather or from excessive pumping for municipal, industrial, and agricultural use, new sinkholes are likely to appear (▶Figure 9-8).

Not only can sinkholes damage houses and roads but they can drain streams, lakes, and wetlands. In low areas, they can channel contaminants directly into underground aquifers, the main source of water used for drinking and other purposes in Florida.

Sinkhole Problems in Central and Eastern Pennsylvania

Widespread limestone in the Pittsburgh area of Pennsylvania is broken by numerous near-vertical fractures. Acidic rainwater percolating down through those fractures slowly dissolves and widens them. Soil above the limestone fractures may intermittently break apart to form a progressively enlarging cavity. Gradual expansion of that cavity eventually leads to formation of a cover-collapse sinkhole at the surface. Fluctuations of the groundwater level with periods

of heavy rainfall tend to loosen soil, enlarge the cavities, and promote sinkhole formation. Most Pennsylvania sinkholes are typically 1 to 6 meters in diameter. Enlargement of fractures to form underground caverns is not common in most parts of Pennsylvania.

In urban areas, the large volume of water carried by storm drains and leaking water mains can enhance the development of sinkholes. A large water flow can quickly flush soil from above fractured limestone to form large sinkholes. Roads and buildings above such sinkholes are vulnerable to severe damage (▶Figure 9-11). In one recent case in Allentown, Pennsylvania, a nearly new eight-story office building broke up after a pair of large sinkholes developed below major support columns. Damage required that the $10 million building be razed (▶Figure 9-12).

Sinkholes in Kentucky

Approximately 55 percent of Kentucky is on limestone weathering to karst, including one of the largest and most famous caverns in the United States, Mammoth Cave National Park. Slow solution of this limestone has left the landscape pockmarked with more than 60,000 sinkholes, some of which line up along lineaments. Just as elsewhere, sinkholes and the limestone fractures and caverns under them cause damage to structures and water supplies and water contamination.

As in Pennsylvania and many other areas, collapse of the limestone roof of a cavern is not common. More common is the collapse of an enlarging soil cavity above the limestone, a cover-collapse sinkhole. Some of these can do considerable damage, as in the February 25, 2002, collapse at a road intersection in Warren County, Kentucky (▶ Figure 9-13). In this case, storm water funneled underground at three corners of the intersection and caused further solution and instability in a limestone cavern underground. Before the road was built, karst experts who knew that the cave passage underground had an unstable roof recommended a different and safe route for the road, but their advice was not followed. Individual collapses such as this cost more than $1 million each to repair.

Collapse over a Salt Mine in Western New York

Early the morning of March 12, 1994, a section of shale roof rock 150 by 150 meters collapsed into a large mining excavation 360 meters below the surface. It caused a magni-

▶**FIGURE 9-13.** This sinkhole formed on February 25, 2002, by cover collapse over a limestone cavern in Warren County, Kentucky, where storm drain water was funneled underground at the road intersection.

tude 3.6 earthquake. Mining had been going on since the 1880s in this, the largest salt mine in the United States and second largest in the world; the underground area covered 25 square kilometers. Annual gross sales had reached $70 million. Groundwater pouring into the mine through the collapsed area gradually filled the mine cavities until mining ceased in September 1995. Consequences of the collapse included sinkholes as large as 250 meters across, deep wells that dried up, salt contamination of groundwater, and release of natural methane and hydrogen sulfide gases from the groundwater.

Land Subsidence

Subsidence of the ground is a serious problem throughout North America (▶ Figure 9-14). The causes are varied. Some subsidence is caused by extraction of groundwater, drainage of organic or clay-rich soils, and collapse into underground mine openings. These are dominantly human-caused.

Mining Groundwater and Petroleum

Many groundwater aquifers consist of thick accumulations of sand and gravel with water filling the spaces between the grains. Withdrawal of water from the pore spaces of loosely packed sand grains permits them to pack more tightly together and take up less space (recall Figure 8-14, page 195). The original groundwater reservoir capacity cannot be regained by allowing water to again fill the pore spaces; those areas have been **mining groundwater.** This is mainly because the sand grains cannot be pushed apart into a more open structure, reversing the reduction in pore space illustrated in Figure 8-12 (page 194).

Randomly oriented marine clays can have an extremely high **porosity,** a house of cards arrangement (recall Figure 8-15, page 195). Collapse of that open arrangement can result in significant land subsidence. Water is not normally pumped from clay layers because, although their porosity or pore space is high, their permeability is extremely low. Water cannot easily flow between pores because the openings are so tiny. The abundant intergrain spaces of some clays—quick clays—can collapse when jarred by an earthquake or heavy equipment. They also can collapse when flushed with fresh water. In both cases, the collapse leads to permanent reduction in porosity.

Prominent examples of subsidence because of long-term groundwater withdrawal are found in Silicon Valley around the southwest side of San Francisco Bay, in the San Joaquin Valley (the south half of the Central Valley of California), the Long Beach area just south of Los Angeles, and the Texas Gulf Coast region near Houston. Large water withdrawals can result in permanent reduction in groundwater reservoir capacity and permanent ground subsidence.

J. D. Kiefer, Kentucky Geological Survey.

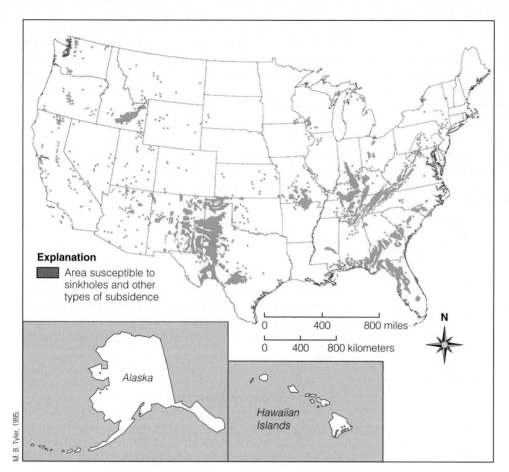

M. B. Tyler, 1995.

FIGURE 9-14. This map shows the distribution of areas that are susceptible to formation of sinkholes and other types of subsidence.

Explanation

▉ Area susceptible to sinkholes and other types of subsidence

0   400   800 miles

0   400   800 kilometers

N

Alaska

Hawaiian Islands

In the Santa Clara Valley at the southern end of San Francisco Bay, rapid depletion of groundwater accompanied the rapid growth in population (▶ Figure 9-15). In the warm and dry climate of the San Joaquin Valley, copious amounts of water were applied to crops, causing more than 8 meters of subsidence. Although the aquifers are partly recharged by water percolating along permeable layers from mountains

to the east and west, the river flows are small (see "Case in Point: San Joaquin Valley of California").

In the Houston area, oil and gas were produced early along with groundwater. Subsidence in parts of Houston has exceeded 1 meter. As a consequence, coastal flooding that was formerly only a problem with the occasional major storm is now a significant problem with many much smaller

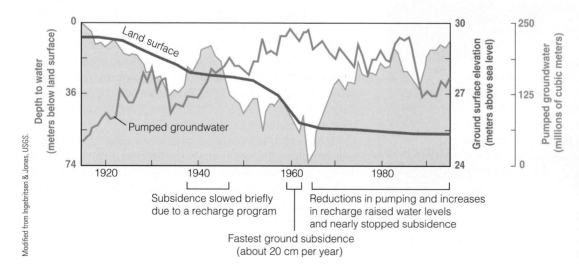

Modified from Ingebritsen & Jones, USGS.

Land surface

Pumped groundwater

Subsidence slowed briefly due to a recharge program

Fastest ground subsidence (about 20 cm per year)

Reductions in pumping and increases in recharge raised water levels and nearly stopped subsidence

FIGURE 9-15. Groundwater pumping in Santa Clara, California (shown in green), increased significantly from 1915 through 1960. As expected, the increases in pumping caused declines in groundwater levels. It also had the unanticipated effect of dropping the land surface by subsidence (shown in brown), until water was imported beginning in 1963 to recharge the groundwater. This recharge and reduced groundwater pumping nearly stopped the subsidence.

The semi-arid Sacramento–San Joaquin valley of central California is noted for its prolific agricultural products, everything from fruit orchards and vineyards to various grains, cotton, and rice. The area produces 25 percent of the food in the United States. Heavy pumping of groundwater for irrigating crops has lowered the water table, causing settling of the ground, in some areas dramatically (▶ Figure 9-16). Pumping groundwater leaves pore spaces between the sediment grains without the water that helps keep them apart. The grains rearrange and compact, causing the ground to settle.

The Sacramento–San Joaquin river delta, just east of San Francisco Bay, is one prominent case where ground subsidence is partly related to decomposition of the organic matter in sediments from agricultural activity. The delta has subsided to almost 5 meters below sea level in the last 100 years, so that the main river channels flow well above the level of the agricultural ground, held there behind dikes. That subsidence threatens, in turn, agricultural production, large water-supply lines to major metropolitan areas, and important large ecosystems.

▶**FIGURE 9-16.** Parts of the San Joaquin Valley settled almost 8 meters in the fifty years between 1925 and 1977.

FIGURE 9-17. This subsidence resulted from oil and water extraction from the basin sediments in the Long Beach area southeast of Los Angeles. Note that the ground has settled around the well stems (white "posts"), leaving their well heads standing high. The levy (left edge of photo) around the subsided ground under the oil well area prevents flooding from the adjacent marina.

FIGURE 9-18. This earth fissure opened up near Picacho Peak, approximately halfway between Tucson and Phoenix, Arizona. Excessive groundwater withdrawals for agricultural and urban uses caused differential subsidence. Such fissures do not tend to open more than 2.5 centimeters (1 inch) per event, but they then widen as blocks slump down into the fissure.

storms. Although much of the subsidence is regional, small faults cause some areas to drop more than others, damaging houses and utilities.

In the coastal areas of Wilmington and Long Beach just south of Los Angeles, subsidence began in the 1940s with pumping of groundwater at the naval shipyard. By 1958, subsidence occurred over 52 square kilometers by as much as almost 9 meters, warping rail lines and pipelines and damaging buildings (▶Figure 9-17). Oil wells sheared off, causing pressure losses. In 1958, the California Subsidence Act was passed, and groups of property owners injected water back underground to repressurized the pore spaces and stabilize the rate of ground subsidence.

Southcentral Arizona pumped almost all of its water from underground sources until the Central Arizona Project began supplying a large amount of water in 1985 from the dwindling Colorado River. In that desert climate, recharge from natural sources is minimal. Water table levels in the main agricultural area between the major population centers of Phoenix and Tucson had dropped 30 to 150 meters (100 to 500 feet) because of groundwater pumping. With evapotranspiration of 1.5 meters per year, precipitation in the mountains of 0.5 meters, and precipitation in some valleys as low as 7 centimeters, the natural recharge of water just does not add up. This mining of groundwater caused significant land subsidence from Phoenix south through Tucson. In some areas, there was significant differential subsidence from one region between adjacent areas, causing large **earth fissures** (▶Figure 9-18) to open to as much as 6 meters deep and 9 meters across.

Ground subsidence resulting from groundwater and petroleum extraction has been significant in many parts of the world. Table 9-1 lists a few of the more significant cases. See also "Case in Point: Venice, Italy."

Mitigation of ground subsidence initially involves stopping the activity that causes the problem. In some cases, governments institute regulations that prohibit or limit the groundwater or petroleum extraction. This can slow or eventually stop the subsidence; however, most of the subsidence or compaction of sediments cannot be reversed once it has occurred. The reduction in **aquifer** porosity and permeability are also mostly permanent. Groundwater can be a renewable resource in many basins, provided that the amount removed is no more than what is replenished, at least as averaged over many years. Appropriate *sustainable use* policies must be designed to maintain that balance. Unfortunately, the problem is not simple. It is important to understand the relationships between extraction of fluids, maintenance of fluid levels, and subsidence. Available wet-year storage for water in pore spaces between sediment grains for use in dry years must be carefully monitored so that the subsurface reservoir capacity is not permanently reduced. The sustainable use or "safe yield" for a basin may approach the natural and artificial replenishment but should take into account the climatic variability that affects

Table 9-1

Table 9-1 Examples of Ground Subsidence Resulting from Groundwater and Petroleum Extraction

Location	Amount of Subsidence (m)	Fluid Extracted	Some Consequences
Sacramento–San Joaquin Valley, California	9 in some places	Groundwater for agricultural use	Sacramento–San Joaquin delta area fields are below sea level
Phoenix to Tucson, Arizona	Up to 5	Groundwater for agriculture and municipal uses	Subsidence and earth fissures
Las Vegas, Nevada	Up to 2	Groundwater for municipal use	Subsidence, ground failure
Long Beach, California	9	Oil and gas	Pipelines and harbor facilities damaged. Dikes needed to prevent seawater flooding.
San Jose, California; and Santa Clara Valley (Silicon Valley), California	>2–2.5	Groundwater for orchards, industrial, and municipal use	Tidal flooding of San Jose by San Francisco Bay
Houston–Galveston area, Texas (a circular area 130 km across)	>2	Oil and gas, groundwater for municipal use	Frequent flooding, sometimes severe; submerged wetlands
Everglades, Florida	2–3.5	Drained for agricultural land and urban development	Wetlands now reduced by 50%; subsidence; saltwater intrusion into groundwater now slowed.
Mexico City, Mexico	8–10	Groundwater for industrial and municipal use	Tilting buildings, broken water, sewer, and utility lines
Pisa, Italy	Several	Groundwater for municipal use	Uneven subsidence endangers Leaning Tower.
Venice, Italy		Groundwater for industrial and municipal use	Settling of buildings below sea level, severe flooding at especially high tides

both the natural and artificial amount of water that can be returned to the groundwater system. Groundwater quality is also a concern, and a minimum rate of flow through the system may be necessary to maintain water quality.

Drainage of Organic Soils

Organic-rich soils, mostly peat, have been drained to plant crops in several large areas. Under their natural state, these peat soils would be saturated with anaerobic (without free oxygen) water, which allows them to maintain a stable state. When groundwater levels drop because of activities such as municipal and agricultural pumping, the peat soils are exposed to aerobic (oxygen rich) water that percolates down from the surface to the water table (See "Case in Point: San Joaquin Valley of California"). This allows bacteria in the sediment to oxidize the organic matter, which causes decomposition largely into carbon dioxide, a gas linked with additional natural hazards through global climate change (as discussed in Chapter 10, "Climate and Weather Related to Hazards"). This decomposition occurs at a rate of as much as 100 times the rate of accumulation of such organic material.

A vast area of the Florida Everglades was drained decades ago for farming, causing almost 2 meters of subsidence. An Everglades restoration plan is currently being implemented to reduce these and other impacts and take the Everglades back toward its natural state.

Ground Subsidence with Drying of Clays

The Leaning Tower of Pisa, near the west coast of Italy, was built as a bell tower for the adjacent cathedral of Pisa. Although construction began in 1173, the tower began settling and leaning, and construction was stopped in 1185 with only three levels completed. Construction resumed in 1274 after 100 years of inactivity. Ten years later, all except the top bell enclosure was complete, but again the rate of leaning accelerated so construction was again stopped. The tower, including its bell chamber, was finally completed in 1372, but by this time it was 1.5 meters off vertical.

The ground under the tower consists of 3 meters of clay-rich sand and more than 6 meters of sand over thick, compressible, and spreading marine clay, not the type of material that today's engineers would choose as a foundation. By 1975, the 55-meter-high Leaning Tower had sunk 2 meters into the ground and was 4.3 meters off vertical. If the photo (▶ Figure 9-21) appears to show slight curvature of the sides of the tower, that is not an illusion. The builders tried to straighten the tower as construction proceeded. Various attempts were made to stabilize the tower. A 600-ton weight added to the base of the upper side of the tower combined with dewatering of the clay and sand by pumping from boreholes. More recently, cables were attached to the upper side to try to straighten the tower somewhat.

A Case History of Water Extraction

General Grant is remembered for many things, perhaps the least of them being his remark upon seeing Venice that the place appeared to have a plumbing problem. It does indeed. The situation has not improved since General Grant was there more than a century ago.

Venice, along the northeastern coast of Italy, is built at sea level on the now-submerged delta of the Brenta River. The delta and the city have since subsided below sea level for a variety of reasons, the most important being extraction of water from the delta sediments. The load of buildings squeezes water out of the soft delta sediments, and pumping groundwater for domestic and industrial use collapses the pore spaces in the sediments. That decreases the volume of the sediment, which sinks the original surface of the delta below sea level. So Venice sinks. Some parts have now sunk as much as several meters. Streets that were above sea level when they were built are now canals. Doorways and the lower floors of homes are now partly submerged (▶Figure 9-19).

The high tides that accompany winter storms compound the problem. They submerge most of the few normally dry walkways under as much as a meter of water (▶Figure 9-20). Maps on the water taxis advise people of the few dry routes available for walking. Pumping groundwater from under the city is now prohibited, and sinking has all but stopped. But the high winter tides that formerly flooded the city twice a year now come several times a year. Ancient drainage pipes installed to carry off rainwater now carry seawater back into the city during times of high winter tides, along with whatever effluent they normally carry.

Tides in most of the Mediterranean Sea range less than 30 centimeters, but where the tidal bulge sweeps north up the Adriatic Sea near the east end of the Mediterranean, they average 1 meter. When low atmospheric pressure of a storm that raises sea level coincides with high tide, sea level may rise even higher. If this rise happens to coincide with a sirocco, a warm winter wind blowing north out of Africa, pushing the tidal bulge ahead of it, the tidal rise can be especially high. Such a series of coincidences happened in November 1966 when three days of rain had already raised both river levels and the lagoon level to cause catastrophic flooding.

Merely raising the sea walls around the city will not cure that problem. Global warming and the rise in sea level undoubtedly contribute to the problem. A few historical highlights illustrate the political background that makes any solution to the problem difficult.

1534   A major dike is built to divert the mouth of the Sile River to the east and away from Venice.

1613   Venice, on the delta of the Brenta River, diverts the Brenta around the city. New sediment is no longer added to that part of the delta.

(a)

(b)

▶**FIGURE 9-19.** **(a)** In Venice, half-submerged doorways along canals emphasize the slow sinking of the city. **(b)** The 1996 Venice flood.

Donald Hyndman photo.

▶**FIGURE 9-20.** High water from the lagoon floods a broad walkway in Venice.

1744–1782 Seawalls are constructed along the offshore barrier islands to protect the Venice lagoon from the Adriatic Sea.

1925 Large-scale pumping of groundwater from under Venice began for industrial use.

1966 A major storm on November 3, compounded by an especially high tide following a full moon, pushes Adriatic water into the Venice lagoon that normally averages only 60 centimeters deep. The water reaches 1.94 meters above normal sea level. The city floods for 15 hours, causing severe damage to buildings and works of art. Heating oil tanks floated, breaking attached pipes, and draining oil into canals and homes. Sinking permanently floods the first floors of most buildings; most families now live on their second floors.

1970 After 45 years of pumping and gradual sinking, the government begins to completely prohibit industrial pumping of groundwater from under Venice.

1981–1982 A technical team of the Public Works Ministry and Venice City Council approves plans for a major flood-control project to protect the shallow lagoon from such flooding. The plan is to build mobile barriers across the mouth of the lagoon that can be raised at times of potential high water.

1984 The Italian Parliament appropriates the equivalent of billions of dollars to build the barriers and raise parts of the urban center.

Given Italian politics, governments that change hands every year or two, and internal bickering, the money remained unspent even by 2002. As in most places, there are those who argue that it is more important to spend available funds on building industrial capacity and promoting new jobs rather than in preserving Venice as one of the world's historic and artistic centers. In this case, the development would involve a third industrial zone at the landward edge of the lagoon in Marguera.

During floods, the impermeable marble bases of most of the old buildings are submerged and seawater seeps into the porous bricks of their walls that then begin to crumble. More frequent submergence of drainpipes that carry rainwater from the city into the lagoon creates another major problem. Those pipes act as the city sewers; they carry untreated sewage including laundry suds and toilet effluent into the canals, then into the lagoon where the tides periodically flush it out to sea. During especially high tides, of course, the drains carry the effluent back into the canals and walkways of the city. Some of the controversy concerning the proposed movable barriers at the mouths of the lagoon involves restriction of the tidal action that flushes the lagoon.

Donald Hyndman photo.

▶**FIGURE 9-21.** The cables that help keep the Leaning Tower of Pisa from falling are just visible between the tower and the church on the left. The tower curves slightly because it began leaning during construction and builders tried to straighten the upper floors.

Some water-rich marine clays are subject to subsidence when they dry out. One case in point is the Ottawa–St. Lawrence River lowland, where seawater invaded lakes ponded at the edge of the retreating continental ice sheet 10,000 to 13,000 years ago. Much of Ottawa, Ontario, is built on the notorious Leda clay, a marine clay up to 70 meters thick. It is especially weak because the natural salts have been leached out. The solid grains of the clay amount to less than one-third of the total volume so that the clay dries and shrinks under old homes and buildings, causing differential settling and distortion (▶ Figure 9-22). Landslides throughout the St. Lawrence River valley and its tributaries are in the Leda clay. In an ironic twist, the new headquarters building for the Geological Survey of Canada, completed in 1911, was built on a site underlain by more than 40 meters

of Leda clay without consulting geologists of the Geological Survey. The heavy outer stone walls immediately began to sink into the clay, causing cracking of the building and separation of its tower from the main building. The tower was finally torn down in 1915 and the building and its foundations strengthened at considerable cost.

An infamous case involving the Leda clay began in April 1971 at St. Jean Vianney, where a new housing development had been built on a terrace of Leda clay 30 meters thick. Heavy rains loaded the clays, and eleven days later it collapsed and flowed downslope toward the Saguenay River in a channel at 25 kilometers per hour for some 3 kilometers. The slide took thirty-one lives and thirty-eight homes (see also Figure 8-16, page 195).

Swelling Soils

The annual cost in the United States, from swelling soils, is more than $4.4 billion. Approximately 50 percent of the damage is to highways and streets and 14 percent to homes and commercial buildings.

Swelling soils and shales are those that contain smectites, a group of clay minerals that swell when they get wet. These minerals are formed by weathering of aluminum silicate minerals and glass in volcanic ash (their behaviors are discussed in Chapter 8, "Landslides and Other Downslope Movements").

As introduced earlier, the swelling of dry clay occurs when water soaks into the interlayer spaces in the mineral structure. This mostly occurs within several meters of the ground surface because the greater pressure at depth typically exceeds the force exerted by the expanding clay, thereby preventing clay expansion; that is, the load of soil above prevents the water from entering the mineral structure and lifting everything above. Thus, these clays do not swell after heavy structures such as large buildings, supports for large bridges, and dams are built on top of them.

Donald Hyndman photo.

▶**FIGURE 9-22.** This is a telltale sign of a brick house settling and being distorted by shrinkage of its clay substrate in Ottawa, Ontario. Note the diagonal zigzag crack in the brick wall (see arrow). The mortar between the bricks tilts over to the left, indicating that the upper part of the wall has moved to the left.

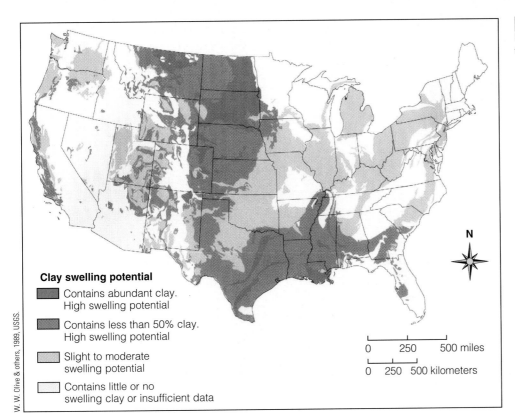

W. W. Olive & others, 1989, USGS.

Clay swelling potential

- ⬛ Contains abundant clay. High swelling potential
- ⬛ Contains less than 50% clay. High swelling potential
- ⬜ Slight to moderate swelling potential
- ⬜ Contains little or no swelling clay or insufficient data

```
0    250    500 miles
0   250  500 kilometers
```

▶**FIGURE 9-23.** These areas in the United States are potentially susceptible to damage from swelling clays.

Much of the United States is affected by swelling soils, including most large valleys in the mountain West, all of the Great Plains, much of the broad coastal plain area of the Gulf of Mexico and southeastern states, and much of the Ohio River valley (▶Figure 9-23, "Case in Point: Denver's Swelling Soils").

You can often spot the presence of swelling clays by the expansion and cracking of surface soils to form "**popcorn clay**" (▶Figure 9-24). Driving on a dirt road on such soils is not a problem until it starts to rain. Then it becomes nothing short of an adventure—as this wet clay is extremely slippery. Mud builds up on tires and shoes—as the "gumbo" that is all too familiar to those who live in areas with smectite clay. In such areas, people create problems for themselves by merely building houses. The area under the house remains drier; that around the house is open to moisture from rain, snow, lawn watering, house gutter drainage, or from a septic tank or drain field. Or prevailing wind-driven rain against one side of a house can cause the ground around that side of the house to swell (▶Figure 9-25). Vegetation concentrated on one side of a house can dry out soil on that side. Even removal of vegetation can increase moisture in the soil by decreasing evapotranspiration. During long dry periods, the soil shrinks; in long wet spells, it expands. The result is the cracking of foundations, walls, chimneys, and driveways of houses.

Donald Hyndman photo.

▶**FIGURE 9-24.** Popcorn weathering is a telltale sign of swelling clays in the Makoshika badlands near Glendive, eastern Montana. View is 60 centimeters across.

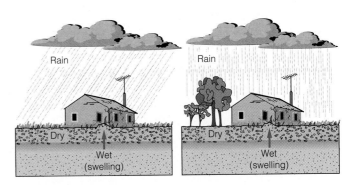

▶**FIGURE 9-25.** The soil under a wet side of a house, either because of a prevailing rain direction or lack of vegetation, can swell and deform the house.

Denver is built on the central High Plains of the United States, which are underlain by flat-lying Cretaceous and older sedimentary rocks that curve up to nearly vertical at the surface just east of the Front Range of the Rockies. Among them is a subtle villain, the Pierre shale. The shale contains smectite-rich clay layers formed by weathering of volcanic ash. When wet, the clays swell, lifting parts of houses and twisting them out of shape. Parts of the southwestern suburbs of Denver, south of U.S. Highway 285 and east of state Highway 470, were built in the 1970s to 1990s over the upturned edges of these formations. In the Denver area, where horizontal smectite-bearing layers under the High Plains tilt up to reach the surface, those layers create differential expansion under houses (▶Figure 9-26).

The problem appears in the growth of long, crudely parallel rises 10 to 20 meters apart and separated by swales. Within a few years of building houses, more than 0.6 meter of differential movement has created broad waves in roads and cracked and offset underground utility lines. Parts of houses rise, causing twisting, bending, and cracking of foundations, walls, and driveways. Windows crack and chimneys separate from houses (▶Figures 9-27, 9-28, and 9-29).

The problem of swelling and deformation of buildings does not extend to larger, heavier structures such as large commercial buildings and bridges. These are so heavy that water cannot penetrate the structure and lift the load.

The obvious solution to damage from expansive soils is to build elsewhere. Where this is not feasible, near-surface effects can be minimized by preventing water access to the soil, removing the expanding near-surface soils, or sinking the foundation to greater depth. The addition of hydrated lime, calcium hydroxide [$Ca(OH)_2$], to an expandable clay can reduce its expandability by exchanging calcium into the molecular interlayer spaces so water cannot enter as easily. How-

Donald Hyndman photo.

▶**FIGURE 9-27.** The driveway and garage doors deformed in response to the swelling clays on which this home is built in a south-western Denver suburb. The brick wall pulled away from the garage door, leaving a gap at the bottom (see arrow).

ever, this should not be done if the soil is a quick clay because that would markedly decrease the cohesiveness of the soil and foster soil collapse and landslides.

Until 1990, engineers attributed the problem to swelling soils. Their mitigation plans called for drilled piers under foundations or rigid floating-slab floors. Most proved unsuccessful. More recent research by the USGS and the Colorado Geological Survey demonstrated that the problem is the Pierre shale bedrock. That "bedrock" is fairly young and about as soft as a soil. Recent work in the Denver area shows that excavation and homogenization of both clay- and non-clay-bearing layers to depths of less than 10 meters in preparation for new subdivisions can increase uniformity of expansion and minimize problems. Regulations since 1995 require new subdivisions to excavate to a depth of at least 3 meters beneath foundations and replace the homogenized fill. Significant reduction in damage has resulted.

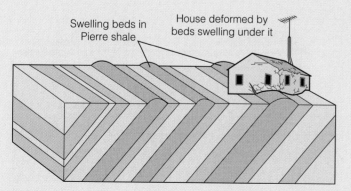

Swelling beds in Pierre shale

House deformed by beds swelling under it

▶**FIGURE 9-26.** Houses in parts of Denver are built over inclined beds of smectite-bearing Pierre shale. Swelling of the shale beds breaks up the houses.

David Noe photo, Colorado Geological Survey.

▶**FIGURE 9-28.** The swelling of Pierre shale bentonite beds that stand on edge caused these waves in Sangre de Cristo Road in southwestern Denver. The bentonite-rich beds bulge near the surface. The interlayered beds do not. This road has since been leveled and repaired, eliminating the evidence.

KLP Consulting Engineers photo, Englewood, CO.

▶**FIGURE 9-29.** Swelling clays in southwest Denver deformed this house's foundation.

What Went Wrong?

New housing developments are built where smectite-rich shale layers under the High Plains turn up to the surface. When the smectite gets wet and swells, houses deform and crack.

http://earthscience.brookscole.com/hyndman

KEY POINTS

✓ Sinkholes, subsidence, and swelling soils cause far more monetary damage in North America than other natural hazards. **Review p. 228.**

✓ Rainwater, naturally acidic because of its dissolved carbon dioxide, can dissolve large cavities. **Review pp. 228–230; Figures 9-1 to 9-3; Sidebar 9-1.**

✓ Soil above those limestone caverns can percolate down, leaving a soil cavity that can collapse to form a sinkhole. **Review pp. 230–231; Figure 9-7.**

✓ Sinkhole collapse can be triggered by lowering the groundwater level that previously provided support for the cavity roof, loading the roof, or increasing water flow that caused rapid flushing of soil into an underlying cavern. **Review pp. 231–233.**

✓ Groundwater or petroleum removal can cause sediments to compact with resulting subsidence of the land surface. In some areas, this can lead to subsidence below sea level, which can result in submergence and saltwater invasion of freshwater aquifers. **Review pp. 233–237.**

✓ Land subsidence can also be caused by drainage of peat or clay-rich soils. Oxygen-bearing surface water invades the peat, causing it to decompose. **Review p. 237.**

✓ Smectite soils, formed by the alteration of soils rich in volcanic ash, swell when water invades the clay mineral structure. Such soils are abundant in the midcontinent region, the Gulf Coast states, and western mountain valleys. **Review pp. 240–243; Figure 9-23.**

✓ Swelling soils can make dirt roads slippery when wet and can crack and deform foundations, driveways, and walls. The problem can be minimized by deep excavation and homogenization of underlying soils. Heavy structures such as bridges and tall buildings are not affected. **Review pp. 241–242; Figure 9-26.**

IMPORTANT WORDS AND CONCEPTS

Terms

aquifer, p. 236
caverns, p. 229
cover collapse, p. 230
cover subsidence, p. 230
dissolution, p. 230
earth fissures, p. 236
karst, p. 230
mining groundwater, p. 233

overburden, p. 230
popcorn clay, p. 241
porosity, p. 233
sinkholes, p. 228
soluble, p. 228
subsidence, p. 233
swelling soils, p. 228

QUESTIONS FOR REVIEW

1. What causes sinkhole collapse—that is, where is the cavity that it collapses into and how does that cavity form? Explain clearly and concisely.

2. What causes caverns in limestone—that is, where does the limestone go and why? Explain clearly.

3. Where in the United States are sinkholes most prevalent? Why?

4. What is karst and how does it form?

5. What are the main characteristics of soil above a limestone cavern that lead to the formation of sinkholes?

6. What weather conditions are most likely to foster the formation of sinkholes? Why?

7. Other than limestone, what other types of rocks are soluble and can form cavities that collapse?

8. What types of human behavior typically lead to widespread ground subsidence?

9. Name three areas in the United States that have seen major subsidence and explain what caused it.

10. What causes swelling soils to swell?

FURTHER READING

Assess your understanding of this chapter's topics with additional quizzing and conceptual-based problems at:

 http://earthscience.brookscole.com/hyndman.

CHAPTER 10

Many factories emit carbon dioxide and other pollutants.

Donald Hyndman photo.

Global Warming

The vast majority of scientists believe that the Earth's average surface temperature has been rising since the Industrial Revolution began in the late 1700s. Most attribute a significant part of the rise to increases in atmospheric carbon dioxide, methane, and nitrous oxide emitted by vehicles, domestic heating, and industry (▶ Figure 10-1). Most of these emissions are derived from the burning of fossil fuels: the coal, oil, and natural gas used to produce heat and electricity. Additional amounts come from forest and range fires, some of which are natural. All of these **greenhouse gases** trap heat in the atmosphere in about the same way that a greenhouse permits the sun to shine through glass but prevents much of the heat from escaping. Surface temperatures thus rise on the Earth as more greenhouse gases enter the atmosphere and insulate our planet.

Intermittent volcanic eruptions also add huge volumes of carbon dioxide to the atmosphere. One might think that this could lead to a significant warming of the atmosphere,

CLIMATE AND WEATHER RELATED TO HAZARDS
Storms on the Horizon

245

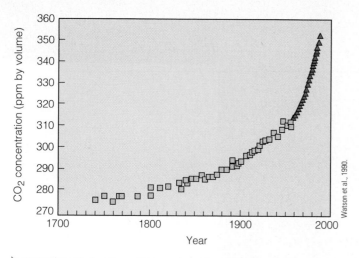

Watson et al., 1990.

▶**FIGURE 10-1.** Since 1700, carbon dioxide levels measured in the atmosphere have risen dramatically. Values from 1700 to 1960 or so were derived from air samples trapped in the Antarctic ice sheet. Values since 1960 were directly measured from the atmosphere on the top of Mauna Kea, Hawaii.

but measurements show that these releases are offset by the atmospheric additions of ash and sulfur dioxide. Both reduce the amount of incoming solar radiation, as noted below and in Chapter 6, "Volcanoes: Materials, Hazards, and Eruptive Mechanisms." The overall effect of a large volcanic eruption appears to be a significant short-term cooling of the climate. Long term, the effect is unclear.

An additional source of greenhouse gas is stored beneath the seafloor in methane-ice compounds called **gas hydrates.** Thin layers of these gas hydrates at depths of 200 to 500 meters in mud under the seafloor on the continental shelves constitute the Earth's largest source of hydrocarbons. At those depths and low temperatures, the compound is stable. When methane hydrate is brought to the surface in a drill core, the methane bubbles out vigorously. Could the gas hydrate decompose suddenly to cause catastrophic re-

lease of methane? Some scientists believe it could happen. If **global warming** were to increase the ocean temperature enough to change the pattern of oceanic circulation, some parts of the shallow sea over the continental shelves could warm by several degrees Celsius. That might release enough methane to significantly increase the rate of global warming.

The Earth's surface temperature increased by 0.4° to 0.8°C (~0.7° to 1.4°F) over the past century, with the most dramatic increase since 1970 (▶Figure 10-2). The United Nation's International Panel of Climate Change in 2000 estimated that the Earth's average surface temperature will likely rise by 1.5° to 4.5°C (35° to 40°F) over the next century and possibly as much as 6°C. This change compares with a rise of 4°C since the peak of the last ice age 20,000 years ago.

The increase in atmospheric temperature has been moderated by the fact that most of our planet is covered with oceans, and heating the oceans uses a tremendous amount of energy. The average change in ocean temperature over the past 40 years was 0.3°C in the top 300 meters and 0.06° in the top 3.5 kilometers. This small warming of the seas is more ominous than it appears because ocean heat contributes significantly to the energy that drives storms. Hurricanes, for example, form and strengthen where sea-surface temperatures are above 25°C (77°F). Warmer seas may cause these storms to be stronger and more frequent in the future. Warming sea temperatures may be responsible for the recent increase in frequency and strength of especially stormy weather in the southwestern United States. Wetter West Coast winters are likely, which would cause more floods, landslides, and debris flows. Snowpacks are expected to be smaller, runoff would generally be earlier, and summer streamflow would be lower on average.

Not all climate changes involve an increase in average temperature. The most significant change would be greater variability of weather, with more extremes of temperature, winds, and precipitation. Atmospheric pressure would show higher highs and lower lows. That would cause winds

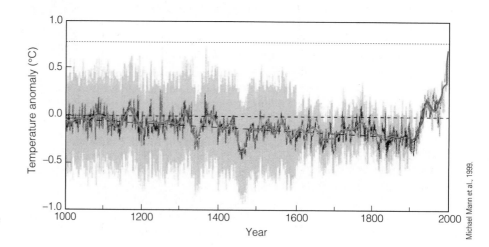

Michael Mann et al., 1999.

▶**FIGURE 10-2.** Changes in temperature from the years 1000 to 2000 have been determined from tree rings and glacial ice. Note the distinct increase in temperature since 1900.

to flow faster between them, with associated stronger and more frequent storms. Natural catastrophes such as hurricanes and tornadoes would be more frequent and stronger. Floods would be more frequent and more intense, even without other influences. Mudflows and coastal erosion would become more severe on the West Coast. Lower rainfall and drier vegetation would lead to more thunderstorms and more fires. Although populations in some northern climates might welcome modest increases in temperature, detrimental effects would include increased incidence of both droughts and floods. More precipitation would fall as rain rather than snow. In western North America, this would mean that less moisture would be stored until late spring, when its runoff would be available for the dry summers. Warmer winters in the Midwest would reduce the seasonal frozen surface area of the Great Lakes, which would lead to more winter evaporation from the open water surface and thus lower the lake levels.

Melting of temperate climate glaciers will continue. Wholesale melting of the Greenland and Antarctic ice sheets, which would cause global sea level to rise a phenomenal 80 meters, is not likely in the foreseeable future. Water expands as it warms; thus the increase in ocean temperature will fill the oceans beyond that expected for the volume of melted ice from the continents. Expansion occurs because heating causes the water molecules to vibrate faster and therefore take up more space. The larger volume of water requires that global sea level will rise, ultimately causing the flooding of coastal areas, including many of the world's largest cities. The current rate of sea-level rise is roughly 2 millimeters per year or 20 centimeters per 100 years, and sea level is expected to rise by between 20 centimeters and 1 meter over the next century. Bangladesh, which occupies the immense near-sea-level delta of the Ganges and nearby rivers, is already subjected to severe flooding during storms. Deaths run into the tens to hundreds of thousands during severe events. The future of its 125 million people looks dismal (see "Case in Point: Cyclones of Bangladesh and Calcutta, India (Bay of Bengal)," pages 389–390 in Chapter 14 on hurricanes).

To reiterate, climate and weather would become more irregular with greater extremes, not only in temperature but also in rainfall. Areas that already have water shortages would be in desperate shape during similar periods in the future. Water access in places such as the Middle East would spark even more conflict, if not wars. Some countries such as Turkey have already diverted irrigation water with dams, depriving arid Iraq and Syria, which are plagued by still lower river levels.

A few scientists, though in decreasing numbers, argue that the temperature increase of the last century is mostly natural; even that group expects a slight temperature increase over the next century. Those who remain unconvinced that humans cause global warming point to satellite measurements of upper atmosphere temperature since 1979 that show a lesser, though significant temperature rise. All agree that carbon dioxide (CO_2) concentration has increased some 30 percent since before the Industrial Revolution, less than 300 years ago.

The new studies have polarized nations over who is to blame and what to do about global warming. One hundred eleven countries, including most industrialized nations but neither the United States nor Russia, signed the **Kyoto Protocol** in 1997 to reduce emissions of six greenhouse gases beginning in 2008. Developing nations such as India and China argued that the industrial nations must first reduce their emissions because they can more afford the associated costs. The United States argued that more research was needed on atmospheric pollutants from developing nations; unfortunately, such research is now politicized. What to do about this pollution is an equally difficult problem. Huge populations in Southeast Asia are so poor that they cannot afford to purchase more efficient fuels. How do you tell a family that has little money for food to stop cooking with the only fuel that it can afford? Finally, seven years later, on November 5, 2004, Russia signed the protocol, bringing the percentage of industrialized nations to do so to 55 and thus meeting the treaty's threshold for activation.

Atmospheric CO_2 is expected to double in the next century, and there is no clear way to stop this trend. Weather patterns will definitely change, and some economists and politicians argue that we need to "learn to live with" those changes. If average temperatures rise as anticipated, predictions indicate that winter rains will dramatically increase in the southwestern United States while snowpack in the Rockies could fall by 50 percent. Both would create severe problems. Northern Europe could see increases in rainfall while southern Europe would likely become drier. Warming of some northern climates would permit the northward migration of insects bearing diseases such as malaria and West Nile virus because fewer insects are killed off in warmer winters.

Positive policy changes that would reduce vulnerability to some of the effects of global warming would include restricting development on floodplains and offshore barrier islands. Significant governmental funds are used to rebuild in areas harmed by floods and mudslides, and the costs of natural hazards will likely increase significantly. However, even greater costs will arise as larger populations move into unsuitable areas. Increased property values in such hazardous areas compound the problem.

Various changes have been proposed to reduce the carbon dioxide levels in the atmosphere. The most obvious solution is to reduce the use of fossil fuels for energy. Burning hydrogen as a fuel sounds good because that produces only water, which is not a pollutant. Producing hydrogen in large quantities, however, requires immense amounts of energy, and most ways of doing so produce carbon dioxide and other pollutants. A big difference is that the emissions would be in large, centralized power plants rather than disbursed through individual vehicles. It would presumably be less expensive to capture harmful emissions at centralized power plants. Another method of reducing atmospheric greenhouse gas concentrations is to remove CO_2 from the air and pump it into underground storage. This is currently done to repressurize oil fields and increase the amount of recoverable fossil fuels. If all of the world's oil fields used such CO_2 reinjection, roughly half of the new carbon dioxide we produce could be confined to the subsurface. An additional strategy would be to react CO_2 with magnesium-rich silicate minerals such as olivine and serpentine to produce magnesium carbonates. That would permanently remove CO_2 from the air; however, it is not yet clear how such a process could be carried out on a large scale.

Atmospheric Cooling

The atmosphere can be cooled by anything that puts particulates such as soot and dust in the atmosphere. Forest fires, both natural and human-made, are a major source of particulate. Industrial smokestacks contribute, especially in underdeveloped countries. Pollution drifting east from coal-fired plants in China has even shifted normal rainfall patterns, causing more rain in the south and worse droughts in the north. Burning wood, coal, and peat provide significant amounts of particulates in poor countries. A 3-kilometer-thick brown cloud of smoke, soot, and dust was recently discovered over part of the Indian Ocean roughly the size of the United States. Part of that pollution is due to the huge population of people in India who use dried cow dung as a cheap fuel source for cooking. Dark soot in the cloud absorbs heat and warms the upper atmosphere. However, it also cools the surface of the Indian Ocean, which reduces evaporation. This in turn increases the intensity of regional droughts because less moisture is available to fall as rain.

As mentioned above, a more intermittent but locally important contributor appears to be volcanic eruptions that produce large amounts of ash (see "Case in Point: Mount Tambora"). Environmental effects have followed many large eruptions of rhyolite ash since 1816. The violent eruption of Krakatau near the island of Java in Indonesia in 1883 injected large volumes of ash into the upper atmosphere, where it spread westward around the Earth. It cooled the climate for several years but did not cause widespread crop failures and famines such as those that devastated populations following the eruption of Tambora in 1816 (see "Case in Point: Mount Tambora").

Any huge eruption of rhyolite ash is likely to develop into a global climatic catastrophe of dimensions beyond easy reckoning. No warning or evacuation scheme can in any way mitigate this danger. People everywhere should dread a major eruption of any large resurgent caldera such as the Yellowstone Volcano or the Long Valley Caldera.

Large eruptions of rhyolite ash also cause spectacular sunsets, which are especially notable for their vivid streaks

CASE IN POINT
Mount Tambora

A gigantic eruption reduced the peak of Tambora volcano on the densely populated Indonesian island Sumbawa from an elevation of 4,300 meters to 2,900 meters in April 1815. The eruption (with a VEI of 7) produced 40 cubic kilometers of ash and pumice, an even larger volume than was produced in the eruption that reduced Mazama to Crater Lake almost 8,000 years earlier (discussed in Chapter 7). The caldera is roughly 7 kilometers wide and more than 600 meters deep. The eruption killed some 10,000 people directly and another 80,000 in the famine and epidemics that followed.

The environmental aftermath developed into a global catastrophe of famine and misery. The rhyolite ash that Tambora injected into the upper atmosphere blocked enough sunshine throughout the northern hemisphere to reduce the average temperature about 0.3°C (~0.5°F) within a few weeks. That seems a small drop, but it was enough to cause agricultural havoc.

Freezing temperatures ravished crops in New England during every month of summer 1816, "the year without a summer," causing widespread famine. Summer frosts ruined crops as far south as Virginia, including, by some accounts, Thomas Jefferson's corn. Meanwhile, abnormally cold and rainy weather caused widespread crop failure and famine in Europe. It was a hard year.

Some historians argue that the eruption of Tambora helped inspire a large migration from New England to the region west of the Ohio River, as well as considerable movement from Europe to North America.

of green, a color not ordinarily observed in sunsets. These typically continue for several years after such eruptions. Weather is mainly confined to elevations below 15 kilometers, so ash lingers high in the stratosphere where rain does not wash it out.

Climatic Cycles?

Many theories have been proposed to explain significant observed cyclic changes in Earth's northern hemisphere climate, and there are clearly multiple overlapping processes that contribute to these long-term climatic shifts. These include changes in ocean currents, cyclic changes in solar radiation because of changes in Earth's orbit, plate tectonic movements, and changes in atmospheric composition. Climate cycles occur across a broad range of time periods from days to 100,000 years and more.

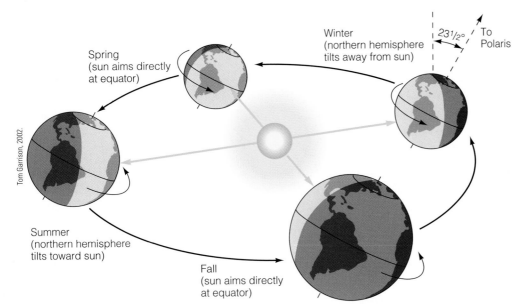

Spring
(sun aims directly
at equator)

Winter
(northern hemisphere
tilts away from sun)

23½° To
Polaris

Summer
(northern hemisphere
tilts toward sun)

Fall
(sun aims directly
at equator)

Tom Garrison, 2002.

▶**FIGURE 10-3.** Seasons: Earth rotates around its north–south axis, inclined at 23.5 degrees to Earth's orbit around the sun. Because Earth rotates around the sun once per year, the northern hemisphere is tilted toward the sun in summer, with the longest northern hemisphere day on the summer solstice (usually June 21). It then tilts away from the sun in winter with the shortest northern hemisphere day on the winter solstice (usually December 21). The arrows from the sun show the solar energy striking the Earth from directly overhead—for example, on the northern hemisphere in summer.

Days

Day and night are the shortest regular cycles in temperature on the Earth. The Earth rotates once per day around its north–south axis, causing any place on Earth to face most directly toward the sun once a day at noon. At that time, the sun's energy is more intense than at any other time of the day, in the same way that the light of a flashlight has more energy in a smaller area when it is pointed directly at a flat surface than when it is pointed at an angle. At midnight, our side of the Earth is facing away from the sun.

Seasons

Annual changes from winter to summer and back to winter (**seasons**) occur because Earth's axis is tilted with respect to its annual orbit around the sun (▶Figure 10-3). When Earth's axis is perpendicular to the sun on both the vernal equinox (spring, usually March 20) and the autumnal equinox (fall, usually September 22), the sun is directly overhead at the equator. At these times, the sun's radiant heat strikes Earth's equatorial region directly, warming it more than at any other time. We are not, as some people seem to believe, closer to the sun in summer. If that were true and the distance change was large enough, the entire Earth would have summer at the same time.

The Earth's axis is tilted 23.5 degrees to the plane of the Earth's orbit around the sun, thus when the sun points directly at 23.5 degrees north latitude on the summer solstice (usually June 21), the northern hemisphere receives its maximum solar radiation and thus the maximum heat from the sun. Although this is midsummer in the northern hemisphere, the warmest temperatures lag the solstice by more than a month because it takes time to heat up the

land and water, just as it takes time to heat up an egg in a frying pan after the stove is turned on. When the axis tilts 23.5 degrees in the opposite direction, the sun is directly above 23.5 degrees south latitude, and it is midsummer in the southern hemisphere. The South Pole gets twenty-four hours of sunlight during the southern hemisphere summer, but despite all of this light, the total solar energy reaching the pole is much less than regions closer to the equator because the solar radiation is at a low angle.

Precession of the Equinoxes: ~26,000 Years

Earth acts somewhat similar to a spinning top with a wobble in the orientation of its axis. Over this cycle, Earth's axis of rotation would point at different positions in space, which results in different amount of solar radiation reaching parts of Earth.

Change in the Tilt of Earth's Axis: ~41,000 Years

This cycle is related to departures from the current 23.5 degree tilt of our axis to the sun (▶Figure 10-4). Earth's axis changes from roughly 22 degrees to 24.5 degrees over this cycle, with stronger seasonal contrasts at the larger tilts.

Change in the Elliptical Shape of Earth's Orbit: ~100,000-Year Cycles

THE ICE AGES Still longer "cycles" include the **ice ages** of the Pleistocene epoch, roughly the last 2 million years. Although only two to four ice ages are readily documented

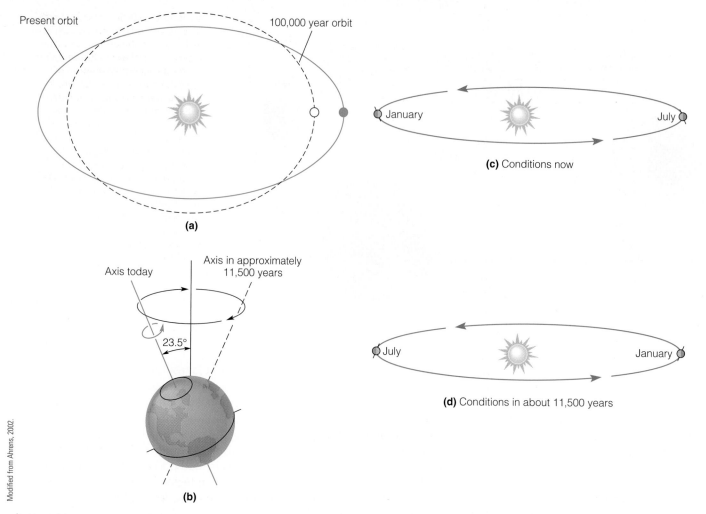

FIGURE 10-4. Three factors contribute to long-term climate change. The difference is much less dramatic than the seasons. **(a)** Earth's orbit changes from nearly circular to elliptical. **(b)** Earth's rotation axis slowly changes its orientation to the sun. **(c)** Earth is slightly closer to the sun in January. **(d)** Earth is slightly closer to the sun in July.

Modified from Ahrens, 2002.

in North America, as many as twenty are recorded in deep-sea cores. These cyclic ice periods average 100,000 years apart but vary somewhat in length. The typical beginning of an ice age shows somewhat irregular advances with lesser retreats to a maximum ice cover, followed by a rapid retreat to minimum ice cover over a few decades to a few thousand years. The atmosphere was 5° to 10°C cooler during the last ice age relative to the present.

Various explanations for the cyclic nature of the Pleistocene epoch include changes in Earth's orbit and rotation. Yugoslavian astronomer Milutin Milankovitch calculated in the 1920s that the Earth's orbit around the sun slowly changes from nearly circular to elliptical over a period of 100,000 years. This corresponds well with the twenty ice ages recorded in the 2 million years of the Pleistocene.

Global temperatures of the past are often estimated from oxygen isotope ($^{18}O{:}^{16}O$) studies of sedimentary materials or glacial ice formed at the time. During cooler periods, more water that is evaporated from the oceans is stored on the continents as ice and snow. During such periods, evap-

oration takes more of the lighter ^{16}O into the air, causing the oceans to be slightly enriched in ^{18}O. Because marine organisms incorporate oxygen from the seawater into their shells as they grow, the shells preserve the $^{18}O{:}^{16}O$ ratio of the seawater at the time. This permits estimates of the water temperature over long periods of time. Higher ^{18}O in a clamshell indicates lower Northern Hemisphere temperatures at the time the shell grew, because more of the water had evaporated from the oceans to form the continental glaciers.

Certainly there are long-term variations in climate. Throughout geological time, for at least the last 2 billion years, there have always been cycles of warm versus cold and wet versus dry. Warm most commonly coincides with wet, though the correlation is not perfect. Major cycles can last for tens of millions of years or even longer. Regardless of the origins of warm and cool or wet and dry cycles, some cycles are unrelated to others. Cycles of different lengths can overlap to give extreme highs or lows that do not come at well-defined regular intervals.

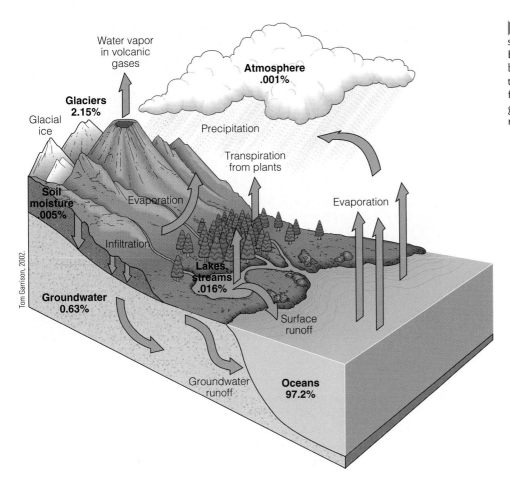

Tom Garrison, 2002.

FIGURE 10-5. The hydrologic cycle shows how, although almost all of the Earth's water is in the oceans, a small but important part evaporates to cycle through clouds and precipitation to feed lakes and streams, vegetation, and groundwater, before much of it again reaches the ocean.

Changes in ocean currents resulting from changes in salinity could rapidly change climate in some areas. Changes in local ocean salinity from factors such as melting glaciers and ice sheets could alter the path and strength of the Gulf Stream, the ocean current that keeps Europe abnormally warm for its northerly location. Any reduction in the flow of this warm current to the north would thus drastically cool Europe and other Northern Hemisphere climates. Many scientists believe that the current warming trend has the potential to rapidly push us into a little ice age for a few hundred years, with average temperature declines of 3° to 6°C (5° to 10°F) in the northern United States and Canada, northern Europe, and northern Asia. This would have devastating effects on humans as well as natural ecosystems, and the economies of most industrialized nations; clearly we should be proactive about our effects on climate change.

Basic Elements of Weather

Hydrologic Cycle and Humidity

Water in the oceans covers more than 70 percent of the Earth's surface. Unfortunately, the oceans are salty; humans need fresh water for life and less than 3 percent of the water on the Earth is fresh. We see fresh water falling as rain, feeding lakes and rivers, and ultimately flowing back into the sea. Consider the **hydrologic cycle** in which fresh water evaporates from the salty oceans; some of this then spreads over the continents before it returns to the oceans via groundwater and stream flow (▶ Figure 10-5).

Air can dissolve water vapor. The sun warms the ocean surface, and the air above it dissolves water. Under most conditions we do not see the water vapor because it is colorless and invisible like most gases. The amount of water vapor that can be dissolved in air depends on the air temperature; cold air can dissolve little water vapor whereas warm air can dissolve a lot. When air dissolves the maximum that it can hold, it is saturated with water vapor, thus its **relative humidity** is 100 percent. If it contains half as much as it can hold, its relative humidity is 50 percent.

If air cools to the temperature at which its relative humidity is 100 percent, the water-saturated air is still invisible. This is the dew point. If it cools further, it will have more water vapor than it can hold and some of the water will condense into tiny water droplets, to form clouds or fog. If enough water droplets form, they coalesce into larger droplets and eventually fall as rain or snow.

As breezes blow across the ocean, the air picks up water vapor. Air cannot dissolve significant amounts of solids, so what happens when warm oceanic seawater evaporates? Water vapor evaporates into the air, but the salt stays in the oceans. Water vapor dissolved in the air travels over the ocean with the prevailing winds until it reaches a continent

▶**FIGURE 10-6.** Air near the ground surface is often not saturated with water. At higher elevations, the air is cooler and can hold less moisture; at some higher elevation, it cannot hold all of its moisture, so some condenses to form water droplets in the form of clouds. Thus, below the base of a cloud layer, the air is unsaturated with moisture; above that, it is saturated.

Base of cloud layer

Donald Hyndman photo.

where some will drop out of the atmosphere as rain or snow (▶Figure 10-5).

Rain or snow falling on the continents can take several different paths. Some falls on leaves of trees or other vegetation; some of that evaporates directly back into air. Some that reaches the ground may run off the surface into streams or lakes, but part soaks into the ground to become groundwater, the water in spaces between soil grains or in cracks in rocks. The roots of trees and other vegetation draw on some of that moisture, taking it up through the plant to carry nutrients to the leaves. The leaves, in turn, transpire water back into the atmosphere because the air generally has less than 100 percent humidity. The direct evaporation and the transpiration by vegetation are both included in the category of evapotranspiration.

Groundwater slowly migrates down the slope of the water table to reach streams or lakes; groundwater is the main source of water for all but dry climate streams. Streams in turn carry the water downslope into larger rivers and ultimately back into the ocean to complete the cycle.

Adiabatic Cooling and Condensation

Adiabatic cooling occurs when rising air expands without change in heat content. Whenever an air mass expands, the available heat is distributed over a larger volume so the air becomes cooler (▶Figures 10-6, 10-7, and 10-8). The rate of cooling with rise in elevation is called the **adiabatic lapse rate.** Recall that warm air can hold more moisture; cold air can hold much less. Thus when warm, moist air cools to its dew point, it cannot hold all of the available water as vapor; the excess condenses into tiny water droplets—fog or clouds (▶Figures 10-6 and 10-8). As the saturated air

mass continues to rise, the droplets get bigger and precipitation falls.

During condensation, the water releases the same amount of heat originally required to convert liquid water into water vapor. In fact, this heat of condensation is large enough to reduce the adiabatic lapse rate from 10°C per 1,000 meters of rise to 5°C per 1,000 meters for saturated air (▶Figure 10-8). In other words, saturated air cools much more slowly as it rises compared to dry air. When an air mass falls, its temperature rises at the dry adiabatic lapse

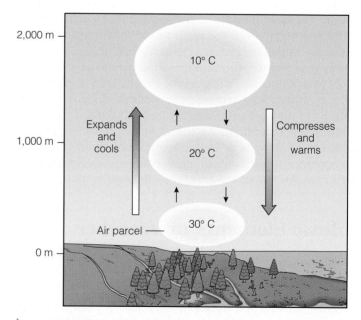

2,000 m — 10° C

Expands and cools — 1,000 m — 20° C — Compresses and warms

Air parcel — 30° C

0 m —

▶**FIGURE 10-7.** Adiabatic cooling: Atmospheric pressure is less at higher altitudes so rising air expands and cools.

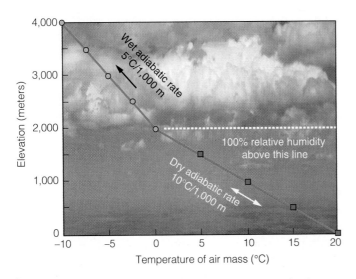

rate (10°C per 1,000) because it would then be below saturation and could thus hold more moisture.

When this moist air reaches land and is forced to rise over a mountain range, it expands adiabatically and cools, in what is called an **orographic effect.** Because cool air can dissolve less water, the moisture separates as water droplets that form clouds, and in turn rain, hail, or snow (▶Figures 10-9 and 10-10). When an air mass moves down the other side of mountains after having much of the moisture fall as rain or snow, the air is generally warm and dry. This rain shadow effect is why deserts often exist on the downwind side of mountain ranges.

Atmospheric Pressure and Weather

Air pressure is related to the weight of the column of air from the ground to the top of the atmosphere. When an air mass near the ground is heated, the air molecules vibrate faster, have more collisions, and thus cause the air mass to expand and its density to decrease. With a lower density than the surrounding air, the mass rises to create a

▶**FIGURE 10-9.** Moist air expands in upslope winds to cool and condense as clouds over a mountain range—in this case, from left to right.

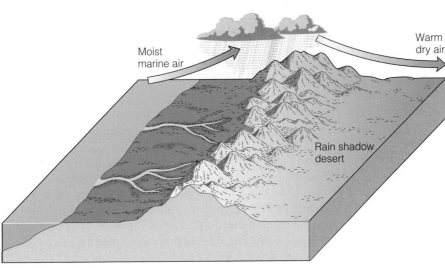
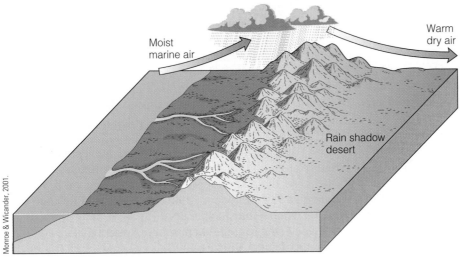

▶**FIGURE 10-10.** This schematic diagram of the orographic effect shows how warm moist air rises, expands, cools, and dumps precipitation. Descending on other side of mountains, it contracts, warms, and dries out.

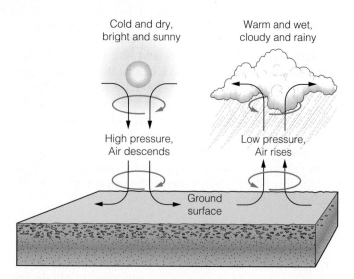

Cold and dry,
bright and sunny

Warm and wet,
cloudy and rainy

High pressure,
Air descends

Low pressure,
Air rises

Ground
surface

▶**FIGURE 10-11.** Differences in atmospheric pressure drive air movements and winds. With high atmospheric pressure, cold dry air descends, compresses, and spreads out; skies are clear. With low atmospheric pressure, air is drawn in from the sides, rises, expands, and condenses to form clouds and rain. In the northern hemisphere, descending air in high pressure cells rotates clockwise (viewed downward); rising air in low pressure cells rotates counterclockwise. The opposite rotation occurs in the southern hemisphere.

low-pressure region (▶Figure 10-11). It has lower pressure because rising air exerts less downward pressure than still or falling air. Air near the ground surface is pulled in toward the low pressure center to replace the rising air.

The opposite occurs in high pressure systems where cool air in the upper atmosphere has a higher density than its surrounding warmer air and thus falls. As it moves toward the ground, it has to spread out or diverge, so air moves away from high pressure systems near Earth's surface (▶Figure 10-11).

Horizontal differences in air pressure between high and low pressure systems create winds. Where these horizontal pressure differences are large, winds are strong. Water acts the same way, but we can visualize the driving force for water movement as the difference in water surface elevations: The greater the difference in elevations, the faster the water moves.

Typical global air pressure patterns for January and July are shown in Figure 10-12. During midsummer, high temperatures over the southwestern United States lead to rising air masses and a thermal low pressure area. The same happens in southwestern Asia. Note that although the location of the big high pressure cell in the Pacific Ocean just off the North American coast shifts its position between winter and summer, that shift leads to a big difference in moisture in those seasons. In January, for example, clockwise circulation around the Pacific high pulls moisture inland from the ocean to keep Oregon, Washington, and southwestern British Columbia wet. In July, circulation around that high

off the coast and the low over the southwestern states pull relatively dry air from the continental interior into the same areas. By contrast, in the eastern United States, clockwise circulation around the big high in the Atlantic Ocean pulls warm, moist air inland, causing frequent rainstorms and sometimes hurricanes. Similarly, examination of high and low pressure cells elsewhere helps us understand their typical winter and summer weather.

Coriolis Effect

Because the Earth rotates from west to east, large masses of air and water on its surface tend to lag behind a bit. It completes a rotation around its axis through the North and South poles once every twenty-four hours; thus, a point at the equator has to travel roughly 40,000 kilometers in a day (the approximate circumference of the Earth), while rotation of the Earth causes no movement for points on either the North or South poles. This large rotational velocity of Earth's surface makes both oceans and air masses near the equator move from east to west because they are fluid and thus are not pulled at the same speed as the rotating solid Earth. As water or air moves away from the equator toward the poles, it moves over part of the Earth that is rotating at a lower velocity and thus the water or air mass shifts off to the east because it is moving faster than solid Earth. The opposite occurs as water or air moves toward the equator. In this case, the air or water is moving slower than the Earth's surface rotation, and the resulting lag causes a curve off to the west. This forced curvature due to Earth's rotation is called the **Coriolis effect.**

Thus, in the northern hemisphere, large currents of air and surface water tend to curve to the right; they curve to the left in the southern hemisphere. Ocean currents move westward as they flow toward the equator, eastward toward the poles. Thus, ocean currents circulate clockwise in the northern hemisphere and counterclockwise in the southern hemisphere.

Airplanes are also affected by the Coriolis effect because they fly above the Earth, which rotates under them. For a plane flying due south from Detroit to Cancún, Mexico, east-to-west rotation of the Earth under it leaves the plane veering off to the west (to the right of its intended due-south flight path if this effect were not considered) (▶Figure 10-13). On its return trip from Cancún to Detroit, the plane on the ground before departure is rotating eastward with the Earth (~40,000 km/day). As it flies to the north, it moves over parts of the Earth with slower rotation speed, because the circumference of the Earth decreases toward the North Pole. Thus its initial eastward ground movement carries the plane farther to the east than its intended target. In either case, the plane veers off to the right compared with its intended flight direction, although pilots take this into account when setting their course. The same thing happens to a plane in the southern hemisphere, except that it veers left instead of right (▶Figure 10-13).

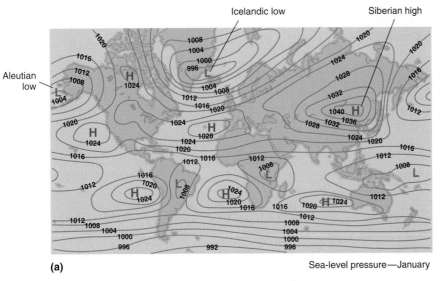

Icelandic low

Siberian high

Aleutian
low

(a)

Sea-level pressure—January

▶**FIGURE 10-12.** These maps show typical global air circulation for **(a)** January and **(b)** July. Note that high atmospheric pressure cells in the northern hemisphere rotate clockwise and spread outward as the air falls; low atmospheric cells rotate counterclockwise and inward as the air rises. During midsummer, high temperatures over the southwestern United States lead to rising warm air and a thermal low pressure area. The same is true in southwestern Asia.

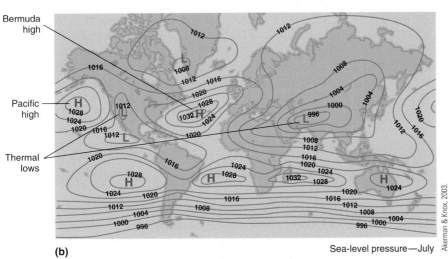

Bermuda
high

Pacific
high

Thermal
lows

Akerman & Knox, 2003.

(b)

Sea-level pressure—July

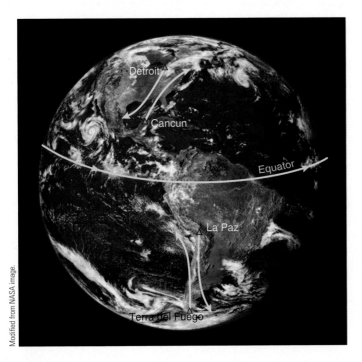

Modified from NASA image.

Detroit

Cancun

Equator

La Paz

Terra del Fuego

Rising and falling air currents also rotate (▶Figures 10-11 and 10-12) because of converging or diverging winds. When air rises in a **low pressure system,** air near the Earth's surface has to converge toward this low to replace the rising air. The Coriolis effect causes these winds to shift to the right of straight in the northern hemisphere, which causes these low pressure systems to rotate counterclockwise. Falling air diverges away from a **high pressure system** as it pushes down on Earth's surface. Again the Coriolis effect

▶**FIGURE 10-13.** The Coriolis effect: In the northern hemisphere, a plane flying due south (red dotted arrow) from Detroit to Cancún would veer well off to the west (yellow arrow) because the Earth is rotating from west to east under it. The curvature to the right of the flight path is the Coriolis effect. The same plane, flying due north from Cancún, would veer off to the east. In the southern hemisphere, a plane flying due south from La Paz, Bolivia, to Tierra del Fuego, would veer off to the left. It would also veer off to the left on its return trip north.

CLIMATE AND WEATHER RELATED TO HAZARDS   **255**

shifts these northern hemisphere winds to the right, which in this case causes a clockwise rotation of these high pressure systems. The opposites are true in the southern hemisphere; low pressure systems rotate clockwise, and high pressure systems rotate counterclockwise. A simple way to remember this rotation is to use the **right-hand rule** in the northern hemisphere (see Sidebar 10-1). Air that rises faster will also rotate faster. Because storms are localized in zones of much lower pressure, the rapidly rising air there rotates rapidly counterclockwise (in the northern hemisphere) (▶Figures 10-11 and 10-12). On a weather map of air pressure, the closer together the pressure contours, the higher the winds.

Circulating storms such as thunderstorms are small-diameter cyclones; tornadoes are still even smaller diameter and higher velocity winds. Severe thunderstorms accompanying fronts can spawn severe circulation and tornadoes. These smaller low pressure systems also rotate counterclockwise (in the northern hemisphere) unless spun off by some uncommon storm effect.

Global Air Circulation

Atmospheric heating is especially prominent near the equator, with cooling at the poles. Less-dense warm air rises near the equator and colder surface air moves toward those low pressure regions.

At higher altitudes in the tropics, the rising air spreads north and south. At approximately 30 degrees north and south latitude, the air sinks to form the subtropical high pressure zone. It warms adiabatically as it sinks; because warm air can hold more moisture, it has far less moisture than it can hold. As a result, the humidity in these regions is generally low and the climate is dry (▶Figure 10-14).

The combination of these global air movements with the Coriolis effect gives rise to the southwest-moving **trade**

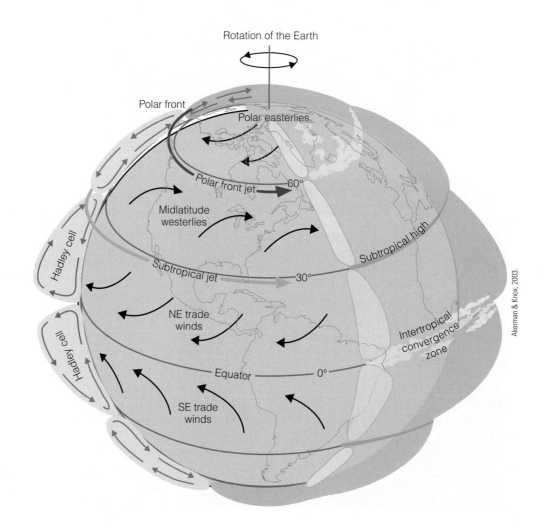

▶**FIGURE 10-14.** This diagram shows global air circulation. See the text for details.

winds between the equator and 30 degrees north latitude and the northeast-moving **westerly winds** between latitudes 30 and 60 degrees north (▶Figure 10-14).

Jet Stream

Meteorologists on television often use weather maps that show the subtropical **jet stream** meandering across North America. This narrow, 3- or 4-kilometer-thick ribbon of high velocity winds blowing from west to east across North America is at altitudes near 12 kilometers (that of many jet aircraft). It forms along the boundary between warm equatorial and colder, dense air to the north, meandering like any other stream (▶Figures 10-14 to 10-16). The upper-level warm equatorial air moves northeast, returning air of the lower-level trade winds. Similarly, the upper-level cold air to the north moves southwest, returning air of the lower-level westerly winds (▶Figure 10-14). Note also that the jet stream moves east along the interface between circulating high and low pressure cells; the high altitude rotation of these cells helps propel the stream (▶Figure 10-15).

When a meander in the jet stream shifts farther south, a huge polar low pressure area in the North Pacific often lies off western Canada. The counterclockwise circulation of the low carries warm moist air from the Pacific Ocean directly into California (▶Figure 10-16). Subtropical storms initiated in the mid-Pacific tend to follow the edge of the polar **cold front** of the Pacific Northwest coast. They carry heat and moisture to the Pacific coast where they can dump intense rainfall for a short time. In some cases, back-to-back storms every day or two provide heavy rainfall over broad areas and cause widespread flooding. Such frequent storms soak open pores in the soil; runoff is magnified and landslides are common.

With more warm air production to the south in the summer, the jet stream shifts northward over Canada; in

▶**FIGURE 10-15.** The region of the subtropical high pressure area near 30 degrees north latitude where warm tropical air meets colder polar air is the area with the greatest contrast in air temperature. Their interaction forms the subtropical jet stream shown in this vertical cross section from the North Pole to the equator.

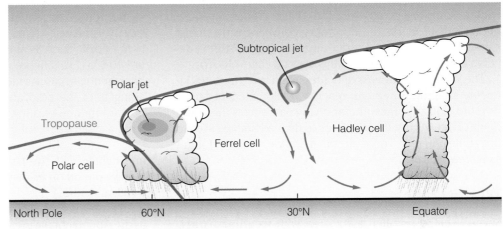

National Weather Service.

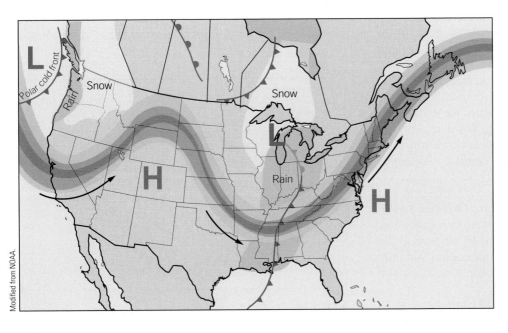

Modified from NOAA.

▶**FIGURE 10-16.** The weather forecast for December 16, 2003, shows the meandering jet stream with a huge polar low in the Pacific off western Canada that funnels rain into coastal California. Another Arctic low over the Great Lakes funnels rain into the Ohio River Valley.

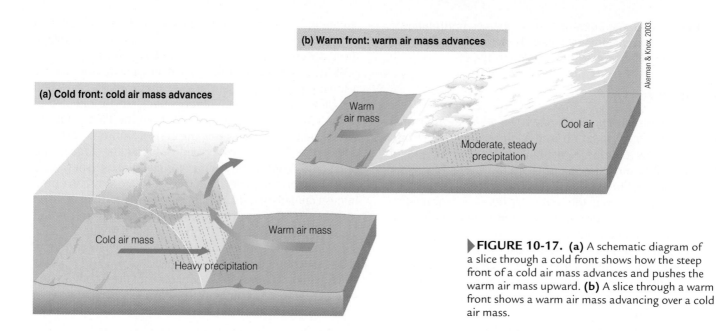

(b) Warm front: warm air mass advances

(a) Cold front: cold air mass advances

Warm air mass

Cool air

Moderate, steady precipitation

Cold air mass

Warm air mass

Heavy precipitation

Akerman & Knox, 2003.

▶**FIGURE 10-17.** **(a)** A schematic diagram of a slice through a cold front shows how the steep front of a cold air mass advances and pushes the warm air mass upward. **(b)** A slice through a warm front shows a warm air mass advancing over a cold air mass.

winter, it moves down over the United States. Because the jet stream marks the boundary between warm, moist subtropical air and cold, drier air to the north, storms develop along weather fronts in this area (▶Figure 10-15).

The jet stream moves at velocities much more than 100 kilometers per hour and can reach 400 kilometers per hour or more. Long-distance aircraft flying east across the continents or oceans use the jet stream to their advantage to save both time and fuel. Those aircraft flying west avoid it for the same reasons.

Weather Fronts

Severe weather is often associated with **weather fronts,** or the boundaries between cold and warm air masses. Thunderstorms develop at such fronts. They are initiated by moist, unstable air rising, cooling, condensing, and raining (▶Figure 10-17). In a cold front, a cold air mass moves more rapidly than an adjacent warm air mass, causing rapid lifting and displacement of the warm air. In a **warm front,** a warm air mass moves more rapidly than the adjacent cold air mass, and thus it rises over the adjacent cold air mass.

Advancing cold fronts rapidly lift warm air to high altitudes, causing instability, condensation, and in many cases thunderstorms (▶Figure 10-17). Air in advancing warm fronts rises over a flatter wedge of cold air, causing widespread clouds and in some cases thunderstorms. Counterclockwise, midlatitude, low pressure winds form where the jet stream dips to the south to form a "trough" (▶Figure 10-18). Cold and warm fronts sometimes intersect, such as at a low pressure center (▶Figure 10-18). The rising air mass in the low pressure center cools, causing condensation and rain. When cold air moves toward the equator, it collides with warm air, which is forced to rise.

A low-pressure center can develop as warm moist air to the south moves more rapidly to the east than cold dry air

to the north. That shear between air masses can initiate a low pressure cell where part of the warm air begins to move to the north, such as along a bulge in the jet stream (▶Figure 10-19a). This figure also shows a cross section of such intersecting fronts. As this occurs, air will be forced upward along both the cold and warm fronts, with the strongest upward motion existing where these fronts come together. Such a system can continue to develop such that the angle between the cold and warm air masses will become tighter and tighter, with the potential for the low pressure cell to finally spin off on its own as a cutoff low (▶Figures 10-19b, 10-20, and 10-21).

In eastern North America, coastal mountain ranges such as the Appalachians increase atmospheric convection. Cold near-shore air can push warmer onshore air upward, causing expansion, cooling, and rain. Storms moving onshore tend to move northward toward the pole. If they do not move far offshore, they can persist for many days, dump heavy

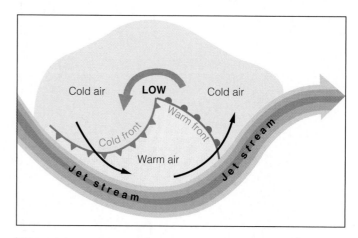

Cold air **LOW** Cold air

Warm front

Cold front

Warm air

Jet stream

Jet stream

▶**FIGURE 10-18.** A schematic diagram of cold and warm fronts intersecting in a low pressure cell. Compare Figure 10-19.

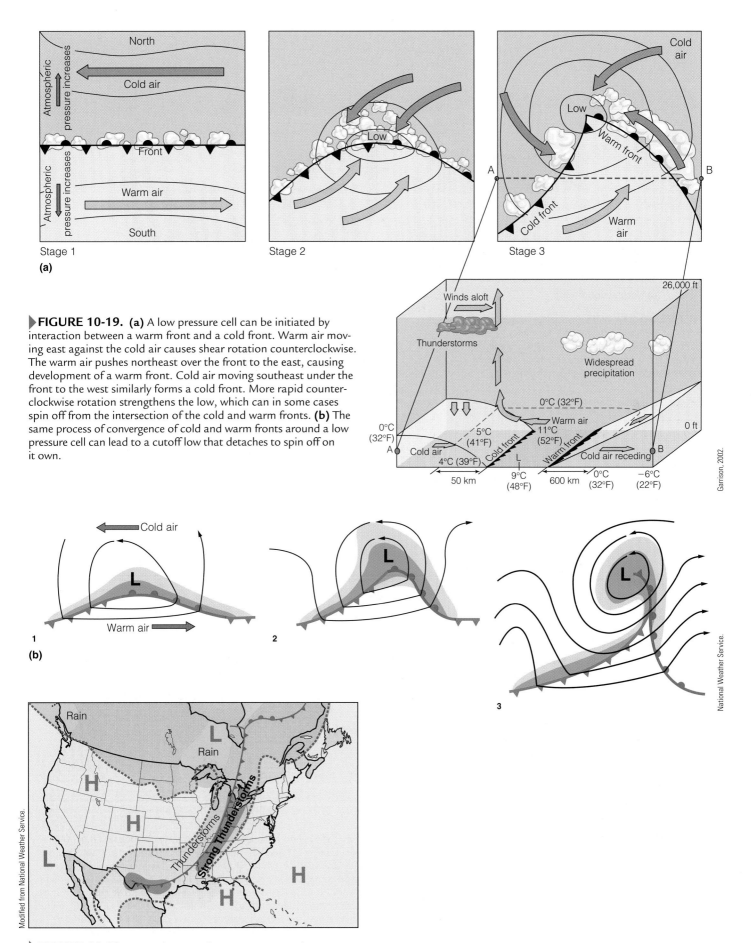

▶ **FIGURE 10-19.** **(a)** A low pressure cell can be initiated by interaction between a warm front and a cold front. Warm air moving east against the cold air causes shear rotation counterclockwise. The warm air pushes northeast over the front to the east, causing development of a warm front. Cold air moving southeast under the front to the west similarly forms a cold front. More rapid counterclockwise rotation strengthens the low, which can in some cases spin off from the intersection of the cold and warm fronts. **(b)** The same process of convergence of cold and warm fronts around a low pressure cell can lead to a cutoff low that detaches to spin off on it own.

▶ **FIGURE 10-20.** A weather map for June 26, 2003, shows a prominent cold front (blue) intersecting a warm front (red in upper right of map), with a cutoff low in Canada north of Lake Superior.

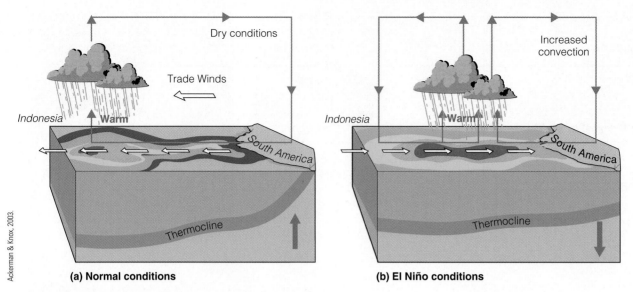

(a) Normal conditions   **(b) El Niño conditions**

Ackerman & Knox, 2003.

▶ **FIGURE 10-21.** Wind and ocean temperature conditions are shown during **(a)** normal conditions with strong trade winds and **(b)** El Niño conditions with weak trade winds. El Niño leads to warm temperatures off the west coast of South America.

rain along the coast, and cause storm surges of 4 meters or more. Because such higher latitude storms can have similar wind strengths and rainfall as tropical cyclones and are more frequent, they can be just as destructive.

El Niño

Normal Pattern

Oceanic circulation in the equatorial Pacific Ocean normally drifts westward, pushed by the trade winds (▶ Figure 10-14). A high-pressure cell exists over Tahiti in the central Pacific, and a major low pressure cell resides over the warm ocean water of Indonesia. Descending dry air over Tahiti (▶ Figure 10-22) and rising moist air over Indonesia circulate strong trade winds that pull warm ocean currents westward to Indonesia.

Farther east, the trade winds blowing to the west across the Andes Mountains of South America descend to the coast, severely drying the coastal desert environment. Warm water at the ocean surface is blown west from coastal Peru and is replaced by upwelling cold water from depth. That cold water carries nutrients to feed the highly productive fisheries off the coast of Peru.

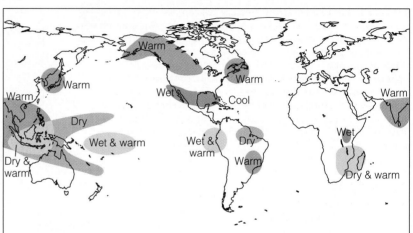

(a)

(b)

▶ **FIGURE 10-22.** **(a)** El Niño promotes greater winter precipitation in some areas, less in others. During El Niño, the southwestern United States is wet, in contrast to its normally dry weather, while the southeastern states are cooler than normal. Western and eastern Canada are warmer. **(b)** During normal, or the opposite extreme, La Niña conditions, southwestern British Columbia and Washington state are wet, as are areas immediately south of the Great Lakes. Western Canada, including Alberta and Saskatchewan, and adjacent states are cool. The southernmost states and Mexico are dry and warm.

El Niño Conditions

Every six years (on average), the Pacific Ocean circulation reverses in a pattern called **El Niño,** especially in winter from November to March. El Niño is the cyclic change in ocean and atmosphere circulation when the subtropical trade winds blowing west over the Pacific decrease in intensity, permitting warm central Pacific water to move east to the coast of Peru. A related term, **southern oscillation,** refers to the alternations in atmospheric pressure between east and west. For this reason, El Niño is also known as El Niño–southern oscillation (ENSO).The accompanying warming of the Pacific Ocean is along the equator between South America and the international date line in the mid-Pacific.

During El Niño, a major high pressure area develops in equatorial latitudes over Indonesia and northern Australia in the western Pacific, and a major low pressure cell develops near Tahiti in the central Pacific (▶Figures 10-21 and 10-22a). Rising moist air over Tahiti, along with descending dry air over Indonesia and northern Australia, weakens the trade winds, allowing warm ocean currents to flow eastward toward Ecuador and Peru.

The weather changes that result from these shifts in warm water locations can be dramatic. The big low pressure zone in the Pacific pulls the high altitude, meandering winds of the subtropical jet stream south before it loops north as it approaches western North America. This keeps cold Arctic air farther north. Eastern Pacific weather systems off Mexico repeatedly hammer southwestern North America. The warm water piles up against the west coast of South America, evaporates, and causes heavy rainfall in

the deserts of Peru. Part of the warm water sweeps north along the coast to bring warm subtropical rains to the southwestern United States, especially Southern California (▶Figure 10-23). In contrast, droughts and widespread fires affect Australia, Indonesia, and Southeast Asia. The mideastern United States tends to be drier than normal. If some areas are abnormally wet, others have to be abnormally dry because El Niño simply involves a large-scale redistribution of moisture rather than a change in the worldwide average.

A recent major event, in late 1997 and early 1998, brought incessant rains to Peru. It caused big problems for peasant farmers. Rains were welcome in this usually arid coastal area of Peru, but not in such torrential amounts. Nor were the rains a big surprise; they always came every few winters. Fields were soggy; rivers rose and flooded. Few people had seen so much rain—as much as 12 centimeters in a day. Instead of prevailing trade winds blowing westward across the Andes, leaving the coast dry, the winds blew eastward, rising against the mountains to bring rain from the Pacific. El Niño was here. Because of the timing, Peruvian fishermen refer to it as El Niño, Spanish for "the child" who typically comes every four to seven years around Christmas. The opposite extreme in weather patterns is called **La Niña,** the "girl child." The usually lush Amazon rain forest, then in the rain shadow of the Andes, would be unusually dry and subjected to widespread wildfires.

North American streams respond to similar El Niño influences. El Niño years in the generally dry Southwest bring more local flooding at lower elevations. Farther east in North America, hurricanes are less frequent and weaker when the jet stream shears off the tops of westward-bound

(a) January 17 1999

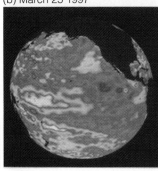

▶FIGURE 10-23. Sea-surface temperatures for the Pacific Ocean are shown for the major 1997–1998 El Niño. The highest sea level in warm water over the eastern equatorial Pacific Ocean is shown in white and red. Darker shades to the west are the coldest temperatures. North America is shown in the upper right. Before an El Niño event, **(a)** warm water (white and red colors) is in western Pacific Ocean north of Australia. As El Niño proceeds **(b, c, d),** warm water moves east across the equatorial Pacific to build up against coastal Peru. At the El Niño maximum **(e),** warm water piles up against coastal Peru and spreads north up the coast of Mexico and sometimes into Southern California.

(b) March 25 1997

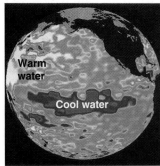

(c) April 25 1997

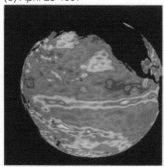

(d) May 25 1997

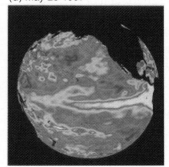

(e) December 10 1997

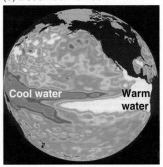

NASA & CNES.

Sea-surface tempera-
ture anomalies in the
mid-Pacific Ocean
are shown from 1900
to 1950 and 1950 to
2003. El Niño shows
abnormally high sea-
surface temperatures.

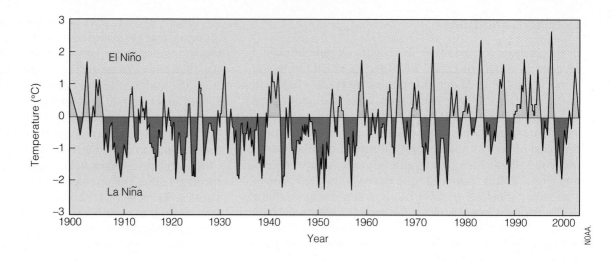

Atlantic storms. Although El Niño minimizes hurricanes, the strong jet stream flow over Florida tends to stir up strong tornadoes. In February 1998, they killed 41 people in the Orlando area of Florida. Tornadoes were also active in early 2003, another El Niño year.

Although the recurrence interval for strong El Niño events since 1930 has averaged six years, these events have been more frequent since 1976 (▶Figure 10-24). Some researchers blame global warming; others disagree.

Prior to 1997–1998, the last major El Niño was in 1982–1983. It killed 2,000 people worldwide, with damage totals of $14.4 billion. More recent events followed in 1986–1987, 1995, and 2003.

Predicting El Niño

El Niño events can be predicted in advance based on monitored sea-surface temperature variations in the equatorial Pacific Ocean (▶Figure 10-24). The issue is obviously important in planning for the likelihood of flooding in Southern California during an El Niño event and the severity of the hurricane season in the southeastern states during normal years. It is also important in planning for droughts and crop losses and for regional changes in heating oil usage and costs; the difference can amount to billions of dollars. Even the stock market responds to such predictions.

Tropical atmosphere and ocean buoys anchored in the equatorial Pacific Ocean monitor surface and deeper water conditions, atmospheric temperature, humidity, and winds. Satellites monitor the ocean surface elevations, circulation patterns, and the huge back-and-forth rhythmic sloshing of the Pacific Ocean between El Niño events and normal conditions.

When sea-surface elevations and temperatures rise in the central Pacific near Tahiti and atmospheric pressures fall, past events show that more rain is to be expected in Peru and southern California—a new El Niño. The latest models

of how the process operates were successful in predicting the massive 1997–1998 El Niño event a year before it occurred. The intensity of El Niño events also varies. In fact, such events appear to be more severe over the last twenty years (▶Figure 10-24). It is still unclear if this is a long-term trend and whether any of these changes reflect the long-term measured increase of approximately 0.5°C in the last century.

North Atlantic Oscillation

As with ENSO in the Pacific Ocean, there is a recurring atmospheric pressure pattern in the North Atlantic Ocean, but there is essentially no correlation between them. In contrast to El Niño, which is recognized by variations in Pacific Ocean conditions, the **North Atlantic Oscillation** (NAO) is defined by variations in winter atmospheric pressure over the North Atlantic Ocean. The greatest weather effects are from December to March. There is, of course, interaction between the North Atlantic atmosphere and the ocean, but changes in the ocean lag behind that of the atmosphere by two years or more.

In a "positive NAO," high pressure sits west of southern Portugal, the Azores, and the eastern Atlantic Ocean; low pressure sits over the North Atlantic southeast of Greenland (▶Figure 10-25). The prevailing westerly winds are strong, and warm weather covers the Atlantic Ocean east of the United States and in northwestern Europe. Europe is warmer and wetter. The trade winds are also strong, and the eastern subtropical Atlantic Ocean off Africa is relatively cold.

In a negative NAO, the high pressure area moves south to the central Atlantic Ocean closer to the equator; low pressure also moves south to east of Labrador and Newfoundland. The North Atlantic off Greenland is warmer, as is the trade wind belt off West Africa. New England east into the

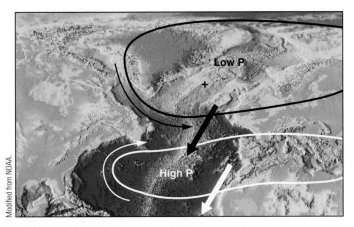

Modified from NOAA.

▶**FIGURE 10-25.** A positive NAO (+) high pressure lies over the eastern Atlantic Ocean (white) while low pressure sits near Iceland (black). Warm weather develops in the western Atlantic and northwestern Europe. The westerly and trade winds are strong. With negative NAO (−), both pressure systems shift to the southwest (bold arrows).

recognized since the early 1800s, but accurate predictions of NAO patterns remain elusive.

Climate Controls on Flooding

Because atmospheric moisture is the main source of water for floods, large floods generally depend upon large areas of humid air. Water stored in snow and ice is a secondary reservoir. Evaporation from oceans, and secondarily from continents, constantly resupplies the atmosphere. Because warm air can hold more water vapor, the atmosphere at low latitudes contains much more moisture. More than 60 percent of water evaporated into the atmosphere is between the equator and 30 degrees north or south. Even with global air circulation, tropical air contains ten times as much moisture as polar-latitude air.

In tropical regions, temperature and pressure gradients are weak, as are regional winds. Rising air, often resulting from localized heating, causes condensation, formation of towering convection clouds, and frequent thunderstorms. Rainfall rates are high but do not last long. In winter, the area of tropical moisture reaches only as far north as Mexico City and Miami (▶Figure 10-26). In summer, it pushes north into Southern California and Arizona. East of the Rockies, the tropical moisture loops north through all of the high plains into north-central Canada, southeastern Canada, and all of the eastern United States.

Tropical cyclones, including hurricanes, move westward in the belt of trade winds. In the Atlantic Ocean, they drift westward, then northward and eastward along the eastern fringe of the United States, where they interact with midlatitude frontal systems. Reaching the Atlantic coastal plain, Hurricanes Camille and Agnes in 1969 and 1972, for example, dumped 25 to 50 centimeters of rain on many areas.

Atlantic Ocean is colder, and so are Spain and the western Mediterranean. The westerly and trade winds are both weaker. Periodicity is not regular, but positive NAO comes along every three to ten years.

Winter storm activity shifts, in general, with the NAO. Positive NAO is associated with more frequent and intense Atlantic storms in northern Europe. Central and southern Europe tend to be drier, with drought conditions prevailing locally. From 1905 to 1970, the NAO index decreased gradually, with a significant increase to record positive values since then. Oscillations in atmospheric pressure in the North Atlantic and their effect on weather have been

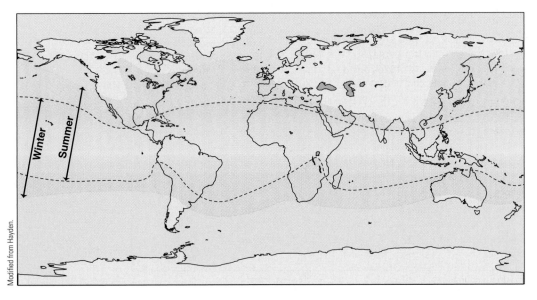

Modified from Hayden.

▶**FIGURE 10-26.** Seasonal boundaries are shown between the warm, moist air of tropics and cooler, drier, higher-latitude air. The limits of warm, tropical air in summer shift northward as shown in the labels at the left.

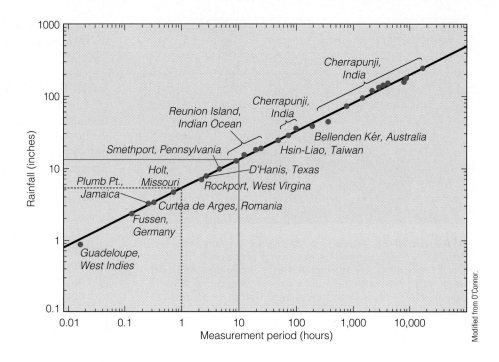

▶**FIGURE 10-27.** Maximum rainfalls are plotted against the duration of rainfall. The dotted red lines show the expected maximum value for a one-hour period, while the solid red lines indicate the expected maximum rainfall for a ten-hour period. Note that on either axis $10^0 = 1$, $10^1 = 10$, $10^2 = 100$, and so on.

Importance of Duration and Intensity of Rainfall

Especially heavy rainfall causes floods almost anywhere. In most cases, such heavy rainfall accompanies thunderstorms, which last for a few minutes or an hour or two. A line of thunderstorms may prolong the deluge for several hours, whereas tropical cyclones may stretch out the rains for several days. Areas such as Southeast Asia, which have seasonal monsoons, may experience extreme rains for several weeks or even months.

The maximum amounts of rain recorded for various periods of time are truly amazing. Most of these result from an extraordinary weather system such as a hurricane, coupled with the orographic effect of the system rising against a mountain range. The maximum recorded amounts are at Cherrapunji in northeastern India, where monsoon-driven winds, laden with moisture from the Bay of Bengal, rise against the eastern Himalayas (▶Figure 10-27 and Table 10-1). Imagine more than a meter of rain of water trying to flow off the ground surface all at once.

Table 10-1 Some Extreme Rainfalls*			
Location	**Date**	**Amount**	**Cause**
Cherrapunji, India	July 1861 (month)	9.3 m	Monsoon
	June 24–30, 1931	3.2 m	Monsoon
	June 9–16, 1876	3.39 m	Monsoon
	June 14, 1876 (one day)	1.03 m	Monsoon
	March 11–19, 1952	4.13 m	
Cilaos, Reunion (Indian Ocean)	March 15–16, 1952 (one day)	1.87 m	
Alvin, Texas (south of Houston)	July 25–26, 1979	1.09 m	Tropical storm
Bowden Pan, Jamaica	Jan 23, 1960 (one day)	1.11 m	Frontal storm
Rapid City, South Dakota	June 9, 1972 (6 hours)	37 cm	
Curtea de Arges, Romania	12 minutes	21 cm	
Barot, Guadeloupe, West Indies	1 minute	3.8 cm	

*Compiled from various sources.

The straight-line relationship on the log–log plot in Figure 10-27 indicates maximum rainfalls for the duration of rainfall. Thus, because 10^0 is equivalent to one hour, the maximum one-hour rainfall to be expected anywhere would be approximately 12.5 centimeters or 5 inches (dotted red lines); the maximum ten-hour rainfall would be approximately 30.5 centimeters or 12.2 inches (solid red lines).

Floods generally occur after prolonged soaking of the ground by either rainfall or snowmelt, when the near-surface soil is water-saturated. Only during uncommon torrential rainstorms will water run directly off the surface to streams as sheet flow.

Except in arid regions, groundwater flows out to streams, keeping them flowing year-round (Figure 11-28a, page 283). Stream flow in nonarid regions varies from high in wet weather to low in dry weather. Groundwater in such areas does not respond rapidly to changes in rainfall, and inflow to streams from groundwater is slow and relatively constant. Thus, wet climates tend to provide streams that flow year-round and flood in prolonged heavy rains. In northern latitudes, they commonly flood during the spring snowmelt period.

Dry Climates

With low annual rainfall, little or no vegetation can grow to soften the impact of raindrops and slow their infiltration. Rain falls directly on the ground, packing it tightly to permit less infiltration and cause more surface runoff. The rain also kicks up sediment from the surface, permitting it to be carried off by streams. During dry seasons, less water gets into the ground, even during the less frequent rainstorms, so less feeds the groundwater (▶ Figure 11-28b, page 283). Dry climates with no year-round streams can see flash floods after any major or prolonged rainfall.

Streams in desert basins generally only flow during and shortly after a rainstorm but then dry up until the next rainstorm. Because more sediment is supplied to the dwindling amount of water, sediment deposits in the gullies (see Figure 11-22, page 280). This progressively chokes the flow, causing some of the water to spill over and follow another path. This gully in turn fills with sediment and so on. The result is a braided alluvial fan that continuously deposits sediment that builds with time.

Strong Winds Not Associated with Storms

Chinook Winds

Along the eastern slope of the Rocky Mountains, strong warm and dry winds occasionally develop in the winters of Alberta, Montana, Wyoming, Colorado, and New Mexico. Strong winds from the west cool and lose their moisture as they rise across the Rocky Mountains. Continuing east, they form strong downslope winds that become warmer adiabatically as the air compresses into a smaller space at lower elevations. At Boulder, Colorado, **Chinook winds** have reached speeds of 63 meters per second (140 mph). The warm, dry winds can rapidly melt any snowpack and cause flooding.

Santa Ana Winds

In Southern California, strong trade winds from the east develop in the fall when dry continental air flows southwest around a high pressure system in Nevada. The air mass flows west, down through mountain passes and toward the coast. In doing so, its temperature rises rapidly because of adiabatic compression and its humidity drops. These hot, dry **Santa Ana winds** create extremely dangerous conditions for wildfires in Southern California (see Chapter 16 on wildfires).

KEY POINTS

✓ Greenhouse gases such as carbon dioxide and methane trap heat in the Earth's atmosphere much as the glass in a greenhouse permits the sun to shine in but prevents most heat from escaping. Atmospheric carbon dioxide and temperature are increasing, especially since 1970. Consequences include more frequent and stronger storms, smaller snowpacks and earlier runoff, drier vegetation and more fires, as well as warming and expansion of the oceans that leads to rise of sea level. **Review pp. 245–248; Figures 10-1 and 10-2.**

✓ The atmosphere can be cooled by particulates from sources such as industrial smokestacks, forest fires, and volcanic ash eruptions. **Review p. 248.**

✓ Earth's climate has cycles from days and seasons to those that come thousands of years apart. The large climatic variations that come with these long-term climate changes would dramatically disrupt life as we know it. **Review pp. 248–250.**

✓ Water continuously evaporates from oceans and other water bodies, falls as rain or snow, is transpired by plants, and flows through streams and groundwater back to the oceans. **Review pp. 251–252; Figure 10-5.**

✓ Rising air expands and cools adiabatically, that is, without loss of total heat. The rate of temperature decrease with elevation—that is, the adiabatic lapse rate—in dry air is twice the rate in humid air. **Review pp. 252–253, Figures 10-7 and 10-8.**

✓ Cool air can hold less moisture; thus, as moist air rises over a mountain range and cools, it often condenses to form clouds. This is the orographic effect of mountain ranges. **Review p. 253.**

✓ Warm air rises, cools, and condenses to form an atmospheric low pressure zone that circulates counterclockwise in the northern hemisphere. Cool air sinks, warms, and dries out to form a high pressure zone that circulates clockwise in the northern hemisphere. Air moves from high to low pressure, producing winds. **Review pp. 253–254, 256; Figure 10-11; Sidebar 10-1.**

✓ West-to-east rotation of the Earth causes air and water masses on its surface to lag behind a bit. Because the rotational velocity at the equator is greater than at the poles, the lag is greater near the equator. This causes air and water masses to rotate clockwise in the northern hemisphere and counterclockwise in the southern hemisphere. **Review pp. 254–255; Figures 10-13 and 10-14.**

✓ The prevailing winds are the westerlies north of 30 degrees north (wind blows to the northeast) and the trade winds farther to the south (blowing to the southwest). Cells of warm air rise near the equator and descend at 30 degrees north and south. **Review pp. 256–257; Figure 10-14.**

✓ The subtropical jet stream meanders eastward across North America, across the interface between warm equatorial air and colder air to the north, and between high and low pressure cells. **Review pp. 257–258; Figures 10-15 and 10-16.**

✓ Warm fronts, where a warm air mass moves up over a cold air mass, and cold fronts, where a cold air mass pushes under a warm air mass, both cause thunderstorms. Such fronts often intersect at a low pressure cell. **Review pp. 258–259; Figures 10-17 to 10-20.**

✓ Equatorial oceanic circulation normally moves from east to west, but every few years the warm bulge in the Pacific Ocean drifts back to the east in a pattern called El Niño, bringing winter rain to the west coast of equatorial South and North America, including Southern California. **Review pp. 260–262; Figures 10-21 to 10-24.**

✓ The North Atlantic Oscillation is a comparable shift in winter atmospheric pressure cells that affects weather in the North Atlantic region. It also shifts every few years, but the times do not correspond to those of El Niño. **Review pp. 262–263; Figure 10-25.**

✓ Because flooding depends on water in the atmosphere and tropical air contains ten times as much water as cold polar air, tropical air masses bring the wettest storms. Warm, moist, tropical air moves to much higher latitudes during the summer. **Review p. 263; Figure 10-26.**

✓ Vegetation is abundant in wet climates, so rain falls on leaves and soaks slowly into the ground to feed groundwater and year-round streams. Lack of vegetation in dry climates permits rain to fall directly on the ground, where most of it runs off the surface. **Review p. 265.**

✓ Especially heavy rainfall can cause floods, as can prolonged rainfall that saturates surface soil to prevent further rapid infiltration. **Review pp. 264–265.**

IMPORTANT WORDS AND CONCEPTS

Terms

adiabatic cooling, p. 252
adiabatic lapse rate (dry and wet), p. 252
Chinook winds, p. 265
cold front, p. 257
Coriolis effect, p. 254
El Niño, p. 260
gas hydrate, p. 246
global warming, p. 246
greenhouse gases, p. 245
high pressure system, p. 255
hydrologic cycle, p. 251
ice ages, p. 249
jet stream, p. 257
Kyoto Protocol, p. 247

La Niña, p. 261
low pressure system, p. 255
North Atlantic Oscillation, p. 262
orographic effect, p. 253
relative humidity, p. 251
right-hand rule, p. 256
Santa Ana winds, p. 265
seasons, p. 249
southern oscillation, p. 261
trade winds, p. 256
warm front, p. 258
weather fronts, p. 258
westerly winds, p. 257

QUESTIONS FOR REVIEW

1. What are the main sources of important greenhouse gases?

2. About how much has carbon dioxide increased in the atmosphere? When did the increase begin?

3. How much has the Earth's atmosphere increased in temperature in the last 1,000 years? When did most of the increase begin?

4. Other than an increase in temperature, what would be the most prominent changes in weather with global warming?

5. Approximately how much is sea level expected to rise in the next 100 years? What country is expected to see the largest loss of life as a result of rise in sea level. Why?

6. What is the main contributor to the rise in worldwide sea level? Be specific as to what makes the level rise—not just "global warming."

7. What causes the Earth's seasons such as winter and summer?

8. An area of low atmospheric pressure is characterized by what kind of weather?

9. Why do the oceans circulate clockwise in the northern hemisphere?

10. Explain the right-hand rule as it applies to rotation of winds around a high or low pressure center.

11. Explain the orographic effect and its effect on weather.

12. What is the main distinction between a cold front and a warm front?

13. What main changes occur in an El Niño weather pattern?

14. Why do streams flow year-round in a wet climate? Explain clearly.

15. If a humid air mass has 100 percent relative humidity and is 20°C at sea level, what would the temperature of this same air package be if it was pushed over a 2,000-meter-high mountain range before returning again to sea level? Explain your answer and show your calculations.

FURTHER READING

Assess your understanding of this chapter's topics with additional quizzing and conceptual-based problems at:

 http://earthscience.brookscole.com/hyndman.

CHAPTER 11

A flash flood overflows the reservoir spillway at Canyon Lake Dam above New Braunfels, Texas, on July 4, 2002, providing this dramatic display of power.

U.S. Army Corps of Engineers photo.

STREAMS AND FLOOD PROCESSES
Rising Waters

Floods in the United States account for one-quarter to one-third of the annual dollar losses and 80 percent of the annual deaths from geologic hazards. Between 1930 and 2000, the long-term average annual costs of flood damages more than doubled. Floods often come as a surprise to those living in the affected areas.

Before Europeans settled North America, rivers flooded as they do now. Local residents of the time merely packed up and moved to higher ground until the water subsided. Then, as now, the amount of water falling on the land surface varied from year to year. Some of that water evaporates from the ground and vegetation surfaces, some is taken up by vegetation, and some soaks into the soil to eventually reach rivers. During torrential rainfall, some will flow across the surface into the rivers.

Most early towns grew along rivers, generally on floodplains (▶ Figure 11-1). With permanent settlement of floodplain areas, people felt the need to protect their structures

State of Massachusetts photo.

▶**FIGURE 11-1.** Lawrence, Massachusetts, lies partially submerged during flooding in October 1996.

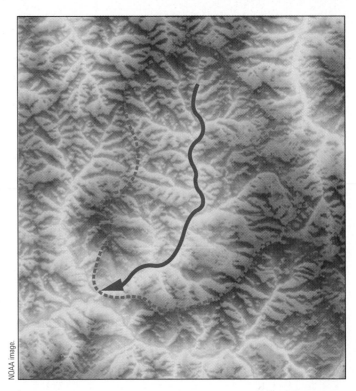

NOAA image.

▶**FIGURE 11-2.** The drainage basin of a river includes all of the slopes that drain water downslope to feed the river. In this shaded relief map from southern Oregon, the drainage basin or watershed above a point is outlined with a blue dashed line. The main channel of the river is shown in dark blue.

from floods, provide irrigation water for crops, and later provide electric power. People began to build protective dikes and dams on rivers, changing the rivers' behavior, especially during floods. They excavated sand and gravel from river channels and floodplains that would otherwise have been carried downstream. They logged forests, often by total clear-cut methods, built roads, and paved large areas. By doing so, they radically increased the rate at which water flows off the surface and the rate of sediment supply from the slopes. Progressive increases in population have intensified all of these influences, as discussed in Chapter 12.

Stream Flow and Sediment Transport Processes

To understand how streams respond to external influences, we first need to know their "normal" behavior. Rivers are complex networks of interconnected channels with many small tributaries flowing to a few large streams, which in turn flow to one major river. They flow in valleys that they have eroded over thousands or millions of years. Most streams are wide and shallow, with nearly flat bottoms. Rivers have little control over their own behavior. They respond to changes in regional climate and local weather through the amount and variability of flow and, as we discuss below, to the size and amount of sediment particles that are deposited in their channels.

Stream Flow and Gradient

Streams in humid regions collect most of their water from groundwater seepage. As a result, flow generally increases in the downstream direction as additional flow from tribu-

tary streams and groundwater enters the channel. Streams accumulate surface water from their **drainage basin** or **watershed,** the upstream area from which surface water will flow toward the channel above a point (▶Figure 11-2).

Most streams begin high in their drainage basins, surrounded by steeper slopes and often by harder, less easily eroded rocks. Those rocks slowly move downslope by processes that include creep and landslides until they reach the valley bottom at a stream. There the streams move the rocks and sediment on down valley.

The **discharge** of a stream, or total volume of water flowing per unit of time, is the average water velocity times the cross-sectional area of the stream (see Sidebar 11-1). Because there is no easy way to measure the average velocity, point velocities are measured at equal intervals across the stream channel, and each such velocity is multiplied by a

---

Sidebar 11-1

$Q = vA$

Q = discharge (e.g., m^3/sec.)

v = velocity (e.g., m/sec.)

A = cross-sectional area (e.g., m^2)
 = width × depth (in cm)

July 31, 1976, was the beginning of a three-day celebration of Colorado's centennial. Motels and campgrounds in the otherwise sparsely populated canyon were filled with more than 3,500 people. The Rocky Mountain Front rises abruptly from the high plains immediately west of Denver, Colorado. Moist summer air masses from the east often rise into the mountains, cool, and dump their moisture in thunderstorms. On July 31, a cool Canadian air mass moved in to become stationary near the Big Thompson basin, while a warm front continued to funnel in moisture.

Thunderstorms formed along the north-trending front just east of Estes Park, west of Fort Collins, and 80 kilometers northwest of Denver (▶ Figure 11-3). It remained there, dumping approximately 30 centimeters of rain in four hours, 75 percent of the total for a typical year. The heaviest rain fell directly over Big Thompson Canyon, just east of Estes Park. For its uppermost 34 kilometers through Rocky Mountain National Park, the river descends more than 1,524 meters. At the town of Estes Park, it enters a narrow, rocky canyon for 40 kilometers and drops another 762 meters before spreading out on the Great Plains.

Heavy rain began to fall at 6:30 P.M., and the danger of a major flood soon became evident; police moved through the canyon telling people to leave. Unfortunately because heavy summer thunderstorms in the area are common, many people did not believe they were in danger and remained. Rapid runoff from the mountain slopes developed into a flash flood, and by 7 P.M. a wall of water more than 6 meters deep was racing down the canyon at roughly 6 meters per second (~22 km/

Robert Jarrett photo, USGS.

▶ **FIGURE 11-4.** This car was caught in the Big Thompson flood.

hour). Flows were between 860 and 1,512 cubic meters per second or between 14 and 24.6 million gallons per minute.

By 8 P.M., the flood was carrying dirt, rocks, buildings, and cars with people in them. Most of Highway 34 through the canyon was washed out. Of the 400 cars on the canyon highway, those who abandoned their cars and ran upslope survived; the 139 who tried to outrun the flood in their cars died (▶ Figure 11-4). More than 600 others were never accounted for. An ambulance crew that drove into the canyon to render aid reported that a huge, choking dust cloud led the wall of water, picked up the ambulance, and slammed it into a wedge on the canyon wall. The crew climbed out of the wrecked ambulance and up to a ledge 15 meters above the highway. The water surged to their level, but they survived to be rescued by helicopter the next morning, along with hundreds of others. In the debris at the mouth of the canyon was a motel register with twenty-three names. The motel disappeared in the flood; none of its guests were found.

The flood destroyed 400 cars, 418 houses, and fifty-two businesses, and totaled $114 million* in damages (▶ Figures 11-5 and 11-6). The flow rate was four times larger than any previously measured flood in the 112 years of record. The recurrence

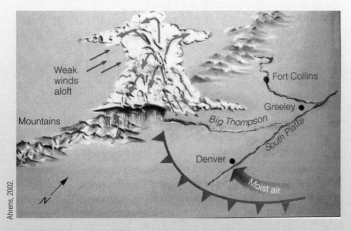

Ahrens, 2002.

▶ **FIGURE 11-3.** Map of the weather pattern that led to the torrential rains associated with the Big Thompson flood.

*All costs are in 2002 dollars.

Robert Jarrett photo, USGS.

▶**FIGURE 11-5.** A house in the bottom of Big Thompson Canyon was washed downstream, coming to rest on a concrete bridge, though somewhat the worse for the experience.

interval for a flood of this magnitude is uncertain, but radiocarbon dating of old flood deposits suggests a recurrence interval of several thousand years assuming no changes in climate. Could it happen again? Not much has changed. The winding highway still follows the canyon bottom, and the patches of narrow canyon bottom are again lined with houses.

Donald Hyndman photo.

▶**FIGURE 11-6.** Big Thompson Canyon, east of Estes Park, Colorado, still preserves the bouldery deposits from the July 1976 flood that killed hundreds of people.

cross-sectional area surrounding that measurement (▶Figure 11-7a). New instruments called Acoustic Doppler Current Profilers (ADCPs) have been developed to more accurately approximate stream flow by measuring water velocity at hundreds of locations using the Doppler shift in sound waves (▶Figure 11-7b). An increase in discharge during a flood involves an increase in water velocity, water depth, and sometimes width of the stream.

The gradient or slope of channels generally decreases down valley as sediment is worn down to smaller sizes and the larger flow is capable of transporting the particles on a gentler slope. Larger rivers often **meander** or sweep back and forth as they drain down valley, eroding the outside of bends and depositing on the inside of the same bends. That process of erosion and deposition, continuing for centuries, gradually moves the river back and forth to erode a broad valley bottom. At high water, the flooding river spills out of its channel and over that broad area—its **floodplain.** Ultimately, the stream will reach a lake or the ocean, a **base level** below which the stream cannot erode. Below

we discuss these features and the processes that form them to provide the context for peoples' interactions with rivers through flood and erosion hazards.

Where tributary streams descend from a steep mountainside onto a broad valley bottom, they leave the narrow valley that they eroded to reach a local base level of a larger valley. The rapid decrease in their channel slope or gradient causes them to drop much of the sediment they were carrying. Thus, the stream changes from an erosional mode to a depositional mode at the point where the slope decreases. The excess sediment spreads out in a broad fan-shaped deposit called an **alluvial fan.** Similarly, where a river reaches the base level of a lake or ocean, the abrupt drop in stream velocity at nearly still water causes it to drop most of its sediment in the form of a delta. The delta is like an alluvial fan except that the delta sediments are deposited under water.

Streams in humid regions adjust their channels to carry the typical annual flows that fill them. Channels in semi-arid regions adjust their channels to less frequent large floods

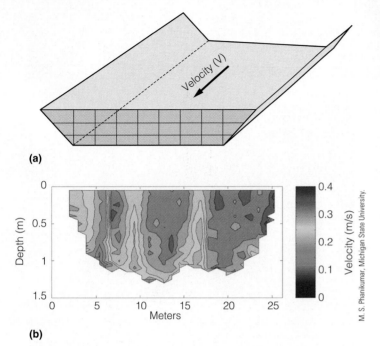

(a)

(b)

M. S. Phanikumar, Michigan State University.

▶**FIGURE 11-7.** **(a)** The cross-sectional area of a simple stream channel can be approximated by dividing it into a rectangular grid. With individual velocity measurements for each such box, the total flow would be the sum of $V_1 \times A_1 + V_2 \times A_2 + \ldots$, where A_1 and A_2 are the areas of individual boxes. **(b)** Stream flow through the Red Cedar River in Michigan was calculated using an ADCP, which measures hundreds of water velocities across a vertical cross section of the channel. The stream flow is then calculated by summing the velocity of each cell by the cross-sectional area of that cell.

because smaller flows do not significantly affect channel shapes. It is not simply the size of the flow that makes a flood, but how unusual such a flow is for that particular channel. Streams generally reach bankfull stage every 1.5 to 3 years on average. They flood and spill over their flood-plains less often (▶Figure 11-8).

Bankfull Channel Width, Depth, and Capacity

A stream floods when it rises above bankfull level (Figure 11-8). The bankfull channel width appears to have a simple relationship to bankfull stream discharge over a large range of channels. Channel widths of 6, 12, and 30 meters generally carry, respectively, bankfull discharges of approximately 0.5 to 1, 14 to 28, and 70 to 250 cubic meters per second. Thus, merely by measuring the average bankfull channel width you can approximate the discharge of a stream just below flood stage.

Damage to bridges, roads, and other streamside structures depends heavily on the depth of channel scour during floods (▶Figures 11-8, 11-9, and 11-10). A classic early study of the extent of channel deepening during floods is illustrated in Figure 11-9. Note the general deepening of the

channel bottom as water levels rise, though the channel width does not change substantially. As the flood wanes, sediment again deposits and raises the channel bottom.

Large rivers can have large flows without being above flood stage. Small streams can flood with fairly small flows. A flood is an unusual event for a stream of any size. Factors that produce unusually high stream flows cause dramatic changes in flood turbulence or energy and thus the shape of the stream channel. These factors include:

- **precipitation,** especially intense or long-duration rainfall;

- **rate of runoff,** with steep slopes, sparse vegetation, thin soils, and closely spaced streams can collect water to collect in larger channels quickly; and

- **bedrock,** with deep channels that foster fast flow and transport coarse bed load.

When peak flood velocity and depth are high enough to develop significant turbulence, stream power on the channel bed, and therefore erosive power, becomes very large.

The ratio between peak discharge and mean annual discharge is critical for assessing flood hazards and the impacts of stream channel modifications. A map of flash flood hazard tendency for the United States shows that high flash flood danger areas are primarily in the semi-arid areas

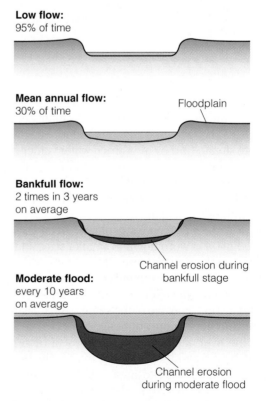

▶**FIGURE 11-8.** Generalized cross sections of a stream channel at various flows.

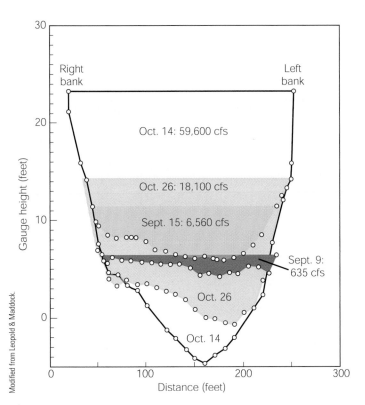

(a)

(b)

▶**FIGURE 11-9.** Channel depth of scour in response to a flood on the San Juan River near Bluff, Utah, in September 1941. Low water on September 9 is outlined with dots. With early water rise (September 15), sediment from upstream deposited to raise the channel bottom. As water rose further to the maximum stage (October 14), the channel eroded to its deepest level. As the flood level waned (October 26), sediment again deposited to raise the channel bottom. Dots are individual channel-bottom measurements. cfs = cubic feet per second

▶**FIGURE 11-10.** **(a)** A flood scour hole eroded next to the downstream end of one of the Alaska Highway bridge support piers in the Johnson River. **(b)** A flood in March 1999 along the Amite River in Louisiana caused failure of this bridge along Highway 10.

of the southwestern United States—Southern California to western Arizona, and west Texas (▶ Figure 11-11). Moderate flash flood dangers exist in areas such as the eastern Rocky Mountains, in the Dakotas, western Nebraska, eastern Colorado, New Mexico, and central Texas. This is not to say that other areas are not prone to flooding, rather that the floods there tend to be less extreme compared with the normal stream flows.

Sediment Transport and Stream Equilibrium

Streams and rivers carry water from the land surface to the ocean. Over millions of years, the gradient or slope of a river and its cross-sectional profile adjust to accommodate the amount of flowing water, the amount and grain size of supplied sediment, and the ease of stream bank erodibility. A river will adjust to accommodate changes in any of these external influences, a fundamental tendency of rivers to adjust to restore their equilibrium or **grade.** Some segments of a graded river may meander back and forth across its floodplain, in response to minor variations in the stream banks or input from tributaries. Other segments may be

relatively straight, or have multiple interconnected channels and deposit large volumes of sand and gravel. Still others may alternate between rapid-flowing **riffles** or rapids and slow-moving pools. Any big external influences on the river will cause changes in its behavior as it adjusts to the change. This state of **dynamic equilibrium** varies over the short term but gradually evolves with long-term changes in climate or mountain building.

Streams are efficient: They adjust their form and behavior to minimize the required work. Coarser grain sizes, such as those supplied from the steeper slopes of mountainous regions, require steeper slopes or faster water to move the grains. During floods, coarser particles are gradually broken down to smaller sized particles in the channel, and then flushed downstream. Any change imposed on the stream by either nature or humans will result in an adjustment of

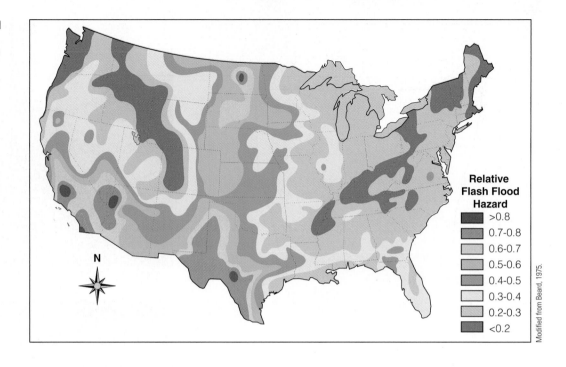

FIGURE 11-11. Flash flood magnitude map for the United States. Higher numbers (orange areas) are most susceptible to flash floods. Higher values are for more extreme floods.

Relative Flash Flood Hazard

- >0.8
- 0.7-0.8
- 0.6-0.7
- 0.5-0.6
- 0.4-0.5
- 0.3-0.4
- 0.2-0.3
- <0.2

Modified from Beard, 1975.

the stream flow and sediment transport. If abundant or unusually coarse sediment is dumped into a river, it adjusts by steepening its slope to an angle at which it can just move that sediment during floods. High water in a flood erodes and transports large amounts of sediment, then deposits it farther downstream as the water recedes, creating deposits with sequences of grains that are finer upward.

The greater discharges and smaller grain sizes that exist downstream lead to gentler slopes in those channels. Thus, slopes in the upstream reaches of a stream are typically steeper, becoming progressively gentler downstream.

The cross section of a river adjusts based on the erodibility of the bottom and banks and the nature of the transported sediment. Almost all of this adjustment occurs during floods when the river erodes and, in turn, deposits sediment. High water typically fills the channel to bankfull stage every 1.5 to 3 years—a level that appears to determine the "equilibrium" form of the channel. This rule of thumb between channel form and discharge seems appropriate for some streams, especially those in humid areas that produce mostly fine-grained sediment. Larger floods fill the channel and spill out over its floodplain. This rule of thumb does not seem to hold for semi-arid environments where normal flows and even modest floods in the following few years typically fail to restore the "damage" to the channel. It may take many years of smaller floods to restore the channel to something resembling its pre–major flood form.

Stream channels with forms characterized by features caused by major floods are sometimes called **flood channels.** Most streams flowing through easily eroded sand and gravel at low water have steep banks and broad, nearly flat bottoms. During flooding, the deepest part of the channel is

eroded even more because of high velocities. As flow velocity decreases, sediment preferentially fills the deepest part of the channel, returning the cross section to a broad and shallow shape. Streams flowing through bedrock or fine silt and clay tend to be narrow and deep, because these materials are less easily eroded.

To reiterate, stream cross sections adjust to accommodate the stream flow, as well as the amount of sediment volume and the grain sizes supplied to the channel. Any change in the stream environment provokes other changes that move the stream back toward equilibrium with its environment. A stream in equilibrium with its environment, called a **graded stream,** adjusts its channel slope or gradient in response to water velocity, sediment grain size, and total sediment load in order to be able to just barely transport the sediment supplied over time.

GRAIN SIZE CARRIED BY A STREAM A stream cannot control the particle size of mud, sand, gravel, or boulders supplied to it from the surrounding slopes. Tributaries bring in whatever they carry. Eroding riverbanks and landslides can supply particles of any size from mud to giant boulders. At low water, virtually all of the material brought into the stream stays put, backing up the water behind it. When water rises in the stream, it has a higher velocity and thus can carry more and larger sediment particles. The grain size a stream can carry is proportional to its velocity; thus rising water first picks up the finest grains, then coarser and coarser particles. Sediment is carried in suspension as long as grains sink more slowly than the upward velocity of turbulent eddies. Fine sediment is first picked up in eddies; at higher flows, pebbles or boulders may tumble along the

FIGURE 11-12. Coarse gravel brought in by the small tributary on the right creates steepening and rapids in the Middle Fork of the Smith River north of Crescent City, California.

FIGURE 11-13. Giant granite boulders dumped by a landslide into the Feather River in the Sierra Nevada range of California can be moved only in an extreme flood.

bottom and even be heard as they collide with one another. This causes more erosion from the stream banks and bottom and deepening of the channel.

As the flood flow wanes, the coarser sand and gravel in suspension progressively drops out, thereby raising the streambed (▶Figure 11-9). Thus, as water level rises, the stream begins eroding and the water gets muddy; as the level falls, the sediment in transit begins to deposit. For this reason, a cross section of a flood deposit shows the largest grains at the bottom, grading upward to finer sediments.

Similarly, if coarser material is added to the channel, such as from a steeper tributary or a landslide, it accumulates in the stream channel until a large flood with sufficient velocity occurs or the gradient of the channel increases sufficiently to move that size material (▶Figures 11-12 and 11-13).

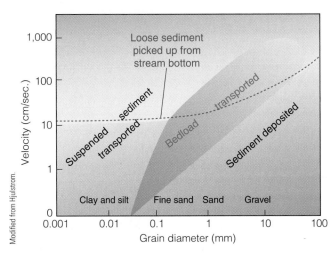

FIGURE 11-14. This diagram shows the approximate velocity required to pick up and transport sediment particles of various sizes. Note that both axes are log scales, so the differences are much greater than it seems on the graph.

Empirical curves have been developed to estimate the approximate maximum particle size that can be picked up or transported for a given water velocity (▶Figure 11-14). Note that the velocity required to mobilize particles is appreciably greater than that to transport the same particles.

The relationship of larger grain sizes requiring higher velocities for movement does not hold where fine silt and clay particles make up the streambed. In that case, the fine particles lie entirely within the zone of smoothly flowing water at the bottom and do not protrude into the current far enough to be moved. Clay-size particles have electrostatic surface charges that help to hold them together.

As water velocity increases, the water drags much more strongly against the bottom. Increase in frictional drag on the stream bottom provides more force on particles on the streambed and thus more erosion. That friction also slows down the water (Sidebar 11-2).

In general, the coarser the particles in the channel, the greater the roughness or friction against the flowing water (Sidebar 11-3). Thus, coarser particles also slow the water velocity along the base of a stream. For this reason, mountain streams with coarse pebbles or boulders in the streambed often appear to be flowing fast but actually flow much

---

Sidebar 11-2

Drag or total friction on the stream bottom is proportional to velocity squared:

$$\tau_0 \, \alpha \, v^2$$

where

τ_0 = friction

v = velocity

FIGURE 11-15. Clark Fork River near Missoula, Montana, spilled muddy water over its floodplain in the June 1964 flood.

more slowly than most large, smooth-flowing rivers such as the Missouri or Mississippi. Note that the velocity depends on water depth and slope.

SEDIMENT LOAD OF A STREAM Although a stream is capable of carrying particles of a certain size, there can be too much of a good thing. Large volumes of sand dumped into a stream from an easily eroded source or sediment provided by a melting glacier will overwhelm its carrying capacity. The excess sediment will be deposited in the channel. Floodwaters do, of course, carry more sediment. As water depth and velocity increase during a flood, shear or drag at the bottom of the channel increases. That extra shear picks up more sediment. Leopold and Maddock (1953) demonstrated that the suspended load of a stream depends upon the discharge (Sidebar 11-4).

Particles picked up during floods clearly come from the streambed and banks (Figure 11-15). The streambed is eroded more deeply as the water depth and velocity rise. For this reason, bridge pilings in a river channel must be set deep enough that major floods will not undermine the pilings and cause a bridge to collapse (see Figures 10a and 10b). Conversely, as the water level begins to fall after a flood crest, the larger particles are dropped first and the streambed is gradually built up again. Again, the volume and velocity of the flow limit both the size and the amount of sediment that can be carried by the stream.

The geometry of the channel is controlled by the flow velocities and the associated ability of the stream to carry sediment. As a stream spills over its floodplain, it moves from a deep channel with high velocity to a shallow broad floodplain with low velocity. This causes sediment to deposit on the edge of the deeper channel in a deposit called a **natural levee** (Figure 11-16). These features often form

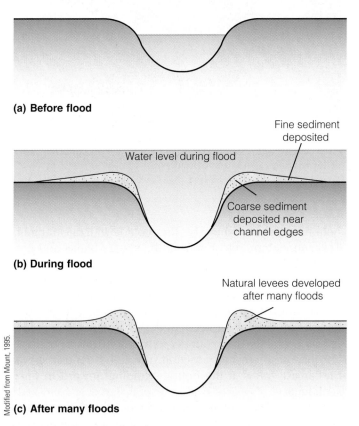

(a) Before flood

Fine sediment deposited

Water level during flood

Coarse sediment deposited near channel edges

(b) During flood

Natural levees developed after many floods

(c) After many floods

FIGURE 11-16. This vertical cross section through a channel shows the development of natural levees in **(b)** as the flood waters deposit sediment on the edge of the channel where the velocity is lower than in the channel. These natural levees are commonly stable after flood waters recede.

a nearly continuous small ridge along the edge of the channel that may keep small floods within the channel.

Sediment Transport Mechanisms

Where large amounts of easily eroded sediment are carried downstream by floodwaters, the nature of the sediment transport also changes. In this chapter, we address floods where water dominates the flowing material. As the proportion of sediment increases to more than 20 percent, a flood becomes a hyperconcentrated flow and, with more than 47 percent sediment, a debris flow (see Sidebar 11-5). The three end-member types are completely gradational. Debris flows are addressed in Chapter 8, pages 212–220.

Stream Types

Characteristics of natural streams vary widely from highly sinuous single-channel paths of some meandering streams, to less arcuate multichannel braided rivers, to nearly straight channels, to the short-lived ephemeral channels on alluvial fans. We focus here on two common yet quite distinct stream types, meandering and braided.

Meandering Streams

Most streams meander, regardless of climate or location. Braided streams are much less common, and naturally straight streams are rare. Eroding streams are not straight but meander from side to side. Any irregularity that diverts

Riffle, between meander bends, where coarse grains are deposited

Low velocity on floodplain during floods; depositing mostly mud

Shallow, slow deposition to form a "point bar" that also migrates downstream

Deep, fast, erosion in "pool" eroding fine bank material and coarse bed material on downstream outside bank

Riffle

Thalweg line along deepest part of channel

Donald Hyndman photo.

▶**FIGURE 11-17.** Meanders in the Smith River south of Great Falls, Montana, nicely illustrate the eroding cut bank on the outside of a meander and the depositional gravelly point bar on the inside of the meander. Flow is toward the lower left.

water toward one bank moves more water to help erode that bank. As the water swings back into the main channel, it sweeps toward the opposite bank, much like a skier making slalom turns. This deep and high-velocity part of the stream is the **thalweg.** Water preferentially erodes the outside of meander bends and deposits sediment as a **point bar** downstream in the slow water along the inside corner of the bend (▶Figures 11-17, 11-18, and 11-19). Erosion occurs on the outside of the bend because the water has forward momentum that drives it into the bank, where the higher velocity can mobilize more sediment. The flow commonly alternates between deep pools and shallow riffles.

Meanders erode outward and slowly migrate downstream. Their bends sometimes come closer together until high water breaks through the narrow neck between. When a stream cuts across a meander bend, either naturally or with "help" from humans, the new stretch of channel follows a shorter path for the same drop in elevation. This creates a short, steeper cutoff and leaves behind the abandoned meander as an oxbow lake (▶Figure 11-19). The steepening of the channel slope causes increased velocity and thus greater erosion.

The migration of meandering channels creates hazards for those who would build their homes next to such a stream. People often build their houses at the outside of a meander bend because the view of the river is best there or the trees are larger. Unfortunately, if you follow this strategy, you may find the stream running through your living room during the next flood.

This was the case for one resident of Plains, Montana, who recently built a home next to the cut bank on the out-

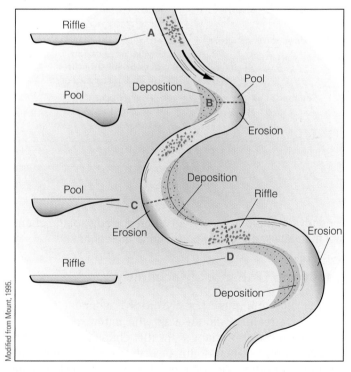

Riffle

Pool

Pool

Riffle

A

Deposition

B

Pool

Erosion

Deposition

C

Erosion

Riffle

D

Erosion

Deposition

Modified from Mount, 1995.

▶**FIGURE 11-18.** Cross sections of typical meandering stream channel from a riffle at A, downstream through pools at B and C, then finally through a riffle at D. Note that the river erodes on the outside of the meander bends where it has a deep channel and deposits on the inside of bends where it has a shallow channel.

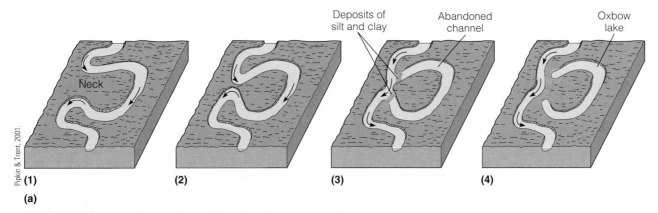

Deposits of silt and clay   Abandoned channel   Oxbow lake

Neck

(1)   (2)   (3)   (4)

Pipkin & Trent, 2001.

(a)

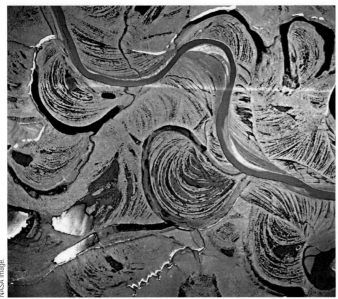

NASA image.

(b)

▶**FIGURE 11-19.** **(a)** Meanders erode the outsides of bends and can migrate until one meander bend spills over to one farther downstream, leaving an abandoned oxbow lake. **(b)** This meandering river on the Arctic Coastal Plain clearly shows how the channel has moved across the floodplain in various episodes. Old inactive channels have the shape of a meander and in some cases are oxbow lakes.

side of a meander bend with a great view of the river and surrounding mountains. The owner was careful to build 10 meters back from the riverbank, thinking that that would be more than enough. It was not (▶Figure 11-20a). The situation was especially embarrassing because the owner was also the local disaster relief coordinator.

In late spring 1997, the annual high water on the river approached bankfull stage, eroding the steep cut bank of the meander bend on the floodplain of the river next to the house. The fast-moving water caused progressive caving of the bank until it undercut the edge of the house. The owner hired a house-moving company but too late. Before they could move the house, erosion had progressed so far that the moving company was not willing to risk loss of its equipment on the unstable riverbanks. Finally, local authorities

Karl Christians photos, Montana Department of Natural Resources and Conservation.

(a)

▶**FIGURE 11-20.** **(a)** No, this house was not built overhanging the cut bank of the river. It was built 10 meters back but on the outside of a big meander bend of the Clark Fork River near Plains, Montana. **(b)** Authorities finally burned the house.

(b)

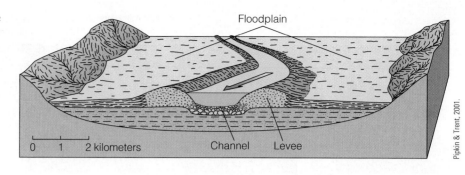

Pipkin & Trent, 2001.

▶**FIGURE 11-21.** This cross section shows the main channel, natural levees, and floodplain of a river.

decided to burn the house rather than find it floating down the river in pieces (▶Figure 11-20b).

Meander size is controlled by the amount of water at bankfull stage. Typically, the meander wavelength is about 12 times the channel width, and the wavelength is also controlled by the total discharge at bankfull stage. Thus, artificial attempts to change a channel by narrowing it or straightening it will be met by the river's attempts to return to a more natural equilibrium channel cross section and meander path.

Natural levees build up at the edges of the main channel where the water spreads out and slows down on the floodplain. The floodplain, with its relatively slow-moving water, is part of the overall river path; it carries a significant flow during floods (▶Figure 11-21).

Braided Streams

Streams overloaded with sediment deposit it in the stream channel, locally clogging it so that water must shift to one or both sides of the deposit. With further deposition, water shifts to form new channels. Such streams do not meander but form broad, multichannel, braided paths. That behavior is promoted by dry climates in which little vegetation grows to protect slopes from erosion. Eroding banks, abundant stream **bedload,** and a steep gradient characterize such braided streams. With waning flow, bedload is deposited in the channel and flow diverges around it, splitting into separate channels. Depending on flow, some channels are abandoned while others temporarily become dominant (▶Figure 11-22).

Meandering and braided river types form end members of a complete range of behavior. Meandering streams are more typical of wet climates with their finer-grained sediments, and **braided streams** are common in dry climates with abundant coarse sediment; however, there are many exceptions. Even within one river, some reaches may meander and others may braid, depending on the erodibility of the banks and the amount of supplied sediment. As with meandering streams, bankfull discharge appears to be responsible for the shape of the channel cross section. Braided channels are also characteristic of meltwater streams flowing from sediment-laden glaciers and of arid

Donald Hyndman photo.

▶**FIGURE 11-22.** This braided river channel of the Wairou River is southwest of Blenheim, New Zealand.

basin-and-range valleys in which heavy, intermittent rains from the mountains carry abundant sediment across valley alluvial fans. When a stream channel moves from an area of high slope to one of lower slope, sediment will generally deposit in a fan-shaped feature because the stream can no longer carry its full load (▶Figure 11-23).

Active alluvial fans are always marked by braided streams. The term *alluvial* implies the transport of loose sediment fragments, and *fan* refers to the shape of the deposit in map view. Alluvial fans are also somewhat arched, with higher elevations along their midline and with lower elevations toward the side edges because they are built by deposition of sediment flushed out of a mountain canyon at the apex of the fan (▶Figure 11-24). Like other braided environments, most alluvial fans tend to be in dry climates and lack significant vegetation. Rain may produce a flash flood, a hyperconcentrated flood, or a debris flow, depending on the proportions of sediment and water (see Table 11-1 and "Debris Flows" in Chapter 8). Any of these events can be catastrophic to those who would build on alluvial fans.

People tend to build on alluvial fans because they are inviting areas with gentle slopes at the base of steep, often rocky, mountainsides. They typically do not realize that the fan is an active area of stream deposition because the

▶**FIGURE 11-23.** Water runs off this steep pile of sand in heavy rains despite its high permeability. Overland flow erodes gullies and carries sediment down onto depositional fans.

▶**FIGURE 11-24.** Flood hazard zones vary on alluvial fans. All of the flow from a canyon concentrates at the apex of the fan, then spreads out, decreasing in depth and intensity downslope. At different times, though, the flow tends to concentrate on different parts of the fan as the flow finds the easiest path down the slope.

streams rarely flow, but watch out when they do! A torrential flood can wash out of the canyon above and with little or no warning. They destroy, bury, or flush houses and cars downslope (see Figures 11-2 and 11-3). Deaths are common, especially during rainstorms at night when people do not hear the telltale rumble of an approaching flood.

Bedrock Streams

Areas where the river erodes down to resistant bedrock tend to abruptly steepen. These abrupt changes in gradient from gentle to steep are called knickpoints. Upstream, the

▶**FIGURE 11-25.** This bedrock channel of the Middle Fork of the Smith River is north of Crescent City, California.

gentle gradient has low energy for the river and may deposit sediment. Downstream, the steeper channel provides high energy and sediment erodes.

Streams with bedrock channels have sufficient energy during floods to transport any loose material in the channel. They do not reach equilibrium. The roughness of bedrock channels is especially high, and erosional drag on the bottom is likewise high (▶ Figure 11-25). Bedrock is resistant to erosion, so only major floods can scour or pluck fragments from the rock.

Floods in bedrock channels have excess energy that cannot be dissipated by changing channel roughness or by sediment transport. Extreme turbulence is typical, and large-scale vortexes or whirlpools can appear. These can be effective in swirling rocks on the bottom to drill potholes in the bedrock (▶ Figure 11-26). Recreational rafting or boating at high water in such channels, although exciting, is especially hazardous because of the extreme turbulence and vortexes. Flotation devices can be insufficient to raise a person against the effect of a whirlpool.

▶**FIGURE 11-26.** Pebbles swirling in whirlpools drilled these cylindrical potholes in McDonald Creek, Glacier National Park, Montana.

Donald Hyndman photo.

The Hydrograph

A **hydrograph** is a plot of the volume of water flowing in a stream versus time. A typical flood hydrograph rises steeply and then falls more gently (▶Figure 11-27). Anything that causes more rapid transfer of water to a stream will heighten and shorten the hydrograph. In areas where the rainfall saturates the soil and is forced to run over the surface to rapidly feed streams, the hydrograph peaks much more quickly and rises to a greater maximum discharge (see "Case in Point: Hurricane Agnes, June 1972" in Chapter 14). Such rapid large discharges develop in urban areas with large areas of pavement or in natural areas that have been deforested. Deforestation can increase the volume of storm runoff by roughly 10 percent. Several factors affect the hydrograph shape.

Precipitation Intensity and Runoff

The intensity of precipitation plays a significant role in the rate of runoff to streams and, in turn, floods. Light precipitation can generally be absorbed into the soil without surface runoff. Heavy precipitation can overwhelm the near-surface permeability of the soil, leading to rapid runoff on the surface. This is called **overland flow.** Even where the permeability of the soil is relatively high, heavy rainfall, especially over a long period or in multiple storms, can saturate near-surface sediments, thus forcing the water to rapidly flow over the surface to streams. In addition, part of the water that infiltrates into the soil will raise the local groundwater level, which thereby forces more water out to the surface within and near the stream.

Rapid flood peaks during large storms are most common in areas with fine-grained soils or desert soils, especially in areas having a tight clay hardpan or soils with a shallow, nearly impervious calcium carbonate–rich layer. Flat areas that contain many depressions can temporarily store enough water to delay runoff and lower and stretch out the flood hydrograph. Decomposed granite soils dominated by coarse sand have high permeability and high in-

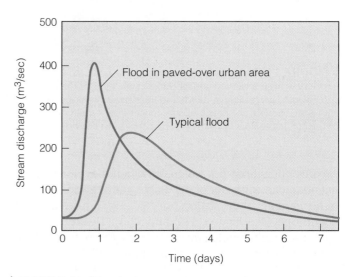

▶**FIGURE 11-27.** This hydrograph is a plot of stream discharge versus time for a similar eighteen-hour rainfall event for the same area before and after urbanization. Note that the area under the two curves is similar, that is, approximately the same total volume of water for both floods. Actually, because less water infiltrates, the flood volume after urbanization will be a little larger.

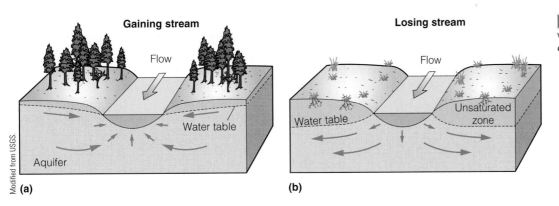

Gaining stream

Flow

Water table

Aquifer

Modified from USGS.

(a)

Losing stream

Flow

Water table

Unsaturated zone

(b)

▶**FIGURE 11-28.** Groundwater flow directions depend on climate.

filtration capacity, which can cause stream levels to rise rapidly.

In areas of moderate to high annual rainfall, groundwater levels stand higher than most streams and thus continuously feed them. During low water periods, the stream surface is generally at the groundwater surface or the exposed water table. In such areas of **gaining streams,** the rate at which water flows from groundwater into a stream is proportional to both the slope of the water table and the hydraulic conductivity of the water-saturated sediments or rocks. Storm discharge includes this groundwater flow as well as any overland flow. New groundwater does not need to flow all the way from its infiltration source during a storm to a river but merely has to raise the water table and displace "old water" into the stream. In semi-arid to arid regions, however, **losing streams** often dry up between storms. When they do flow, they lose water into the ground and raise the water table (▶Figure 11-28).

Stream Order and Hydrograph Shape

The number of tributaries of a stream (its **stream order**) has a significant effect on the rate of rise of floodwaters during and following a storm. Small streams that lack tributaries are designated first-order streams (▶Figure 11-29). First-order streams join to form a second-order stream, and second-order streams join to form a third-order stream, and so on.

Low-order streams tend to have a rapid response to a storm and steep hydrographs because water only has to travel short distances to the stream. Such streams provide less flood warning time for downstream residents. They have smaller catchment basins and carry coarser and larger amounts of sediment for a given area.

High-order streams with numerous tributaries have longer lag times between a storm and a downstream flood; their hydrographs are less peaked and cover longer time periods. Such streams, because of water distribution through many larger basins and temporary storage of water on floodplains, take longer to respond to storm precipitation.

A storm in a headwaters area may cause flooding in several first-order streams. As the flood crest of each moves downstream, the length of time it takes to reach the second-

order stream varies, so each first-order flood crest arrives at somewhat different times. The flood peak for the second-order stream will therefore begin later and be spread over a longer period. The same goes for several second-order streams coming together in a third-order stream; its flood peak will again begin later and be spread over a still longer period.

Flood Crests Move Downstream

Even with local intense rainstorms on a small drainage basin, there will be a lag between a storm and the resulting flood peak. A torrential downpour may last for only ten minutes, but it takes time for the water to saturate the surface layers of soil and to percolate down to the watertable. More time is required for sheet flow to collect in small gullies, and for water in those gullies to flow down to a stream. In turn, it takes time for the water in small streams to combine and cause flooding in a larger stream. The length of the lag time depends on many factors, including basin area and shape, spacing of the drainage channels, vegetation cover, soil permeability, and land use.

Flood intensity depends on similar characteristics. If a storm only occurs in an upstream portion of a watershed, the peak height of the hydrograph will be lower and the

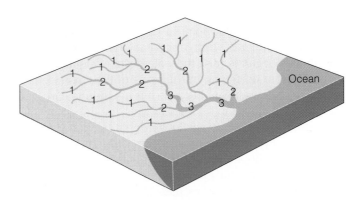

Ocean

▶**FIGURE 11-29.** Stream order: First-order streams have no tributaries but join to form a second-order stream and so on.

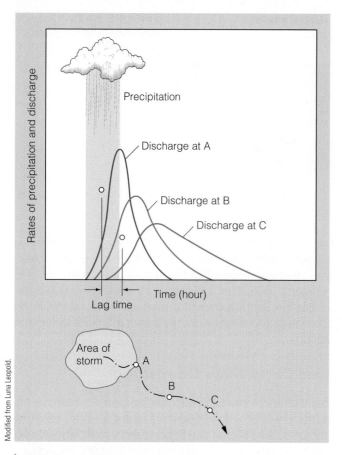

Rates of precipitation and discharge

Precipitation

Discharge at A

Discharge at B

Discharge at C

Time (hour)

Lag time

Area of storm

A

B

C

Modified from Luna Leopold.

▶**FIGURE 11-30.** Storm rainfall entering the stream precedes the flood crest that it causes. The flood hydrograph nearest the rainfall area is highest and narrowest. Farther downstream at B, the flood hydrograph crests at a lower level but lasts longer. Still farther downstream at C, the flood crests at an even lower level and lasts longer.

flood duration will be longer farther downstream from the storm area (▶Figure 11-30).

At the downstream edge of the main rainfall area, water levels rise as water flows in from slopes, tributaries, and upstream. The level often continues to rise until about the time the rainfall event stops (station A in Figure 11-30). At that time, water levels on the upstream slopes begin to fall, and flood level at that point in the stream crests and begins to fall. Farther downstream, some of the earlier rainfall has already begun to raise the water level. Water continues to arrive from upstream, but the size of the stream channel is larger downstream because it is adjusted to carry the flow from all upstream tributaries. Thus, the flood flow fills a wider cross section to shallower depth (station B in Figure 11-30). Even farther downstream, the channel is still larger and the flood crest still lower (station C in Figure 11-30). In cases where the precipitation occurs over the whole drainage basin, flows will increase downstream.

For flood hazards, we are normally more concerned with the maximum height of the flood crest than when the first

water arrives from a storm. The flood crest moves downstream more slowly than the leading water in the flood wave. Note in Figure 11-30 that the first rise in flood water from the storm appears downstream with time at stations A, B, and C. Note also that the flood crest or peak discharge takes longer to move downstream. A rule of thumb is that the flood crest or flood wave moves downstream at roughly half the speed of the average water velocity.

Floods on Frozen or Water-Saturated Ground

Floods are common when heavy rain falls on ground that is either saturated from earlier storms or frozen. In both of these cases, most of the water that reaches the ground surface runs off directly to the stream. This can cause large floods, such as occurred when Hurricane Agnes dumped 5 to 7 centimeters of rain in Pennsylvania in June 1972 on ground that was already saturated with water. Large floods occurred in that event from Pennsylvania to Maryland and Virginia as discussed in Chapter 14 (see "Case in Point: Hurricane Agnes, June 1972").

The largest floods in cold climate regions are generated during rainstorms, just as they do in warmer climates. Regions north of the influence of warm, moist tropical airflow also are affected by snowpack on the ground for a month or more in most years. Floods in these areas can result from melting snow or glaciers, a rainstorm, or a combination.

At high elevations, much of the winter precipitation falls as snow. This stores the water at the surface until snowmelt in the spring. If the snowpack melts gradually, much of the water can soak into the ground. If it melts rapidly, either with prolonged high temperatures or especially with heavy warm rain, much will flow off the surface directly into streams. When the snowpack warms to 0°C, a large proportion of meltwater remains in the pore spaces and the snowpack is "ripe." Water draining through the pack concentrates in channels at the base of the snow. The largest floods develop when heavy warm rain accelerates melting of a ripe snowpack. Water has a large heat capacity, so dispersal of warm rainwater and the consequent melting of snow is quite effective. Under these conditions or if the ground is already saturated with snowmelt, water from a rainstorm may entirely flow off the surface, as it did in the February 1986 floods from the Sierra Nevada mountains in California. Or if the ground is frozen, as is sometimes the case with minimal snowpack, heavy rains quickly run off the surface because they cannot soak into the ground. These circumstances lead to rapid rise of stream levels and flooding.

During winter and early spring, ice and snow may block stream channels. Ice coating rivers during winter breaks up and moves as the ice begins to melt. A sudden warm spell that causes ice to break up can dam the channel

▶**FIGURE 11-31.** An ice jam built up at a river constriction threatens a bridge at Gorham, New Hampshire. The power shovel tries to get the river flowing again.

at constrictions such as at bridges (▶Figure 11-31). Large rivers flowing northward have the additional problem that the ice upstream thaws before that downstream. As upstream meltwater flows north, it encounters ice jams and restricted downstream flow, causing floods (see "Case in Point: Spring Thaw from the South on a North-Flowing River").

Destructive Energy of Floods

The destructive effect of a flood depends primarily on the flood intensity, which is related to the shear stress or stream power exerted on the stream channel. Therefore, floods in small, narrowly confined drainage basins are typically much more violent than those along major rivers such as the Mississippi (▶Figures 11-4, 11-5, and 11-6).

Stream channel changes are insignificant with normal flows or even small floods, regardless of their frequency. Significant changes occur only when flow reaches a threshold level that mobilizes large volumes of material from the streambed and channel sides.

Stream Power

For a given area of stream channel, the erosive power of the stream is proportional to:

hydraulic radius × slope × velocity

(See Sidebar 11-6.)

The Mississippi River, with a drainage area of about 1.5 million square kilometers has a **stream power** of 10 watts per square meter. By contrast, small streams with drainage areas of less than 10,000 square kilometers tend to have stream powers between 1,000 and 10,000 watts per square meter.

Flood Frequency and Recurrence Intervals

Flood frequency is commonly recorded as a recurrence interval, the *average* time between floods of a given size. Larger flood discharges on a given stream have longer recurrence intervals between floods. A **100-year flood** is used by the U.S. Federal Emergency Management Agency (FEMA) to establish regulations for building purposes near streams. A 100-year flood has a 1 percent chance of happening in any single year, although a 100-year flood still has a 1 percent chance of happening the year following a similar magnitude event.

Bridges and other structures should be designed to accommodate the runoff in at least the expected life of the structure. Those who design structures such as dikes, bridges, and buildings for a given life span generally consider the probabilities of large floods.

Complicating matters is the fact that extreme precipitation events are likely becoming more frequent, perhaps from global warming or from upstream changes or from some cyclic aspect of climate. All-time records were set, for example, in the Susquehanna River of Pennsylvania in 1972, the Santa Cruz River in Tucson in 1983, the upper Mississippi River in 1993, the Red River in North Dakota in 1996, and the American and San Joaquin rivers of California in 1997.

Recurrence Intervals and Discharge

For a given stream, the statistical average number of years between flows of a certain discharge is the **recurrence interval** (T). The inverse of this (1/T) is the probability that a certain discharge or flow will be exceeded in any single year.

Sidebar 11-6   Stream Power

$$P = S_s \times V$$

where

P = stream power in watts per m^2

S_s = boundary shear stress: $w \times R \times s$

w = specific weight of the fluid (including suspended sediment),

  = 1 g/cm^3 for pure water with no sediment

R = hydraulic radius (i.e., average depth) = wetted cross-sectional area / wetted perimeter

s = energy slope of the channel (often first approximated as the water surface slope)

V = velocity in m/sec.

(Text continues on page 290.)

The Red River, North Dakota

The winter of 1996–1997 brought numerous heavy snowstorms to the northern plains, a snowpack two or three times the depth of the previous record. In early April, a major blizzard dumped up to 1 meter of snow on parts of the area; this was the culmination of eight major blizzards that provided as much as 3 meters of snow in Fargo.

The Red River that flows through Fargo, North Dakota, and Winnipeg, Manitoba, freezes in winter (▶Figures 11-32

▶**FIGURE 11-32.** The Red River flows north along the border of North Dakota and Minnesota, through southern Manitoba, and into Lake Winnipeg.

▶**FIGURE 11-33.** Ice jams fill the channel of the Red River below Grand Forks, North Dakota, caused this flood in April 1997.

U.S. Army Corps of Engineers photo.

and 11-33). Come spring, it thaws first in North Dakota while it is still frozen farther north, a recipe for flooding beyond the control of the local residents. As the thaw progresses, ice floes drift north to pile up in ice jams that cause widespread flooding. In North Dakota and adjacent Manitoba, the river flows in a shallow, meandering valley across subdued topography that includes the broad flat floor of former glacial Lake Agassiz. Grand Forks, North Dakota, is entirely within a 500-year floodplain; its highest elevation lies less than 10 meters above flood stage.

People in the area are used to floods, but not on the scale of 1997. On April 17, the Red River broke its 100-year record at Fargo by cresting at 6.9 meters above flood stage (▶Figures 11-32 to 11-38). Discharge reached some 3,400 cubic meters per second compared with this river's average of 75 cubic meters per second. On April 19, it crested at Grand Forks 7.9 meters above flood stage and covered 90 percent of the city for several days. Flooding of the water and sanitation systems not only left the city without drinking water and sewage treatment but also contaminated the floodwater, making the city uninhabitable. Fire broke out in downtown Grand Forks on April 19 and spread to eleven buildings, creating severe access problems for firefighters.

Sixty thousand people in the Grand Forks area were evacuated, and damages and cleanup costs exceeded $1 billion. Fortunately, no one died in Grand Forks. Overland flooding from snowmelt from farm fields on April 3 worsened the problem. To lessen damage to the larger population areas, authorities built a dike in the outskirts of town to block overland flooding from the fields. The Minnesota River also flooded, though it produced less damage than the Red River. At least seven people died in North Dakota and four in Minnesota. In contrast to many rainfall floods, the hydrograph for this flood is rather broad, remaining above flood stage from March 26 to May 20. The federal government purchased many homes in

Gary Gorham photo, North Dakota State University.

▶**FIGURE 11-34.** In this aerial view of the Red River flood in North Dakota, April 27, 1997, thin white lines around some "islands" are sandbags. The view is near Harwood, north of Fargo.

FIGURE 11-35. Downtown Grand Forks was mostly submerged during the 1997 flood. The water level rose high enough to submerge some buildings to their rooftops as well as the bridge, which is visible near the right side of the photo.

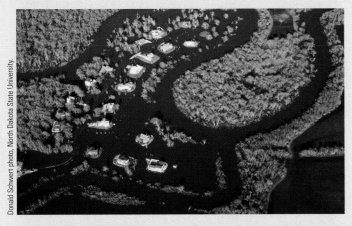

FIGURE 11-36. This flooded subdivision lay near a river meander south of Fargo, North Dakota. Homes in the area are built on mounds to keep them above floods, but the flood blocked access.

FIGURE 11-37. Several different approaches were taken to hold back the Red River flood in Fargo in April 1997.

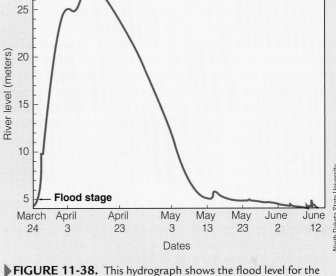

FIGURE 11-38. This hydrograph shows the flood level for the Red River at Fargo, North Dakota, March 24 to June 30, 1997.

the lower parts of Grand Forks and tore them down to minimize damage from future floods. The city is now considering either moving levees farther from the river or constructing a flood-diversion channel to bypass the city.

So what does the future hold? Climate predictions suggest that the average precipitation and temperature in the northern Great Plains will both increase during the next century. If correct, floods on the Red River may become more frequent and larger in future.

What Went Wrong?

The spring thaw on the ice-covered Red River begins first in the south. Broken ice then piles up as the river tries to flow north where ice remains. As the unusually heavy snowpack of 1997 melted, the ice dams blocked the overland flow. The region's flat topography caused the floods to spread over a broad area.

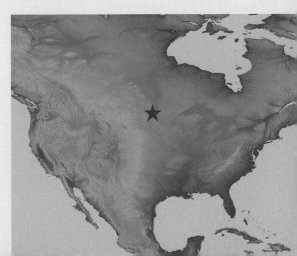

July 4, 2002

Like much of south-central Texas, the area around Austin and San Antonio is fairly dry except for heavy rainstorms delivered by warm moist air funneled from the Gulf of Mexico. One such storm lingered over the Guadalupe River watershed upstream from New Braunfels for the first week of July 2002. It began as a typical Texas summer storm, a tropical low pressure system fed by moisture from the gulf. However, instead of moving east it stalled over central Texas for almost a week because no other weather systems helped push it eastward. In that time, it dumped more than three-quarters of a meter of rain on thin soils draped over widespread limestone bed-

rock that is incapable of rapid infiltration. The amount of precipitation was unprecedented, surprising even the regional meteorologists.

The river 16 kilometers upstream from New Braunfels is dammed for a combination of flood protection, water supply, and recreation. Unfortunately, storm inflow from the upstream drainage raised the Canyon Lake reservoir level 12 meters in four days, then topped the spillway of the broad containment dam by more than 2 meters beginning on July 4 (▶Figure 11-39a, b).

The flood raced down the narrow, tree-lined valley of the Guadalupe River, gouging a new rock-bound canyon through the thin soils and bedrock (▶Figure 11-40). It ravaged the

▶**FIGURE 11-39.** **(a)** Water level in Canyon Lake on the Guadalupe River rose about 40 feet (12 meters) in four days during the flood in early July 2002 (blue line). **(b)** The flood crested at 7 feet (2.1 meters) above the spillway.

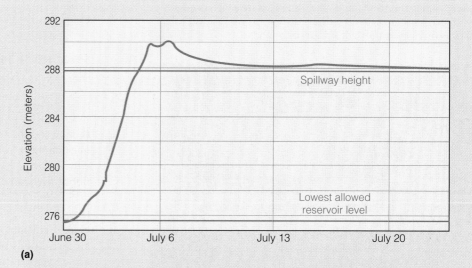

(a)

(b)

Comal County, Texas, Engineers Office photo.

(a)

Comal County, Texas, Engineers Office photo.

▶**FIGURE 11-41.** This kitchen was submerged during the Canyon Lake–Guadalupe River flood of July 4, 2002, leaving behind heavy deposits of mud.

Comal County, Texas, Engineers Office photos.

(b)

▶**FIGURE 11-40. (a)** Water recedes in the Guadalupe River flood of July 4, 2002, below Canyon Lake after washing out this bridge. **(b)** The Guadalupe River eroded a new gorge in bedrock below the Canyon Lake spillway after it was overtopped in the July 4, 2002, flood.

additional urban development cause more rapid runoff and higher flood peaks. Any climate changes such as global warming could cause more moisture to be picked up from the Gulf of Mexico and carried inland.

What Went Wrong?

Central Texas is commonly dry enough for several years that reservoirs fall to dangerously low levels; in other groups of years they fill to overflowing. Low pressure cells commonly funnel warm moist air north over this region from the Gulf of Mexico. In 2002, an especially moist storm stalled over the Guadalupe River watershed for the first week of July. This was not so much a combination of circumstances as an extreme event in a climate with significant variability.

homes of people who built near the stream, seemingly protected by the dam upstream (▶Figure 11-41).

The flood crest reached New Braunfels six hours after topping the dam. It reached the Gulf of Mexico seven days later. So why wasn't the dam built higher in the first place, high enough to withstand a flood of this magnitude? Some large dams and bridges are built to withstand only a 100-year flood. To build them for a larger flood costs significantly more money, especially given that the dam would have to be much wider if it were higher.

In addition, a dam may prove inadequate for future flood control because any upstream changes in land use such as

Table 11-2 Chance of a Flood of a Given Size in a Certain Period of Time

Flood (Recurrence Interval)	Occurrence (%) in			
	Any Single Year	10 Years	25 Years	100 Years
50-year	2		40	86
100-year	1		22	63
1,000-year			1	9.5
10,000-year				1

To determine recurrence intervals for floods on a stream, the peak annual discharge is recorded for each year on record. The largest of these discharges is then ranked number 1, the next largest 2, and so on. Each discharge can then be plotted against its calculated recurrence interval using the Weibull formula (Sidebar 11-7).

Exceedence probability, the chance of a certain size flood being exceeded in any single year, is the inverse of recurrence interval.

For example, on Squaw Creek, a tributary of the Mississippi River at Ames, Iowa, the largest recorded flood in seventy-five years of record was 14,500 cubic feet per second in 1992. This record flow ranked as number 1. Thus, $T_{1992} = (75 + 1)/1 = 76$ years. The new record 1993 flood was 24,300 cubic feet per second, which was a 77-year flood. The corresponding exceedence probability (P) is 0.013 [$P_{1993} = 1/(76 + 1) = 0.013$]. The expected discharges or flood volumes (in thousands of cubic feet per second) are typically plotted against log of recurrence interval in years (▶ Figure 11-42). Each stream is different.

In general, there is a higher ratio of fifty-year to five-year flood magnitudes for smaller catchment basins (see Table 11-2). It is dramatically higher in monsoon and semi-arid climates such as Sri Lanka and the southwestern United States as compared with tropical environments such as Congo and Guyana because the semi-arid regions have

much higher variations in precipitation. Similarly, this ratio is much higher in arid New Mexico than in Maine.

Problems with Recurrence Intervals

With any calculation of recurrence interval, there are several built-in assumptions:

- The data on which it is based must cover a long enough time interval to be representative. A quantitative example of how dramatic the change in recurrence interval can be with a single slightly larger record flood is shown in Sidebar 11-8.

- The use of **paleoflood data** can dramatically extend the total time interval on which the record is based, which may result in significant changes in estimated recurrence intervals. The main qualification here is whether the paleofloods had a distinctly different cause than more recent floods. Floods during the ice ages of the Pleistocene epoch, for example, originated in a colder, wetter climate; the conditions were different

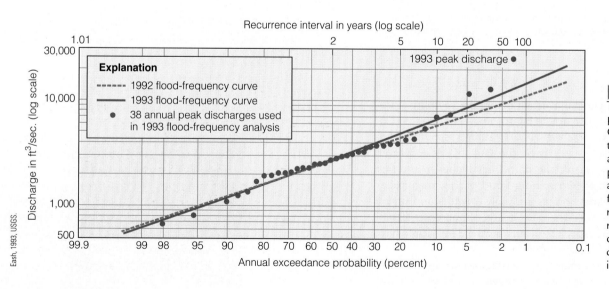

▶ FIGURE 11-42. The flood frequency plot for Squaw Creek, a tributary of the Mississippi River at Ames, Iowa, is plotted for before and after the largest flood of historic record in 1993. The recurrence interval or inverse of exceedence probability is also shown.

than at present, probably both for amounts and frequency of precipitation.

- The conditions that affect stream flow upstream in the watershed must be similar through time. This is generally not the case because of human impacts such as urbanization, channelization, the building of dikes and dams, deforestation by any process, and overgrazing. Population growth clearly causes changes, especially in areas where residential communities continue to expand the area of impermeable ground with pavement and concrete. Rapidly growing cities such as Bellevue, just east across Lake Washington from Seattle, removed vegetation and added huge areas of pavement and rooftops. The estimated 100-year recurrence interval for Mercer Creek, which drains the area, was estimated in 1977 as 420 cubic feet per second. In 1994, following seventeen years of rapid growth, the 100-year flood was estimated as 950 cubic feet per second (▶Figure 11-43)!

- A third assumption is almost never met. All of the floods plotted to determine recurrence interval or exceedence probability must belong to a single group that originates from similar flood causes so that the distribution of flood sizes should be random. For example, all must

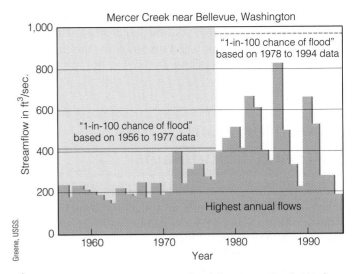

▶**FIGURE 11-43.** The 100-year flood for Mercer Creek, Washington, near Seattle, increased dramatically following rapid urbanization from 1977 to 1994.

result from an El Niño event or all from storms originating at a warm front. Combinations of causes such as a hurricane colliding with a cold front would create aberrations that would not fall in the same group and should ideally not be plotted on the same frequency distribution. Unfortunately, the data sets are so small—the period of record so short—that all floods on a river are typically lumped together, regardless of size or origin.

Even worse is a simplistic public perception of a 100-year flood as a flood that comes along every 100 years. People quickly forget most major floods that have affected an area and believe that similar or larger floods are unlikely to recur anytime soon. Although a flood of a given size will recur on average every so many years, it may recur anytime, including the next year. Similarly, if people have not seen their homes or businesses flooded in their lifetime—or that of their parents—they also believe that such a flood is not likely to happen to them. In both sets of experience, people underestimate their risk.

A **100-year floodplain** is the area likely to be flooded by the largest event in 100 years—*on average*. Such an event is based on extrapolation from a few large events over a short and incomplete hydrologic record. The 100-year flood level does not account for huge probable changes, including later upstream alterations to the drainage basin. FEMA maps show the limits of a 100-year floodplain with no indication of uncertainty in this level in spite of the fact that the level is a simple estimate based on limited data, is always changing, and may not have been updated in decades. We know that almost all upstream human activities will decrease the average number of years for floods to reach the former 100-year level and raise the height of the average 100-year

---

Sidebar 11-8   How a Recurrence Interval Can Change

The importance of having a long enough record can be emphasized with data from Squaw Creek at the Ames, Iowa, gauging station.

In 1992, the largest flood in thirty-seven years of record had a recurrence interval of:

$$T_{37} = \frac{n+1}{m} = \frac{37+1}{1} = 38 \text{ years}$$

Following the new record flood in 1993, that same previous record flood (now the second largest in thirty-eight years of record) has a recurrence interval of:

$$T_{37} = \frac{n+1}{m} = \frac{38+1}{2} = 19.5 \text{ years}$$

Note that it does not matter, for this analysis, how much larger the 1993 flood was, just that it was larger.

Another way of looking at the recurrence interval problem can be seen from Figure 11-42.

In 1992, a flood flow of 10,000 cubic feet per second would have a projected 80-year recurrence interval. Following the record 1993 flood, that same 10,000 cubic foot per second flood flow would now be calculated as only a 42-year recurrence interval!

Similarly, we can consider the expected volume of the 100-year flood in 1992 (blue dashed line) versus that after the new record flood in 1993. In 1992, you would estimate the flow of a 100-year flood to be 10,200 cubic feet per second. After the record 1993 flood (red solid line), you would estimate the 100-year recurrence interval flood to be 13,500 cubic feet per second, a 32 percent increase.

event. We also know that we cannot prevent flooding during major storm events, in spite of some people's belief in safety above the 100-year floodplain. Compounding the problem, building levees induces people to build in the "protected" area behind them.

Floodplains and 100-Year Floodplains

As discussed above, the area outside the low-water channel but covered by the stream during moderate and larger floods is the floodplain. During floods, water that would otherwise flow rapidly downstream spreads out over the floodplain and slows down. Although that water eventually continues downstream, its temporary storage on the floodplain lessens flood levels downstream. Constructed levees eliminate that storage, causing higher flood levels downstream.

Many people do not recognize that the floodplain is part of the natural pathway of a stream or river (▶Figure 11-44). Typically, a flooding stream drops its coarsest suspended sediment just outside the main channel to form natural levees, giving the impression of additional flood protection. A stream may not flood across its floodplain for many years, but it will eventually.

Many people build on rural floodplain areas because the land there is often inexpensive. Governments then pay millions to build and maintain levees to protect such developments and to provide disaster relief following a flood. The cycle continues when governments allow relief funds to be used to rebuild in the same unsuitable places. Federal government policies are now beginning to change. Disaster relief funds are now commonly provided in a manner that encourages people to move out of floodplains. The government also now purchases many homes on floodplains with the requirement that building new structures in the floodplain is prohibited (IFMRC, 1994).

Floods in the United States cause an average of $5 billion per year in property damage, more than all other natural hazards combined. Federal flood relief amounted to somewhere between $15 billion and $30 billion between 1990 and 1998. In spite of significant moves to keep people out of high hazard zones, the percentage of federal to nonfederal cost sharing actually increased from 75 percent to 90 percent between 1992 and 1994 for Hurricane Andrew and the 1993 Mississippi River flood. In 2003, participants in the National Flood Insurance Program (see "Flood Insurance" below) were paying only 38 percent of actuarial risk rates. However, attitudes are changing. Individuals must take responsibility for their own actions. Natural hazards expert, R. H. Platt, commenting on a 1994 U. S. Supreme Court case, noted that "If private owners are unwilling to accept limits on unsafe building practices in known hazardous areas, why should the nation hold them harmless from the

▶**FIGURE 11-44.** In this aerial view of the Kentucky River flood at Hazard, Kentucky, in 1963, the floodplain is underwater.

results of their own free choice?" In 2003, the U.S. Congress began to move toward providing only mitigation funds to flood policyholders with a record of repeated losses. Mitigation would require relocation outside the flood zone, elevation of the home above flood level, flood proofing, or demolition. Those who reject the mitigation offer would be charged insurance rates based on standard actuarial costs for properties with severe repetitive losses.

Floodplains in the United States are often inappropriately used. They should be reserved for agriculture and related uses, not housing and industry. Parks, playing fields, and golf courses are reasonable uses in urban areas. Natural rivers spill over their floodplain every couple of years. When permanent structures are built on floodplains and levees are built to protect those structures, the river is confined to a more restricted area and must then crest at a higher level. The problem is not that the river floods but that we encroach on the natural behavior of the river.

Unfortunately, many people do not know that they are living on a floodplain. A survey of Missouri residents conducted seven months after the end of the catastrophic 1993 Mississippi River flood (see the discussion in Chapter 12) concluded that approximately 70 percent of floodplain residents thought they did not live on a floodplain. If people are not made aware that they are living on a floodplain when water is rising against their doorway, they typically learn when they attempt to sell the property or apply for a permit to modify or develop the land or buildings. This is often when the property loan must be secured by a flood insurance policy if it is within the floodplain. Most experts believe that real estate agents should be required to disclose flood risks when properties are offered for sale and that lending institutions should make similar disclosures at

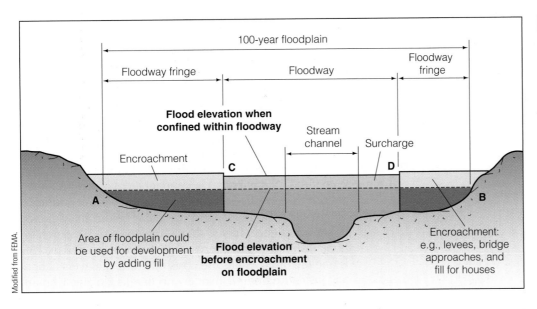

FIGURE 11-45. This schematic river cross section shows FEMA's definitions for flood insurance purposes. Line AB is the flood elevation before encroachment. Line CD is the flood elevation after encroachment. The *surcharge* is the rise in flood water level near the channel as a result of artificial narrowing of the floodplain. The surcharge is not to exceed 1.0 foot (FIA requirement) or lesser amount if specified by a particular state.

the time of a loan application. In many areas, they are required by law to do so. Only since 2002 have virtually all banks and mortgage companies required that a residence be surveyed to see if it is in the floodplain. If it is in the floodplain, it needs to be covered by flood insurance before the mortgage company or bank will provide a mortgage or offer a loan on a property.

Some people feel that regulations on building in a floodplain infringe on their rights to use their own property as they choose. However, an individual's choice to build in a floodplain often infringes on many other individuals. Huge sums of public tax dollars are spent every year to fight floods, build flood-control structures, and provide relief from flood damage for structures that should not have been built on the floodplain. Streams and rivers generally pass through many people's property. How one person or one town affects, restricts, or controls a stream commonly influences stream impacts to others both upstream and downstream. Developers and builders can make profits by building on floodplains, leaving homeowners and governments to pay the price of poor or insufficient planning. A coordinated approach is necessary to protect everyone. In many places, floodplains are not adequately zoned to minimize damages. Newly developed structures should generally be prohibited, and in some cases entire towns should be moved. Expensive as this may be, it is less expensive in the long run.

Flood Insurance

The **National Flood Insurance Program** (NFIP) for the United States was established by the National Flood Insurance Act of 1968 and the Flood Disaster Protection Act in 1973. Insurance for floods is provided by the federal government but purchased through private insurance companies.

Ratings and insurance premiums are actuarial, that is, they are based on flood risks and the existence of certain mitigation measures.

Guidelines for federal flood insurance stipulate several definitions (Figure 11-45). The floodplain that lies below the level that is likely to be flooded on average once every 100 years is formally separated into a *floodway* and a *flood fringe*.

The **floodway** includes the stream channel and its banks. During flooding, this zone carries deeper water at higher velocities. Most new construction is prohibited in this area, including homes, commercial buildings, waste disposal, storage of hazardous materials, and soil-absorption sewage systems. Also prohibited are structures, fills, and excavations that will significantly alter flood flows or increase 100-year flood levels.

The **flood fringe** zone is farther from the stream channel but still below the 100-year flood level. It includes the stream channel and banks. During flooding, it may be underwater but would consist of backwater and flood-storage areas. Water there is generally shallower and flows more slowly. Construction is typically permitted there as long as the lowest floor is on compacted fill and at least 2 feet above the 100-year flood level. Building or encroaching on the floodway must not raise the level of a 100-year flood by more than 1 foot (Figure 11-45). Some states set more stringent restrictions of no more than a 0.5 foot rise. Bridge approaches and levees, however, commonly encroach on the channel. Prohibited uses include waste disposal, storage of hazardous materials, and soil-absorption sewage systems, including septic tank drain fields.

Flood-hazard boundary maps (FHBMs) and flood insurance rate maps (FIRMs) are available for communities under the regular program of FEMA (Figure 11-46). To be

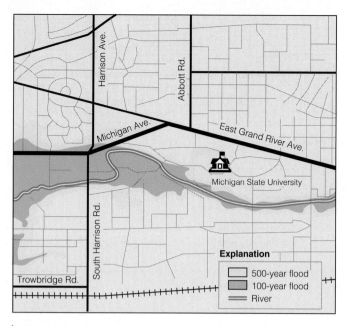

FIGURE 11-46. This example of a National Flood Insurance flood hazard map is for part of East Lansing, Michigan.

eligible for flood insurance, the community must complete the required studies to designate floodplain zones and enforce its regulations. Larger amounts of insurance are available at actuarial—that is, "true risk"—rates (see Chapter 1). Flood insurance is required by law under the following circumstances:

- where a building lies below the FEMA-identified 100-year floodplain; and

- where financing is required to buy, build, or improve property in a community where flood insurance is available; and

- where the loan for the property is subject to federal insurance or guarantees, or the lender is subject to federal rules or guarantees.

For people's behavior with respect to a river to be appropriate, flood insurance premiums for a given location should be proportional to the risk of damage caused by flooding. Premiums are currently set too low to cover the actual cost of flood insurance. Although insurance can be purchased up until thirty days before flood damage occurs, few people whose property is damaged by flooding have purchased such insurance. We can only conclude that most people are not aware that their property could be flooded or that flood insurance is available until thirty days before an event, thus until the flooding danger is imminent. However, insurance cannot be cancelled after a flood for the one-year duration of the policy. The short, thirty-day delay before an insurance policy takes effect means that a homeowner can purchase flood insurance even after the initiation of some floods. This is one factor that shifts the costs onto the taxpayers nationally instead of the relative few on the floodplains who are reimbursed.

Flood insurance rates depend on the likelihood and severity of flooding and are designated in mapped flood zones. Because building in areas of 100-year floodplains is now generally restricted, most existing homes in those areas were built before 1974. For a pre-1974 home in the 100-year floodplain (flood insurance zone A), typical rates in 2003 for a single-family dwelling with no basement were $520 per year for a $100,000 home and $800 for a $200,000 home. The federal insurance covers only the building, not the land. Rates for homes with basements and for home contents are somewhat higher. Above the 100-year level but within the 500-year, rates for a similar $100,000 home are $360 or so per year. Deductibles and additional base costs apply to initiate coverage.

Other areas in flood insurance zone A are those on alluvial fans or slopes that have been subjected to sheet flow in the past. Flow depths are typically 1 to 3 feet (0.3 to 1 meter) and insurance rates are as on the 100-year floodplain.

Human alterations to the landscape can have significant effects on the magnitude of future floods, as discussed in Chapter 12.

KEY POINTS

✓ Floods in the United States cause one-quarter to one-third of the monetary losses and nearly 90 percent of the deaths caused by natural hazards. **Review p. 268.**

✓ In humid regions, streams are fed primarily by groundwater. A stream's flow or discharge is proportional to the average water velocity and the cross-sectional area of the flowing part of the channel. **Review pp. 269, 272; Figure 11-7; Sidebar 11-1.**

✓ Streams commonly reach bankfull stage every 1.5 to 3 years, on average, and spill over floodplains less often. **Review p. 272; Figure 11-8.**

✓ During floods, not only does the water surface rise higher but also the channel erodes more deeply. Channel scour strongly affects damages. **Review pp. 272–274; Figures 11-8 and 11-9.**

✓ A flood does not have to be a large flow; rather, it is an unusually large flow for the stream. **Review pp. 272–273; Figure 11-11.**

✓ Steam equilibrium or grade is adjusted to accommodate grain size and amount of sediment in the channel, and the amount of water. Coarser grains and less water lead to steeper channel slopes. Similarly, larger particles can be moved only by higher water velocities. **Review pp. 273–275; Figure 11-14.**

✓ The sediment load that can be carried depends on the total flow or discharge. Thus, floods can carry more sediment and erode more from the channel. Where they spill over onto a floodplain and the flow slows, sediment is deposited as natural levees. **Review p. 276; Figures 11-16 and 11-21; Sidebar 11-4.**

✓ Meandering streams erode on outside bends and deposit on inside bends, so the meanders gradually migrate over the whole floodplain. The meander wavelength is controlled by the discharge at bankfull stage and is roughly 12 times the channel width. **Review pp. 277–281; Figures 11-17 and 11-18.**

✓ Braided streams develop where they are overloaded with sediment because of either a dry climate with too little water or overabundant sediment supplied to the channel such as below melting glaciers. **Review pp. 280–281; Figures 11-22 to 11-24.**

✓ The stream hydrograph, a plot of stream discharge against time, becomes higher and steeper if less of the precipitation infiltrates into the ground and runoff moves more rapidly over the surface because of urbanization, forest fires, or overgrazing. **Review p. 282; Figure 11-27.**

✓ Some streams are gaining, that is, they are fed by groundwater and flow year round; others are losing, in which case they flow primarily after a rain and may dry up in between rainstorms. Other streams have both gaining and losing segments. **Review p. 283; Figure 11-28.**

✓ In stream order, the smallest tributaries are designated first order, those where first-order tributaries join are second order, and so on. Low-order streams have short lag times between rainfall and flood. The flood crest moves downstream slower than the leading water in the flood wave. **Review pp. 283–284; Figures 11-29 and 11-30.**

✓ When the ground is frozen or water-saturated, most of the new rainfall must flow off the surface. **Review pp. 284–287.**

✓ Because stream power or erosive energy depends on the cross-sectional area of the channel times the slope of the channel, the Mississippi River has low power, but a steep mountain stream has high power. **Review p. 285, Sidebar 11-6.**

✓ Average flood frequency or recurrence interval is based on the past record for that stream. A 100-year flood has a 1 percent chance of occurring in any single year, regardless of the date of the last major flood. The recurrence interval for a stream is calculated as the total number of years of flood record for that stream + 1, divided by the rank of the flood under consideration, where the largest flood on record has a rank of 1, the second largest has a rank of 2, and so on. **Review pp. 285, 290–291; Sidebars 11-7 and 11-8.**

✓ The 100-year floodplain is the area flooded by the largest flood in 100 years on average. Federal flood insurance maintains restrictions on building within the 100-year floodplain. For insurance purposes, the stream channel is designated as the floodway; the area farther away, out to the 100-year floodplain, is designated as the flood fringe. **Review pp. 292–294; Figure 11-45.**

IMPORTANT WORDS AND CONCEPTS

Terms

alluvial fan, p. 271
base level, p. 271
bedload, p. 280
braided streams, p. 280
discharge, p. 269
drainage basin, p. 269
dynamic equilibrium, p. 273
flood channels, p. 274
flood fringe, p. 293
floodplain, p. 271
floodway, p. 293
gaining stream, p. 283
grade, p. 273
graded stream, p. 274
hydrograph, p. 282

losing stream, p. 283
meander, p. 271
National Flood Insurance Program, p. 293
natural levee, p. 276
100-year flood, p. 285
100-year floodplain, p. 291
overland flow, p. 282
paleoflood data, p. 290
point bar, p. 278
recurrence interval, p. 285
riffles, p. 273
stream order, p. 283
stream power, p. 285
thalweg, p. 278
watershed, p. 269

QUESTIONS FOR REVIEW

1. A river slope or gradient adjusts to what three factors that it has no control over?

2. How often, on average, does a stream reach bankfull stage just before spreading over its floodplain?

3. Does water move faster in a mountain stream or in a large smooth-flowing river like the Mississippi? Why?

4. If a large amount of sediment is dumped into a stream but nothing else changes, how does the stream respond? Why?

5. Why does a stream bottom erode more deeply when its water level rises in a flood?

6. What aspects of weather cause a flood? (Be specific: not merely "more water.")

7. In what climate are flash floods most common? (Provide an example location.)

8. Stream power or the destructive capacity of a stream depends on what factors?

9. Why does removal of vegetation by any mechanism cause more surface runoff and thus more erosion?

10. People should not build homes on floodplains because of the danger of flooding. What are better uses for floodplains?

11. What is the simple formula for calculating the recurrence interval for a certain size flood on a stream in 1999 if there are sixty-nine years of record? Give a numerical example.

12. With changes in characteristics of the upstream drainage area (e.g., urbanization), what changes can be expected in the height or extent of the 100-year floodplain?

13. Why do floods in first-order streams provide only a short time of warning for people downstream but high-order streams may provide days of warning?

14. Does a flood crest move downstream at the same speed as the water flow? Explain why or why not.

15. Why, in parts of Canada or northern United States, do north-flowing rivers often flood in spring? Explain clearly.

FURTHER READING

Assess your understanding of this chapter's topics with additional quizzing and conceptual-based problems at:

 http://earthscience.brookscole.com/hyndman.

CHAPTER 12

Missouri Highway and Transportation Department.

In 1993, the Missouri River flooded this major interstate freeway near Jefferson City, Missouri.

FLOODS AND HUMAN INTERACTIONS

Effects of Development on Floodplains

Long before Europeans arrived in North America, Native Americans built their houses on high ground or, lacking that, built mounds to which they could retreat in times of flood. European descendants, however, built levees to hold back the river. In spite of extensive high levees, more than 70,000 homes in the Mississippi River basin flooded in 1993, predominantly those of poor people living on the floodplains where the land is less expensive. Outside the area covered by the floodwaters, life went on almost untouched, but those nearby worried about flooding if the levees failed.

Levees

Most **levees** are constructed from fine-grained sediments dredged from the river channels or the floodplains. Different materials have different advantages. Compacted clay resists erosion and is nearly impermeable to floodwater but can fail by slumping. Crushed rock is permeable but less prone to slumping. During the1993 floods across the Mississippi River basin, locals, National Guard personnel, and others dumped loads of crushed rock and filled sandbags to raise the height of critical levees across the region. For days and weeks on end, it seemed that the work would never stop. As the higher levees raised river levels, they became saturated, causing slumping. Crushed rock in the levees minimized that. People inspecting a levee would sound an alarm if they found a leak. Cloudy water indicated that soil was being washed out of the levee through a process called **piping.** Wave erosion, slumping, or piping damaged or caused the failure of more than two-thirds of the levees in the upper Mississippi River drainage.

Levees did save some towns from flooding, but it is the levees themselves that created much of the problem. Every levee keeps floodwater from spreading out over its floodplain. All of the floodwater that should have spread over the floodplain is confined between the levees, causing the flood flow to be many times deeper and faster. A few towns on the floodplain are not protected by levees. When the water rises, people merely move out, then clean up afterward. Grafton, Illinois, at the confluence of the Illinois River with the Mississippi, is such a place, but after six big floods in twenty years, some people began to think about a change. When finished, they will have moved much of the town to higher ground with the help of funds from various federal agencies. Small towns such as Cedar City and Rhineland, Missouri, accepted a governmental buyout of their flood-damaged homes and moved off the floodplain. The homes were bulldozed to create public parkland.

People's typical response to unwanted action by a stream is to treat the symptoms and not consider the consequences of such treatments. Individuals, municipalities, states, and even the U.S. Army Corps of Engineers commonly try to "protect" floodplain areas from floods by building levees and then higher levees. Historically, levees were typically built on top of original **natural levees,** which are at the edge of the stream channel. Although a levee or dike may be initially of sufficient height to constrain a 100-year flood, the stream will eventually overtop the levee in a larger flood. Changes such as urbanization, levee construction, or channelization in the upstream drainage basin can cause flooding both upstream and downstream of the constricted river stretch that is "protected" by levees (Figure 12-1). Upstream of the channelized section, water levels rise because the flow is constricted, causing flooding. Immediately downstream, the water level is higher because it is deeper within the constricted area. Flooding will thus commonly occur both upstream and downstream where it would not have occurred before the levees were built.

Most people feel safe when they live on floodplains that they believe are protected by levees. They think flooding can only occur behind a levee if it is overtopped. Overtopping or **breaching** of levees does frequently occur in major floods. Other common failures are caused by bank

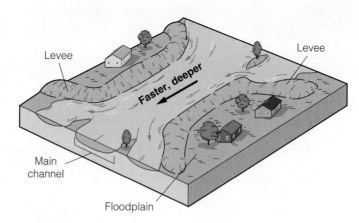

FIGURE 12-1. Levees that constrict the flow of a stream cause water to flow faster and deeper past levees. This causes flooding both upstream and downstream.

erosion from river currents or waves, slumps into the channel, piping, or seepage through old gravels beneath the levee (Figure 12-2). Or, if a flood is prolonged, lateral seepage beneath the levee may raise the groundwater level, which then floods surrounding areas behind the levee.

Levees are always built on floodplains, which often are composed of old permeable sand and gravel channels surrounded by less permeable muds. The floodplain muds beyond the current channel are sediments that were left behind by the river where it spilled over a natural levee to flow on the floodplain.

The river migrates across all parts of that floodplain over a period of hundreds or thousands of years. Under a mud-capped floodplain, the broad layers of sand and gravel deposited in former river channels interweave one another (Figure 11-19b, page 279; and Figure 12-3). These permeable layers provide avenues for transfer of high water in a river channel to lower areas behind levees on a floodplain.

Rising floodwater in a river increases water pressure in the groundwater below; this can push water to the surface

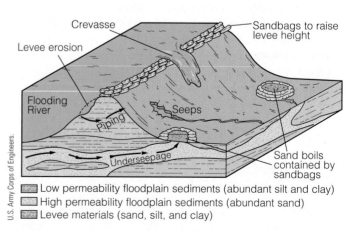

Low permeability floodplain sediments (abundant silt and clay)
High permeability floodplain sediments (abundant sand)
Levee materials (sand, silt, and clay)

FIGURE 12-2. A levee may fail by overtopping, seeping through, or piping.

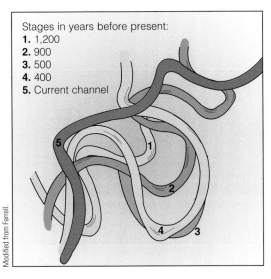

Stages in years before present:
1. 1,200
2. 900
3. 500
4. 400
5. Current channel

Modified from Farrell.

▶**FIGURE 12-3.** The Mississippi River at False River, Louisiana, shows gradual migration of meanders before the formation of a cutoff in an oxbow lake. The river flows from upper right to lower left.

California Department of Water Resources.

▶**FIGURE 12-4.** A California Conservation Corps crew places sandbags around sand boils at the Sacramento River, at north Andrus Island, in the delta area southwest of Sacramento on January 4, 1997.

on the floodplain, where it can potentially rise to nearly the water level in the river. With prolonged flooding, that water often reaches the surface as **sand boils** (▶Figures 12-2 and 12-4). The water, under pressure, gurgles to the surface to build a broad pile of sand a meter or more across. Workers defending a levee generally pile sandbags around sand boils to prevent the loss of the piped sand (▶Figure 12-4). They leave an opening in the sandbags to let the water flow away and reduce water pressure under the levee.

Artificial levees damage not only homes and businesses when they fail but also fertile cropland on floodplains. A flooding river without artificial levees spills slowly over its floodplain, first dropping the coarser particles next to the main channel to form low natural levees. Farther out on the floodplain, mud settles from the shallow, slowly moving water to coat the surface of the floodplain and replenish the topsoil. When artificial levees fail, cropland on the floodplain is often damaged by both deposition and erosion. The floodplain areas adjacent to a levee breach are commonly buried under sand and gravel from the flood channel. Farther away, the faster flow from the breach may gully other parts of the floodplain and carry away valuable topsoil.

The Great Mississippi River Basin Flood of 1993

August 1993 had the largest upper Mississippi flood on record, larger than a 100-year event (▶Figures 12-5, 12-6, and 12-7). River flow reached 29,000 cubic meters per second at St. Louis, 160 percent of the average flow at New Orleans near the mouth of the river. Twenty-five thousand square kilometers of floodplain was underwater. Fifty people died,

and damages exceeded $22 billion,* the worst flood disaster in U.S. history. More people died in the previous record flood on the lower Mississippi River in 1927, primarily because flood forecasting and warning systems then were less advanced. In 1993, highways, roads, and railroads were submerged for weeks on end, along with homes, businesses, hospitals, water-treatment plants, and factories (▶Figures 12-8 and 12-9). Almost 100,000 square kilometers, much of it productive farmland, lay underwater for months. Wells for towns and individual homes were flooded and contaminated, requiring boiling of domestic tap water. The U.S. Army Corps of Engineers halted all river barge traffic on the Mississippi north of Cairo, Illinois, in late June because it could no longer operate the locks and dams along the river.

Rivers across the basin overtopped or breached numerous levees as the flood crest reached them. Of 1,576 levees on the upper Mississippi River, 1,043 of the 1,347 that were built by local or state agencies were damaged. Only thirty-nine of the 214 that were federally constructed were damaged. Levees failed from north of Quincy, Illinois, to south of St. Louis, Missouri, and on the Missouri River from Nebraska City, Nebraska, south through Kansas and Missouri to St. Louis. Once a river breached a levee, it would flush sand and gravel over previously fertile fields, flood the area behind the levee, and flow downstream outside the main channel (see Figures 12-7 and 12-8). Each levee breach would lower the river level, sparing downstream levees, at least for a while. Even large cities were affected. Des Moines, Iowa, had no drinking water and no electric power for almost two weeks. The Mississippi River at St. Louis was

---

*Note: All costs are in 2002 dollars.

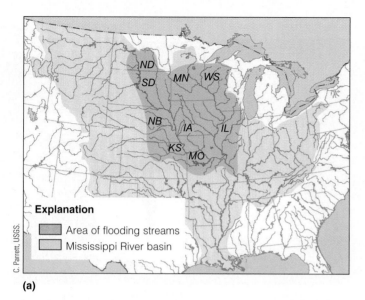

(a)

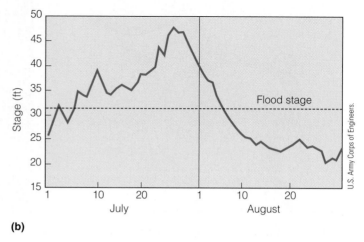

(b)

▶**FIGURE 12-5.** **(a)** This map shows the Mississippi basin and the general area of flooding streams during the summer of 1993. Heavily affected states are labeled with their abbreviations. **(b)** The flood hydrograph for the 1993 flood event shows that the Mississippi River was above flood stage for more than one month.

above flood stage from June 26 to late August, then again from September 13 to October 5. Clearly, the upper Mississippi River levee system only provided limited control of the 1993 flood.

Every time we build a new levee to protect something on a floodplain or to facilitate shallow-water navigation, we reduce the width of that part of the river and raise the water level during flooding (▶Figure 12-10). (See also Sidebar 12-1.)

Sidebar 12-1   Discharge Estimated from Cross Section and Water Slope

Specifically,

$$Q = 4.0A^{1.21}S^{0.28} \text{ for bankfull stage,}$$

or

$$= 3.4A^{1.30}S^{0.32} \text{ (to } \pm 20\%),$$

where

  Q = water discharge (in m³/sec.)

  A = average cross-sectional flow area (in m²)

  S = water-surface slope (along the channel)

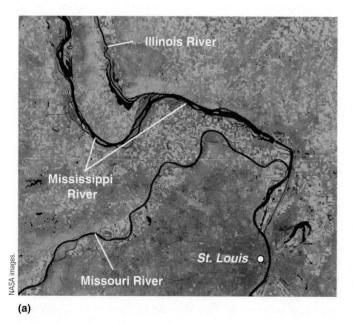

(a)

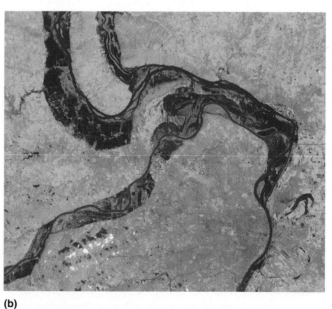

(b)

▶**FIGURE 12-6.** The Mississippi River flood of 1993 can be clearly seen by comparing satellite images taken **(a)** before and **(b)** during the flood. Note that the river fills its floodplain except in the channelized St. Louis reach.

FIGURE 12-7. In the breach of this levee during the 1993 flood on the Mississippi, the river (on the left) spills through two breaches to cover the floodplain (right). A second levee at the right edge of the photo provides temporary relief.

FIGURE 12-8. The Missouri River flooded, filled its floodplain, and submerged this freeway near Jefferson City, Missouri, in late July 1993.

FIGURE 12-9. The 1993 flood left a large number of homes deeply submerged in floodwaters for an extended period of time.

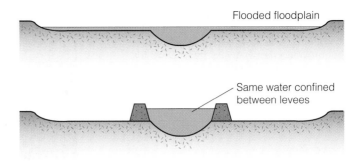

FIGURE 12-10. Levees confine at least the same amount of water between levees as could spread out over the floodplain.

A study conducted for the U.S. Congress in 1995 showed that if levees upriver had been raised to accommodate the 1993 flood, the water level in the middle Mississippi would have been 2 meters higher than it was. In fact, the increase in flood stages, for constant discharge, has increased 2 to 4 meters in the last century in the parts of the Mississippi River with levees. This rise in flood level is mostly the result of building the levees. The upper Missouri and Meramec rivers do not have levees and show no increase in flood levels.

Navigational dikes or **wing dams** constrict the river channel at St. Louis and other locations to increase river depth for barge traffic at low discharges (▶Figure 12-11). Even at high discharges when the river flow tops the wing dams, these structures increase resistance to flow near the river banks, which slows the velocity and raises the water level. As a result, for the same flood discharge, the stage or water height increases. This artificial increase in stage clearly affects the inferred recurrence interval for any huge flood such as the 1993 event on the Mississippi. The 1993 peak stage lies well above the recurrence interval curve adjusted for river flow without the wing dams.

The adjusted recurrence interval for the 1993 flood at St. Louis is 100 years or less, rather than previous estimates of up to 500 years. Recall that any new record flood changes previous recurrence intervals because the previous largest flood is now the second largest in the **recurrence interval** formula. This can have significant consequences for 100-year-event floodplain maps. Properties that were outside the 100-year flood hazard area are now within it; flood-protection structures that appeared to provide an adequate margin of safety no longer do so.

Some houses lay for weeks in water as much as halfway up the second story. After the lengthy flood, putrid gray mud coated floors and walls, and plaster was moldy to the ceiling because it and the insulation wicked it up. Belongings that could not be moved in time had to be thrown out and carted away. For those who expected the levees to hold, losses were higher because they made less effort to move their belongings. Not all the damage occurred within the floodplain. Many other areas with saturated soils

U.S. Army Corps of Engineers.

FIGURE 12-11. Wing dams on the Mississippi River, halfway between the Missouri and Ohio rivers, slow the flow at flood stages but provide a deepwater central channel for shipping at lower water.

Deeper flow in mid-channel

Wing dams

had flooded basements and backups of sewers and drain fields.

As in some previous major floods, the preceding winter and the first half of 1993 were 50 to 100 percent wetter than average, so most of the ground was saturated with water. Along with the usual melting of winter snow in the upper parts of the Mississippi basin, water levels in rivers of the basin rose as they usually do. However, 1993 was different, especially in late June and July. The jet stream moved farther south, bringing cool, dry air from Canada and circulating around a low pressure system in southwestern Canada. That part of the jet stream swept northeastward from Colorado toward northern Wisconsin. At the same time, warm, moist air pulled into the central United States from the Gulf of Mexico collided with the cool, dry air to produce persistent low pressure cells and northeast-trending lines of thunderstorms centered on Iowa and the surrounding states (▶Figure 12-12). Even this was not a particularly unusual weather pattern. What was unusual was its coincidence with a persistent high pressure system that stalled over the Southeast coast. That high kept the storms from moving east as they would normally have done.

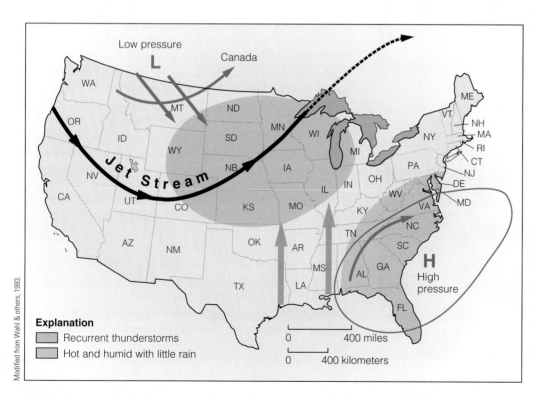

Modified from Wahl & others, 1993.

Explanation

- Recurrent thunderstorms
- Hot and humid with little rain

0   400 miles

0   400 kilometers

FIGURE 12-12. The dominant weather pattern for June and July 1993 that created the Mississippi River floods included a stationary low over western Canada and a persistent high off the Southeast coast.

For the July 5 storm, for example, a stationary front extended northeast from northern Missouri to southeastern Wisconsin (▶Figure 12-12). A series of cold fronts rotated counterclockwise around a low in southwestern Manitoba. Each cold front colliding with the warm, moist air over Iowa lifted the warm air to cause condensation and thunderstorms. A strong vortex in the middle atmosphere, some 5 kilometers above the surface, created additional lift for the thunderstorms. Strong upper-atmosphere winds of the jet stream created a chimney effect that created additional updraft.

By late June, flood-control reservoirs in the upper Mississippi basin were full or nearly so. The storms kept forming in the same area; it rained and rained, literally for months. The soil was still saturated from heavier than normal winter and spring precipitation, but torrential rains continued as high water arrived from upstream. Most of the area was drenched with 0.6 meter of rain between April and August; areas in central Iowa, Kansas, and northern Missouri received more than 1 meter. On June 17–18, 7 to 18 centimeters of rain fell in southern Minnesota, Wisconsin, and northern Iowa. In late June, flooding in Minnesota was the worst in thirty years. The flood crest moved downstream while it continued to rain. Compounding the problem, storms in Iowa on July 4–5 and 8–9 dumped 5 to 12, and 20 centimeters, respectively. Other major storms, on July 15–16, and 22–24, dumped an additional 12 and 33 centimeters of rain on various parts of the region.

The flood wave moved downstream at approximately 2 kilometers per hour, so it was relatively simple to predict when the maximum height of the flood would reach any area. In reality, however, other factors came into play. Tributaries added to the flow and broad areas of floodplain that were not blocked by levees removed flow to release it more gradually to the river. Tributaries backed up and locally even flowed upstream. Most places had multiple flood crests. Between Minneapolis and Clinton, Iowa, high bluffs confine the river to a narrow floodplain. From Clinton down to St. Louis, the floodplain is wider, and various agencies have built levees of different heights. From St. Louis down to Cairo, Illinois, they channelized the Mississippi River to maintain dikes for an average river depth of 7 meters. The river crested at nearly 6 meters above flood stage in St Louis.

In most places in the upper Mississippi River drainage, the recurrence interval of the flood was thirty to eighty years. In the lower Missouri River drainage basin in Nebraska, Iowa, and Missouri, the peak discharge was greater than a 100-year event. Large portions of the upper Mississippi River basin were declared federal disaster areas.

Intentional Levee Breaks

Individuals, towns, and government agencies go to great lengths to maintain and raise levees to protect towns from floods. So why would levee district officials and the U.S. Army Corps of Engineers intentionally sever a Mississippi River levee during the flood of 1993? They did just that in southwestern Illinois on August 1, 1993, to save the town of Prairie du Rocher (▶Figure 12-13). Fifteen kilometers north of Valmeyer, Illinois, a levee failed, permitting Mississippi River water to flow onto the floodplain on the east side of the river. There it began flowing south, soon overtopping a pair of levees on a tributary stream and continuing south behind the main levee to flood Valmeyer (▶Figures 12-13

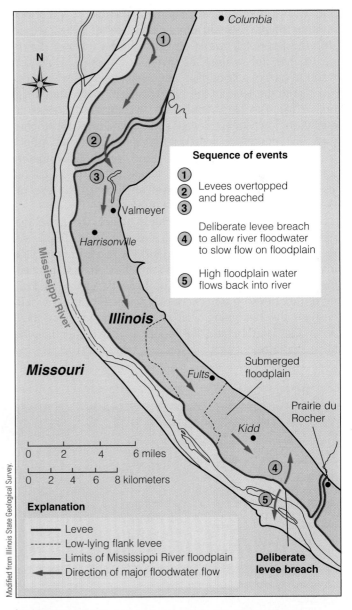

▶**FIGURE 12-13.** This map shows the Mississippi River, levees, and floodplain in the Valmeyer–Prairie du Rocher area of Illinois. The sequence of events can be followed from the breach of the main levee, flow south along floodplain, to deliberate breach to release water near Prairie du Rocher. This diagram shows the flow through a breach and downstream on a floodplain outside the channel levee.

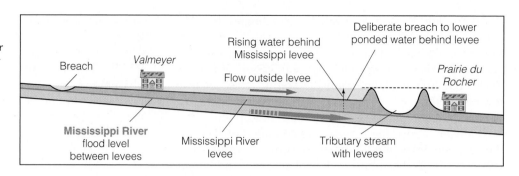

▶**FIGURE 12-14.** This vertical cross section through the area shown on Figure 12-13, shows that the water will gradually rise to the level of water at the breach upstream if that flow is blocked downstream.

and 12-14). Twenty-seven kilometers south of Valmeyer, another tributary stream flanked by levees of its own would stop the floodplain water and protect the town of Prairie du Rocher unless they also were overtopped. However, water in part of a floodplain enclosed by levees will gradually rise to the level of the inflow breach upstream (▶ Figure 12-14). The Mississippi River, of course, decreases in elevation downstream; levees protecting Prairie du Rocher were high enough to keep out the advancing flood but not as high as the Mississippi River at the breach 42 kilometers upstream. That means that the "lake" behind the main Mississippi River levee would rise well above the flood level on the Mississippi and flood Prairie du Rocher.

The solution was to deliberately breach the main Mississippi River levee 1 kilometer upstream from Prairie du Rocher to permit the "lake" behind the levee to flow back into the Mississippi River. Officials and the Corps of Engineers breached the levee before the floodplain flood reached Prairie du Rocher so the backflood of Mississippi water would cushion the oncoming wall of water on the floodplain. When the "lake" behind the floodplain rose higher than the Mississippi River, it again flowed back into the river.

Levees, Safety, and Costs

More and more people built homes and businesses in the Mississippi floodplain behind the supposed safety of the levees. The 1993 floods, however, completely submerged seventy-five towns. Are the costs of more levees and dams more than the value of the buildings they protect? Should taxpayers pay for flood losses for people who build in the floodplain? As noted above, federal relief funds are increasingly restricted to only those who move to higher elevations outside the floodplains or raise the floor level of their homes.

The **National Flood Insurance Program** (NFIP) made insurance available to those living on designated floodplains at modest cost (see "Floodplains and 100-Year Floodplains" in Chapter 11, page 292). In spite of that, only 5.2 percent of households in the Mississippi flood-hazard area had purchased flood insurance; those who did not probably believed that their floodplain property was protected by the levees. Some communities were again flooded in 2001 (▶ Figure 12-15).

One important outcome of the widespread failure of levees and flooding behind them was a general consensus

Andrea Booher, FEMA.

▶**FIGURE 12-15.** Davenport, Iowa, also on the Mississippi River floodplain, was again underwater during the 2001 event.

that rebuilding in flood-prone areas was not acceptable. Where flood insurance funds were used for reclamation or rebuilding, the work had to conform to NFIP standards. For insured buildings, the structure had to suffer loss of more than 50 percent of its value, and the rebuilt lowest floor had to be above the level of the 100-year flood level. Some 10,000 homes and businesses were approved for removal or non-rebuilding from more than 400 square kilometers of floodplain. A 1994 committee of federal experts recommended that levees along the lower Missouri River be moved back from the river by 600 meters to give the river room to meander and spill over its floodplain during high water.

Avulsion

What happens if a stream flooding outside its levees never returns to the main channel? This process is called **avulsion.** The most famous and catastrophic cases have occurred on the Yellow River in China many times over the past several thousand years. Details of these cases are described in "Case in Point: Yellow (Huang-Ho) River of China" below. Meandering streams follow paths through their floodplains, gradually shifting position as each meander erodes its outside bank and deposits on the point bar of its inside bank. If the meander belt remains in one place for a long time, such as between artificial levees, deposition of sediment gradually raises the channel elevation. Large floods often breach levees.

Initially, water crossing a levee breach is relatively clear and below its sediment load capacity, so it erodes vigorously. As the breach erodes deeper, the floodwater carries more sediment and the water-surface slope decreases, ultimately bringing the breach flow into equilibrium and limiting further erosion. If floodplain flow is ponded locally, the breach flow will slow, sediments will deposit, and the breach or crevasse will stop flowing. Levee breaches near Quincy, Illinois, along the Mississippi River in 1993, ranged from 100 meters to 1 kilometer in width and continued for seven to fifteen hours before deposited sediment stopped the breach flow.

If the floodplain flow is unrestricted downstream to lower elevations and the flood level remains high, the breach flow may continue to erode, ultimately diverting the main channel through the breach and causing avulsion. The stream moves to a new lower-elevation path on the floodplain. Because the new path is lower, the stream does not return to its original path.

At high flood levels, water may break through a crevasse in a levee where the floodplain lies lower than the streambed. The stream may abandon its channel and form an entirely new one. The Mississippi River did this locally in the 1993 flood. In the 1870s, the Bear River in California shifted to a new path following channel aggradation caused by hydraulic gold mining. The social and economic consequences of avulsion on a major river such as the Mississippi can be severe (see "Case in Point: New Orleans").

Channelization

Confining a stream within a concrete-lined channel causes the flow velocity to increase by making the stream deeper and straighter, as well as by reducing channel roughness. For hundreds of years, European cities such as Florence and Pisa, Italy, have sought protection from floods by building walls lining rivers that run through them (see "Case in Point: A Channelized Old World River"). In other cases, such as the upper Rhine valley of Germany, levees were constructed to deepen the flow to permit year-round navigation. The intended results were as anticipated. Those walls raised flood water levels and increased the water velocity. Unfortunately, such changes also enhanced erosion, both in the channeled reach and downstream where the river is not confined to the channel. Floods also pass through more rapidly, resulting in much higher downstream flood crests. Urban Los Angeles provides an extreme example of **channelization** (▶ Figure 12-16).

David Gatley photo, FEMA.

▶**FIGURE 12-16.** The Los Angeles River runs through a straight section of concrete channel with angled energy dissipation structures.

(Text continues on page 311.)

The Future of the Mississippi?

The 4,845-kilometer-long Yellow River (Huang Ho) drains most of northern China, an area approximately 945,000 square kilometers. That is a similar length but less than one-third the drainage area of the Mississippi River (▶Figure 12-17). The upper reach, flowing generally east from the Tibetan highlands, carries relatively clear water through mountains and grasslands for more than half the river's length. The middle reach south from Baotou and the deserts and plains of Inner Mongolia drains a broad region of yellowish wind-deposited silt or loess originally blown from the Gobi Desert in Mongolia to the northwest.

Sediment supply to the river from the loess plateau is vigorous because of vast arid to semi-arid, hilly areas of easily eroded silt. Before heavy agricultural use of the loess plateau beginning in 200 B.C., the plateau was mostly forested and the sediment load fed to the river would have been one-tenth of the current load. After the tenth century A.D., agriculture had largely destroyed the natural vegetation. Silt supplied by sheetwash and gully erosion (▶Figure 12-18) is so abundant that floods often carry hyperconcentrated loads of yellow-colored sediment that gives the river its name. Average sediment

George Leung photo.

▶**FIGURE 12-18.** The largest remaining area of loess tableland at Dongzhiyuan in Gansu, China, is being rapidly eroded. The size of the view can be inferred from the road around the end of the deep canyon.

load during a flood is generally greater than 500 kilograms per cubic meter, which is some 50 percent by weight or 20 percent by volume.

The lower reach of the Yellow River, downstream from Zhengzhou, flows across a densely populated and cultivated alluvial plain, one affected repeatedly by flooding for more than 4,000 years. As the river gradient decreases and the river spreads out over a width of several kilometers, sediment deposits and progressively raises the channel bottom; that regularly requires raising the levees. Near Zhengzhou, siltation raises the river bottom at an average of 6 to 10 centimeters per year, aggravates flooding, and rapidly fills the reservoirs behind dams.

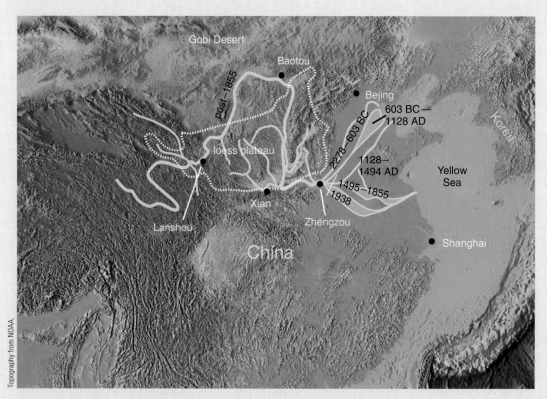

Topography from NOAA.

▶**FIGURE 12-17.** The middle reach of the Yellow River drains the easily eroded loess plateau. During the last 4,000 years, the lower Yellow River of China changed its course many times.

As early as 4,000 years ago, Emperor Yu dredged the channel and dug nine separate diversion channels to divert floodwaters. In 7 B.C., Rhon Gia advised evacuating people rather than fighting the river, but people did not follow his wise suggestion. After a disastrous flood in 1344 A.D., people used a combination of river diversion, river dredging, and dam construction. After each flood, they plugged breached dikes and raised existing dikes. With the channel raised by siltation, some breaches drained the old channel and followed an entirely new path to the sea—that is, by river avulsion. Unfortunately, as with most rivers, levees here are built from the same easily eroded silt that fills the channel; the river erodes the levees just as it does the loess plateau. Downstream from a breach, the riverbed between levees is left dry, a serious problem for those who are dependent on its water.

In 1887, the river topped 20-meter-high levees and followed lower elevations to the south to reach the East China Sea at the delta of the Yangtze River at Shanghai. The flood drowned people and their crops; the flood and resultant famine killed more than 1 million people. Official government policy since 1947 has been to contain floods by flood-control dams and by artificial levees along the channel. However, these structures are only designed to control a flood with a recurrence interval of sixty years, clearly not a long-term solution. The bed of the river is now as much as 10 meters higher than the adjacent floodplain (▶Figures 12-19 and 12-20). Because the riverbed is well above the surrounding landscape, the lower 600 kilometers of the river receives no water from either surface runoff or groundwater.

In 1960, the Chinese completed the San-men Gorge Dam, 122 meters high and more than 900 meters wide. Its 3,100 square kilometer reservoir that is designed to col-

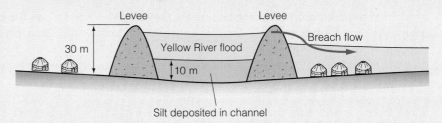

▶**FIGURE 12-20.** Siltation in the channel of the Yellow River, between levees, has raised the channel some 10 meters. Even with 30-meter levees, breaches often lead to avulsion and abandonment of the main channel.

lect water and silt probably will be filled with sediment by 2006. At the same time, a major effort was launched to plant trees and irrigate huge areas of the silt plateau to reduce the amount of silt reaching the Yellow River. Unfortunately, most of the trees died. As with droughts, construction of reservoirs and extraction of water for irrigation and other uses leads to low downstream flow, especially since 1985. Flood stage in 1996 was an all-time high even though the discharge was much lower than the 1958 and 1982 floods.

Repeated construction of levees over the last 2,500 years has not prevented catastrophic floods and river course changes. The lower 800-kilometer stretch of the Yellow River has repeatedly shifted its course laterally by hundreds of kilometers (Figure 12-17). Although the Mississippi drainage basin does not include either desert or active wind-blown loess in its headwaters, its riverbed rises higher with each year. Is this the future of the Mississippi River as people continually raise levees in response to each major flood?

What Went Wrong?

China has lived with the Yellow River and its floods for thousands of years and have tried to control the floods with levees just as we do. It has not worked. Levees broke and killed thousands of people. Avulsion dramatically changed the river course several times. With each flood, they built the levees higher and the levees still failed. There should be a message here; our levees fail just about as frequently.

▶**FIGURE 12-19.** Levees of the Yellow River stand high above the surrounding floodplain.

In addition to permanently flooding and eroding large flood-plain areas, the disruptions to a city that depends on river shipping for its economic livelihood can be severe. Avulsion of the Mississippi River above New Orleans would be economically catastrophic for the city. The Yellow River avulsion in 1855 took only one day. Avulsion of the Saskatchewan River in the 1870s turned more than 500 square kilometers of its floodplain into a belt of braided channels and small lakes. Avulsion there occurred at the outside of a large meander and apparently evolved over a period of several years.

New Orleans Flood History

Originally settled in 1718 as a French colony on a natural levee of the Mississippi River and 4.5 meters above sea level, New Orleans soon built low artificial levees to protect itself in times of flood. By 1812, the levees extended 114 kilometers upstream. New Orleans has battled the river ever since. With each large flood, the levees were built higher to protect the town. Major levees built in 1879 to contain floods around New Orleans broke in 1882 in 284 places. In 1927, with 2,900 kilometers of levees, the flood broke through in 225 places. In the 1973 record flood, the river submerged 50,000 square kilometers of floodplain.

By 1900, the city began to spread north into the swamps of the floodplain by building canals and draining them with pumps. Soggy peat on the floodplain began to compact because of dewatering and the weight load of roads and build-ings, so the city began to settle. Parts of it are now almost 4 meters below sea level, even farther below the Mississippi River that flows along the dikes right next to the core of downtown (▶Figure 12-21). Ships on the Mississippi look down on the city. High capacity pumps capable of pulling 1,100 cubic meters per second keep the groundwater at bay and the ground free of water, even after torrential rains.

Each additional dike further confines the river, which aggravates its tendency to flood downstream. Each time, the river rises higher during flood and flows faster. Instead of permitting the river to spread over its floodplain to shallow depths during flood, levees along the Mississippi and tributaries exacerbate the problem by raising the river level well above the natural floodplain.

The Atchafalaya River Problem

As the Mississippi River carries sediment to the Gulf of Mexico, it builds its delta seaward, thereby decreasing the river's slope in the depositional delta area. At the same time, the river builds both its bed and natural levees higher. During a major flood, the river breaches the natural levees, and its water heads down a steeper, shorter, lateral path to the sea. Because water flows faster on the steeper slope, it will begin to carry more of the river's flow, gradually taking over to become the new main channel. Sixty to seventy years ago, a new distributary channel, the Atchafalaya River, formed and began to sap some of the flow from the main channel (▶Figure 12-22).

▶**FIGURE 12-21.** New Orleans, the major shipping center on the lower Mississippi River, is protected by levees so that the river now stands 4 meters above the downtown area.

Marti Miller photo.

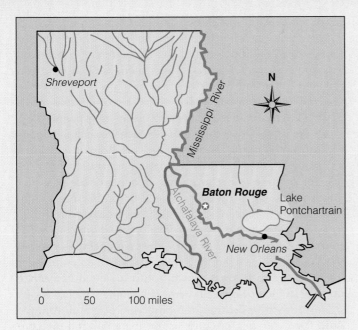

▶**FIGURE 12-22.** In Louisiana, the Mississippi River takes a long gentle path to the Gulf of Mexico. The Atchafalaya River drains part of the flow of the Mississippi upstream from Baton Rouge, taking a shorter and therefore steeper path.

nance of the Atchafalaya River over the current Mississippi course through New Orleans. Or the U.S. Army Corps of Engineers could decide to deliberately breach the barrier at the Old River control structure to save New Orleans. Either way the city loses. If the Atchafalaya takes most of the flow, the city is left high and dry without the river that is critical to its future.

How long can the U.S. Army Corps of Engineers keep the Mississippi confined and prevent it changing course to the straighter, steeper path to the ocean along the Atchafalaya River? The long-term effect of using levees to confine a river such as the Mississippi can be tragic as seen by comparing the history of the Yellow River in China.

Unfortunately New Orleans is located on the main channel downstream from where the Atchafalaya splits off. For New Orleans to be left high and dry without its river would destroy its key role as a shipping center for the whole Mississippi basin.

In the 1960s, recognizing the problem, the U.S. Army Corps of Engineers built the Old River control structure to permit 30 percent of the flow to enter the Atchafalaya, keeping 70 percent in the main channel through New Orleans. The idea was to regulate flow so that floodwaters could be channeled into the Atchafalaya to save New Orleans and other towns from flooding. Some of the sediment carried by a flood could also be channeled off to minimize further siltation of the main channel through New Orleans. The 1973 Mississippi River flood almost destroyed the Old River control structure, thereby channeling the main flow into the Atchafalaya River.

New Orleans is now at the mercy of the river. If a catastrophic flood should breach levees in the area, much of the city would be drowned in as much as 7 meters of water, along with a thick layer of sand and mud brought in through the breach. Could it happen? It was not supposed to happen in the upper Mississippi in 1993. A catastrophic flood could also destroy the Old River control structure, leading to domi-

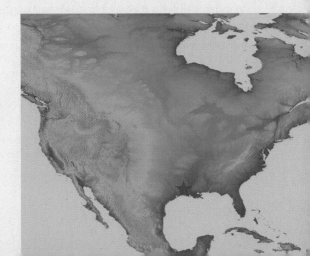

The Arno River, Florence, Italy, November 4–5, 1966

Like many other old European towns, multistory buildings are built right next to the channel of the Arno River. The river itself is channelized through the city, with vertical walls raised only a little above street level. In the seventeenth century, walls almost 9 meters high were built on the riverbanks, reducing the bankfull channel width from 300 meters to its current 150 meters. Even at low water, the river reaches both walls in many places. Although the surrounding region is hilly, the old part of town, including most of its most famous museums and churches, is on a broad, flat floodplain 1 to 2 kilometers wide, primarily on the north bank of the river.

In Florence, the Arno is only 160 kilometers from its headwaters but has a record of catastrophic floods including those in 1117 and 1333 that decimated the city and its bridges. Two years before the disastrous flood of 1547 that killed more than 100 people, Bernardo Segni pointed out that cutting so many trees for timber in the mountains upstream permitted water to erode the soil and to silt up the beds of the rivers. Thus, humans had contributed to the flood. In spite of major floods averaging one per twenty-six years, Florence remained unprepared.

The largest flood ever on the Arno River was recorded on November 3, 1966. Coincidentally, the largest previous flood was on the same day in 1333. Following an exceptionally wet October that saturated soils, a heavy storm on November 3 dumped 48 centimeters of rain on Florence and the Arno headwaters, a third of the average annual rainfall. Discharge reached 2,580 cubic meters per second and overfilled reservoirs. The flood tore through small towns upstream, continued down river at almost 60 kilometers per hour into its narrow, concrete-lined river channel within Florence. At 2:30 A.M. on November 4, floodwaters rapidly rose to a depth of 6.2 meters, 2 meters higher than the 1333 flood. By 4 A.M., water invaded the main square and was soon 1.5 meters deep in the Piazza Duomo.

The raging floodwaters rose to the roadway of the famed Ponte Vecchio, threatening the famous bridge that has spanned the Arno River in one form or another since Roman times; it was previously destroyed by floods in 1117 and 1333 and rebuilt each time. A bus carried downriver by the raging torrent crashed into the bridge during the 1966 flood, opening a huge hole. Ironically, that permitted water to pass, easing pressure on the bridge, and probably saving it from complete destruction. By 7 A.M. on November 4, water 1 to 2 meters deep completely covered the central part of the city. Water was 6 meters deep in parts of town—up to third-floor levels. Heating oil from thousands of basement tanks flushed

to the surface and was carried along with the floodwaters to contaminate everything it touched. For a city whose claim to fame is one of Europe's most valuable centers of culture and art, the effect was disastrous, with damage from mud and polluted water and the destruction of many priceless medieval and Renaissance paintings, sculptures, and books, many of which were stored in basements. Twenty-nine people died.

A significant part of the blame for these devastating floods was attributed to the residents. Since pre-Roman settlement in the region, they had stripped natural vegetation from the hills for centuries. That produced an annual cycle of winter floods and summer droughts. The 1966 problem was compounded by the failure to gradually release water from two hydroelectric dams upstream from Florence during heavy October rains. Late on November 3, the dam operators realized they had a problem, so they released a huge mass of water from the upstream dam, which in turn required immediate opening of the downstream dam, unleashing a wall of water.

In the last few decades, the extraction of gravel from the Arno River channel for construction materials and the construction of reservoirs upstream have caused increased channel erosion. Little has been done to rectify the basic causes, but most vulnerable art works are now kept on the upper floors of buildings and out of range of future floods.

What Went Wrong?

The Arno River, channelized since the seventeenth century, is bordered by multistory buildings. Deforestation permitted erosion of the drainage basin and deposition of sediment in the river channel. A major storm dumped 48 centimeters of rain on already saturated soils. Operators of two dams upstream failed to gradually release water to lower reservoirs during the rainy period that preceded the major storm, so they were not able to collect water behind the dams during the flood.

In another example, part of the Tia Juana River drains northwestward from Mexico to the ocean through a small corner of San Diego. A 4.3-kilometer section of the river is channelized to protect the city of Tijuana from flooding just before it enters San Diego. To protect agricultural land on the California side from severe flooding and erosion, and to permit groundwater recharge, the U.S. Army Corps of Engineers built an energy-dissipating structure and sedimentation area to slow the water and deposit excess sediment. They used levees to protect built areas on both flanks.

Flood Control and Multipurpose Dams

Some dams are built to trap floodwaters upstream and then release that water more slowly once the flood has passed (▶Figure 12-23). One of the most serious problems with dams is a more subtle one. Many are designed and financially justified as **multipurpose dams** to provide electric power, flood control, water for irrigation, and recreation.

These are desirable attributes, but the timing of their needs often clashes. For example, winter is California's wet season, so reservoirs and rivers fill and flood then or in spring as rainfall overlaps snowmelt. Summer is dry, but because that is the growing season, irrigation water is in high demand. Electricity use is also highest in summer, primarily because of air conditioning. Dam officials try to fill reservoirs as full as possible in the winter to prepare for summer irrigation and electric power needs. In contrast, reservoirs need to be kept as low as possible to provide maximum capacity to contain a potential flood. Thus, the dilemma: predict rainfall well enough in advance to fill the reservoirs but keep them low enough to contain possible floods. Unfortunately, predicting the weather is not entirely reliable, especially weeks in advance. Immense floods that inundated much of the Central Valley of California, including several cities,

in the winter of 1997–98 demonstrated the consequences of multiuse management based on unreliable weather predictions (see "Case in Point: Sacramento–San Joaquin Valley, California").

Floods Caused by Failure of Human-Made Dams

Federal agencies or states own only 7.8 percent of dams, local agencies and public utilities own 18.5 percent, and private companies or individuals own 59 percent. Thus, care in siting, design, construction quality, and maintenance of dams is highly variable.

Many dams serve their intended purpose, in theory, until the reservoir behind the dam is filled with sediment or, fortunately not often, the dam fails. When the U.S. Army Corps of Engineers studied more than 8,000 U.S. dams in 1981, they found that one-third of them were unsafe. The hazard to people living downstream depends on the volume of water released, the height of the dam, the valley topography, and the distance downstream. Calculations show that a dam-failure flow rate in a broad open valley would likely drop to half of its original rate in 60 kilometers or so, but in a steep narrow valley this level of drop in flow rate requires 130 kilometers.

Flooding on Rapid Creek in the Black Hills of South Dakota provided dramatic evidence of why it is dangerous to live downstream of a multiuse dam (▶Figure 12-24). In just six hours in June 1972, 37 centimeters of rain fell over the Rapid Creek drainage basin. Southeast winds carrying

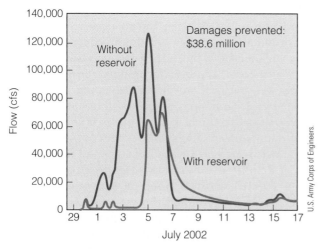

▶**FIGURE 12-23.** This flood hydrograph shows the Guadalupe River with Canyon Lake Reservoir (blue) and a calculation of what it would have been without the reservoir (red).

▶**FIGURE 12-24.** Damages from the June 10, 1972, flood at Rapid City, South Dakota, were extensive. Cars were mangled, and all that is left of a nearby house is the tangle of boards in the lower right.

Flood of January 5, 1997

When the State of California proposed limiting new development in flood-prone areas on low ground near the confluence of the Feather and Yuba rivers north of Sacramento,

Norm Hughes photo, California Department of Water Resources.

▶**FIGURE 12-25.** Failed Feather River levees left only rooftops of homes showing above floodwaters on the floodplain north of Sacramento on January 4, 1997.

Norm Hughes photo, California Department of Water Resources.

▶**FIGURE 12-26.** A levee of the Feather River was eroded and breached at a country club north of Sacramento, California, on January 3, 1997.

the California Building Association and California Association of Realtors fought the restrictions, arguing that it would take away people's property rights. The State Department of Natural Resources responded that such building is not in the public interest because property owners expect the government to use public tax dollars to bail them out when disaster strikes.

From December 29, 1996, through January 4, 1997, a "pineapple express" accompanying El Niño brought one of a series of five warm, subtropical rainstorms from the central Pacific Ocean (▶Figure 12-25). Heavy, warm rains fell on a Sierra Nevada snowpack that was almost double the average, causing record stream flow, but torrential rains made up 85 percent of these flows. Ground saturated from earlier rains and snowmelt amplified the runoff, causing streams to rise to record levels.

Levees failed on the Cosumnes, Mokelumne, Tuolumne, and Feather rivers (▶Figure 12-26). The Mokelumne River levee failure swept a marina and 230 boats downstream to crash into a bridge. Multipurpose reservoirs filled up, and erosion and mudslides severed U.S. Highway 50 across the Sierra Nevada in five places. Floodwaters covered 650 square kilometers, killed at least eight people, damaged or destroyed 16,000 homes, and caused damages of more than $1.8 billion (in 2002 dollars).

The worst event of the flood was failure of a problem-plagued levee on the lower Feather River, out on the broad Sacramento River valley, where water rose to a depth of 23.5 meters during the night of January 2. This flooded 39 square kilometers of farmland, killed three people, displaced 80,000 more, and caused $225 million in damages (▶Figures 12-26 and 12-27). The disaster would have been much worse if the new city of Plumas Lakes that was approved in 1993 had been built. Developers and politicians planned it on the floodplain as a Sacramento bedroom community for as many as 30,000 people. By 2001, the Plumas Lakes development had all the basic approvals for an initial 700 homes. All of these sites were partly or entirely flooded in the 1997 storm.

On January 3, authorities deliberately breached the levee downstream near the confluence with the Bear River to permit some of the floodplain water to flow back into the Feather River and relieve pressure on the levee. Parts of the site were still under 2.5 meters of water two months after the flood.

Voters rejected another group of planned towns for 100,000 people just upstream from Sacramento in 1993, but the developers continue to challenge the rejection in the courts. Downstream near Stockton, other developments for 29,000 people have been approved by city councils but also are being fought in court by opponents who are concerned about future flooding and long-term costs to the public.

▶**FIGURE 12-27.** Flooding from the Feather River extended far off to the horizon in some places.

Flooding on the Sacramento River downstream was ranked as approximately a fifty-year event. Damages reached almost $2.2 billion, including flood-control sites, homes, small businesses, private irrigation systems, highways and other roads, buildings, fences, and crops. This figure did not consider the large losses in income from businesses and tourism. Some 775 square kilometers flooded as levees failed in more than eighty places. In some cases, water flowing through old river gravel channels under the levees eroded the base of the levees. Developers and local officials still insist that the planned cities will be safe from floods when levees are upgraded. Developers are lured by the huge development profits; cities want the tens of thousands of new jobs and tens of millions of dollars in new annual tax revenues. Unfortunately, taxpayers pay for the consequences when things don't go as planned.

The recurrent arguments on one side are that people should not be denied the right to build on their own property, that development will bring new jobs and new tax revenues. They also insist that flood-control dams, levees, and levee improvements, many as yet not built, will provide ample safety. Opponents argue that levees and dams have proven inadequate and unsafe again and again and that property owners expect the government and therefore the public to pay for cleanup, repairs, and rebuilding after any natural disaster. Flood experts doubt that levee upgrades will prevent flooding. Even if levees do not fail, water will flow through subsurface gravels to flood areas behind the levees (Figures 12-2 and 12-4).

Multipurpose Dams

The demands on many multipurpose dams in the Sierra Nevada compounded the flooding problem in 1997. Engineers rationalized building of the dams by adding up the total benefits of irrigation, electric power, recreation, and flood control. Unfortunately, some of these are mutually incompatible.

On a positive note, a broad consensus among federal, state, and county agencies noted that rivers need more space to "do their own thing." Even the U.S. Army Corps of Engineers, which built most of the levees and dams, is having second thoughts. Instead of building levees adjacent to river channels, the corps suggests a 30-meter corridor on either side of the channel where development is not allowed and where the river is permitted to meander and develop riffles. This would minimize flood damage, reduce construction and maintenance costs, and provide a small area of natural flood plain and a "nice river corridor."

What Went Wrong?

Large dams were built to serve several distinctly different purposes that were not compatible. The water level in reservoirs behind the dams needed to be low enough to provide protection from floods in wet springs but high enough to provide water for irrigation and electric power in hot, dry summers. Weather prediction was not accurate enough weeks in advance to adjust the reservoir heights and prevent overtopping of the dams.

Norm Hughes photo, California Department of Water Resources.

warm, moist air from the Gulf of Mexico banked up against the Black Hills where it encountered a cold front from the northwest. Pecola Dam, 16 kilometers upstream, was built on Rapid Creek just twenty years earlier for irrigation and flood control after an earlier flood. Building of this and other dams made people feel secure from floods, so they built homes along the creek downstream.

During the intense flood of 1972, the creek's typical flow of a few cubic meters per second became a torrent of 1,400 cubic meters per second within a few hours. With rising water, authorities began ordering evacuation of the low-lying area close to the creek at 10:10 P.M., and the mayor urged evacuation of all low-lying areas at 10:30. The spillway of a dam just upstream from the city became plugged with cars and house debris from upstream, raising the lake level by 3.6 meters. At 10:45 P.M. the dam failed, releasing a torrential wall of water into Rapid Creek that flows through Rapid City. The flood just after midnight killed 238 people, destroyed 1,335 homes and 5,000 vehicles, and caused $690 million in damages. More than 2,800 other homes suffered major damage (▶ Figures 12-24 and 12-28).

In this case, the lesson was learned—at least for now. The city used $207 million in federal disaster aid to buy all of the floodplain property and turned it into a greenway, a park system, a golf course, soccer and baseball fields, jogging and bike paths, and picnic areas. Since then, building in the floodplain has been prohibited. However, decades later, pressure increases to build shopping centers and other structures in the greenbelt. The decision rests in a politically divided city council—the usual struggle between developers or "jobs" versus long-term costs, aesthetics, and safety.

More than 3,300 high and hazardous dams are located within 1.6 kilometers of a downstream population center. Few local governments consider the hazard of upstream dams when permitting development. Major floods from dams in narrow valleys have occurred for a variety of reasons:

- Overtopping a reservoir after prolonged rainfall, as in the 1972 flood in Rapid City, South Dakota. Although most dams could be built higher, the cost increases rapidly with height, in large part because a dam's length also increases rapidly with its height.

- Seepage of water under a dam leads to piping and erosion of the dam foundations, resulting in catastrophic dam failure. The Teton Dam in eastern Idaho that failed on June 5, 1976, took eleven lives and caused more than $3.2 billion in damages (see "Case in Point: Failure of the Teton Dam, Idaho").

- Subsurface erosion along faults or other weak zones in the foundation rock below the dam.

- Poor design and engineering standards of a privately owned slag-heap dam at Buffalo Creek, West Virginia, resulted in a 1972 failure that drowned 125 people.

- Improper maintenance of a dam, including failure to remove trees, repair internal seepage, or properly maintain gates and valves.

- Negligent operation, including failure to remove or open gates during high flows, as in the 1966 flood on the Arno River in Florence, Italy.

- Landslides into reservoirs that cause a surge and overtopping of the dam, as in the Vaiont Dam in northeastern Italy that killed 2,600 in 1963. Filling the reservoir behind the dam increased pore-water pressure in sedimentary rocks sloping toward the reservoir. A catastrophic landslide into the reservoir displaced most of the water to drown more than 2,500 people downstream (see "Case in Point: The Vaiont Landslide" in Chapter 8, pages 210–211).

- Earthquakes that weaken earth-fill dams or cause cracks in their foundations. The Van Norman Dam, owned by the city of Los Angeles and less than 10 kilometers from the epicenter of the 1971 San Fernando Valley earthquake, is immediately upstream from thousands of homes. It is an earthen structure that was thirty years old when the earthquake struck. The earthquake caused a large landslide in its upstream face and so drastically thinned the dam that it seemed likely to fail. Operators were fortunately able to lower the water to a safe level so that the dam did not fail, but authorities evacuated 80,000 people from the area downstream until they could lower the water level.

Urbanization

Flash floods also materialize wherever roads or buildings cover the ground and where drainage channels are built to handle the runoff, because water cannot soak into materials such as pavement. Increasing **urbanization** in many parts of the world causes increasing numbers of flash floods and higher flood levels. In urban areas, floods often create problems for those who think they can wait out the rising

▶**FIGURE 12-28.** This house was carried off its foundation onto the road by the 1972 Rapid City flood.

The Teton Dam, near Rexburg in eastern Idaho, was built by the U.S. Bureau of Reclamation to provide not only irrigation water and hydroelectric power to east-central Idaho but also recreation and flood control. After dismissal of several lawsuits by environmental groups, construction began in February 1972, and filling of the reservoir behind the completed dam began in October 1975. The dam was an earth-fill design 93 meters high and 945 meters wide, with a thin "grout curtain" or concrete core to prevent seepage of water through the dam.

On June 3, 1976, workers discovered two small springs just downstream from the dam. On June 5 at 7:30 A.M., a worker discovered muddy water flowing from the right abutment (viewed downstream). Although mud in the water indicated it was carrying sediment, project engineers did not believe there was a problem. By 9:30 A.M., a wet spot appeared on the downstream face of the dam and quickly began washing out the embankment material. The hole expanded so rapidly that two bulldozers trying to fill it could not keep up and were themselves lost into the hole. At 11:15 A.M., project officials told the county sheriff's office to evacuate the area downstream. At 11:55, the crest of the dam collapsed; two minutes later, the reservoir broke through and rushed downstream (▶Figure 12-29). The flood obliterated two small towns and spread to a width of 13 kilometers over Rexburg, with a population of 14,000, and continued downslope at 16 to 24 kilometers per hour.

The flood killed eleven people and 13,000 head of livestock, and the federal government paid more than $530 million (in 2002 dollars) in claims. The cause of failure was never settled, but numerous flaws came to light. After construction started, U.S. Geological Survey geologists expressed concern about pressures from a filled reservoir and loading that could cause movement around the dam, as well as internal shearing, endangering the dam.

What Went Wrong?

An expert's review panel blamed design and construction flaws for the dam's failure. A modern dam was built on bedrock with open joints that were not cemented as specified. Fill settled away from the cracked rock, causing internal deformation of the structure. A concrete wall inside the dam, which was designed to block water flow, was built too thinly and cracked under the high water pressure at the base of the dam.

(a)

(b)

(c)

▶**FIGURE 12-29.** Progressive failure of Teton Dam, eastern Idaho, June 5, 1976: **(a)** At 11:20 A.M., muddy water pours through the right abutment of the dam. **(b)** At 11:55 A.M., the right abutment begins to collapse and a large volume of muddy water pours through the dam. **(c)** In the early afternoon, the dam fails and the reservoir floods through it.

FIGURE 12-30. This flood in the Midwest in June 1994 certainly made roads impassable.

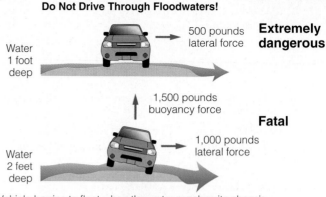

Do Not Drive Through Floodwaters!

Vehicle begins to float when the water reaches its chassis, which allows the lateral forces to push it off the road.

Washed-out roadway can be hidden by muddy water, allowing a vehicle to drop into unexpected deep water.

FIGURE 12-31. Even shallow floodwaters can lift a vehicle and wash it downstream.

waters. When they realize they really do have to evacuate, the roads have become impassable, preventing evacuation (Figure 12-30).

Streams in all but arid regions are fed mostly by groundwater. Even during floods, a large proportion of the flow commonly moves through the subsurface. The saturation of near-surface soils is necessary before significant surface runoff begins. Exceptions include conditions with especially heavy rains or ground that has been frozen or sealed by wildfires. Small drainage basins have higher, narrower hydrographs than those of large basins, and their floods arrive sooner after a major rainstorm. Paved streets and parking lots, houses, and storm sewer systems carry runoff quickly to streams and result in a significantly higher hydrograph flow in a shorter length of time. The stream discharge rises and falls much faster than in rural areas.

Flash Flood Hazards

Any type of flood can be dangerous, but flash floods are especially so because they often appear unexpectedly and water levels rise rapidly. Deaths are frequent because of little warning and because of their violence. Even under a clear blue sky, floodwaters may rush down a channel from a distant storm. On many occasions, people have been caught in a narrow, dry gorge because they were not aware of a storm upstream. At night, people in their homes have been swept away.

Driving through a flooded roadway can be dangerous or even fatal. Even though water may appear shallow, the force of its flow against the wheels, or worse against the side of the car, can wash it downstream. Shallow fast-moving water can erode a deep channel that may not be visible. Fast-moving water only 1 foot deep exerts more than 500 pounds of force against the wheels and is extremely

dangerous (Figure 12-31). Water 0.6 meters deep and above the vehicle floorboards can exert 700 kilograms of buoyancy and 450 kilograms laterally to push the vehicle off the road and potentially drown the occupants. Driving into apparently shallow muddy water that hides a washed-out roadway can drop the vehicle into deep water and drown the occupants. Even where water is still or hardly moving, liquefaction or settling of part of a roadway can cause unexpected deep water. Cars seem heavy, but their weight is generally much less than the volume of water to fill a car. Most cars will float in water—that is, until water seeps in and they sink. If trapped in a partly submerged and sinking car, it will be difficult to open the doors to escape. Either lower a window or kick out the windshield to climb out.

Even after a storm passes, stay away from downed power lines. Be wary of broken tree limbs that may fall unexpectedly. If you smell gas in a building, extinguish any open flames, turn off the main gas valve, open windows and doors, and evacuate the building.

Coarse material in a stream is supplied by erosion of bedrock or from landslides that enter the channel. In most cases, debris flows on steeper slopes supply that coarse debris to the streams; steep stream gradients are also needed to transport the coarse material. High-gradient bedrock channels generally have deep, narrow cross sections that carry deep, turbulent flows that are highly erosive during floods. Their turbulence makes them capable of transporting large boulders as bedload and even in suspension. Bedload boulders impact and abrade the channel sides, severely damaging roads and bridges.

Changes Imposed on Streams

Increased stream sediment load requires a steeper slope for transport. With excess sediment supplied to a stream, more than the stream can carry, sediment drops out to choke the channel. The stream generally becomes braided, and the channel becomes steeper. Examples include a landslide or mudflow filling the channel, or deforestation by fire, heavy logging, or overgrazing of a watershed, all of which cause excessive erosion and lead to the dumping of large amounts of sediment into a channel.

Forest Fires and Range Fires

In vegetated areas, rain droplets impact leaves rather than landing directly on the ground. Rich forest soils soak up water almost like a sponge, providing a subsurface sink for rainwater that can then be used by the vegetation. Fire removes that soil protection, permitting the droplets to strike the ground directly and run off the surface. Intense fire also tends to seal the ground surface by sticking the soil grains together with resins developed from burning organic materials in the soil. This decrease in soil permeability reduces water infiltration, forcing a large proportion of the rain to directly run off the surface. This can carve deep gullies into steep hillsides and feed large volumes of sediment to local streams (see Chapter 16 for further discussion and pictures of fire-related gullies).

Logging and Overgrazing

Logging by clear-cut methods can promote significant erosion. Perhaps the most destructive method involves "tractor yarding" in which felled trees are skidded downslope to points at which they are loaded on trucks. The method leaves skid trails focused downslope to a single point like tributaries leading to a trunk stream. Skidding logs along the ground removes brush and other vegetation, leaving the ground vulnerable to erosion. Logging roads tend to intercept and collect downslope drainage, permitting the formation of gullies, increased erosion, and the addition of sediment to streams.

Cattle and sheep grazing on open slopes similarly remove surface vegetation that formerly protected the ground. Rainfall running off the poorly protected soil is more likely to erode gullies, thereby carrying more sediment to the streams. Once gullies begin, the deeper and faster water causes rapid gully expansion and destruction of the area for most uses (▶ Figure 12-32). Animals grazing near streams also break down stream banks, causing more rapid bank erosion and heavier stream sediment loads.

Donald Hyndman photo.

▶**FIGURE 12-32.** Heavy sheep grazing has encouraged deep erosion in a steep slope south of Lake Wakatipu, New Zealand. The broader valley in the upper right is a more extreme stage of the same process.

Vegetation removal also decreases evapotranspiration, the evaporation of rain from leaves and the transpiration of moisture from the living cells of leaves, which naturally pulls water from the soil via roots. This reduction in evapotranspiration permits more water to soak into the ground and to run off the ground surface. Evapotranspiration can drop to 50 percent after clear-cutting. More water penetrating a slope tends to promote landslides, which in some areas contributes as much as 85 percent of the sediment supplied to a stream. Increased runoff and erosion on slopes also carries more sediment into streams. That upsets the balance of the stream, causing increased sediment deposition in the channel and increased flooding downstream.

Hydraulic Placer Mining

The historic practice of hydraulic placer mining in California during the 1860s and 1870s provided a classic case history of how rivers respond to large volumes of added sediment load. Gold miners originally panned gold or separated it from sand and gravel in streambeds with small sluice boxes fed by water diverted from the stream. As those river gravels became depleted, the miners discovered gold in high-level terrace gravels above the streams. To separate that gold, they used high pressure jets of water from higher elevations to hose down the gravels into large sluice boxes at stream level (▶ Figure 12-33). The accumulated loose gravel was picked up by the streams during flood and washed downstream. The Bear River in the western Sierra Nevada, for example, built up its bed by as much as 5 meters in response to hydraulic gold mining upstream.

The first big flood from heavy rainfall in the Sierra Nevada in January 1862 flushed much of the placer gravel from

▶**FIGURE 12-33.** Hydraulic placer gold mining in California in the 1860s contributed to heavy gravel accumulations in the rivers of the western Sierra Nevada mountain range.

tributaries into the main rivers, and in turn out through the mouths of their canyons into the edge of California's Central Valley. The rivers became choked with sediment because they could not carry it all; channels filled with gravel, and the flood spread far beyond where it should have. Previously productive farmland was covered with gravel, making it unusable. Cities were not much better off. The next catastrophic floods, in 1865, turned the Central Valley into an "inland sea" 20 miles wide by 250 miles long, submerging farms and towns. Similar floods occurred in the following thirty to forty years.

Hydraulic mining was finally outlawed in 1884, but by then the damage was done. Landowners and governments tried to deal with the floods by the usual means of treating the symptoms. Build levees near the river to contain the flood; when those are topped during a subsequent flood, build them higher. Channelize and straighten the river to carry the water through more quickly and to prevent the water from backing up to form a lake. Build dams to contain the floods. These actions, however, made matters worse downstream. Floodwaters raced right down the channel rather than spreading out across floodplains to slowly drain back into channels as the flood waned. Flood levels between the levees were much higher, so water flowed faster. All that water created greater flood levels downstream.

Although downstream towns built levees to protect themselves, the sediment-choked channels contained so much new sediment that their beds in some cases built up higher than the towns behind the levees. On January 19, 1875, a modest flood along the Yuba River breached a levee at Yuba City north of Sacramento, sending a flood of gravel through the town, all but destroying it.

The hydraulic placer-mining fiasco may be past, but other landscape alterations can lead to similar results. Deforestation from fire or vegetation stripping by overgrazing

may lead to rapid erosion in hilly terrain. Sediment from such erosion is carried to stream channels, choking them and leading to heavy sediment deposition. In some areas, the result is braided streams.

Dams and Stream Equilibrium

Building dams removes sediment from streams because the velocity in reservoirs behind the dams drops to virtually zero. Downstream of the dam, the stream carries little or no sediment, so it erodes its channel more deeply during flood (▶Figure 12-34).

Dams are built for different competing purposes, as noted below in the discussion of flood control and multipurpose dams. According to the 1994 National Inventory of Dams,

■ more than 31 percent of dams were built to provide recreation,

■ 23.5 percent were built to collect water for water supply or irrigation use during dry seasons,

■ 14.6 percent were designed to control flooding downstream from storm or high-water runoff, and

■ 2.9 percent were built to generate hydroelectric power.

Flood-control dams are built to stop flooding of floodplains. The intent is not to allow people to build on floodplains, but that is often the effect. People may feel protected by a dam, but that is a false sense of security.

Flood-control dams are built high enough to contain a certain magnitude of expected flood, perhaps a 100-year flood. The ability of a dam to contain such a flood depends on how full the reservoir is just before the flood, as well as changes in management of the upstream land such as deforestation or urbanization. In addition, the reservoirs behind dams eventually fill. At some point, the dam may not be adequate, and in extreme cases the dam may fail.

Most of the largest dams are built by states, the U.S. Army Corps of Engineers, or the Bureau of Reclamation with funds supplied by the federal government. Most of the money comes from public taxes. For Congress to appropriate funds for a large dam, the corps must justify the cost weighed against perceived benefits. The benefits might include flood control, hydroelectric power generation, water

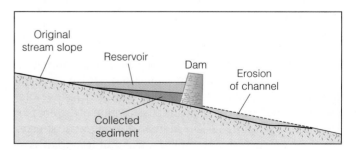

▶**FIGURE 12-34.** Trapping of stream sediment in the reservoir behind a dam causes erosion downstream.

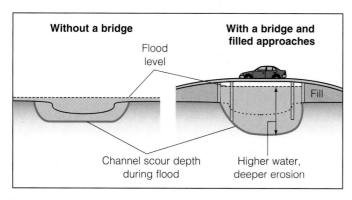

▶ **FIGURE 12-35.** This diagram shows the effect of a bridge on a flood.

Donald Hyndman photo.

▶ **FIGURE 12-36.** Now flooded, these gravel pits along the South Platte River near Denver, Colorado, are only separated from the river by thin gravel barriers.

stored for irrigation, and recreation. Often the cost is justified by adding together all of the perceived benefits even though some are often incompatible with others. Because the public pays for the dam through its taxes, Congress must be convinced of the importance of the benefits.

Bridges

The road or railroad approaches to bridges commonly cross floodplains by raising the roadway above some planned flood level. Most commonly this involves bringing in fill and effectively creating a partial dam across the floodplain so that river flow is restricted to the open channel under the bridge (▶ Figure 12-35). Where roadway approaches restrict flow, floodwater upstream is slowed and becomes deeper. Deeper water flowing under the bridge flows faster, causing erosion of the channel under the bridge. In major floods, the channel deepening may undermine the pilings supporting the bridge, causing the bridge to fail. Where more enlightened planners or engineers design the bridge—or where better funding is available—the approaches may be built on pilings to permit floodwater to flow underneath the roadway. Even where planners are aware of the problem, fill may be used to reduce construction cost.

Mining of Stream Sand and Gravel

Large amounts of sand and gravel are used in construction materials for roads, bridges, and buildings. Much of that sand and gravel is mined from streambeds or floodplains. At first glance, that might seem like a harmless thing to do. After all, won't the stream merely bring in more gravel to replace that which was mined? Actually, removal of sand or gravel from a streambed has the same downstream effect as a dam; supply of sediment in the stream decreases. Because the water flow in the stream is unchanged, the decrease in sediment load leaves the stream downstream from the mining area with excess energy that it uses to erode its channel deeper. Deepening a channel can severely damage roads and bridges. It also typically lowers the water table because more groundwater flows into the deeper stream channel; as

a result, water supplies are damaged. Where gravel is mined from pits on the floodplain, temporary barriers are often used to channel the stream around the pit (▶ Figure 12-36). Later, rising water may erode the barriers. Water entering the pit from upstream slows and deposits sediment in the upper end of the pit, eventually filling it.

Although such mining may appear, therefore, to exploit a renewable resource, the gravel removed from flood flow by deposition is not being carried farther downstream. The increased energy of the stream downstream amplifies the erosion. In one prominent court case, it was shown that mining of gravel from stream bars has resulted in deepening of the channel by as much as 3 meters for many kilometers downstream from Healdsburg, California (▶ Figure 12-37).

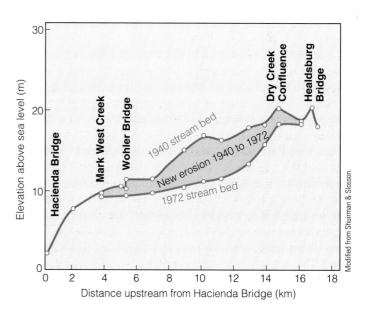

Modified from Shurman & Slosson.

▶ **FIGURE 12-37.** Downcutting of part of the Russian River channel between 1940, before gravel mining began, and 1972 as a result of gravel-mining operations downstream along the Russian River in northern California.

The change threatened several bridges, destroyed fertile vineyard land, and lowered groundwater levels.

Streambed mining can affect nearby and upstream structures such as roads and bridges. As water nears a deepened gravel pit, its velocity increases, in many cases eroding the upstream lip of the pit and washing that gravel down into the pit. That lip migrates upstream. The deepening channel undercuts roads, bridge piers, and other structures, destroying them. The cost to repair or replace such structures often exceeds the value of the mined gravel. (See "Case in Point: Channel Deepening and Groundwater Loss from Gravel Mining.")

EFFECT OF RIVER GRAVEL MINING ON COASTAL EROSION Removal of sediments from a stream often has far-reaching consequences where a river dumps its sediment at the coast. Normally, much of the sediment deposited in the river delta is carried up or down the coast by longshore drift. The sands and gravels that form beaches are supplied by such longshore drift and may gradually move along the coast for hundreds of miles. When the sediment supply is reduced because of gravel mining or an upstream dam, the longshore movement of sediment along the coast continues but is not replenished. Beaches erode and may even disappear. Waves that normally break against the beach then break against the beach-face dunes or sea cliffs, causing severe erosion (See Chapter 13 for a detailed discussion of coastal erosion.)

Paleoflood Analysis

A major problem in estimating the sizes and recurrence intervals of potential future catastrophic floods in North America is that we only have a short record of stream flow data; the measurement of flood magnitudes is not much more than 100 years, even less in the West. The record is better in civilizations such as China and Japan that have written records extending a few thousand years, and to a lesser extent in Europe. We can project graphical data on magnitudes and their recurrence intervals to less-frequent larger events. However, as outlined above, any record-sized event can dramatically change the estimate of recurrence intervals. **Paleoflood analysis** uses the physical evidence of past floods that are preserved in the geological record to reconstruct the approximate magnitude and frequency of major floods in order to extend the record into the past and to recognize larger floods. Even where the paleoflood magnitudes cannot be determined reliably from the evidence, the flood height can often be fairly well determined. By itself, this can provide critical information on the minimum hazard of a past flood.

Early Postflood Evidence

The nature and magnitude of a flood is most obvious immediately after it occurs. Streams in different environments and different climates, however, are highly variable. Most usable

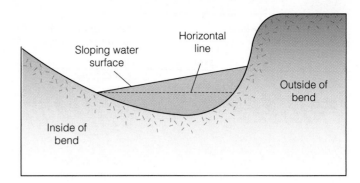

▶**FIGURE 12-38.** This diagram shows water-surface inclination at a bend.

evidence comes from meandering streams, not braided or straight streams. Unfortunately, most studies have been in single regions such as the west-central United States that may not apply well to other regions.

The best reaches of a stream for paleoflood analysis are those with narrow canyons in bedrock, slack water sites, and areas with high concentrations of suspended sediment. Useful features include the following (see also Figure 12-41):

- **high-water marks,** which can provide the elevation and width of the high-water surface;
- **cross-sectional area,** if a cross section is exposed (assuming no post-peak scour or channel fill);
- **mean flood depth** (equal to cross-sectional area divided by water-surface width);
- **estimated water velocity,** which can be calculated (± 50%) from the inclination of the water surface as it flows rapidly around a bend (▶Figure 12-38);
- **mean flow velocity,** which can also be estimated (± 25–100%) from the size of the largest boulders that were moved;
- water **discharge** for high within-bank flows, which can be estimated from cross-sectional flow area and water slope along the channel (Sidebar 12-1);
- **meander wavelength,** which approximately equals 1.63 × meander belt width or 4.53 × meander radius of curvature (▶Figure 12-39);

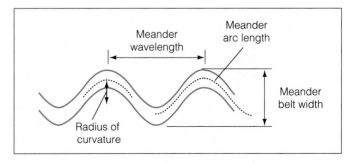

▶**FIGURE 12-39.** This diagram shows the relationship between meander characteristics.

The Russian River, Northern California

The Russian River downstream from Healdsburg, California, north of San Francisco, provides an excellent example of the consequences from river-channel gravel mining. Beginning in 1946, both the region's population and large construction projects increased the need for sand and gravel mined from the Russian River watershed. By 1978, private companies were mining 4.06 million metric tons per year of sand and gravel along the Russian River, mostly downstream from Dry Creek. Of this, 80 percent came from stream terraces and 10 percent by draglines from within the stream channel. Draglines excavated gravel in pits as deep as 18 meters below water level in a stream that is typically less than 1 meter deep.

Recall that the slope of a stream strives to remain in equilibrium with the amount and grain size of supplied sediment, and the erodibility and cross section of the channel (see "Dams and Stream Equilibrium," page 318). If any of these factors change, the others will adjust to help bring the stream back toward equilibrium. In the early 1970s, Byron Olson and other fruit farmers along Dry Creek, upstream from Healds-burg, noticed that riverbanks were eroding, steepening, and collapsing more rapidly in storms than they had in the past. The channels eroded more deeply, and the streams became wider at the expense of adjacent orchards and vineyards.

Local farmers blamed the gravel miners for the increased erosion and filed suit for damages. Detailed stream studies followed that eliminated large storms, fire or flood events, and land use changes as causes for the observed increase in discharge and erosion. The total deepening of the Russian River streambed upstream from its confluence with Dry Creek ranged, by 1972, from 2 meters at the confluence to 5.7 meters some 5 kilometers upstream and 2 meters some 9 kilometers upstream. Dry Creek eroded its bed 3.3 meters deeper for some 10 kilometers upstream from the confluence (▶ Figure 12-37). It also undercut and threatened bridge supports (▶ see Figure 12-40).

The removal of so much gravel left Dry Creek with much less bedload to carry, locally increased the channel gradient into the deep mining pits, and increased turbulence. This resulted in more aggressive downcutting all the way from the mining areas into Dry Creek.

▶**FIGURE 12-40.** Dry Creek bridge supports near Healdsburg, California, have been severely undercut by stream-bed gravel mining downstream.

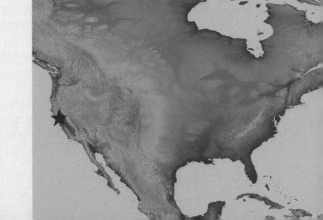

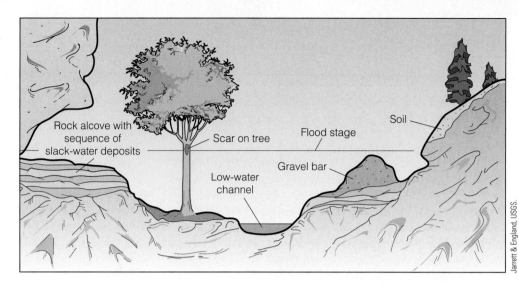

FIGURE 12-41. This idealized sketch shows the paleoflood features that are used to determine flood level.

Rock alcove with sequence of slack-water deposits

Scar on tree

Flood stage

Soil

Low-water channel

Gravel bar

Jarrett & England, USGS.

- **meander belt width,** which approximates 2.88 × meander radius of curvature (▶Figure 12-39); and
- **length of a meander arc,** which is related to the bankfull channel width and depth (▶Figure 12-39).

Note that many numerical relationships hold for interrelated variables across all stream sizes. This emphasizes the need to maintain those relationships if any artificial changes are imposed on a stream. These relative proportions of meander wavelength, radius of curvature, and meander belt width hold regardless of the stream size. It does not matter whether it is a small stream only 2 meters across or the lower Mississippi River 1,000 meters across.

Paleoflood Markers

DRIFTWOOD AND SILT LINES Organic debris including leaves, twigs, logs, and silt carried in floodwaters tends to collect at the edges of the flow, including in back eddies (▶Figure 12-41). These provide perhaps the best evidence for maximum flood height, though they may not be preserved long after the flood and driftwood is likely to be well below the maximum water level. Debris may pile up on bridges, providing a minimum height of the flood (▶Figure 12-42). More commonly, floods will leave behind drift lines that show the high-water mark for a short period of time until it erodes away with the next rainstorm (▶Figure 12-43).

TREE RING DAMAGE Individual trees may preserve effects of damage during a flood, indicating the number of years since the flood (▶Figure 12-44). Scars on a tree trunk or branch remain at their original height during tree growth. The height of the damage generally indicates the minimum stage of a flood, though it could be somewhat above the flood height if vegetation piles up.

The age of trees growing on a new flood-deposited sand or gravel bar indicates the minimum age since the flood that produced the bar. When all of the oldest trees on the

Donald Hyndman photo.

▶FIGURE 12-42. A catastrophic flood on this small stream, Shoal Creek, in Austin, Texas, on November 15, 2001, overtopped the bridge deck and left branches tangled in its railings, evidence of the previous flood height. The high water also deeply eroded the channel and undercut the supports, causing the bridge deck to sag.

Donald Hyndman photo.

▶FIGURE 12-43. This organic drift line marks the high water along Shoal Creek in Austin, Texas, from a flood that followed an intense rain storm that dropped 7.5 centimeters of rain in one and one-half hours.

D. Usher photo, USGS.

▶**FIGURE 12-44.** The number of tree growth rings after the point of damage indicates the number of years since damage occurred.

Donald Hyndman photo.

▶**FIGURE 12-45.** These Lake Missoula slack water sands were deposited over coarse, darker cross-bedded Lake Missoula flood deposits near Starbuck in eastern Washington state.

deposit are of roughly equal age, then that age is probably close to the age of the deposit.

SLACK WATER DEPOSITS During a flood, silt and fine sand can be deposited on sheltered parts of floodplains, the mouths of minor tributaries or shallow caves in canyon walls in bedrock canyons, or downstream from major bedrock obstructions (▶Figure 12-45). These provide information on the maximum water height in a flood.

Organic material in silt and mud layers on floodplains can be dated with radiocarbon methods to indicate the dates of former floods. Bounds can be placed on the heights of former floods even if the specific heights cannot be determined.

Boulders are often deposited where flood velocity decreases, such as where a channel abruptly widens or gradient decreases. These provide a minimum height for a flood.

Review of Factors that Influence Floods

- Floods appear to be more frequent and severe with time, partly because our measurement record is so short. The longer the record, the greater chance of recognizing larger floods. Gauging stations in the United States were not used until 1895. Global climate change is another factor, but human influences such as urbanization are in many cases much more important.

- Floods are generally natural, although they are commonly aggravated by changes imposed by humans. Rivers cover their floodplains during flooding. "Damage" reflects injury to humans and their structures. That damage is not caused by nature but by peoples' choices of where to live. If you build on a floodplain, your house will eventually be flooded.

- Flood damage costs rise year after year because of population growth, more expensive property, and people moving into less suitable areas.

- The probability of a 100-year flood is the same every year, regardless of how many years it has been since the last 100-year flood.

- However, following any new record flood, the 100-year floodplain is likely to get larger; the recurrence interval for the existing mapped floodplain will get shorter. Therefore, the 100-year floodplain area mapped some years ago is *not* correct—it is almost always larger because of human activity. In other words, the correct 100-year average flood level is higher and covers a greater area than the mapped level.

Hydrologic conditions and basin characteristics that can lead to rapid rise and high level of floodwaters include:

- moderate to heavy rainfall over an extended period over a large area (a series of storms or a stalled storm);

- extremely heavy rainfall as a major storm carries heavy moisture onshore because water cannot soak in quickly enough;

- heavy rainfall on frozen or already water-saturated ground;

- rapid snowmelt from prolonged high temperatures or heavy rainfall on a warm snowpack;

- large drainage area above a site;

- low-order streams so that there are few tributaries to spread out the flow;

- steep channel gradients;
- narrow, deep channels and narrow or missing floodplains;
- lack of vegetation as in an arid climate, after an intense fire, or after clear-cut logging with numerous roads;
- thin or fine-grained, nearly impervious soils that minimize the ability of water to soak in;
- increased runoff because of upstream urbanization, including paving, houses, storm drains, and levees or channelization of upstream valleys leading to constricted and rapid flow;
- failure of a human-made dam or a landslide dam; or
- some combination of the above.

Complexity, or "Coincident Criticality" and Floods

With all of the influences that can lead to flooding, it is overlapping, generally unrelated events that often lead to the most extreme floods. Examples include:

- heavy rainfall on already saturated or frozen ground;
- heavy warm rainfall on a deep snowpack;
- rapid melting of snow over frozen ground;
- heavy rainfall filling a reservoir and causing failure of its dam; and
- earthquake-caused failure of an earth-fill dam, leading to catastrophic flooding downstream.

KEY POINTS

✓ Most stream levees are built on top of the natural levees, adjacent to the stream channel from fine-grained sediments dredged from the stream channel. Because levees confine the stream to the main channel rather than permitting floodwaters to spread over the floodplain, the levees dramatically raise water levels during a flood. **Review pp. 297–298; Figures 12-1 and 12-10.**

✓ Although people feel safe behind a levee, levees fail by overtopping or breaching, bank erosion, slumps, piping, or seepage through old river gravels below the levee. **Review pp. 298–299; Figures 12-2 and 12-4.**

✓ The Mississippi River in 1993 remained above flood level for more than one month—in some places, for two months. Approximately two-thirds of the levees on the upper Mississippi River were damaged, and many were breached. **Review p. 299.**

✓ Mississippi River flood levels, for a constant discharge, have become higher each time levees were constructed or raised. **Review pp. 300–301.**

✓ Some Mississippi River levees were deliberately breached to decrease flood levels and protect critical areas downstream. In one case, the breach permitted rising water in the "lake" outside the levee to flow back into the river. **Review pp. 303–304; Figure 12-13 and 12-14.**

✓ Many people living on floodplains are eligible for national flood insurance but are not aware that they live on a floodplain; nor are they aware that they may be flooded even though there is a levee between them and the river. **Review p. 304.**

✓ Avulsion occurs when a breach flow does not return to the river but follows a new path. The consequences for some river-dominated cities such as New Orleans would be catastrophic. The Yellow River of China is an excellent example of repeated avulsion over more than 2,000 years and its devastating consequences. **Review pp. 304–309; Figures 12-17 and 12-22.**

✓ Rivers are often channelized to protect adjacent towns, pass floods through more quickly, and increase water depth for shipping. **Review pp. 305, 310–311; Figure 12-16.**

✓ Multipurpose dams are built on rivers to provide electric power, flood control, water for irrigation, and recreation. Unfortunately, the demand to keep reservoir levels low enough for flood control often does not coincide with demands to keep levels high enough for electric power generation, irrigation, and recreation. **Review pp. 311–313.**

✓ Floods caused by the failure of human-made dams are worst in steep, narrow valleys. Some fail during floods, others because of seepage and erosion under the dam, some by poor design and construction, and others by major landslides into the reservoir upstream. **Review pp. 311 and 314; Figure 12-29.**

✓ Urbanization aggravates the possibility of flash floods because it hastens surface runoff to streams. Cars driven into a flooded roadway with water above their floorboards are often pushed off the road, which can cause their occupants to drown. Most vehicles will float until they fill with water and sink. **Review pp. 314 and 316; Figure 12-31.**

✓ A dam on a stream or mining of stream gravel removes sediment so that the excess stream energy causes erosion downstream. **Review pp. 318–320; Figures 3-34 and 3-37.**

✓ Paleoflood analysis, the study of the magnitude and timing of past floods, includes indications of high-water marks, cross-sectional area, meander wavelength, among other factors. **Review pp. 320–323; Figures 12-38 to 12-45.**

IMPORTANT WORDS AND CONCEPTS

Terms

avulsion, p. 305
breaching, p. 298
channelization, p. 305
deliberate breach, p. 312
evapotranspiration, p. 317
levee, p. 297
multipurpose dams, p. 311
National Flood Insurance Program, p. 304

natural levee, p. 298
paleoflood analysis, p. 320
piping, p. 298
recurrence interval, p. 301
sand boil, p. 299
streambed mining, p. 320
urbanization, p. 314
wing dams, p. 301

QUESTIONS FOR REVIEW

1. Roughly what depth of flowing stream is dangerous to drive through?

2. What non-natural changes imposed on a stream cause more flooding and more erosion?

3. Aside from protecting the adjacent stream bank, what effects do levees have on a stream?

4. What are the negative effects of mining sand or gravel from a streambed?

5. Aside from storing water for irrigation or water supply, flood control, hydroelectric power, and irrigation, what negative physical effects do dams have?

6. What negative physical effect do most bridges have on the streams they cross?

7. What process can lead to the failure of a river levee?

8. What process can lead to flooding of the floodplain behind a levee (of a flooding river) if the levee does not fail?

9. What is a common sign of seepage under a levee?

10. Under what circumstance (or for what purpose) might a levee be deliberately breached?

11. What catastrophic problems may arise with a multi-purpose dam that a single-purpose dam should not have? Why?

12. How would a hydrograph for a drainage basin change if major urban growth were to occur upstream? Be specific.

13. What specific evidence can be used to estimate the maximum water velocity in a prehistoric flood?

FURTHER READING

Assess your understanding of this chapter's topics with additional quizzing and conceptual-based problems at:

 http://earthscience.brookscole.com/hyndman.

CHAPTER 13

Waves breaking against this rocky coast along the south coast of France maintain its steepness by intermittent cliff collapse.

David Hyndman photo.

WAVES, BEACHES, AND COASTAL EROSION
Rivers of Sand

Living on Dangerous Coasts

Hurricanes and other major storms affect beaches and the people who live on them. To understand coastal hazards, we need to understand wave processes and the formation of beaches and sea cliffs. We also need to understand how human activities affect wave action, beach response, and sea-cliff collapse. As populations grow in numbers and affluence, more people move to the coasts, not only to live but also for recreation. But beaches and sea cliffs constantly change with the seasons and progressively with time. When people build permanent structures at the beach, coastal processes do not stop; the processes interact with and are affected by those new structures. Storms, hurricanes, and their dramatic aftermath are often viewed as abnormal or "nature on a rampage." In fact, they are normal for a constantly evolving landscape. What is abnormal is how human actions and structures cause natural processes to impose unwanted damages.

Waves

Winds blowing across the sea push the water surface into waves because of friction between the air and the water. Gentle winds form small ripples. As the wind speed increases, ripples grow into waves. Two other factors that increase wave height are **fetch,** which is the length of water surface over which the wind blows, and the *amount of time* the wind blows across the water surface. Ocean waves are generally much larger than those on small lakes, and prolonged storms often build large damaging waves (▶ Figure 13-1).

The timing and size of waves approaching a shoreline varies by the location and size of offshore storms. Waves move out from major storm centers, becoming broad, rolling swells with large **wavelengths** (▶ Figure 13-2). A constant mild onshore wind will produce smaller and shorter-wavelength waves.

As waves approach the beach, water in a wave itself does not move onshore with the wave; it merely moves in a circular motion within the wave, otherwise staying in place (▶ Figure 13-3). You can see that motion by watching a stick or seagull floating on the water surface. It moves up and down, back and forth, not approaching the shore unless blown by the wind or caught in shallow water where the waves break. Waves in this circular motion are not damaging because the mass of water is not moving forward. Watch waves moving past the pilings of a pier or against any kind of vertical wall in deep water. The waves have little forward momentum and do not splash against the vertical surface; they merely ride up and down against it. A steep "wall" of coral reef facing offshore from some tropical islands has a similar effect, thereby helping protect such islands from the impact of storm waves. When waves approach shore they begin to "feel bottom," and thus gain forward motion,

NOAA.

▶ **FIGURE 13-1.** In addition to shore damage, huge storm waves can topple boats.

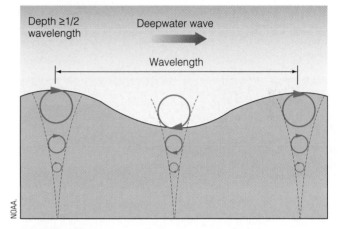

NOAA.

▶ **FIGURE 13-2.** Individual water particles rotate in a circular motion but do not travel in the direction of wave travel. The water motion fades out downward until a depth of approximately half a wavelength.

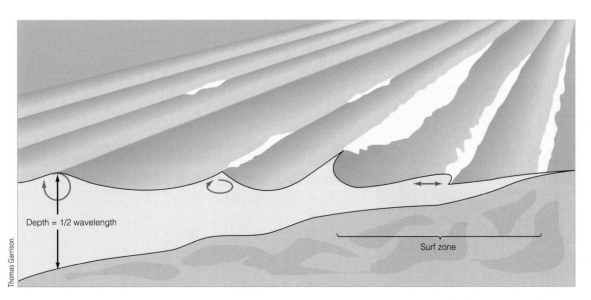

Thomas Garrison.

▶ **FIGURE 13-3.** Characteristics of waves change on approaching a shore.

which causes the waves to break. These are conditions under which waves gain the potential for serious damage.

Waves begin to feel the bottom when the water depth is less than approximately half the wavelength. Because the size of the circular motion is controlled by wave size, the depth at which wave action fades out is controlled by wavelength. In shallower water, the crest of the wave moves forward as the base drags on the bottom. The waves slow in shallow water but rise in height (Sidebars 13-1 and 13-2). The momentum of the upper mass of water carries it forward to erode the coast (▶Figures 13-3 and 13-4).

Big waves are more energetic and cause more erosion (▶Figures 13-4 and 13-5). **Wave energy** is proportional to the mass of moving water. This can be approximated by multiplying the density of water by the volume of water in a wave, which is approximately the wave height (H) squared times wavelength (L). Because the height term is squared, waves that are twice as high have four times the energy; those that are four times as high have sixteen times the energy (Sidebar 13-3).

▶**FIGURE 13-4.** As waves approach the shore, they drag on bottom and their crests lean forward to break as seen here along the southern Oregon coast. The water is brown from stirred sand.

▶**FIGURE 13-5.** Storm waves pound the seawall at Galveston, Texas. The seawall is in the lower left, under the breaking wave.

Sidebar 13-1

Wave velocity in deep water is proportional to the square root of wavelength:

$$V = \sqrt{\frac{gL}{2\pi}} = 1.25\sqrt{L}$$

where

  V = velocity (in m/sec.)

  L = wavelength (in m)

  g = acceleration of gravity (9.8 m/sec²)

  π = 3.1416

Thus, a group of waves with a wavelength of 5 meters moves at 1.25 × 2.24 = 2.8 m/sec. or 10.1 km/hr. Waves with a wavelength of 100 meters move at 1.25 × 10 = 12.5 m/sec. or 45 km/hr. We can also show this relationship graphically, as shown below. Note that if we measure the period (the time between wave crests), we can easily determine both the wavelength and wave velocity using the graph below.

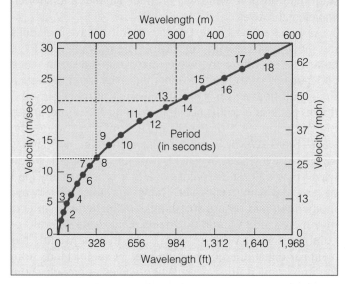

Sidebar 13-2

In shallow water, wave velocity is proportional to square root of water depth:

$$V = \sqrt{gD} = 3.1\sqrt{D}$$

where

  V = velocity

  D = water depth (m)

Sidebar 13-3

Doubling the wave height quadruples the energy:

$$(E_w) = 0.125\sigma gH^2L$$

where

  E_w = energy of the wave

  σ = water density (close to 1)

  g = acceleration of gravity = 9.8 m/sec.²

  H = wave height

  L = wavelength (in m)

FIGURE 13-6. This cliff at Pismo Beach, California, has been undercut by waves. Large chunks of rock break off and are pounded into sand by the waves.

FIGURE 13-7. A sandy summer beach covers the lower part of a bouldery upper beach left by winter waves north of Newport, Oregon.

Beaches

Beaches are accumulations of sand or gravel supplied by sea cliff erosion and by river transport of sediments to the coast. The size and number of particles provided by sea-cliff erosion depends on the energy of wave attack, the resistance to erosion of the material making up the cliff, and the particle size into which it breaks. Waves often undercut a cliff that collapses into the surf, then break it into smaller particles (▶ Figure 13-6). The size and amount of material supplied by rivers depends similarly on the rate of river flow and the particle size supplied to its channel.

Most of the sand and gravel supplied to the coast is pushed up onto the beach in breaking waves; it then slides back into the surf in the backwash. Big waves during winter storms carry sand offshore into deeper water, leaving only the larger gravels and boulders that they cannot move. The gentle breezes and smaller waves of summer slowly move the sand back onto the beach (▶ Figures 13-7 and 13-8). As a result, some cliff-bound beaches show sand near the water's edge with gravel or boulders upslope. In such areas, the beaches are commonly sand in summer but more steeply sloping gravel or boulders in winter. The active beach extends from the high-water mark to some 10 meters below sea level.

Wave Refraction and Longshore Drift

Waves often approach shore at an angle. The part of each wave in shallower water near shore begins to drag on the bottom first and thus slows down; the part of the wave that is still in deeper water moves faster, so the crest of the wave curves around toward the shore (▶ Figure 13-9). This is called **wave refraction** because waves bend or refract toward shore.

(a)

(b)

FIGURE 13-8. This beach north of Point Reyes, California, was eroded by waves during the 1997 El Niño event; in **(a)** much of the sand has been removed (October 1997), but in **(b)** it has been naturally rebuilt by April 1998.

▶**FIGURE 13-9.** Waves curve by refraction in the shallow water near shore in a New Zealand bay.

When wave crests approach a beach at an angle, the breaking wave pushes the sand grains up the beach slope at an angle to the shore. As the wave then drains back into the sea, the water moves directly down the beach slope perpendicular to the water's edge.

Thus, grains of sand follow a looping path up the beach and back toward the sea. With each looping motion, each sand grain moves a little farther along the shore (▶Figure 13-10a). The angled waves thus create a **longshore drift** that essentially pushes a river of sand along the shore near the beach. Over the period of a year or so, with high and low tides, large and small waves, and storms, most of the sand on the beach moves farther along shore.

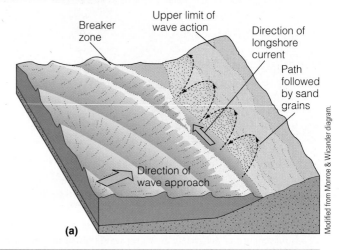

Breaker zone • Upper limit of wave action • Direction of longshore current • Path followed by sand grains • Direction of wave approach

(a)

▶**FIGURE 13-10. (a)** Sand grains are pushed up onto the beach in the direction of wave travel. Gravity pulls them back directly down the slope of the beach. The combination is a loop that moves each sand grain along the shore with each incoming wave. **(b)** Wave crests bend to conform to the shape of the shoreline; wave directions bend to attack the shoreline more directly. Thus, rocky headlands are vigorously eroded and bays collect the products of that erosion.

Wave crests • Wave energy converging on headlands • Wave energy diverging • Quiet beach • Sediment movement • Quiet beach

(b)

Longshore drift also moves sand from headlands to bays (▶Figure 13-10b).

Along both the west and east coasts of the United States, longshore drift is dominantly toward the south. Longshore drift in parts of coastal California averages a phenomenal 750,000 cubic meters of sand past a given point per year, more than 20,000 cubic meters per day. If a large dump truck carries 10 cubic meters, that would be equivalent to 2,000 dump truck loads per day. Along the East Coast, it is much less but still averages 75,000 cubic meters per year, more than 2,000 cubic meters per day.

Waves on Rocky Coasts

Waves approaching a steep coast such as those along much of the Pacific coast of North America or the coast of New England or eastern Canada encounter rocky points called **headlands** that reach into deeper water with shallower sandy bays in between. Waves bend or refract toward the rocky points, causing the energy of the waves to break against the headlands (▶Figure 13-10b). Thus, wave refraction dissipates much of the wave energy that would otherwise have impacted sandy bays. Sand blasted off the rocky point migrates along the shore to be dumped along the beach in the bay because the currents on both sides carry sand to the center of the bay (▶Figures 13-10b and 13-11).

Beach Slope: An Equilibrium Profile

As waves move into shallow water and begin to touch bottom, they move sediment on the bottom, stirring it into motion and moving it toward the shore. Most sediment moves at water depths of less than 10 meters. Long wavelength storm waves, however, with periods of up to twenty seconds, reach deeper; they touch bottom and move sediments at depths as great as 300 meters on the continental shelf.

Whether the sediment moves shoreward or not depends on the balance between shoreward bottom drag by the waves, size of bottom grains, and downslope pull by gravity. Thus, the slope of the bottom is controlled by the energy required to move the grains, which is related to the water depth, wave height, and grain size. Shallower water, smaller waves, and coarser grains promote steeper slopes offshore, just as in rivers. In the breaker zone offshore, waves can easily move the sand and the beach surface is gently sloping. As breakers sweep up onto the shore, the water is shallower, their available energy decreases, and the **shore profile** steepens. Sand there can move back and forth only on such a steeper slope (▶Figure 13-12).

Grain size strongly controls the slope of the beach. Just as in a stream, fine sand can be moved on a gentle slope, coarse sand or pebbles only on a much steeper slope (▶Figures 13-7 and 13-13).

On shallow gently sloping coastlines, such as those in much of the southeastern United States, the beach both onshore and offshore becomes steeper landward (the onshore

David Hyndman photo.

▶**FIGURE 13-11.** Wave refraction has eroded the former series of headlands along Drakes Bay, Point Reyes National Seashore, California, into a straighter coastline.

portion visible in Figure 13-12). This is because the waves use energy stirring sand on the sea bottom so that they slow as they ride up onto the beach. Most of the wave energy is used in waves breaking and moving water and sand upslope. Water flowing back off the beach carries sand with it. The active beach slope is controlled by the grain size being moved and the amount of water carrying the grains. To reiterate, larger grains or less water need a steeper slope to move the grains.

At high tide, the waves reach higher on the beach, so the change in slope continues landward. At low tide, winds pick up the drying sand and blow it landward to form **dunes.** The combination of an upward-curving shoreline, steeper than the average coastal slope, with dunes grown above sea level, builds an offshore barrier island or barrier sandbar.

Donald Hyndman photo.

▶**FIGURE 13-12.** The beach slope steepens shoreward where the breaking waves reach their upper limit. The result is a ridge or berm as in this case at Positano, Italy. The second berm to the left formed earlier by larger waves.

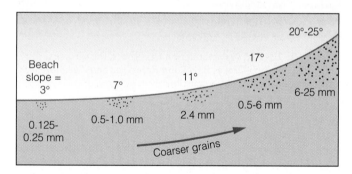

▶**FIGURE 13-13.** This graph shows the general relationship between grain size and beach slope.

Farther shoreward, the area below sea level is a coastal lagoon (see also Figure 13-3).

Rip Currents

Some high-energy beaches show a prominent scalloped shoreline with cusps 5 to 10 meters apart. These erode as **rip currents** carry streams of water back offshore through the surf after groups of strong incoming waves pile too much water onto the beach. These streams of muddy-looking water can be dangerous to swimmers not familiar with them because the currents are too fast to permit swimming directly back to shore. You can escape a rip current by swimming parallel to shore and then back to the beach. Rip currents are sometimes called undertows, but this is a misnomer because these currents would not drag someone under the surface. The danger comes from a swimmer becoming overtired if they try to swim directly toward shore against the current, possibly ending in drowning.

Loss of Sand from the Beach

Sand in the surf zone moves with the waves; however, where it goes changes with the tides and with wave heights. Larger waves, especially those during a high tide or a major storm, erode sand from the shallow portion of the beach and transport much of it just offshore into less-stirred deeper water. Much of the eroded material comes from the surface of the beach, eroding it to a flatter profile. More comes from the seaward face of dunes at the head of the beach where waves either break directly against the dunes or undercut their face.

Storms bring not only higher waves but also a local rise in sea level—called a **storm surge.** High atmospheric pressure on the sea surface during clear weather holds the surface down, but low atmospheric pressure in the eye of a major storm permits it to rise by as much as a meter or more. The stronger winds of a storm also push the water ahead of the storm into a broad mound several kilometers across. The giant waves of a hurricane and the higher water level of storm surges take these effects to the extreme. They cause massive erosion and decrease the beach slope on barrier islands (see Chapter 14 for further discussion). On gently sloping coasts, they may level the dunes.

As waves break onto a beach, some of the water soaks into the sand and is thus not available to carry sand offshore. This effect is most significant with smaller waves, which have a larger proportion of their water and therefore wave energy soaking into the beach than large storm waves that still have most of their water available to flow offshore and carry sand with them. Small waves thus tend to leave more of their sand onshore. During low tide, winds pick up drying sand on the beach and blow it landward into dunes. The next strong storm may erode both the beach and the dune face and carry the sand back offshore.

People Move to the Beach

People have always lived along the shores of inlets and bays and fished in nearby streams and lagoons, but their structures along the open coast were temporary shelters that could be moved, or low-value ramshackle summer cabins that could be replaced after storms. Coastal tourism was not

Beach Erosion and Hardening

Reduction of Sand Supply

Anything that hinders supply of the sand to the beach from "upstream" on the coast or removes sand from the moving longshore current along the beach results in less sand to an area of coast and erosion of the beach. Dams on rivers trap sand, keeping it from reaching the coast, and mining sand from river channels or from beaches for construction has a similar effect. In many industrial countries, major dam-building periods on rivers began in the 1940s to generate electricity, store water for irrigation, and provide flood control. That and better land use practices caused dramatic reductions in the amount of sediment carried by rivers and supplied to beaches. Beach erosion again accelerated. Shoreline recession of 5 to 10 meters per year was common but was as much as 200 meters per year—for example, at the mouth of the Nile River. The resulting erosion of beaches and coastal cliffs is clear in California.

People with beachfront or cliff-top homes are commonly affected by storms that cause beach erosion or threaten the destruction of their property. Until something bad happens to their beach, beach cliff, or home, however, many do not really think about the constant motion of sand along the beach from waves and offshore movement in storms. Unless tragedy strikes near home, they do not realize that soft sediment beach cliffs gradually retreat landward as they erode at their base and collapse or slide into the ocean.

Artificial Barriers to Wave Action

The distribution of homes and accommodations in North America, where private cars dominate, differs from that in Europe where mass transit is more common. In North America, the main access road leading to the coast feeds a shore road behind beachfront residences and other accommodations. In Europe, the main beach access leads to a road along the beachfront itself. In both cases, a promenade or boardwalk may front the beach near the main access road and local business district. Seawalls and other supposed beach protection or **hardening** tends to front the boardwalk and spread out along the beach from there.

So what do beachfront dwellers do to protect their property from coastal erosion? Before large-scale tourism, coastal residences and even small communities fell to the waves or moved inland as the beach gradually migrated landward. Shore-protection projects followed the catastrophic Galveston, Texas, hurricane of 1900. These included building the massive **seawalls** (▶ Figures 13-15 and 13-16). Individuals during this period reacted to try to stop the erosion threat with boulder piles or walls built of either timber or concrete at the back of the beach and in front of their property. The thought was that the waves will beat against the boulders or walls rather than eroding their property. Construction of seawalls accelerated until the 1960s, when scientists and

▶**FIGURE 13-14.** These huge interlocking concrete pieces are designed to reduce wave energy and protect the harbor at Nazare, Portugal.

David Hyndman photo.

important because of the difficulty of access through local brush and the incidence of malaria. By the 1700s, people began building more costly and more permanent structures and even complete towns on the protected landward side of some barrier islands. The old-timers understood beach processes and built homes on stilts on the bay side of the bar with temporary structures at the beach.

By the 1850s, reduction in working hours, formation of an urban middle class with money, and expanded transportation via railroads and steamship service changed the ground rules. Coastal tourism and resorts expanded, especially after the late 1940s. As populations and affluence grew, people installed utilities, paved roads, and built bridges to the islands, along with more expensive permanent homes, hotels, and resorts along the same beaches. More recently, second homes for summer use have become popular, some used for only a few weeks per year. Others have become year-round dwellings for urban retirees.

When hurricanes and other storms damaged these structures, people placed protective **riprap** (▶ Figure 13-14) and seawalls and demanded that local governments protect them from the "ravages of the sea." Instead of living with the sea and its changing beach, they tried to hold back the sea and prevent natural changes to the beach.

For awhile, sediment supply to some beaches increased and erosion decreased. The advent of steam locomotion in the early 1800s, followed by railroads and a large increase in population in the continental interior, also led to deforestation, land cultivation, and overgrazing on a large scale. Invention of the internal combustion engine continued the trend. This removal of protective cover from the land led to heavy surface erosion and large volumes of sediment delivered to the coasts. Along the steep Pacific coast, longshore drift of these sediments caused widespread enlargement of beaches.

FIGURE 13-15. The massive seawall at Galveston, Texas, was built after the catastrophic hurricane of 1900. Beach sand eroded by waves at the base of the seawall is continually replaced by truckloads of sand brought in from coastal dunes.

Donald Hyndman photo.

FIGURE 13-16. The Nor'easter of December 29, 1994, undermined this vertical steel wall at Sandbridge Beach, City of Virginia Beach, Virginia, and toppled into the surf.

Carl Hobbs photo, Virginia Institute of Marine Science.

governments began to recognize that much of these activities have long-term disadvantages.

Waves breaking against coastal cliffs reflect back offshore, carrying smaller particles farther offshore. The same is true when waves break against seawalls or piles of riprap

boulders used to protect the area in front of houses. The effect is often just the opposite from what people intended. Although the structure may slow the direct wave erosion of a beach cliff for awhile, the waves reflect back off the barrier, stir sand to deeper levels, and carry the adjacent beach sand farther offshore. On a gently sloping beach, a wave sweeps up onto the beach and the return swash moves back on the same gentle slope. A wave striking a seawall, however, is forced abruptly upward so the swash comes down much more steeply and with greater force, eroding the sand in front of the seawall. The beach narrows and becomes steeper, the water in front of the barrier deepens, and the waves reach closer to shore before they break. Thus, instead of a protected beach, bigger waves approach closer to shore. The result often hastens erosion and removal of the beach. When the water in front of the barrier becomes sufficiently deep, the beach is totally removed, and the waves may undermine the barrier, which then topples into the surf (▶Figures 13-16 and 13-17). If the beach lies at the base of a sea cliff, the supposedly protected cliff succumbs more rapidly as well.

For those who understand that sand grains on a beach tend to migrate along the coast, another approach has been to try to keep the sand from migrating. Groins and jetties are built out into the surf or offshore. **Groins,** the barriers built out into the surf to trap sand from migrating down the beach, do a good job of that. They collect sand on their upstream side (▶Figures 13-18 and 13-19). Unfortunately, that reduces the sand that continues along the beach, which causes beach erosion on the downstream side of the groin. Effectively, they displace the site of erosion to adjacent areas downstream.

Riprap walls or **jetties** are sometimes used to maintain navigation channels for boat access into bays, lagoons, and marinas. Jetties that border such channels extend out through the beach but typically require intermittent dredging to keep the channel open (▶Figures 13-20 and 13-21).

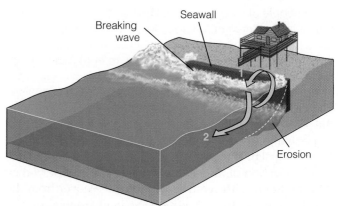

FIGURE 13-17. This diagram shows how waves break against a sea wall, causing erosion that may result in the collapse of the seawall as occurred in Virginia Beach.

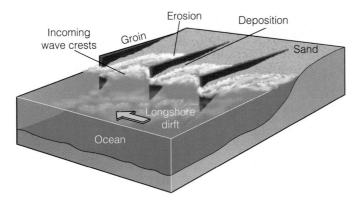

▶**FIGURE 13-18.** This diagram shows the effects on beach sand as groins stop longshore drift.

▶**FIGURE 13-19.** Groins at Cape May, New Jersey, also trap sand. Longshore drift is from the lower right to left.

They also block sand migration along the beach. **Break-waters,** built offshore and parallel to the shore, have a similar effect, causing deposition in the protected area behind the barrier and erosion on the downstream or "downdrift" sides (▶Figure 13-22). Both groins and breakwaters trap sand moving along the beach and cause erosion farther down the beach. Without continued supply, those beaches are starved for sand and thus erode away. So, once someone builds a groin or jetty, people downstream see more erosion of their beaches and are inclined to build groins to protect their sections (see Figure 3-18). New Jersey shows these effects to the extreme. Except where replenished, its once sandy beaches are now narrow or nonexistent and lined

(a)

(b)

▶**FIGURE 13-20.** **(a)** Jetties bordering an estuary block southward drift of beach sand. That starves the beach to the south, leading to its erosion. Note that the beach south of the estuary is much recessed from the straight coastline. Lighthouse Point, Pompano Beach, Florida. **(b)** Dredging of sand from a river outlet between jetties in Pompano Beach, Florida. The dredge is the yellow boat in the middle left. Sand from this dredging is piped to the downstream side of the jetties to replenish the beach seen in the foreground.

▶**FIGURE 13-21.** Longshore drift is interrupted at jetties just as it is at groins. The beach on the right side of this jetty north of San Diego has been completely eroded by longshore drift of sand from left to right.

with groins and seawalls (▶Figure 13-23). The impacts of inlets are dramatic at Ocean City, Maryland, and St. Lucie in Martin County, Florida.

Most knowledgeable people view groins as inherently bad news, and in many cases they are. The state of Florida now requires removal of groins that adversely impact beaches or are simply nonfunctioning. Groins should not stop sand migration permanently—this should only be a remedy until sand is fully deposited on the upstream side. The effects, of course, remain.

However, in some cases groins can work where the drift of sand farther down the coast is undesirable. For example, to keep an inlet open, dredges must remove sand moving down the coast into an inlet. Sand drifting into a submarine valley is generally carried far offshore and lost perma-

nently from the coast (▶Figure 13-24). In such cases, well-engineered groins can provide a useful purpose. They take a wide variety of forms and sizes that depend on the specific purpose. Some even include artificial headlands.

It might seem that continual erosion of sea cliffs and erosion by rivers would add more and more sand to beaches, making them progressively larger. This does happen in some areas. Beaches are actually gaining sediment near the mouth of the Columbia River between Washington and Oregon, between Los Angeles and San Diego, along much of Georgia and North Carolina, and along scattered patches in Florida and elsewhere on the Gulf of Mexico coast.

However, many coastlines are continuing to erode. What happens to the sand? Nature gets rid of some of it. Some is

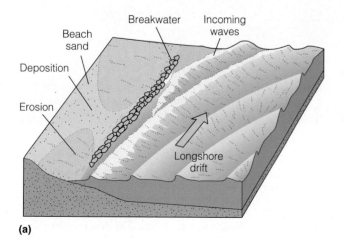

(a)

(b)

▶**FIGURE 13-22.** **(a)** This diagram shows the effects of a breakwater on beach sand. **(b)** Note the deposition of sand behind these breakwaters along the Mediterranean coast of Israel.

FIGURE 13-23. Houses crowd a narrow barrier bar at Manasquan, New Jersey. A groin in the foreground traps sand moving from left to right. The area downstream to the right, having lost its entire beach, is temporarily protected by a rock seawall.

blown inland to form sand dunes, especially in areas without coastal cliffs. Other processes can permanently remove sand from the system. Some is beaten down to finer grains in the surf and then washed out to deeper water. Rip currents form when waves carry more water onshore than returns in the swash. That current flows back offshore in an intermittent stream that carries some of the beach sand back into deeper water. Huge storms such as hurricanes carry large amounts of sand far offshore.

Some sand drifts into inlets that cross barrier islands, where dredges remove it to keep the inlets open for boat traffic. Some migrates along the coast for a few hundred kilometers until it encounters the deeper water of a **submarine canyon** that extends offshore from an onshore valley. One prominent example of a submarine canyon extends offshore from the Monterey area of central California. These valleys extend across the continental shelf to where the sediment intermittently slides down the continental slope as turbidity flows onto the deep ocean floor. Thus, much of the longshore drifting sand of beaches is permanently lost to the beach environment (Figure 13-24).

Areas of Severe Erosion

The main factors in cliff erosion are wave height, sea level, and precipitation. All three factors are heavily influenced by intermittent events such as El Niño and hurricanes. When storms come in from the ocean, their frequency and magnitude strongly affect the rate of erosion. Along the coast of Southern California, erosion is amplified during El Niño events, such as those of 1982–83 and 1997–98 (Figure 13-25). Farther north along the coasts of Washington and Oregon, storms are more common during years when El Niño effects are weakest. Much of the West Coast would be eroding back more rapidly except for the presence of beach cliffs almost everywhere. Where these beach cliffs consist of soft Tertiary-age sediments, waves undermine them to cause landslides and collapse. Disintegration of the collapsed material supplies sand to beaches down the coast. That part is good for the beach, but if your home is perched at the top of that cliff where there is a magnificent view of the ocean, it may not be so good. As the cliff erodes, your house gets closer to the edge and the view gets better, but your house eventually collapses with the cliff.

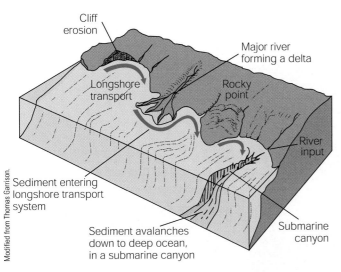

FIGURE 13-24. The longshore drift of beach sediments often leads to loss of the sediment in a submarine canyon.

FIGURE 13-25. Large storm waves pound the shoreline in Eureka, California.

FIGURE 13-26. Storm waves undercut this East Coast parking lot, causing it to collapse into the ocean.

Along the Gulf Coast and southeast coast of the United States, storms and heavy erosion are most likely to occur in the hurricane season of late summer to early fall (▶Figure 13-26). The most severe area of coastal erosion in United States, with more than 5 meters of loss per year, is in the area around the Mississippi delta, where dams upstream trap sediment and the compaction of delta sediments lowers their level. One to five meters are lost annually along much of the coasts of Massachusetts to Virginia, South Carolina, and scattered patches elsewhere (▶Figure 13-26). **Nor'easters,** the heavy winter storms that hit the northeastern coast of the United States with similar ferocity, do similar damage.

As mentioned earlier, one of the most extreme cases of **beach hardening** and severe erosion is along the coast of New Jersey (▶Figure 13-27). Early development was promoted by extending a railroad line along more than half of the New Jersey shoreline by the mid-1880s. Buildings first clustered around railroad stations, dunes were flattened for construction, natural dense scrub vegetation was removed, and marsh areas were filled. The arrival of private cars accelerated development. Marsh areas along the back edge of the offshore bars were both filled and dredged in the earliest 1900s to provide boat channels. Inlets across barrier bars were artificially closed, and jetties built after 1911 stabilized others. People did not understand then that barrier bars were part of the constantly evolving beaches, that nature would resist human attempts to control it.

The large population nearby, and the demand for recreation, stimulated the building of high-rise hotels and condominiums in Atlantic City; it is now so packed with large buildings next to the beach that it strongly affects wind flow and the transport of sand. Large-scale replenishment of sand on beaches is widespread in some areas. In other areas, narrow and low artificial dunes are built in front of separate detached houses, as are barriers to provide backup protection from erosion. Groins are numerous. A prominent seawall and continuous groins leave no resemblance either to the original barrier island environment or the beaches that attracted people in the first place.

Beach Replenishment

Are there better ways to protect a beach? Beginning in the 1950s, replacing sand on beaches became popular in the United States. To replace sand on a severely eroded beach, individuals sometimes contract to have sand brought in dump-truck loads (▶Figure 13-28). A simple calculation shows that this is generally a losing proposition.

A large dump truck carries roughly 10 cubic meters of sand. If a person's lot is 30 meters wide and the beach extends 65 meters from the house to the water's edge at midtide (roughly 2,000 square meters), one dump-truck load would cover that part of the beach to a depth of only

FIGURE 13-27. This is the sad ending for a mishandled beach at Cape May, New Jersey. The building of jetties at Cape May Inlet just up the coast trapped the sand that previously replenished the beach. A concrete and rock seawall now lines the "beach."

▶**FIGURE 13-28.** A truck dumps rusty, fine-grained sand excavated from the lagoon side of a barrier bar at Holden Beach, North Carolina, following a small storm in March 2001. Such sand is more easily eroded by smaller waves than the natural sand on the beach.

half a centimeter or so; it would take 200 loads to add about a meter of sand to the beach in front of one house. Thus, if sand costs $30 per cubic meter, it might take 200 truckloads in front of every home to replace the sand removed in one moderate storm—for a cost of roughly $60,000 for each home. Another problem is that the active beach actually extends well offshore into shallow water. If that part of the beach is not also raised, waves will move much of the onshore sand offshore to even out the overall slope of the beach.

A moderate storm the next month or next year could remove the added sand. How many moderate storms would it take for that type of beach replenishment to reach the value of the house?

And where would the contractor get all that sand? For obvious reasons, mining sand from other beaches is generally not permitted. In some areas, significant sand is obtained from maintenance dredging of sand from navigation channels (▶ Figure 13-20b). Mining sand from privately owned sand dunes well back from a beach or dredging sand from a lagoon or other site behind a barrier island is sometimes possible, though expensive. A significant drawback to such sources is that such sand in dunes and lagoons is generally more distinctly fine-grained than that eroded from the beach. Because the storm waves were able to move the coarser-grained sand from the beach, slightly finer-grained sand can be removed by even smaller waves.

Another solution used in many areas, especially along the southeastern coasts, is to dredge sand from well offshore and spread it on the beach. Because much larger equipment is required, regional or federal governments, often under the direction of the U.S. Army Corps of Engineers, normally undertake such projects. If the sand were taken from near shore, that would deepen water near shore and

make the beach steeper. Because waves shift sand into an equilibrium slope, that depends especially on the size of the sand grains and the size of the waves. An artificially steeper beach will be eroded down to the equilibrium slope. If the sand is taken from well offshore, below wave base, such so-called **beach replenishment** or **beach nourishment** can work better—at least until the next major storm. The bigger waves of storms erode the beach down to a lower slope farther offshore.

Because many people do not readily understand the processes involved in sand movement in the beach environment, there is no easy solution to where to place the sand during replenishment projects. Most people want to see the sand they pay for placed on the upper dry part of the beach rather than offshore. It is also easier to calculate the volume of sand added to the upper beach and therefore the appropriate cost. Because that location steepens the beach, the next storm will carry much of the new sand offshore, where people view it as lost to the beach and think the replenishment was a waste of money. If half of the sand added is offshore in shallow water where it provides a more natural equilibrium profile, people have a hard time understanding that that sand is not being wasted. Either scenario leads to criticism of beach replenishment as a viable solution to beach erosion. As a result of such problems, some beach experts suggest using the expression "shore nourishment" rather than "beach nourishment" to emphasize that the shallow, underwater part of the beach is equally important. Because subsequent storms remove sand from the beach, part of the cost of a sand renourishment project involves continued small-scale nourishments at intervals of two to six years. Small trucking operations may need to be repeated every year, especially if the added sand is finer-grained than that on the original beach (▶ Figure 13-28).

Nourishment operations generally try to minimize long-shore transport and loss of sand from the nourishment site. In some cases, however, a nourishment operation chooses to minimize costs by adding sand up the coast and permitting longshore sand drift to spread sand into the area needing nourishment. Dumping sand at a cape, for example, can lead to sand migration into adjacent bays where it is needed. Similarly, sand drifting into inlets is often lost to the system if inlet dredges dump it far offshore. Some areas such as Florida now require that dredged sand be dumped next to down-drift beaches (▶Figure 13-20). The operation thus permits the sand to bypass the inlet. Other bypass operations use fixed or movable jet pumps, most commonly along with conventional dredges.

Along cliff-bound coasts, sand, sediment, or easily disintegrated fill material is sometimes merely dumped over an eroding bluff to permit breakup by waves. Instead of rapidly eroding the cliff base, the waves gradually break up the dumped sediment to form new beach material.

Sometimes, groups of residents, towns, or counties on barrier islands lobby the local, state, and federal governments to replenish the sand on a severely eroded beach. Because large sand-replacement projects typically run into millions of dollars and homeowners do not want their taxes to increase, the local governments lobby their state and federal representatives to have governments foot most or all of the bill. Politicians want to be reelected and to bring as much money back to their communities as possible, so they lobby hard for state and federal funding. What this means, of course, is that the cost of replenishing sand to benefit a few dozen beachfront homeowners is spread statewide or countrywide among all those who derive little to no benefit from the work. For example, the federal government paid approximately 50 percent of the cost of beach replenish-ment projects in Broward County, Florida, from 1970 to 1991, while local communities paid only 4 percent. To add insult to injury, beachfront communities often try to inhibit beach access by the hoards of mainlanders who flock to the beaches on warm summer days. Although the beach area below high tide is legally public, access routes are often poorly marked or illegally posted for no access. Some communities also make beach access difficult by severely restricting parking along nearby roads.

Where the federal government agrees to foot a large part of the cost of a major beach replenishment project, the U.S. Army Corps of Engineers becomes the responsible agency. Engineers, geologists, and hydrologists with expertise in beach processes design a replenishment project and oversee the private contractors who actually do the work. Common sand sources are shore areas in which sand shows net accumulation or sources well offshore and below wave base. Sometimes sand is dredged off the bottom and transported to the beach area on large barges. Elsewhere sand is suction-pumped from the source and pumped through huge pipes as a sand-and-water slurry. From there, the sand is spread across the beach using heavy earthmoving equipment (▶Figures 13-29, 13-30, and 13-31).

In fifty-six large federal beach projects in the United States between 1950 and 1993, the Corps of Engineers placed 144 million cubic meters of sand on 364 kilometers of coast. Enormous volumes have been placed in some relatively small areas, such as the 24 million cubic meters on the shoreline of Santa Monica Bay, California. More than 28,000 kilometers of coast continue to erode. Sources of usable sand in many areas are being rapidly depleted. Florida's sources of economically recoverable sand, for example, were almost depleted by 1995. They are now looking at tens of billions of tons available from the Great Bahama Banks.

▶**FIGURE 13-29.** To replenish a beach, the U.S. Army Corps of Engineers pumped a high volume sand-and-water slurry from up the coast in 30-inch pipes. Heavy earth-moving equipment spread more than a meter of sand across the new beach at Ocean Isle Beach, North Carolina, in March 2001.

Donald Hyndman photo.

(a)　　　　　　　　　　　　　　　**(b)**

▶**FIGURE 13-30.** These views show Miami Beach, Florida, **(a)** before and **(b)** after beach nourishment by the U.S. Army Corps of Engineers.

▶**FIGURE 13-31.** As part of the massive beach replenishment at East Rockaway, New York, in March 1999, the groin at left minimizes the loss of the replenished sand by longshore drift.

Erosion of Gently Sloping Coasts and Barrier Islands

During the ice ages of the Pleistocene epoch more than 12,000 years ago, sea level dropped as water was tied up in continental ice sheets; the shoreline receded far out onto the continental shelf. In fact, wave-base erosion and deposition to form the continental shelves may be related to thousands of years at which sea levels stood some 100 meters lower than at present. Following the last ice age, as continental ice sheets melted 12,000 to 15,000 years ago and sea level rose, waves gradually moved sands on the continental shelf landward, piling them up ahead of the advancing waves. The Atlantic and Gulf of Mexico coastal plains are gently sloping. The gentle upward-curving **equilibrium profile** of a sandy bottom produced by the waves is steeper than the coastal plain, so it tops off landward in a ridge, the barrier island (▶Figure 13-32).

Waves sweep sands in from the continental shelf. Offshore **barrier islands** are typically 0.4 to 4 kilometers wide and stand less than 3 meters above sea level. Winds picking up dry beach sands may locally pile dunes as high as 15 meters above sea level. Offshore barrier islands and the lagoons behind them are products of dynamic coastline processes: erosion, deposition, longshore sand drift, and wind transport. Barrier island communities live within this constantly changing environment.

Dunes

Sand dunes help maintain the barrier islands. Major storms or hurricanes wash sand both landward into the lagoon and seaward from the barrier bar. Sand moves back up onto the

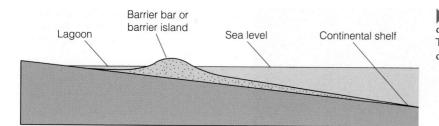

▶**FIGURE 13-32.** This cross section shows an offshore barrier bar with a sheltered lagoon behind. The waves create a steeper profile for the sand than the overall slope of the coastline.

higher beach in milder weather, and wind blows some dry sand into the dunes again. When people modify the dunes, they upset this long-term equilibrium. They level the dunes to build roads, parking areas, and houses, or to improve views of the sea (▶Figures 13-33). True, they improve the view, but with each storm the beach gets closer until the view gets too good. The house is the next to go. People remove vegetation deliberately to provide views or inadvertently by trampling underfoot to reach the beach or by using off-road vehicles. They move sand onto lagoon-margin marsh areas for housing and marinas. They remove sand from dunes for construction or for reclaiming beaches damaged by erosion elsewhere.

In many places, sand mining is now permitted only offshore at depths greater than 18 to 25 meters. Monitoring of such activities, however, is sometimes lacking. And some communities tacitly condone mining by purchasing beach sand for use on roads. Where dunes are managed at local levels with strong input from landowners, the dunes are often low and narrow to permit easy access to the water and direct views of the ocean. Such low dunes provide, of course, minimal protection from storms.

Wind-blown sand not only builds dunes but also piles against houses and collects as drifts around and downwind of them. In areas of lower sand supply, the wind may funnel between houses to cause areas of local scour. It covers roads, driveways, and sidewalks (see also Figure 13-47, page 347). Where buildings are elevated 4 meters or so on pilings, as is commonly the case across the southeastern or Gulf coasts, the effect on wind-blown sand is much reduced.

After storms, the sand is routinely removed from roads by either individuals or municipal employees. The sand is sometimes placed back on the upper beach, but often it is pushed onto vacant lots where it is lost to the surface of the beach. It may be used to partly rebuild artificial dunes or close gaps opened through the dunes.

Sand can be trapped by placing sand fences across the wind direction to slow the wind and promote deposition of sand on their downwind sides. Fences can help keep sand on the upper beach or in dunes; they can also be used to prevent drifts from forming where they create problems for roads and driveways. Beach nourishment projects also often involve dune nourishment, the sand being scraped from the beach or from overwash sediments.

Coastal vegetation was and in some areas still is burned or physically removed to provide building sites, views, or landscaping. Native vegetation on coastal dunes is effective in reducing sand drift and stabilizing the dunes. Where lost, vegetation can be replanted to help stabilize the sand, though new vegetation may be difficult to establish if the sand is salty or mobile. Straw or branches from local coastal shrubs can be strewn on the sand surface to slow sand movement and help establish the growth of grasses.

Dune vegetation that is diverse and native to the area is best for its likelihood of survival. Natural revegetation of dunes may be accomplished using cuttings taken from nearby mobile dunes, but this can be a slow process. European beach grass introduced to the Pacific coast of the United States has been even more successful in trapping sand than native vegetation, but it creates dunes that are less natural and higher than the originals. It now dominates dunes in coastal Washington state and parts of Oregon (▶Figure 13-34). Residents, because of appearance, often plant exotic vegetation, but that often requires artificial watering. The resulting rise in the water table can lead to the

David Scott photo.

▶**FIGURE 13-33.** On this completely built-up barrier island at North Myrtle Beach, South Carolina, the beach in front of the houses has lost its protective dunes. The lagoon behind separates the barrier island from the mainland.

▶**FIGURE 13-34.** Beach sand blown into these dunes along the Oregon coast has been stabilized by European beach grass.

▶**FIGURE 13-35.** This diagram shows typical features of barrier bars served by a bridge across the lagoon.

formation of surface gullies and increase the chance that coastal cliffs will have landslides.

Barrier Bars at Estuaries and Inlets

Offshore barrier bars or barrier islands are a part of the active beach, built up by the waves and constantly shifting by wave and storm action. **Barrier bars** form primarily along gently sloping coastlines such as those of the East and Gulf coasts of the United States, but they also form across the mouths of shallow bays and estuaries along the West Coast. The sea level rise after the last ice age drowned the mouths of these river valleys.

Where high tides or storms carry the sea through low areas in the barrier bars, the water returns from the lagoon to the sea at low tide through the same inlets, eroding them and keeping the channels open (▶Figure 13-35).

Over time, inlets through barrier bars naturally shift in position; some close and others open. Longshore drift of sand at times closes some gaps and storms open others, so they intermittently change locations. When a storm overwashes and severs a beach-parallel road or a storm cuts a new inlet across a barrier bar, some homes and businesses are isolated on part of the bar (▶Figure 13-36). Generally, people fill the new inlet and rebuild the road. Unless they do so immediately, the inlet typically widens rapidly over the following weeks or months as tidal currents shift sand into the lagoon and back out. Therefore, the scale of the repair project can quickly get out of hand. A storm at Westhampton, New York, in December 1992, for example, opened a breach inlet 30.5 meters wide. Within eight months, the inlet widened to 1.5 kilometers.

Shifting sand closing an existing inlet commonly hinders access to marinas and boating in the protected lagoon behind the bar. It also hampers sea access to any coastal industrial sites on the mainland (▶Figure 13-35). Thus, existing

▶**FIGURE 13-36.** **(a)** This breach of Hatteras Island on the North Carolina outer banks followed the appearance of Hurricane Isabel in September 2003. The ocean is to the right. **(b)** The breach of Hatteras Island following Hurricane Isabel in September 2003 severed the only road along the length of the island and isolated many homes from the mainland. The arrow in **(a)** and the yellow line in **(b)** mark the centerline of the road.

FIGURE 13-37. This walkway at Sunset Beach, North Carolina, once reached over a protective beachfront dune, which has since been removed by hurricane waves.

Donald Hyndman photo.

inlets are often kept open by dredging and by building jetties along the edges of the inlet. Where significant populations or large industrial sites are affected, the U.S. Army Corps of Engineers can be involved in constructing or maintaining an inlet. Where significant settlement has occurred on or behind a barrier bar, the maintenance of inlets severely hinders natural evolution of the barrier bar and beach (▶Figures 13-19 and 3-37).

Many barrier islands are now so crowded with buildings that they bear little resemblance to their natural state. Most distinctively, the former broad beaches erode in front of the buildings (▶Figures 13-27 to 13-31 and 13-33). Although the barrier islands help protect low-lying coastal areas from damage from storm waves and floods, the islands themselves are hazardous places to live. Through past experience with hurricanes and other big storms, people living on barrier bars learn to build homes on posts, raising them above the higher water levels and bigger waves of some storms. In many areas, building codes require such construction. Codes also require preservation of beachfront dunes to help protect buildings from wave attack. Unfortunately, most dunes along much of the coasts of the southeastern and Gulf Coast states have disappeared in spite of such efforts. Beach erosion is severe (▶Figures 13-37 to 13-43).

Barrier islands migrate with the gradual rise of sea level. Because the equilibrium slope of the beach and the position of the barrier island are linked to the size of the waves and the depth of water, the beach and barrier island must shift landward as the water level rises. Current rates of sea-level rise are approximately 30 centimeters per century. On especially gentle-sloping coasts like those of the southeastern United States, that 30-centimeter rise can move sand of the beach and barrier island inland by 100 to 150 meters or more (▶Figures 13-39, 13-40, and 13-41). **Barrier island**

(a)

(b)

USGS photos.

(c)

FIGURE 13-38. Shoreward migration of beach at North Topsail Beach, North Carolina, is well shown by a series of three views of the same area. (a) Use the colorful condominiums in the upper left here as a reference and note those in the center of the photo taken after Hurricane Bertha on July 16, 1996. (b) This is the same location following Hurricane Fran on September 7, 1996. Finally, (c) shows the same location after Hurricane Bonnie on August 28, 1998. Note that the series of photos spans only about two years.

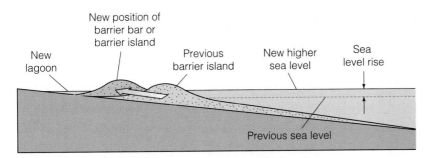

migration happens over decades, but all of the significant movement is in stages during hurricanes and other major storms (▶Figure 13-41).

Oysters grow in the quiet waters of lagoons on the coastal side of barrier islands. If you find oyster shells in sand of the front beach, one possibility is that the barrier bar gradually migrated landward, over the mud of the lagoon. Beach waves winnow out the fine mud of the lagoon, leaving the heavier oyster shells embedded in the beach sand (▶Figure 13-42). Migrating barrier bars also overwhelm trees growing along lagoons; their stumps reappear later along the upper edge of the beach as the bar gradually moves landward after a storm (▶Figure 13-43).

A coastal construction control line (CCCL), established in the 1980s by the Department of Protection of Florida, imposed higher standards for land use and construction in high hazard coastal zones. It restricts the purchase of flood insurance to those communities that adopt and enforce National Flood Insurance Program construction requirements governed by the standard building code. On the coastal side of the CCCL, requirements are more stringent for foundations, building elevations, and resistance to wind loads (FEMA, 1998).

There are a few barrier bars at the mouths of drowned river estuaries along the West Coast; people build on those just as they do on barrier islands on the East Coast (▶Figure 13-44). On gently sloping areas of some Western beaches, as in parts of Southern California, people build right on the

▶**FIGURE 13-40.** The Morris Island Lighthouse, near Charleston, South Carolina, was on the beach in the 1940s. It is now 400 meters offshore because the sand of the barrier island on which it once stood gradually migrated landward.

▶**FIGURE 13-41.** Storm waves reduced the level of sand by almost 4 meters during one storm at Westhampton, New York. The barrier island migrated landward, leaving these houses stranded offshore.

▶**FIGURE 13-42.** These oyster shells embedded in beach sand at Myrtle Beach, South Carolina, probably originated in a back bay or lagoon behind the barrier bar.

Donald Hyndman photo.

▶**FIGURE 13-44.** As viewed across a lagoon, homes cover a bay mouth bar south of Lincoln City, Oregon.

Carl Hobbs photo, Virginia Institute of Marine Science.

▶**FIGURE 13-43.** Peat blocks and stumps in the surf at Sand-bridge Beach, City of Virginia Beach, originally formed in the lagoon behind the barrier island decades ago. The island has since migrated landward over the lagoon. Note the imposing wall supposedly protecting buildings from the waves.

Donald Hyndman photo.

▶**FIGURE 13-45.** These beachfront homes are protected, albeit temporarily, by heavy riprap, northwest of Oxnard, California.

beach (▶Figures 13-45 and 13-46). Perhaps they purchase homes or build them in good weather, not realizing that big winter waves come right up to the house. Some pile heavy riprap on the beach in front of their homes, hoping to protect them (▶Figure 13-45). The protection is temporary because big waves reflect off the boulders, washing away the beach sand in front and steepening the remaining beach. Eventually, the boulders will slide into the wave-eroded trough, leaving the homes even more vulnerable. But then the beach is gone.

Some houses are built on pilings to let waves pass underneath. Bigger storm waves will impact such structures because any protective dunes have long since disappeared. Sand still blows off the beach, but it drifts around the houses instead of forming dunes in the back beach. It piles against the houses and covers sidewalks and streets. Shoveling

California Coastal Records photo.

▶**FIGURE 13-46.** These huge houses are right on the beach at Oxnard Shores in Southern California.

FIGURE 13-47. Sand blown off a beach (out of sight on the right) piled as drifts around houses and covered roads, sidewalks, and driveways in Oxnard in Southern California. Shoveling snow in northern climates is heavy work, but this is ridiculous. "For Sale" signs are often numerous along this beach-parallel street.

Donald Hyndman photo.

snow in some northern areas can be hard work but imagine having to shovel sand, which weighs many times as much and will not melt (▶ Figure 13-47).

Erosion Along Cliff-Bound Coasts

Hard, erosionally resistant rocks, granites, basalts, hard metamorphic rocks, and well-cemented sedimentary rocks mark some steep coastlines. Such coastlines often consist of rocky headlands separated by small pocket coves or beaches. Much of the coast of New England, parts of northern California, Oregon, Washington, and western Canada are of this type. The rocky headlands are subjected to intense battering by waves and drop into deep water. Sands or gravels pounded from the headlands are swept into the adjacent coves, where they form small beaches (see Figures 13-10b and 13-11).

Raised marine terraces held up by soft muddy or sandy sediments less than 15 million to 20 million years old mark other coastlines, including much of the coasts of Oregon and central California. These terraces, standing a few meters or tens of meters above the surf are soft and easily eroded. They were themselves beach and near-shore sediments not long ago. Some terraces rose during earthquakes as ocean floor was stuffed into the oceanic trench at an offshore subduction zone. Elsewhere they may rise by movements associated with California's San Andreas Fault.

Along these coasts, cliffs line the head of the beach, except in low areas at the mouths of coastal valleys. The beaches consist of sand, partly derived from erosion of the soft cliff materials and partly brought in by longshore drift. Waves strike a balance between erosion and deposition of beach sands. Larger waves flatten the beach by taking

sand farther offshore, and smaller waves steepen it. Where streams bring in little sand or the cliffs are especially resistant, the beach may be narrow. Where rivers supply much sand or where the coastal cliffs are easily eroded, the beach may be wide. A broad, sandy beach hinders cliff erosion because most of the waves' energy is expended in stirring up sand and moving it around.

With European settlement of North America over the past 200 years and our attempts to control nature, we have upset that balance. We have severely reduced the supply of river sediment to the coasts by building dams that trap the sediment and by mining sand and gravel from streams. We have even mined sand and gravel from beaches themselves. With less sand and gravel, beaches shrink and the waves break closer to, and more frequently against, coastal cliffs. Waves undercut the cliffs, which collapse into the surf (▶ Figures 13-48 to 13-52).

(a)

Donald Hyndman photo.

(b)

David Hyndman photo.

▶**FIGURE 13-48. (a)** Beach erosion and cliff collapse endangered homes in Pacifica (south of San Francisco) in March 1998. Collapsed parts of houses litter the base of the cliff. **(b)** Taken from the same viewpoint as **(a)** in December 2003, this photo shows that the seven homes nearest the camera are gone. Only two of the original ten houses remain, and one of these sold in 2004 for $450,000! (See Figure 18-2, page 446.)

▶**FIGURE 13-49.** Heavy equipment tried desperately to pile heavy riprap boulders at the base of the same rapidly eroding cliff shown in Figure 13-48 before the next high tide in January 1998. Note the house debris at the base of the cliff.

▶**FIGURE 13-50.** Collapsing sea cliffs at Ocean Beach near San Diego destroyed some homes and severely threatened those remaining in March 1998. Huge boulders were placed at the base in an attempt to arrest the erosion.

▶**FIGURE 13-51.** These large homes sit atop a sea cliff without a beach in Pismo Beach in Southern California. Note that a new house is being built on the right, even though established houses have lost most of their yards and have spray cemented their cliffs to slow cliff loss.

▶**FIGURE 13-52.** Timbers were installed in an attempt to protect an eroding beach cliff below houses in Pismo Beach, California.

Unfortunately, people also choose to build houses on those same cliff-tops so they have a view of the sea, typically not realizing how hazardous the sites are. Developers promote building in such locations as prime view lots, charging a premium for land that may be gone in a few years or decades. Vertical cliffs made of soft, porous sediments are a recipe for landsliding and cliff collapse (▶ Figures 13-48 to 13-52). Homeowners themselves exacerbate the problem by clearing the beach of driftwood that would help to break the force of the waves. They unwittingly become agents of erosion by making paths down steep slopes, cutting steps, and excavating for foundations next to the cliffs. They irrigate vegetation and drain water into the ground from rooftops, driveways, household drains, and sewage drain

▶**FIGURE 13-53.** These homes were built on pilings on the beach west of Malibu, California.

Donald Hyndman photo.

fields. Adding water to the ground further weakens it and promotes landslides.

When such cliff-top dwellers see their property disappearing and recognize that part of the problem is waves undercutting the cliff, the typical response is to dump coarse rocks—riprap—at the base of the cliff (▶Figures 13-48b, 13-49, and 13-50) or to build a wood, steel, or concrete wall there (▶Figures 13-51 and 13-52). Waves reaching such a resistant barrier tend to break against it, churn up the adjacent sand and sweep it offshore. The new deeper water next to the barrier undercuts the barrier, which then collapses into the deep water. The barrier has provided short-term protection to the cliff but after a few years that "cure" is worse than if they had done nothing. A bigger slab of the cliff collapses into the deeper water in the next big storm. Some people spray **shotcrete,** a cement coating, on the surface of the cliff to minimize the loss of the cliff surface (▶Figure 13-51).

In a few places, people even build at the base of cliffs, on the beach itself (▶Figure 13-53). They must know something that we do not! CalTrans, the California highway department, had enough sense to build the coastal highway 6 or so meters above high tide; these houses, on the beach side of the highway, are on pilings sunk into the beach but have their lower floors more than 3 meters below the highway.

Letting Nature Take Its Course

A less expensive, more permanent, alternative to beach hardening or beach replenishment is advocated by many coastal experts. They suggest moving buildings and roads on gently sloping coasts back landward to safer locations after major damaging storms. Sand dunes behind the beach can be stabilized with vegetation in order to provide further protection for areas behind them. The cost of moving buildings may be high, but it is less than the continuing long-term cost of maintaining beach-hardening structures that tend to destroy the beach or continually bringing in thousands of tons of beach sand that is removed by the next storm. And it does provide a way of living with the active beach environment rather than forever trying to fight it.

On cliff-bound coasts, buildings should not be placed close to cliffs. Those that are too close should be moved well back from them, and foot traffic and other activities should be restricted to areas away from the cliff tops and faces. Beaches should not be cleared of natural debris such as driftwood. Mining sand and gravel from beaches and streams should be prohibited. Additional dams should not be built on rivers that discharge in coastal regions with erosion problems; removal of old dams would eventually bring more sediment back to the beaches and help to protect the cliffs.

KEY POINTS

✓ Most waves are caused by wind blowing across water. The height of the waves is dictated by the strength, time, and distance of the wind blowing over the water. **Review p. 327.**

✓ Water in a wave moves in a circular motion rather than in the direction the wave travels. That circular motion decreases downward to disappear at an approximate depth of half the wavelength. Waves begin to feel bottom near shore in water less than that depth. **Review p. 327; Figures 13-2 and 13-3.**

✓ Wave energy is proportional to the wavelength times wave height squared, so doubling the wave height quadruples the wave energy. **Review p. 328; Sidebar 13-3.**

✓ The grain size and amount of sand on a beach depends on wave energy, erodibility of sea cliffs and the size of particles they produce, and the size and amount of material brought in by rivers. Larger winter waves commonly leave a coarser-grained, steeper beach. **Review p. 329.**

✓ Waves approaching the beach at an angle push sand parallel to shore as longshore drift. **Review pp. 329–330; Figures 13-9 and 13-10.**

✓ A beach at its equilibrium profile steepens toward shore. Larger waves erode the beach and spread sand on a gentler slope. **Review pp. 331–332; Figures 13-12 and 13-13.**

✓ Beach hardening to prevent erosion or stop the longshore drift of sand includes seawalls, groins, and breakwaters. All have negative consequences either after a period of time or elsewhere along the coast. In some cases, they result in the complete loss of a beach. **Review pp. 333–336; Figures 13-14, 13-15, 13-16, 13-21, and 13-23.**

✓ Beach replenishment or nourishment involves replacing sand on a beach, an expensive proposition that needs to be repeated at intervals. **Review pp. 338–341; Figures 13-20, 13-28 to 13-31.**

✓ Offshore barrier bars or barrier islands develop where the landward-steepening beach profile is steeper than the general slope of the coast. Wind blows sand shoreward into dunes. Dune sand spreads over the beach, helping to protect it from erosion during storms, but most dunes have disappeared by excavation, trampling underfoot, or wave action. **Review pp. 341–343; Figures 13-28 to 13-30, 13-37 and 13-38.**

✓ Estuaries and inlets through barrier bars are kept open by tidal currents into the lagoon behind the bar. Storm surges and waves open, close, and shift the locations of such inlets. **Review pp. 343–344; Figures 13-35 and 13-36.**

✓ Barrier islands maintain their equilibrium profile by migrating shoreward with rising sea level. Because many barrier islands are now covered with buildings, the island cannot migrate but is progressively eroded. **Review pp. 344–345; Figures 13-39 to 13-43.**

✓ Along cliff-bound coasts, sand is eroded from headlands and deposited in bays. Where the cliffs are made of soft materials, decreasing amounts of sand fed by rivers results in more rapid cliff erosion with disastrous consequences for buildings atop the cliffs. **Review pp. 345–347.**

IMPORTANT WORDS AND CONCEPTS

Terms

barrier bar, p. 343	jetty, p. 334
barrier island, p. 341	longshore drift, p. 328
barrier island migration, p. 344	Nor'easter, p. 328
beach hardening, p. 333	rip current, p. 332
beach nourishment, p. 339	riprap, p. 333
beach replenishment, p. 339	sand dune, p. 341
breakwater, p. 335	seawall, p. 333
dune, p. 331	shore profile, p. 331
equilibrium profile, p. 341	shotcrete, p. 349
fetch, p. 327	storm surge, p. 332
groin, p. 334	submarine canyon, p. 337
hardening, p. 333	wave energy, p. 328
headland, p. 331	wavelength, p. 327
	wave refraction, p. 329

QUESTIONS FOR REVIEW

1. Sketch the motion of a stick (or a water molecule) in a deep-water wave.

2. What force creates most waves?

3. How can there be big waves at the coast when there is little or no wind?

4. Why are ocean waves generally larger than those on lakes?

5. Which side of a beach groin collects sand?

6. How do sand dunes affect stability of a beach?

7. Where does the sand go that is eroded from a beach during a storm?

8. What factors cause growth of wind waves?

9. Where would a sand spit form on a barrier bar island relative to the direction of longshore drift?

10. What happens to wave energy and erosion when riprap or seawalls are installed?

11. Draw the shape of a barrier bar island before and after a significant rise in sea level.

FURTHER READING

Assess your understanding of this chapter's topics with additional quizzing and conceptual-based problems at:

 http://earthscience.brookscole.com/hyndman.

After Hurricane Hugo, pleasure boats moored in the lagoon behind the barrier island lay stacked like fish in a basket.

HURRICANES AND NOR'EASTERS
The Big Winds

Hurricanes, Typhoons, and Cyclones

Hurricanes in the North Atlantic and eastern Pacific; **typhoons** in the western Pacific, Japan, and Southeast Asia; and **cyclones** in the Indian Ocean are all major subtropical cyclones. They have wind speeds of more than 120 kilometers per hour and can exceed 260 kilometers per hour. They have killed more people annually than any other natural hazard. The worst disasters have occurred in heavily populated poor countries in Southeast Asia, in events resulting in hundreds of thousands of deaths in a single event.

The word **hurricane** is derived from a Caribbean native Indian language and means "big wind." A tropical cyclone develops, by definition, as a large, warm-core, low-pressure system over tropical or subtropical waters with a temperature of at least 25°C; cyclones circulate counterclockwise in the northern hemisphere, clockwise in the southern hemisphere. This definition excludes severe thunderstorms at frontal systems that may have hurricane-force winds.

Rebuilding in the aftermath of 1989's Hurricane Hugo was fraught with problems. The same is true after most other major hurricanes. Poorly trained and untrained workers, hasty and shoddy work, unsuitable materials, and unscrupulous contractors were everywhere. Inspection was inadequate or absent.

Many beaches were heavily eroded and dunes flattened, and damage to homes was extensive (▶ Figures 14-1 and 14-2). The storm caused 105 deaths, eleven of them in the Charleston area and twenty-nine in all of South Carolina; property damage was $12.3 billion (in 2002 dollars). Although the largest insured payments were for flooding caused by **storm surge,** the dramatic rise in sea level from low atmospheric pressure and push of the sea ahead of the prolonged winds (discussed in Chapter 13, "Waves, Beaches, and Coastal Erosion"), flooding did less than 10 percent of the total dollar damage. Wind damage was much greater. Depending on coastal location, the storm probably had an 80- to 200-year recurrence interval.

Hurricane Hugo initially lashed the Leeward Islands and Puerto Rico with sustained winds of 225 kilometers per hour, leaving more than 30,000 homeless, and causing damages of $9.9 billion. Although satellite and aircraft tracking permitted the populace to be kept informed of the hurricane's track and intensity in 1989, most communities did not take advantage of the information.

▶**FIGURE 14-2.** This house was not well attached to its foundation. Hurricane Hugo's surge carried it inland, where it was deposited on a roadway.

South Carolina lacked restrictions over the use of coastal sites and the quality of construction. That was left to local communities, some of which imposed controls while others provided few or none. Most adopted a building code only when they wanted to participate in the National Flood Insur-

▶**FIGURE 14-1.** Hurricane Hugo caused rampant destruction of beachfront homes on Sullivan's Island, South Carolina. It destroyed homes behind the two in the foreground, which were also damaged. It even wrecked nearly new homes on 3-meter-high pilings.

FIGURE 14-3. Surges quickly cover escape routes. This warning sign is on the barrier bar road to Sunset Beach on the offshore barrier island of South Carolina.

Donald Hyndman photo.

ance Program (NFIP). That program required raising dwellings above the 100-year flood level of 4 meters or so but provided no design requirements for wind effects. The standard building code did not require appropriate standards for coastal wind loads until 1986. Well-designed buildings generally survived Hurricane Hugo with little damage, but others lost roofing and wall siding, or were completely destroyed. The high level of damages was caused by the lack of standards held by many groups, including governments, developers, builders, lenders, insurers, and owners.

Growth on the barrier islands and coast of South Carolina, in the thirty years after the previous hurricane, was dramatic. Sparsely populated areas became major beach resorts with high-rise hotels and condominiums. Many of the newcomers were not aware of the long history of hurricanes and their effects along the coast. Much of the rural population in South Carolina lives in mobile homes. The governor ordered the evacuation of all such homes, and many people complied. The widespread pine forests helped protect many homes, but with winds near 112 kilometers per hour, those trees began falling. Fortunately, peak winds in the deadly northeast quadrant of the storm hit sparsely populated areas. Landfall 30 kilometers to the south would have heavily impacted Charleston and multiplied the damage. In both residences and commercial buildings, much of the damage was initiated when wall coverings and roofs failed because of winds.

Electric power lines and poles were undermined on the barrier islands. Even hundreds of kilometers inland to the north-

west along Hugo's track, power lines were almost completely destroyed by wind, wind-blown debris, and falling trees. Only 23 percent of customers in the Charleston area had power eight days later; some had none for two to three weeks. The main bridge to the mainland from Sullivan's Island and the Isle of Palms failed.

Hugo came ashore on a northwesterly track, just north of Charleston near midnight on September 21. It weakened progressively so that by 1 P.M. the next day, it entered western North Carolina as a Category 2 hurricane. A huge surge of 6.1 meters along the South Carolina coast caused extensive flooding (▶Figure 14-3), and 0.5 to 1.2 meters of sand disappeared from under most buildings, less than most storms because of the wide, flat beaches and abundant sand. The waves flattened most oceanfront dunes (▶Figure 14-1). Fifteen to thirty meters inland of these dune fronts, overwash deposited as much as 0.6 meter of sand. Many of the destroyed houses were built on columns sunk only 30 centimeters into the sand and on 60-square-centimeter footings. Even on the landward side of the lagoon behind the barrier island, homes were severely damaged, and pleasure boats in marinas were stacked up on shore like piles of fish (see the photo at the beginning of the chapter).

On September 21, 1989, residents in and around Charleston, South Carolina, braced for the onslaught of Hurricane Hugo (▶ see Figure 14-4). The National Weather Service forecast a 5.2-meter-high surge accompanying sustained winds of 215 kilometers per hour. That would completely in-

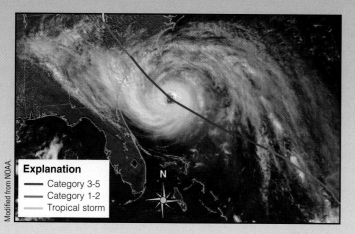

Modified from NOAA.

Explanation
—— Category 3-5
—— Category 1-2
—— Tropical storm

▶**FIGURE 14-4.** Hurricane Hugo's path brought it directly onshore in South Carolina, before weakening and turning north though West Virginia and eastern Ontario. The colors of the storm track on this satellite image relate to the intensity of the storm winds. Note that the winds weakened soon after the hurricane came onshore.

undate most barrier islands in South Carolina because most are less than 3 meters above sea level. In fact, the storm, co-incident with high tide, was a Category 4 with minimum pressure of 934 millibars and a forward speed of 40 to 48 kilometers per hour.

In March 1988, Gwenyth and William Reid completed their home 30 kilometers northeast of Charleston, overlooking the intracoastal waterway which was a long boat channel excavated from lagoons behind the barrier bar (▶Figure 13-20a). They took special care to build with storm-resistant design and materials. The year-old house was built with its floor 6 meters above sea level, well anchored onto dozens of concrete pilings pounded 3 to 4 meters into the ground. In anticipation of the storm, the Reids attached precut plywood over the windows and doors. Most neighbors evacuated, but they elected to stay with one neighbor in another well-built house with 300 meters of woods between the house and the water.

By 7 P.M. the night of the storm, the wind whipped the trees over, and by 8 P.M. the area lost power. By 11:30 P.M. the water rose to 20 to 30 centimeters above the ground and the house trembled and shook in the fierce wind. Feeling that

the house might not survive, the Reids and their neighbor tied lengths of rope around each of them to lash to trees in case they found themselves out in the water. By 11:45, the water was 1 meter deep and the new car parked under the house between the pilings floated free; they opened its doors to keep it from rising and bashing through the floor above it. A strong smell of gasoline fumes from the cars filled the air. By 12:30 A.M., water rose above the floor inside the house. As the water continued to rise, they checked the one small uncovered "escape" window in the bathroom to see rough water, lumber, and branches floating by; all of the trees are broken off above water level. They were terrified that they would not survive until high tide at 2:13 A.M. when they expected the water to begin dropping. Fortunately, the water began to drop at 1 A.M., and the wind eased a little.

The next morning, the Reids left their neighbor's house to find almost every pine tree broken off and huge live oaks with no leaves and few of their branches, lumber, and pieces of houses and docks. Back at their own house, the Reids found gaping holes in every side, furniture and the refrigerator hanging halfway out, the decks and lower stairs gone, the flooring and joists of the great room gone, along with some inner walls, sliding doors, and the bay window on the east. Upstairs there was little damage.

The only other neighbors to remain through the storm, in their house about a kilometer farther west, were forced upstairs to their den, and then with water rising to chest height, they climbed onto cabinets. In the morning, when they tried to go downstairs, they realized that the den in which they found shelter was resting on the ground 10 meters from where it belonged. The rest of the house was gone.

Of the forty-two homes in the immediate area, all less than six years old, twenty-one were destroyed and seventeen were severely damaged. The four remaining homes were above the level of the 6-meter surge. The beach was washed away. Those who stayed said they would never do so again. They emphasized that you must leave, take a chain saw so you can clear trees to return quickly, keep records in pencil because ink generally runs, and spread family pictures and vital papers around to other parts of the country. They would build still higher above base flood elevation and build even more sturdily.

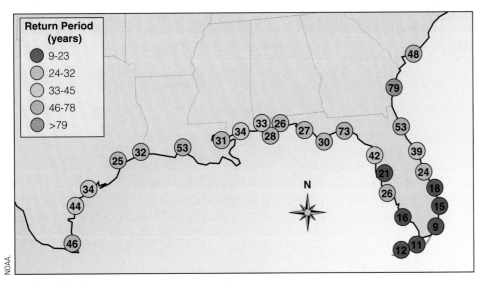

▶**FIGURE 14-5.** Category 3 hurricanes are strong enough to do major damage to coastal areas. They can be expected to strike almost any individual part of the Gulf of Mexico coast on average every 25 to 35 years. Southern Florida is even more vulnerable, with average return periods as low as 9 to 10 years.

Some eighty to ninety tropical storms and forty-five hurricanes affect the Earth every year, or 130,000 such storms in a thousand years. An average of six named hurricanes per year form in the Atlantic and Gulf of Mexico. Thus, in a short period of geological time, cyclones are likely to have a major effect on erosion, deposition, and overall landscape modification, especially in exposed areas such as barrier islands.

For the United States, subtropical cyclones or hurricanes are a major concern along the Gulf of Mexico and southern Atlantic coast. More than 44 million people live in coastal counties susceptible to such storms, roughly 15 percent of the total population of the United States (~294,000,000 in 2005). Because of the rapid growth of these areas, more than 85 percent of the population has never directly experienced a hurricane and thus is poorly informed about the risks. These residents and equally inexperienced developers and builders choose to work in hazardous sites such as beachfront dunes and offshore barrier bars. Rising property values amplify the damages when a major storm does hit.

Although most U.S. hurricanes affect lower-latitude areas, many of those that remain off the coast with northward trajectories reach as far north as New England. Only a few of these have intensities greater than Category 3, but those few can be highly destructive, especially where Long Island, Rhode Island, and eastern Massachusetts protrude into the Atlantic Ocean and across their paths. Based on historic accounts and prehistoric overwash deposits of sand covering layers of peat, such intense hurricanes wreaked havoc in New England in 1635, 1638, 1815, 1938, and 1954. At least twenty-seven hurricanes of Category 1 and 2 have hit the same area in the last 400 years, on average one every fifteen years.

Large hurricanes (Category 3—see Table 14-1) strike most locations from Florida through the Texas Gulf Coast on average once every fifteen to thirty-five years (▶Figure 14-5).

Saffir-Simpson Hurricane Scale

The Saffir-Simpson Hurricane Scale is based on barometric pressure and average wind speed. The lower the barometric pressure, the stronger the hurricane (Table 14-1). We discuss below the various aspects of hurricanes, their barometric pressure, wind speeds, storm surge heights, and damages.

Although people often focus their attention on Categories 4 and 5 hurricanes, lower-category storms can sometimes do almost as much damage and in some cases cost even more lives. A comparison of pairs of Category 5 and Category 2 storms can be instructive (see Table 14-2 and the Cases in Point for Hurricane Camille and Hurricane Isabel).

Formation of Hurricanes and Cyclones

In tropical latitudes, temperatures are high and air-pressure gradients are weak. Air rises by localized heating, which causes condensation that can build into towering convective "chimneys" with frequent thunderstorms. Where one of

Table 14-1 The Saffir-Simpson Hurricane Scale

Category	Example	Barometric Pressure*		Storm Surge		Average Wind Speed		Damages**
		mbar	in.	m	ft	kph	mph	
Normal (no storm)		1,000	29.92	0	0	0	0	
Tropical storm				<1.2	<4	62–119	39–74	
1	Danny, 1997	980	>28.94	1.2–1.5	4–5	119–153	74–95	Minor to trees and unanchored mobile homes
2	Bertha, 1996; Isabel, 2003	965–979	28.5–28.93	1.8–2.4	6–8	154–177	96–110	Moderate to major damage to trees and mobile homes, windows, doors, some roofing. Low coastal roads flooded 2 to 4 hours before arrival of hurricane eye.
3	Alicia, 1983; Fran, 1996	945–964	27.91–28.49	9–12	2.7–3.3	178–209	111–130	Major damage: Large trees down, small buildings damaged, mobile homes destroyed. Low-lying escape routes flooded 3 to 5 hours before arrival of hurricane eye. Land below 1.5 meters above mean sea level flooded 13 kilometers inland.
4	Hugo, 1989	920–944	27.17–27.9	4–5.5	13–18	210–249	131–155	Extreme damage: Major damage to windows, doors, roofs, coastal buildings. Flooding many kilometers inland. Land below 3 meters above mean sea level flooded as far as 10 kilometers inland.
5	Camille, 1969; Gilbert, 1988; Andrew, 1992; Mitch, 1998	920	27.17	>5.5	>18	>249	>155	Catastrophic damage: Major damage to all buildings less than 4.5 meters above sea level and 500 meters from shore. All trees and signs blown down. Low-lying escape routes flooded 3 to 5 hours before arrival of hurricane.

*Standard atmospheric pressure at sea level = 29.92 inches (in.) = 1,000 millibars (mb) = 1 bar or 1 atmosphere pressure.
**These are highly variable and depend on many factors as discussed in the text.

Table 14-2 Comparison of Hurricane Categories 5 and 2

Characteristic	Hurricane	
	Category 5	Category 2
Surge	High	Lower
Winds	High	Lesser
Diameter	Smaller (affects smaller area)	Larger (affects larger area)
Speed	Crosses coastline rapidly (thus shorter life span on coast; dumps less rain on smaller area and less flooding)	Crosses coastline slower (thus longer life span on coast; often dumps more rain on larger area and more flooding)
Example	Gilbert, 1988; Andrew, 1992	Agnes, 1972; Isabel, 2003

August 14–22, 1969, Mississippi: Category 5

Over western Cuba on August 15 Camille was a Category 3 hurricane, strengthening as it moved west-northwest over warm waters of the Gulf of Mexico. Four hundred kilometers south of Mobile, Alabama, this small-diameter but powerful hurricane registered a central atmospheric pressure of 901 millibars and winds of more than 320 kilometers per hour. The actual maximum wind speeds are unknown because the hurricane destroyed the wind-recording instruments in the landfall area. This Category 5 hurricane was the second strongest to ever hit the United States. Reaching Mississippi on August 17, Camille produced a 7.6-meter storm surge, becoming a tropical depression upon moving from Mississippi into western Tennessee. After landfall, it weakened rapidly but caused severe flash flooding when it shed 68 centimeters of rain in Virginia, mostly in a three- to five-hour period. A total of 143 people died on the Gulf Coast, and another 113 died from floods in Virginia. Storm damage reached more than $7.1 billion (in 2002 dollars).

these convective spirals coalesces into larger systems, it develops into a tropical storm that may intensify into a hurricane. Convection may strengthen when the air rises to high elevations without strong winds aloft that can shear off the tops of the convective chimneys and cause the hurricane to dissipate.

Hurricanes begin to develop over warm seawater, commonly between latitudes 5 and 20 degrees. Vertical wind shear, the variation in wind speed at different elevations, must be minimal to permit a column of air to rise and maintain strong circulation. A low pressure zone with cyclonic winds (rotating counterclockwise in the northern hemisphere) becomes hurricane force when the winds exceed a sustained velocity of 119 kilometers per hour (74 mph) where the highest wind speeds exist along the edge of the eye wall. The warm, moist air over the ocean rises and spreads out at the top of the "chimney." The rising warm air expands, cools, and releases latent heat; the air rises faster as the storm strengthens. The center or **eye** of the cyclone, clearly visible in many satellite views (▶e.g., Figure 14-6), is as much as 20°C warmer than surrounding air. This convective rise pulls more moist air into the eye. Coriolis forces initiate rotation in the rising air, the highest winds and lowest air pressures focusing toward the core of the storm. The whole storm may be from 160 to more than 800 kilometers in diameter. Once formed, the storm moves across the ocean with the prevailing trade winds, a forward motion averaging 25 kilometers per hour.

In the eye—generally 20 to 50 kilometers in diameter and bounded by a wind "wall"—the winds drop abruptly from, for example, 220 kilometers per hour to 15 kilometers per hour in the eye. The air pressure drops from normal

▶**FIGURE 14-6.** As Hurricane Ivan roared into the Gulf Coast on September 15, 2004, its eye crossed the west end of the Florida panhandle while its strongest onshore winds, waves, and storm surge pounded areas just east of the eye. State boundaries are superimposed on this natural color image.

NCDC-NOAA

September 18, 2003, North Carolina and Virginia: Category 2

Five days before landfall at the Carolina coast, Isabel loomed as a Category 5 hurricane with a pressure of 920 millibars and maximum winds of more than 250 kilometers per hour. People expected the worst. Fortunately, however, the storm weakened dramatically before landfall, arriving at the coast as a Category 2 hurricane with maximum winds of approximately 160 kilometers per hour. Tropical storm-force winds extended to more than 500 kilometers from the hurricane's eye. Storm-surge levels reached 2 to 3 meters. One hundred thousand people were ordered to evacuate in North Carolina

Mark Wolfe photo, FEMA.

▶ **FIGURE 14-8.** Heavy equipment fills the breach in Hatteras Island, North Carolina on October 21, 2003, after Hurricane Isabel.

Cynthia Hunter photo, FEMA.

▶ **FIGURE 14-7.** Hurricane Isabel severed Highway 12 along Hatteras Island, North Carolina. The continuation of the highway is visible directly across the breach to the right of the power poles.

and another 100,000 in Virginia. In spite of the lower category rating, damages were severe. The storm covered a broad area from part of South Carolina to New Jersey and made clear that lower category large-diameter storms can sometimes be almost as damaging as more energetic storms (▶ Figures 14-7 through 14-10).

Making landfall early the morning of September 18, Isabel severely eroded sand from beaches, destroyed roads (▶ Figure 14-7), toppled trees, blew the roofs off houses, and collapsed others (▶ Figure 14-10). It opened a broad breach that severed Hatteras Island, leaving a lot of people isolated from their bridge to the mainland until the breach could be filled by heavy equipment (▶ Figures 14-7 and 14-8). Many areas were without water or power for weeks. The Federal Emergency Management Agency (FEMA) reported $1 billion in damages and forty-seven deaths, mostly from flooding.

Isabel continued northwest, dumping more than 12 centimeters of precipitation on already saturated ground and causing extensive flooding. It rapidly moved inland at between 28 and 38 kilometers per hour. If its progress over the eastern states had been slowed or stalled by other weather systems, damages from flooding would have been much worse.

▶**FIGURE 14-9.** Sand eroded from the beach under this house during Hurricane Isabel, exposing most of the length of the posts that were buried in the sand.

Mark Wolfe photo, FEMA.

Mark Wolfe photo, FEMA.

▶**FIGURE 14-10.** This house on the beach in Kitty Hawk, North Carolina, lost its support posts during Hurricane Isabel.

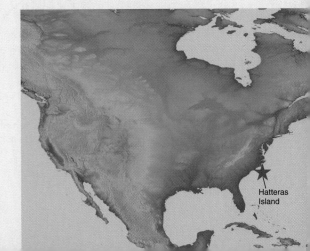

Hatteras Island

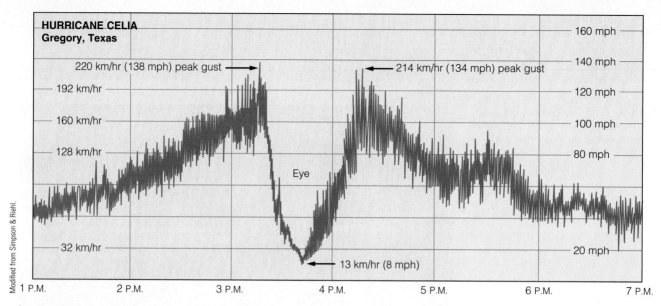

HURRICANE CELIA
Gregory, Texas

220 km/hr (138 mph) peak gust →

192 km/hr

160 km/hr

128 km/hr

Eye

32 km/hr

← 13 km/hr (8 mph)

160 mph

140 mph

← 214 km/hr (134 mph) peak gust

120 mph

100 mph

80 mph

20 mph

1 P.M.   2 P.M.   3 P.M.   4 P.M.   5 P.M.   6 P.M.   7 P.M.

▶**FIGURE 14-11.** This graph for August 3, 1970 shows wind speeds in a typical hurricane in both kilometers per hour and miles per hour.

atmospheric pressure of approximately 1,000 millibars (1 bar) to 960–970 mb in the eye. The air in the sharply bounded eye sinks, causing skies to clear as the air warms and can hold more moisture (▶Figures 14-11, 14-12, and 14-13).

Tropical cyclones rotate counterclockwise but track clockwise in northern hemisphere ocean basins (▶Figure 14-14). Cyclones rotate clockwise and track counterclockwise in the southern hemisphere, the same direction as their ocean currents. Northern hemisphere tropical storms begin in warm waters off the west coast of Africa, then move westward across the Atlantic Ocean with the trade winds. They warm, pick up wind speed and energy, and often develop into hurricanes before they reach the Americas. They generally track west, northwest, and then north, either off the southeastern United States or sometimes into the continent (Figure 14-14). Typhoons in the western Pacific and cyclones in the north Indian Ocean have similar tracks.

Hurricanes generally curve northward as they approach the southeastern coast of the United States because of the Coriolis effect, and they commonly strengthen as long as they remain over the warm water of the Gulf Stream that flows northeasterly along the east coast of Florida and continues up the coast past eastern Canada (▶Figure 14-15). Westward-moving storms therefore tend to track in arcs to the northwest and then north as they approach the coast. As they move over cooler waters or continue over land, they gradually lose energy and dissipate. Some track fairly straight; others take erratic paths (▶Figure 14-16).

An average of five hurricanes develop in the Atlantic Ocean every year, two of them major (Category 3 or greater). Especially vulnerable are southern Florida, the Carolinas, and the Gulf Coast (▶Figure 14-17). Although hurricane season spans late June or July to November, most develop in August or September because this is the period with highest ocean surface temperatures. An average of nine hurricanes appear in the eastern Pacific basin each year, four of them major.

Hurricanes in the United States cost an average of $700 million annually; from 1989 to early 1993, the cost was $50 billion in hurricanes and storm surge. Rainfall of 10 centimeters (roughly 4 inches) per day can fall over a path 400 kilometers wide. If the storm stalls over an area, it can dump tens of centimeters of rain per day.

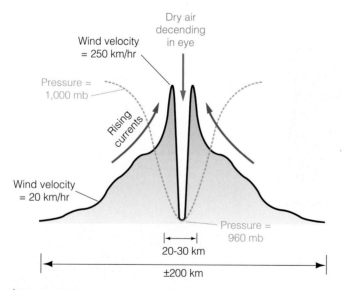

Dry air decending in eye

Wind velocity = 250 km/hr

Pressure = 1,000 mb

Rising currents

Wind velocity = 20 km/hr

Pressure = 960 mb

20-30 km

±200 km

▶**FIGURE 14-12.** This schematic cross section of a typical hurricane shows atmospheric pressure (orange dashed line), wind velocity (heavy black line), and air motions.

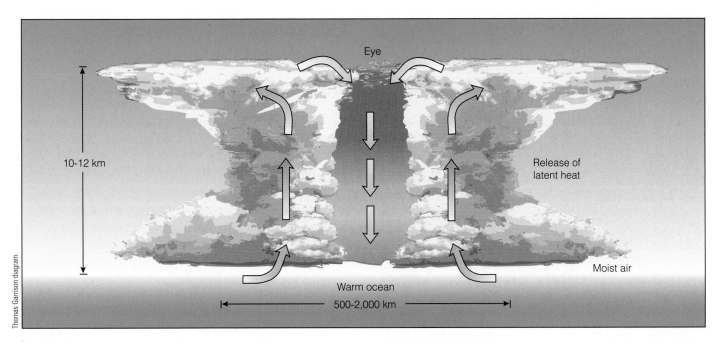

FIGURE 14-13. This cross-section of a typical hurricane shows both its width and height.

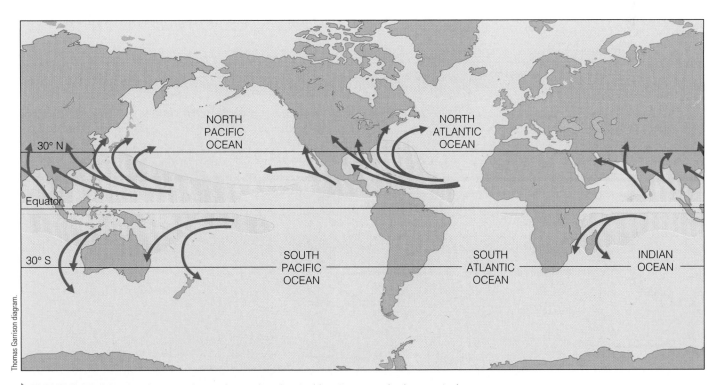

FIGURE 14-14. This diagram shows the tracks of typical hurricanes and other tropical cyclones. Hurricane breeding grounds are shown in orange.

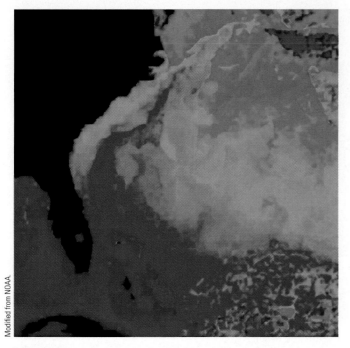

FIGURE 14-15. Warm Gulf Stream temperatures are shown here in red. Sequentially cooler temperatures run from yellow to green.

Modified from NOAA.

Extratropical Cyclones, Including Nor'easters

These storms appear any time in fall, winter, or spring, but especially in February. They behave much like hurricanes and can cause as much damage. **Nor'easters** that strike the northeastern parts of the United States can be huge. The most historically prominent Nor'easters were in 1723, 1888, 1944, 1953, 1962, 1978, 1991, and 1993. They differ from hurricanes in several ways:

1. They are most common from October through April, especially February (rather than late summer for hurricanes).

2. They build at fronts where the horizontal temperature gradient is large and the air is unstable. They often form as low pressure **extratropical cyclones** on the east slopes of the Rocky Mountains such as in Colorado or Alberta when the jet stream shifts south during the winter months. Similar storms can arise in the Gulf of Mexico, near Cape Hatteras, North Carolina, or near the Bahamas or east coast of Florida.

3. They lack distinct, calm eyes and are not circular in form but can spread over much of the northeastern

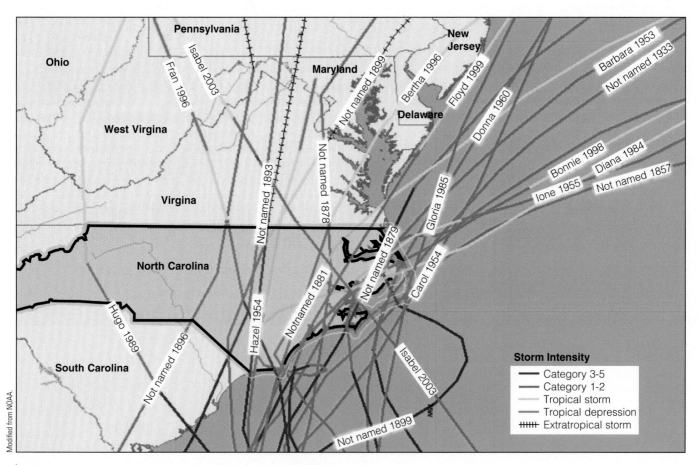

Modified from NOAA.

FIGURE 14-16. This map shows hurricane tracks across North Carolina from 1857 to 2003. The colors of the storm tracks indicate the category of the hurricane near the eye.

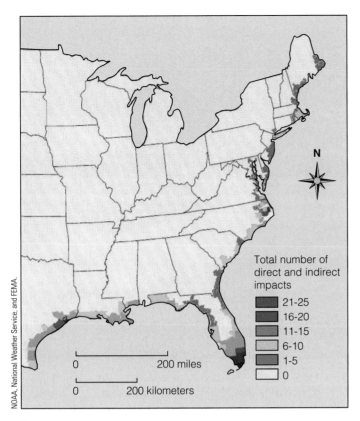

▶**FIGURE 14-17.** In this map, the total number of hurricanes to strike coastal counties from Texas to Maine can be seen for the period 1900–94. More intense hurricanes also concentrate along the coasts with the largest number of hurricanes.

Total number of direct and indirect impacts

- 21-25
- 16-20
- 11-15
- 6-10
- 1-5
- 0

NOAA, National Weather Service, and FEMA.

center of the storm, this drops the surface pressure and increases storm strength.

6. Damage is concentrated along the coast, whereas much of the damage from hurricanes is farther inland. With strong winds from the northeast, they typically batter northeast-facing shorelines.

Nor'easters can build when prevailing westerly winds carry these storms over the Atlantic Ocean and if the jet stream is situated to allow the storms to intensify. The annual number of Nor'easters ranges from twenty to forty, of which one or two are typically strong or extreme.

In addition to the direct damage from high winds, a Nor'easter also generates high waves and pushes huge volumes of water across shallow continental shelves to build up against the coast as a storm surge. Low barometric pressure in the storm permits the water surface to rise, creating a higher storm surge. Extreme surge heights can result from coincidence with especially high tide or movement of surges into bays. As with the surges that accompany hurricanes, these flood low-lying coastal plains and overwash beaches, barrier islands, and dunes. As with hurricanes, the greatest damages occur when a major storm moves slowly at the coast or hits a coast already damaged by a previous storm.

A classification scale for Nor'easters by Davis and Dolan (1993) approximately parallels that of the five-category Saffir-Simpson hurricane scale, except that the emphasis is on beach and dune effects rather than wind speeds and surge heights. It infers a "storm power index" based on the maximum deepwater significant wave height (average of the highest one-third of the waves) squared multiplied by the storm duration (Table 14-3).

Nor'easter wave heights are commonly 1.5 to 10 meters, with energy expended on the coast being proportional to the square of their height. Thus a 4-meter wave expends four times as much energy as a 2-meter wave. Wave height depends on **fetch,** or the distance a wave travels over the water. Waves with a long fetch and constant wind direction in a slow-moving storm can therefore be much more destructive than those in a stronger, fast-moving storm with variable wind directions. Where the storm center is well

United States. As recognized in the late 1700s by Benjamin Franklin, smaller, counterclockwise-rotating cyclonic weather systems are embedded in the broader overall flow.

4. Named for the direction from which their winds come, they bring heavy rain and often heavy snowfall. Franklin also noted that precipitation begins in the south and spreads northward along the coast.

5. They are cold-core systems that do not lose energy with height. If jet stream winds move mass away from the

Class	Maximum Deepwater Significant Wave Height (meters)	Average Duration (hours)	Most Common Site of Formation	Example Storms
I (Weak)	2	10		
II	2.5	20		
III	3	35		
IV	5	60	Bahamas or Florida	December, 1992; March 1993 ("Storm of the Century")
V (Extreme)	6.5	95	Bahamas or Florida	March 7, 1962 (Ash Wednesday); Halloween 1991 ("The Perfect Storm")

Table 14-3 Dolan-Davis (1993) Nor'easter Scale

offshore, the highest waves may be just reaching the coast after the clouds and rain have passed.

When high waves are stacked on top of a storm surge, the effects are magnified. As with hurricanes, a storm surge depends on the lower atmospheric pressure of the storm permitting water level to rise, coupled with the mound of water pushed ahead of the wind. A severe Nor'easter can remain in place for several days and through several tide cycles; high tide superimposed on a high storm surge can be extremely destructive. Examples of Class V storms include the Ash Wednesday storm of 1962 and the Halloween storm of 1991.

The Ash Wednesday Storm of March 7, 1962, was a high latitude Nor'easter along the Atlantic coast of the United States; it stayed offshore approximately 100 kilometers, paralleling the coast for four days. The storm began east of South Carolina and migrated slowly north and parallel to the coast before moving farther offshore at New Jersey. Its slow northward progress was blocked by a strong high pressure system (clockwise rotation) over southeastern Canada. It affected 1,000 kilometers (620 miles) of coast and caused more than $1.8 billion in damages (in 2002 dollars). Sustained winds over the open ocean were 72 to 125 kilometers per hour (45 to 78 mph) and produced waves as high as 10 meters. Storm waves of 4 meters on top of a 1- to 2-meter storm surge washed over barrier sandbars that had been built up for years. A series of five high tides aggravated the height of the surge. Damage included extensive beach and dune erosion and the creation of dozens of new tidal inlets. Almost all of heavily urbanized Fenwick Island, Delaware, was repeatedly washed over by the waves. The coastline moved inland by 10 to 100 meters.

The huge Halloween Nor'easter of October 1991, also called "The Perfect Storm," appears to have had 10.7-meter high deepwater waves and lasted for almost five days. Its wave crests were especially far apart, with intervals ranging from ten to eighteen seconds between crests, so they moved much faster than most storm waves. Accompanied by a major storm surge, it caused heavy damage from southern Florida to Maine, especially in New England (▶ Figure 14-18).

Winter Windstorms and Heavy Snow

Snow depth on the ground is one measure of the severity of a winter storm because of snow loading and accompanying building collapse. Snow depths are typically greatest in Maine, northern Michigan and Wisconsin, and western New York, occasionally exceeding a meter. Heavy snow can even fall in southern states such as Alabama and Georgia (see "Case in Point: Superstorm, March 1993").

Hazards initiated by both tropical cyclones and extratropical storms include the following:

- Storm surges are well above normal tides and cause coastal flooding, salinization of land and groundwater, coastal erosion, damaged crops and structures, and drowning.

Peter Shugert photo, U.S. Army Corps of Engineers.

▶ **FIGURE 14-18.** Storm waves batter this seawall at Sea Bright, New Jersey, during the 1991 Halloween Nor'easter.

- Huge storm waves may overwash dunes and impact coastal structures.
- Heavy rain causes river flooding, flash floods, landslides, structural damage, and overflow of storm sewers and sewage systems. Fifty-nine percent of the people who die in hurricanes drown because of freshwater floods. One-quarter of hurricane deaths are from people who drown in their cars or while trying to abandon them during floods. The heaviest rainfalls are from slower-moving storms and those with larger diameters (see "Case in Point: Tropical Storm Allison" and "Case in Point: Hurricane Agnes"). The National Hurricane Center predicts the total rainfall in inches by dividing 100 by the forward speed of the storm in miles per hour.

$$\text{Inches of rainfall} = \frac{100}{\text{storm's forward speed in mph}}$$

For example,

$$\text{Inches of rainfall} = \frac{100}{20 \text{ mph}} = 5 \text{ inches of rain}$$

- High winds, including tornadoes, damage windows, roofs, and entire buildings. They disrupt transportation and utilities and create large amounts of debris. Tornadoes can form near the eye wall or well away from the center of the hurricane. Although more tornadoes are reported than confirmed, they are difficult to confirm because hurricane winds can reach speeds similar to those in tornadoes. Most of the physical damages in hurricanes are caused by winds.

"Storm of the Century"

On March 12 to 13, a springtime storm off the coast of Texas moved east off the Louisiana coast, its central pressure dropping to 984 millibars. Moving rapidly toward Florida at between 64 and 80 kilometers per hour, the northern part of the squall line then accelerated to 112 kilometers per hour. It generated a storm surge that raised sea levels by 2.7 to 3.5 meters along the Florida panhandle near Apalachicola, decreasing to 2 meters at Tampa. It also spawned at least fifteen tornadoes and numerous downbursts. It moved north-northeast through Georgia on March 13 with a central pressure of 971 millibars, then north along the Atlantic coast. Wind gusts reached 175 kilometers per hour west of Key West and later 231 kilometers per hour on Mount Washington, New Hampshire. Precipitation was only 2.5 to 7.5 centimeters from the Gulf Coast to Maine, but snowfall generally ranged from 15 to 60 centimeters and up to 1.5 meters along the North Carolina–Tennessee border. In Alabama and northern Georgia, areas not used to such conditions, hundreds of roofs collapsed under heavy, wet snow. Fallen trees and high winds left more than 3 million people without electrical power.

Central pressure dropped to 960 millibars over Chesapeake Bay before rising to 966 millibars over Maine. Based on storm surge height and minimum atmospheric pressure, the "superstorm" was equivalent to a Category 3 hurricane. Its total circulation area was the size of the contiguous forty-eight states. All interstate highways north of Atlanta in the eastern United States were closed. Some 270 people died, with forty-eight more missing at sea. Along the Outer Banks of North Carolina, 200 homes were heavily damaged, and on Long Island at least eighteen homes fell into the sea. Total damage amounted to $5.7 billion.

June 5 to 11, 2001: Tropical Storms Can Be as Damaging as Hurricanes

Like most other tropical storms or hurricanes, Allison rose off the coast of Africa and moved west across the Atlantic. It crossed Mexico into the eastern Pacific on June 1, doubled back into southeastern Mexico and north through the western Gulf of Mexico. It made landfall on the east end of Galveston Island, Texas, late on June 5. The cyclonic storm drifted slowly inland to 200 kilometers north of Houston over the next thirty-six hours, dropping more than 25 centimeters of rain in southeast Texas and adjacent Louisiana by the morning of June 8.

Weakening of a high pressure center off Florida and strengthening of a high over New Mexico caused Allison to loop east, then southwest to Houston a second time. Another 51 centimeters of rain fell in twelve hours to cause the worst flood on local record. Twenty-two people drowned in the Houston area. In total, the storm dumped 94 centimeters of rain, almost 1 meter, on the Port of Houston in less than one week. Twenty-one centimeters of that came within two hours in this area, which has notoriously poor drainage. Damages in the Houston area alone reached $5 billion. The Texas Medical Center was especially hard hit. Its below-ground floors were flooded. Backup generators were above ground, but unfortunately switches between the two systems were below ground and destroyed by the flooding.

Central Louisiana did not fare much better with up to 75.8 centimeters of rain at Thibodaux. From there the storm strengthened as it moved east, dumping 30 centimeters of rain on Gulfport, Mississippi; 25 centimeters on Tallahassee, Florida; and more than 30 centimeters near Columbia, South Carolina. It dumped 53 centimeters of rain on Morehead, North Carolina, and then continued up the coast to dump more heavy rain, including more than 25 centimeters in some places in eastern Pennsylvania. It finally moved offshore in Maine.

■ Immediately following the storm, communication lines are lost, power lines are down, roads are out, and a broad range of urgent needs overwhelm local governments. Salt, sewage, various chemicals, and bacteria contaminate surface water and groundwater. Many local officials are inexperienced in dealing with large-scale disasters and with the programs available for assistance. Coordination between all levels of government and teamwork is critical to recover from such disasters.

The number of deaths in the United States because of tropical cyclones before 1910 was in the thousands every decade. Since 1920, the early warning systems and evacuations of populations have reduced the number dramatically but it still amounts to hundreds every decade. In spite of the warnings and evacuations, the costs have increased six- to tenfold because of increasingly heavy development in coastal areas. Less obvious costs not generally recorded in damage reports include loss of business, cleanup costs to individuals, and job and wage interruption. Governments, communities, and insurance companies are finally beginning to react to damages and costs. They are beginning to push for safer communities that are less vulnerable

June 1972: Major Flooding from a Minor Hurricane along the Susquehanna River

Agnes was not much of a hurricane—actually, it was only a tropical storm by the time it reached New England, though it had an immense diameter of 1,600 kilometers. On June 15, 1972, Hurricane Agnes grew in the Gulf of Mexico, moved north though the Florida panhandle on June 19, spun off fifteen tornadoes, then weakened to the intensity of a tropical depression as it continued over Georgia, the Carolinas, and out into the Atlantic Ocean. There it again strengthened to a tropical storm and continued north.

On June 22, its center moved inland over southern New York and Pennsylvania, where it stalled from June 20 to June 25 and dumped 10 to 48 centimeters of rain, the largest amounts in the Susquehanna Basin (Figures 14-19 to 14-21). June 1972 was the wettest month on record for both Pennsylvania and New York, at 21.6 and 28.5 centimeters, respectively, both amounting to more than twice the normal annual precipitation. Agnes arrived in New England when the ground was already saturated from earlier rains. More would have to run off the surface without soaking in. Floods on the Susquehanna and Schuylkill rivers reached 200-year levels, the greatest flood since 1784 when records began. Peak flows on the Susquehanna were 1.5 times larger than the previously known maximum flood. The peak discharge at Harrisburg, Pennsylvania, was 28,900 cubic meters per second; that at Conowingo, Maryland, was 32,000 cubic meters per second. Peak flows on many rivers in Virginia also exceeded 100-year levels.

Susquehanna River Basin photo.

▶**FIGURE 14-19.** In 1972, Hurricane Agnes flooded Wilkes-Barre, Pennsylvania.

Pennsylvania Army National Guard photo.

▶**FIGURE 14-20.** The 1972 Susquehanna River flooding of Wilkes-Barre, Pennsylvania, was extensive.

▶**FIGURE 14-21.** Flood waters surge under the Occoquan River bridge in the 1972 flood.

Flood-control reservoirs on the west branch of the Susquehanna helped reduce flood heights by as much as 1 to 2 meters, but even there the peak flows set records. For the 25,000-square-kilometer area east of the Appalachians to the Atlantic coast, precipitation from Agnes exceeds all others except for two Florida hurricanes in 1916 and 1950. The amount of sediment carried by the Susquehanna and Schuylkill rivers during this single flood was three times the average annual load. Overbank deposition of sand and mud was widespread on major streams.

Until 2001, Agnes was tied for third costliest among major hurricanes. Only Hugo in 1989 and Andrew in 1992 caused more damage. However, except for increases in populations and property values, Agnes would be second only to Andrew. Fifty people died in Pennsylvania, and damages were $9 billion (in 2002 dollars), mostly from flooding. Twenty-four died in New York, and damages were more than $3 billion. Nineteen died in Maryland, and damages came to $4.75 billion. Thirteen died in Virginia. Agnes ranks fifth among twentieth-century hurricanes in damages to the northeastern states, almost entirely from flooding. Clearly, a weak (low category) hurricane can do as much damage as an especially strong one. A large-diameter weak hurricane wreaks havoc primarily because of freshwater flooding rather than high winds, waves, and surge.

The Susquehanna River that drains much of Pennsylvania and adjacent New York and Maryland drains less than 1 percent of the continental United States but averages 6 percent of its annual flood damage. As described for other hurricanes, damages included flooding of homes and businesses, public water and sewage facilities, and industrial and public utility plants; crop loss; and the disruption of jobs and commerce. Water was rationed in some towns. Fires that broke out in many places were left to burn because floods made them inaccessible.

What Went Wrong?

A slow-moving tropical storm with a huge diameter followed previous rains that saturated the ground. That low pressure storm cell pulled huge amounts of Atlantic Ocean moisture into Pennsylvania and adjacent areas.

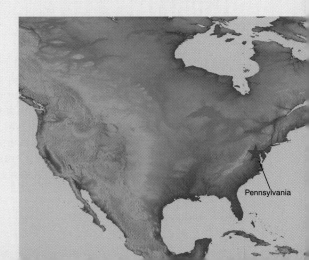

Pennsylvania

and incur reduced costs from coastal hazards—"disaster-resistant communities." Policies used include land use planning, building codes, incentives, taxation, and insurance. However, because tourism is the largest source of income for most coastal areas, most governing bodies are reluctant either to publicize their vulnerability or place many restrictions on anything related to tourist income.

Storm Surges

Storm surges, also called *storm tides,* occur when the low atmospheric pressure over an area of a major storm permits the sea level to rise. Because pressure of the atmosphere pushes down on the water surface, the height of the storm surge rises with lower atmospheric pressure (Sidebar 14-1).

In addition, prolonged high winds push seawater into huge mounds as high as 7.3 meters and 80 to 160 kilometers wide. The surge wave moves more or less at the speed of the storm (▶Figure 14-22). The pileup of water ahead of the wind is higher with greater wind speed and fetch length and with shallower water (Sidebar 14-2). The mound of water ahead of the storm wind slows down and piles up as it enters shallow water at the coast. Shallower water of the continental shelf forces the offshore volume of water into a smaller space, causing it to rise. A bay, inlet, harbor, or river channel that funnels the flow of water against the coast also causes the surge mound to rise. Even the Great Lakes are large enough to build storm surges 1 to 2 meters high.

The recurrence of a particular surge height is generally estimated from past experience in the same way that recurrence intervals are estimated from floods on streams (see Chapter 11, "Streams and Flood Processes," pages 285 and 290). Recurrence intervals for surge heights are estimated from the historic surges measured from tidal gauges in a given time period. A major problem, as with recurrence intervals of any hazard such as earthquakes or floods, is having a sufficiently long time to record the largest events. The calculation of surge height recurrence interval (R) is shown in Sidebar 14-3.

Surge height can be plotted against recurrence interval on semilog paper with values falling on a nearly straight line that can be projected to larger values that have not yet been recorded (▶Figure 14-23).

Thus, as with other hazards, there are numerous small surges, fewer large surges, and only rarely a giant surge. They are subject to a power law. Inverting the recurrence interval equation provides the probability of a surge of some height being exceeded. The probability (P) of a surge of a given height being exceeded (the exceedence probability) is shown in Sidebar 14-3. In this case, the probability can be plotted (on probability paper) against the surge height.

As with any 100-year recurrence interval event or other infrequent event, the timing of the next event of similar size is an average. It may not be far away. It could be the next month or next year. In fact, the largest events tend to cluster

National Weather Service photo.

▶**FIGURE 14-22.** As the storm surge moves inland, the sea level rises to cover large low-lying areas.

Sidebar 14-1

$$h = 43.3 - 0.0433\, P_0$$
where
   h = height of surge (in m)
   P = pressure in eye of hurricane (in hPa or millibars)

Sidebar 14-2

(height)^2 α wind speed × (fetch length − water depth)
Specifically
$$h = \sqrt{\frac{0.0625 \times w \times L \times D}{d \times g}}$$
and
h α 1/atmospheric pressure
where
   w = wind speed (m/sec)
   L = fetch length (m)
   D = water depth (m)
   d = density of salt water
   g = gravitational constant (980 cm/sec^2)

Sidebar 14-3

R = (n + 1) / m = Recurrence interval
P = m × 100 / (n + 1) = Exceedence probability
where
   m = rank of the surge event in the total record
   n = number of years of record

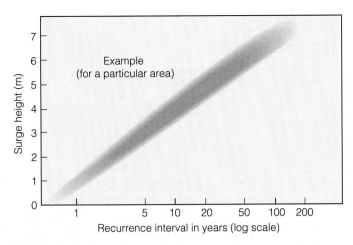

in time, perhaps because of a "forcing mechanism" that influences or causes the hazard.

The level of surge hazard depends on a variety of factors (► Figure 14-24). Because sea level rises rapidly during the onset of a storm surge, low-lying coastal areas are flooded and people drown. In fact, 90 percent of all deaths in tropical cyclones result from storm surge flooding (see Figures 14-22). Contrary to most people's expectations, the highest surge levels are not at the center of the hurricane but in the north to northeast quadrant of the hurricane eye wall (► Figure 14-25). Because hurricanes rotate counterclockwise, winds in that quadrant point most directly at the shore and cause the greatest effect there. The forward movement of the storm also enhances these winds because they blow in nearly the same direction that the storm is moving. Winds south of the eye of the hurricane are moving offshore and have the least effect.

The path of the hurricane compared with orientation of the shore also has an effect. A hurricane arriving perpendicular to the coast can lead to a higher storm surge because the whole surge mound affects the shortest length of coast. One arriving at a low angle to the coast is spread out over a greater length of shoreline but can build higher waves from a longer fetch.

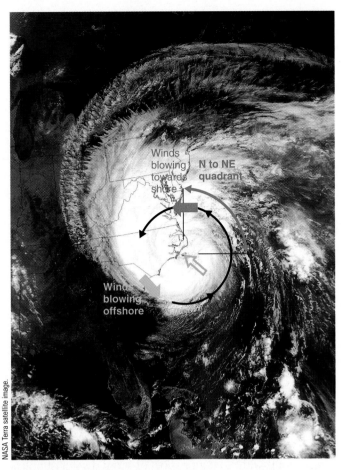

► **FIGURE 14-25.** The northeast quadrant of hurricane winds are directed toward the shoreline and inflict more damage; southeast quadrant winds are directed offshore and inflict least damage. Hurricane Isabel, 2003, is shown here as an example. State boundaries are superimposed on this natural color image taken at 11:50 A.M. EDT on September 18, 2003.

The shape of the shoreline (in map view) also has a major effect on the height of the storm surge. Bays or inlets, for example, can focus the storm surge mound into a smaller area, causing it to rise higher with resulting greater damage. Although people may feel more protected by living along the side of an inlet rather than along the open coast, the surge height in such a location may actually be higher.

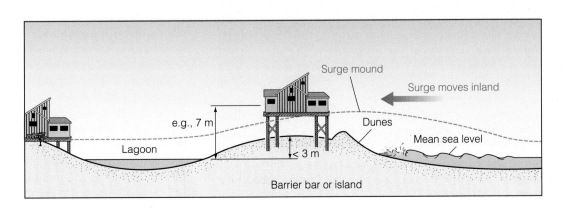

► **FIGURE 14-24.** This diagram shows some of the effects of a storm surge on a barrier island.

The forward speed of the storm center can have mixed effects. A faster storm movement pushes the storm surge into a higher mound that submerges the coast in deeper water, but slow-moving hurricane cells often inflict more overall damage because they remain longer over a region to dump greater rainfall and cause more landward flooding. Similarly, a larger-diameter hurricane also dumps greater rainfall because it impacts a larger area. It can also build higher waves because of longer fetch and build a wider surge mound because it pushes the mound of water over a larger area.

The inland reach of storm surge waters depends on a variety of factors. Most coastal areas of the southeastern United States and Gulf of Mexico stand close to sea level, so there is little to prevent surge flooding inland. However, inland progress of storm surge waters is slowed dramatically by the presence of vegetation and dunes. Because surge waters flow inland like a broad river, they are slowed by the height of coastal dunes, especially those covered with brush, and by near-shore forest. Areas where dunes are low or absent or where vegetation has been removed permit surge waters to penetrate well inland. Thus, people who lower dunes or remove vegetation to improve their view or ease of access to the sea invite equally easy access from any significant storm surge, along with the severe damage that it brings. In many cases, pedestrian paths across dunes to get to the beach foster sites of erosion and overwash. That damage includes the destruction of buildings, roads, bridges, and piers, along with contamination of groundwater supplies with saltwater, agricultural and industrial chemicals, and sewage. Saltwater can invade aquifers and corrode buried copper electrical lines.

▶**FIGURE 14-27.** Hurricane Fran (Category 3) in September 1996 moved the beach shoreward and carried huge amounts of sand (white) across the narrow barrier island and into the lagoon behind North Topsail Beach, North Carolina. Dark ridges bordering the road are railings on small bridges across low areas that provided easy access for storm waters.

A storm surge, finding a low area through dunes, or an area of little or no vegetation, will flow faster through the gap, eroding it more deeply. Subtle low areas invite overwash of sand eroded from the beach toward the lagoon (▶Figures 14-26 and 14-27). Deeper gaps may sever the barrier island when a breach forms. That not only destroys roads and houses but isolates homes and towns so that they have no road access to the mainland (▶Figure 14-8).

Buildings, bridges, and piers can be washed away by surge currents and by waves, or they can float away if not well anchored to foundations, or if the foundation is undermined by waves. Because larger waves feel the bottom at greater depth, they stir bottom sand and erode more deeply, undermining pilings and foundations. For this reason, most low-lying coastal homes are raised high on posts to be above the most frequent storm surge heights (▶Figure 14-28). Coastal houses are often built 4 meters off the ground on posts to raise them above the level of storm surges and waves. Aside from being a good idea, building codes that address flooding generally require it. Unless well braced, these posts may snap because of lateral forces on the walls above. Even if structurally sound, the house may settle when waves accompanying high storm surges undermine the piers on which the house is built.

Because a storm surge locally raises sea level, moves inland as a swiftly flowing current, and raises the height of wave attack, it amplifies all of the erosional aspects of a storm. Anything that can float will do so more readily; anything loose will be moved more easily by the waves. Houses are essentially big boxes full of air, so whole houses can be floated off their foundations and transported inland, often breaking up in the process. Thus, the nature and qual-

▶**FIGURE 14-26.** Overwash tongues of sand spread behind a vegetated dune crest at Camp Lejeune, North Carolina, after Hurricane Bertha (Category 2).

Donald Hyndman photo.

▶**FIGURE 14-28.** Houses in hurricane-prone areas are built on posts 4 meters high to minimize damage from storm waves and surges. These are in North Myrtle Beach, South Carolina.

M. Wolfe photo, FEMA.

▶**FIGURE 14-31.** This roof was lifted off a house in Kitty Hawk, North Carolina, by the storm surge from Hurricane Isabel in September 2003.

Dave Gatley photo, FEMA.

▶**FIGURE 14-29.** This house in Princeville, North Carolina, was not well anchored to its foundation. The surge from Hurricane Floyd lifted the house and deposited it on top of the family car.

Jimmy Oland photo, U.S. Army Corps of Engineers.

▶**FIGURE 14-32.** Hurricane Camille's surge lifted this sailboat and carried it inland to collide with a house. Apparently the boat sustained less damage than the house.

Orrin Pilkey photo.

▶**FIGURE 14-30.** This house, at Carolina Beach, North Carolina, was not well anchored to its slab. The surge accompanying Hurricane Fran (Category 3) in 1996 carried it off its slab and onto a road.

U.S. Army Corps of Engineers, Galveston office.

▶**FIGURE 14-33.** Surge flooding during Hurricane Alicia (low Category 3) in August 1983 drowned houses to their eaves in Baytown, 51 kilometers inland from Galveston, Texas.

ity of construction is important in minimizing damage. The building foundation should be well anchored to the ground such as through deeply embedded strong piles. The floors and walls should be well anchored to the foundation and the roof well anchored to the walls with "hurricane clips" (▶Figures 14-29, 14-30, and 14-31).

When the surge comes at high tide, the resulting sea level rises still higher, with correspondingly greater reach inland and greater damage (▶Figures 14-32 and 14-33). Although flooding compounded by water rise from storm surge can be catastrophic, wind damage is often ten times as great (see "Case in Point: Galveston Hurricane").

September 8, 1900

In 1900, the science of meteorology was in its infancy. Weathermen in Galveston, Texas, were crudely tracking the hurricane as it moved northwest over Cuba to south of Louisiana. Because there were no planes or satellites, much of the tracking was provided by radio reports from ships at sea. They expected that it would swing northeast to move up the eastern United States as most of them did. However, a stationary high pressure cell over Florida forced it to swing west instead—toward Galveston. The city was a prosperous center of shipping and resort activity, and huge 4.5-meter-high sand dunes along the edge of the city had been removed to provide better beach access. High water had spilled into the city in 1875 and 1886 but without doing much damage. The highest part along the crest of the island was only 2.5 meters above sea level.

The local weather bureau had been receiving storm warnings from headquarters in Washington, D.C., for a couple of days, but the first sign of a change was a heavy swell from the southeast, beginning in the afternoon of September 7. By 4 A.M. the morning of September 8, an especially high tide flooded the low parts of town, but the barometer had dropped only a little. Later that morning, people enjoyed watching the big surf until skies began to darken, winds strengthened, heavy rain fell, and the barometer dropped. The water rose to cover bridges to the mainland, and the rough weather precluded using boats. By the time people realized that it was a hurricane, they were trapped on the island. Houses crumbled in the waves, debris filled the streets, and roofing slates, whole roofs, and pieces of houses flew through the air. A wood railroad bridge broke loose and moved inland, sweeping up buildings and debris like a giant bulldozer. Injured people waded and climbed through debris to reach sturdy buildings on the highest ground. Those in sturdy buildings retreated to the upper floors as the water rose. An estimated 1,000 people survived in the elegant Tremont Hotel.

Early the morning of September 7, the steamship *Pensacola* left Galveston harbor, heading east for Pensacola, Florida. Captain Simmons was not aware that storm warnings were issued for Galveston Island thirty minutes after departure. Shortly after leaving the harbor, he ran into high seas and winds but continued east. That evening he dropped anchor, but the chain broke in the huge waves so the ship was adrift for all of the next day. When Simmons finally regained control of the ship he headed back, reaching Galveston on September 9, only to find battered ships and thousands of dead bodies.

The hurricane, with winds up to 193 kilometers per hour during a high tide, created a 6-meter storm surge (▶Figure 14-34). It was a Category 4 storm upon landfall. The offshore barrier island on which the city is built was completely covered with 3 to 5 meters of water. Waves and winds destroyed boats, bridges, more than half of the wooden buildings, and even some brick ones. The only thing that saved some of the remaining business district six blocks back from the beach was a wall of debris from the shattered buildings closer to the beach (▶Figure 14-35). At 8:30 P.M., the barometer hit bottom at 28.5 inches (723 mm). Thousands evacuated before the storm, but 8,000 to 12,000 died, mostly

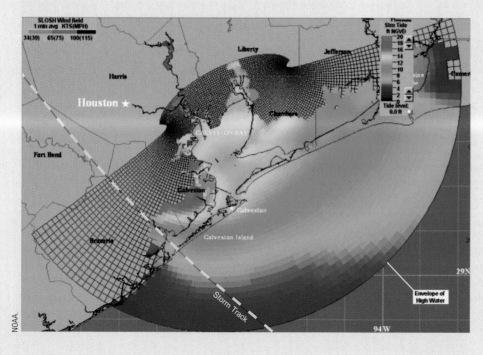

▶**FIGURE 14-34.** This map shows the storm surge that overwhelmed Galveston Island in 1900. The northwest trend of the storm track is shown as a dashed white line in the lower left.

> **FIGURE 14-35.** The 1906 Galveston hurricane piled up debris from buildings closer to the coast into a huge ridge that partly protected those behind.

in the storm surge. Survivors marooned with the loss of their only bridge to the mainland lacked water, food, medical supplies, and electricity. The third of the city closest to the Gulf completely disappeared. Along its landward edge was a ridge of debris that included the remains of 3,600 houses. Innumerable bodies were found in the jumbled debris; seventy a day turned up for a month. In the hot, humid weather, the stench of rotting bodies was overwhelming. Survivors doused the bodies with oil and burned them on the spot because they could not be buried in the water-saturated ground. Looting was rampant until martial law was imposed five days later. Damages, in 2002 dollars, amounted to $14.4 billion.

After destroying Galveston, the hurricane weakened as it continued north through the Great Plains, then turned northeast across the Great Lakes, New England, and eastern Canada before heading out to the Atlantic Ocean on September 15. Then again on August 16, 1915, another Category 4 hurricane, with tides 3.6 meters above normal, surged over Galveston's new seawall (see Figure 13-15) and flooded the business district to a depth of 1.8 meters. Two hundred seventy-five people died from flooding and high winds.

Before 1900, people believed that the greatest dangers from a hurricane were the huge waves and that those would be minimized by shallower water offshore. They were not aware that most people drown from the high water of the storm surge or that the huge mound of water is caused by low atmospheric pressure amplified by motion of the storm and restriction of water depth on the continental shelf. Nor did they understand how destructive a hurricane could be, so most did not try to evacuate until it was too late.

To provide protection from future hurricanes, Galveston and the U.S. Army Corps of Engineers raised the island surface by some 3 meters and built a giant seawall along the beachfront of the city (see Figure 13-15, page 334). The seawall has grown in length over the years and is now 16 kilometers long. As would be expected, the beach in front of the wall largely disappeared. Galveston regularly dumps truckloads of sand over the seawall to provide a narrow beach. West of the west end of the seawall, rapid development in the 1970s permitted homes to be built on the beach at an average elevation of only 1.5 meters. Hurricane Alicia, a Category 3 storm that arrived in 1983, destroyed or severely damaged ninety-nine of 207 homes on the beach. Although Texas law claims all land seaward of the vegetation line, lawsuits and pressure from real estate groups led the state and county to relent and provide building permits to repair and reoccupy damaged homes. Permits have also been provided for additional new homes on the beach near the west end of Galveston Island so that there are now 100 or so more beachfront homes than before Hurricane Alicia. Those houses were built on lots that sold for an average of $200,000 each. Beachfront dunes built since Alicia were washed away by Tropical Storm Frances in 1998. New developments on the west end of Galveston Island are now as vulnerable as Galveston was before the 1900 hurricane.

A storm equivalent to that of 1900 would likely bring a storm surge 4.6 meters high at the seawall to 7.6 meters at the north (inland) edge of Galveston Island. Seawater would be 1.5 to 3 meters deep in buildings in the center of town. West of the seawall, all 300 beachfront homes would be destroyed, their wreckage becoming battering rams that would destroy homes landward of them. Destruction of the built environment is certain; the number of people who die depends on the number who evacuate in time. It will take almost thirty-six hours to evacuate those at risk. The population of coastal counties in the area was 1.6 million in 1961; it is now almost 5 million, a large proportion being new residents who have no experience with hurricanes.

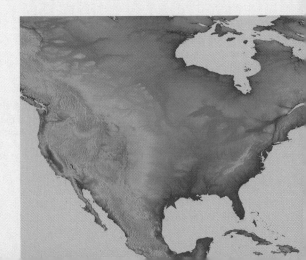

Waves and Wave Damages

Higher waves impact the coast with much more energy. They stir up sand to greater water depths both offshore and onshore, moving the sand farther offshore on gentler slopes. Thus, they erode sand from beaches and dunes, undermining any structures there (▶Figures 14-36 to 14-38). Loose debris carried by waves amplifies the damage. Boards, branches, logs, and propane tanks act as battering rams against buildings. Loose debris collected against building pilings strongly enhances the wave forces against the sides of a structure (▶Figures 14-39 and 14-40).

The level of damage at a particular site can be quite dramatic as shown in photographs taken before and after individual hurricanes. Many beachfront homes are completely destroyed when the sand beneath them is washed offshore. Others nearby sustain major wind and wave damage (▶Figures 14-41 and 14-42).

The width and slope of the continental shelf has a significant effect on wave damage. On a wide, gently sloping continental shelf offshore, the waves drag on bottom and stir up sand; this uses up more wave energy before the waves reach land, decreasing the damage. Large amounts of moving sand on a shallow bottom offshore reduce the wave energy available for coastal erosion. On a narrow continental shelf, as in the outer banks of North Carolina, or a steeper slope offshore, larger waves and breakers maintain their energy as they approach closer to shore, thereby causing more damage.

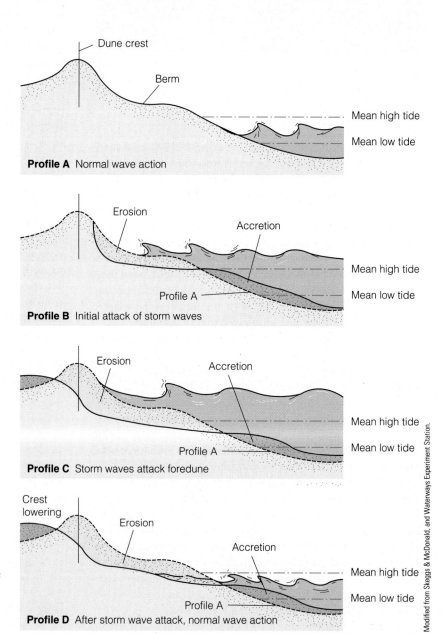

▶**FIGURE 14-36.** This series of diagrams shows the effects of a storm wave attack on a beach and dune. Note the progressive erosion of original shore Profile A, in part because of waves on higher sea level of storm surge.

Modified from Skeggs & McDonald, and Waterways Experiment Station.

FIGURE 14-37. Hurricane Irene on November 5, 1999, eroded much of the sand from the beach in front of Breakers Resort, Melbourne Beach, Florida. It came close to undercutting and destroying the buildings.

FIGURE 14-38. Hurricane Dennis on September 1, 1999, eroded the beach out from under these beachfront homes in Kitty Hawk, North Carolina.

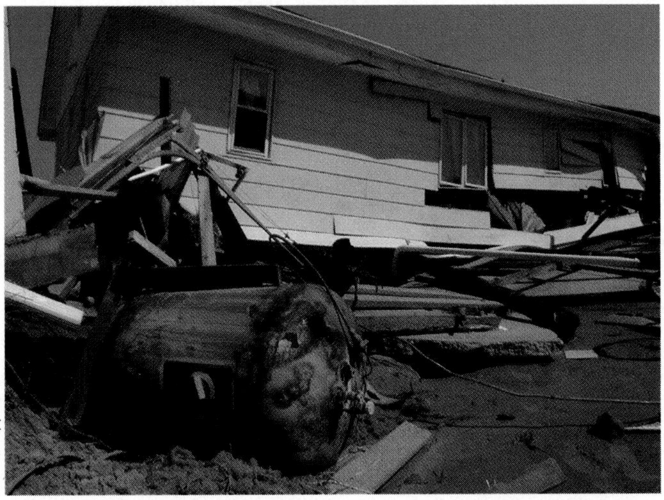

FIGURE 14-39. An unsecured propane tank that ripped loose acted as a battering ram during Hurricane Floyd on September 17, 1999. Aggressive erosion of sand under the house supports dropped this house at Long Beach, Oak Island, North Carolina, onto the sand, effectively destroying it.

FIGURE 14-40. This beachfront hotel at Panama City, Florida, was severely damaged by Hurricane Opal in 1995.

Winds and Wind Damages

Although the Saffir-Simpson Scale is designed to indicate the intensity of a hurricane, it can be misleading as an indicator of the level of expected damage. A hurricane impact scale is more appropriate for that purpose based on the factors below. In addition to the wind speed, the amount of damage from a hurricane is heavily dependent on the height of water rise from the storm surge, how large an area is covered, the duration of high water and high winds, and how recently another storm has affected the area.

Wind can blow down trees, power lines, signs, and wreck weak buildings; it can fan wildfires and destroy crops. It can blow in windows, doors, and walls and lift the roofs off houses. Damages by winds are greatly magnified by flying debris (▶ Figures 14-43 to 14-48). Wind velocity also has a major effect on wave height.

Wind does by far the greatest damage to buildings, much more than flooding, although significant damage results from rain entering the structure after minor wind damage. The lowest-level hurricane winds at 119 kilometers per hour apply approximately 73 kilograms per square meter (15 lb/ft^2) of pressure on the wall of a building. Thus, a force of 1,360 kilograms (3,000 lbs) would press against a wall 2.4 meters high and 7.6 meters long. If the wind speed doubles, the forces would be four times as strong; so in a Category 4 hurricane at 240 kilometers per hour, the force on the same wall would be 5,400 kilograms (12,000 lbs). Reduced pressure caused by the same winds on the downwind side of the building adds to the problem because the winds pull against the wall (▶ Figure 14-47). Roofs often fail before walls because of additional factors. The largest wind forces are caused by suction. Where a roof slopes toward the wind, the air is forced up and over the roof, lifting it in the same way that air flowing over an arched airplane wing lifts the

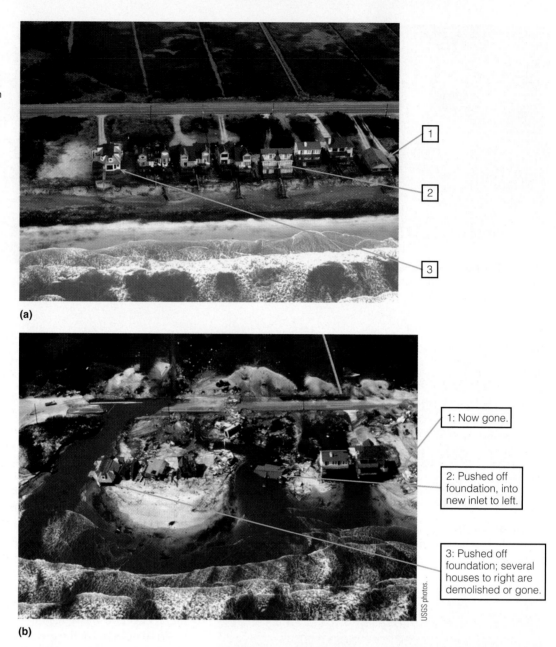

FIGURE 14-41. These photos show part of North Topsail Beach, North Carolina, **(a)** after Hurricane Bertha on July 16, 1996, and **(b)** after Hurricane Fran on September 7, 1996. The area is hardly recognizable.

(a)

(b)

1: Now gone.

2: Pushed off foundation, into new inlet to left.

3: Pushed off foundation; several houses to right are demolished or gone.

USGS photos.

plane. A steeper-sloped roof actually performs better in high wind than one sloped more like an airplane wing. The lifting forces under overhanging eaves tend to cause failure there first. Hip roofs sloping to all sides and lacking overhangs deflect the wind best. Larger roof spans are more vulnerable. Roof material also makes a difference. Shingles fare poorly because they can easily blow off during high winds. Metal, slate, or tile roofs are good; a single-membrane roof is better; and a flat concrete tile roof is much better. Concrete or steel beams supporting the roof are better than wood.

Sidewalls and windows are commonly sucked out rather than blown in. Keep this in mind when anchoring roofs, walls, and window coverings. When windows or doors fail because of flying debris, pressure increases inside the building, and the whole roof may be pulled off (▶Figures 14-47 and 14-48). Shutters or plywood well anchored over

windows and exterior doors help more than anything else because those are the vulnerable points of entry. Impact resistance is also important for windows and doors. Tempered and laminated glass windows are significantly better than ordinary window glass. Where more than 50 percent of a wall is windows or exterior doors, the wall is especially vulnerable. Skylights are especially susceptible to penetration. Double-wide garage doors, particularly if overhead doors, are unusually susceptible because the track holding the door may fail or the wind may force the door out of the track. Unreinforced brick chimneys fail in large numbers, even where against a backup wall, because of poor attachment to the wall. Brick, stone, or reinforced concrete block walls, however, are much stronger than wood-sheathed walls. Because much of the damage is from flying debris, gravel or roof-mounted objects on nearby buildings are dangerous.

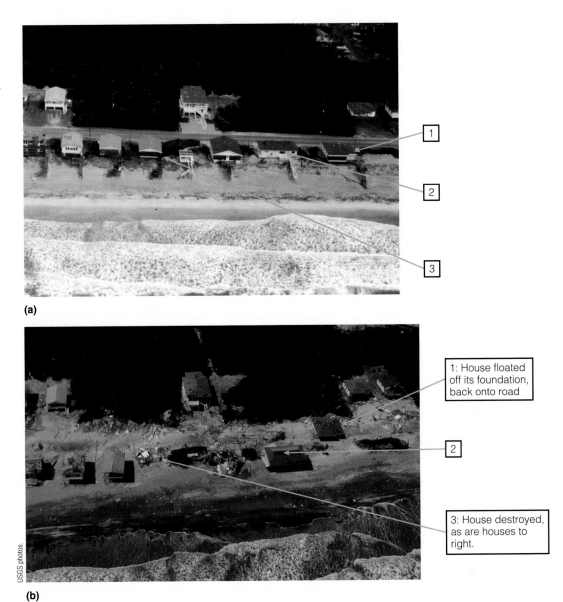

▶**FIGURE 14-42.** These houses on Topsail Beach, North Carolina, show extensive damage **(a)** after Hurricane Bertha on July 16, 1996, and **(b)** after Hurricane Fran on September 7, 1996.

1
2
3

(a)

1: House floated off its foundation, back onto road

2

3: House destroyed, as are houses to right.

USGS photos.

(b)

U.S. Army Corps of Engineers.

▶**FIGURE 14-43.** Hurricane Alicia (Category 3) in October 1983 removed the beach sand from under the concrete parking area below this house on the Texas Gulf Coast, leaving it suspended on the pilings that raised the house above surge level. It also peeled off much of the roof and the front half of this house.

FEMA photo.

▶**FIGURE 14-44.** In some places, entire roofs were lifted off these Florida houses during Hurricane Andrew on August 24, 1992.

▶**FIGURE 14-45.** Yes, hurricane winds can be high! The danger of wind-blown debris is vividly illustrated by a palm tree near Miami, which was impaled by a sheet of plywood during Hurricane Andrew in August 1992. This may have been caused by an embedded tornado.

▶**FIGURE 14-46.** Debris flying in hurricane winds can be dangerous. A sheet of plywood used to cover a window came loose to become a guillotine in Hurricane Frederick along the Gulf of Mexico coast in 1979.

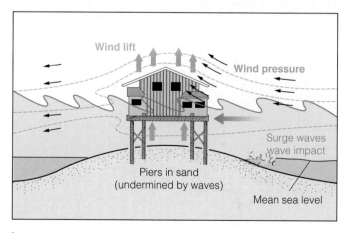

▶**FIGURE 14-47.** This diagram shows the wind, wave, and surge effects on a coastal building on a barrier bar.

▶**FIGURE 14-48.** Hurricane Alicia (Category 3) in October 1983 destroyed most houses near this beach along the Texas Gulf Coast. The house at top, just left of center, survived well because of its nearly flat roof with no overhangs and few windows, as well as its location (in the third row of houses back from the beach).

Rainfall and Flooding

Hurricanes not only cause severe beach and dune erosion by waves and associated surge action but also generally cause significant flooding because the storm is a strong low pressure weather system. Small-diameter or fast-moving storms cover a small area and pass over an area quickly, so they do not generally cause major flooding. Large-diameter or slow-moving hurricanes spend much more time over an area and typically drop large amounts of rain over large parts of a drainage basin, and for a longer period, so they cause more extensive and more prolonged flooding. Hurricane Agnes, only a tropical storm when it came back on land in Pennsylvania and New York in 1972, spread over a diameter of 1,600 kilometers and provided the largest rainfall on record. Heavy rain can not only cause flooding but also wash out structures from flooding streams, drown people, contaminate water supplies, and trigger landslides. Several centimeters of rain dumped over a few hours collect rapidly and run off slowly because of the gentle slopes of coastal plains. Draining downslope, the water collects to even greater depths to cause catastrophic floods in low areas. A good example is Hurricane Floyd in 1999 following on the heals of Hurricane Dennis, which had already saturated the ground (see "Case in Point: Hurricanes Dennis and Floyd, 1999").

Although the circumstance may be unusual, the additional load of a large surge mound on the Earth's crust may be sufficient to trigger an earthquake in an area already under considerable strain—as in the Tokyo earthquake of 1923 in which 143,000 people died, mainly in the cyclone-fanned fire that followed the rupture of gas lines.

Hurricane Dennis, a Category 3 storm, moved up the East Coast offshore beginning August 30, wandered back and forth erratically for a few days, then made landfall as a much weaker tropical storm on September 4 (▶ Figure 14-49). Unlike many hurricanes that move quickly across the shoreline, minimizing the time available for wave damage, Dennis remained 125 kilometers off the North Carolina coast for days. It generated big waves that progressively eroded the beaches through a dozen high tides. Sand overwashed the Carolina Outer Banks at numerous points, and some islands moved a little farther landward. Erosion was equivalent to that of a Category 4 hurricane.

A large frontal dune built in the 1930s for erosion control along 150 kilometers of beachfront had been progressively eroded by storms in the past decade. Hundreds of buildings now rest on the seaward-sloping beachfront where they are directly exposed to storm attack, but the houses would have to be removed to replace the old dune. Hurricane Dennis removed sand from beneath some buildings and stranded others below high tide level.

Ten days later, from September 14 to 18, Hurricane Floyd hit the East Coast from South Carolina to New Jersey (▶ Figure 14-49). However, it was only a strong Category 2 hurricane by the time it reached the U.S. mainland near Cape Fear, North Carolina, on September 16. Buildings on Oak Island, largely destroyed in a hurricane in 1954, had been rebuilt in their original locations well back from the beach. After forty-five years of landward beach migration, some 20 to 45 meters in total,

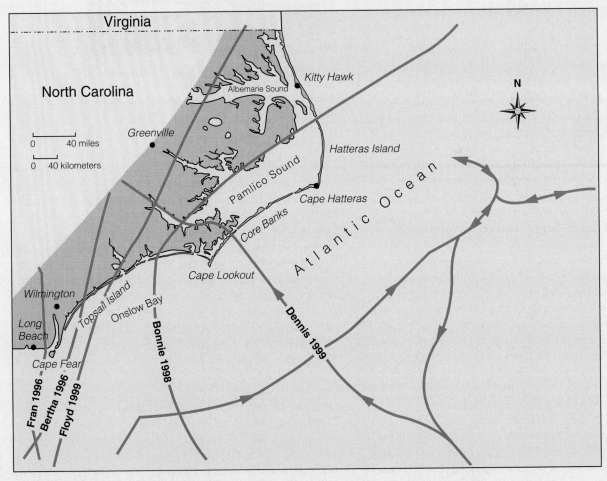

▶**FIGURE 14-49.** This map shows the erratic path of Hurricane Dennis and the nearly straight path of Hurricane Floyd, back-to-back storms in 1999.

Dave Gatley photo.

▶**FIGURE 14-50.** Sand eroded from under the shallowly anchored posts under this house, causing it to topple during Hurricane Floyd on September 17, 1999, at Long Beach, Oak Island, North Carolina.

U.S. Army Corps of Engineers,

▶**FIGURE 14-51.** Hurricane evacuation traffic comes to a standstill during evacuation for Hurricane Floyd. Note that a few people are outside their stopped cars.

half of them were again destroyed by Floyd (▶Figure 14-50). Its big waves came on top of a 1.5- to 2.5-meter storm surge.

On Topsail Island, a sheet of overwash sand covered the low, thin island. The artificial frontal dune along much of the island, constructed from sand bulldozed from the beach, largely disappeared. The dune had been replaced at least five times in the previous ten years. After Hurricane Fran destroyed the roads, utility lines, and septic systems in 1996, Floyd again largely destroyed them. They will again be replaced—primarily at federal taxpayer expense.

At least seventy-seven people died, most of those in flooding. Damages were in the billions of dollars. Some 2.6 million people evacuated from Florida to the Carolinas, the largest such evacuation in U.S. history. At the peak of the evacuation, almost every east-to-west highway was jammed with traffic, some almost at a standstill (▶Figure 14-51). On Interstate 26 west of Charleston, South Carolina, both directions were converted to only westbound traffic to handle the vehicles. Some people took two and one-half hours to cover fifteen miles. Most gas stations and restaurants were closed.

At least 110,000 hogs drowned, along with more than a million chickens and turkeys in this agricultural region. Peanuts, cotton, tobacco, and sweet potato crops were severely damaged. Water everywhere was a stinky mess. Drinking water was contaminated, and flooded water-treatment plants were

inoperable. Military helicopters and emergency personnel in boats rescued thousands of people from flooded houses, rooftops, and even trees. Almost every state from Florida to Connecticut was declared a major disaster area.

Floyd covered a larger area and lasted longer than many larger hurricanes, so its heavy rains lasted much longer. The storm dumped more than 50 centimeters of rain on coastal North Carolina, an area where Dennis had saturated the ground only two weeks earlier. Floyd's torrential rains had nowhere to go but to run off the surface. Flood levels from eastern North Carolina to New Jersey rose above the 100-year flood stage (▶Figure 14-52).

The maximum surge was 3.1 meters on Masonborough Island, North Carolina. The Tar River crested 7.3 meters above flood stage, and most roads east of Interstate 95 flooded. Twenty-four-hour rainfall totals reached 34 centimeters in Wilmington, North Carolina, and 35.5 centimeters at Myrtle Beach, South Carolina (▶Figure 14-53). Total rainfall along part of the coast reached 53.3 centimeters. As if that was not enough, with rivers still high, 15 more centimeters

USGS photo.

▶**FIGURE 14-52.** Heavy rains accompanying Hurricane Floyd submerged much of eastern North Carolina. The U.S. 64A bridge at Tarboro, North Carolina, is isolated by floodwaters. Think about trying to evacuate after you finally realized that the rivers were really going to flood your house. You might make it to the local highway, or even to the freeway—but then what?

of rain fell on eastern North Carolina, causing still more flooding. Heavy runoff carried sediment, organic waste, and pesticides from farms; hazardous chemicals from industrial sites; and raw sewage into coastal lagoons and bays and onto beaches.

According to FEMA, this was the worst flood disaster ever recorded in the southeastern states. For many people, the trauma and mess of the storms were just the beginning. Many thousands of dollars of appliances, computers, stereo equipment, clothing, and other belongings were ruined beyond repair. Far more than half of the people flooded out did not have flood insurance. A common reaction was that "No one told me I was in a flood area" or "I didn't know my insurance didn't cover floods."

Dave Gatley photo.

▶**FIGURE 14-53.** This family returns to its almost completely submerged home in a community near the Tar River just north of Greenville, North Carolina.

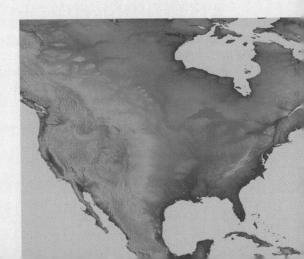

Deaths in Hurricanes

Deaths in hurricanes depend not only on strength of the hurricane—that is, the wind velocity and surge height—but also on several unrelated factors. Obvious problem areas are buildings that are close to the coast, buildings that are close to sea level, and weakly constructed buildings such as mobile homes. Less obvious problems are people's lack of awareness of the level of hazard involved in a hurricane and the wide range of hazards involved in such events. Finally, a large population in the path of the storm cannot evacuate quickly or efficiently, especially if an accident blocks a heavily used highway.

If people fail to evacuate before the storm arrives, high surge levels can flood roads under meters of water, making escape impossible. Fallen trees, power lines, and other debris can block roads, or the surge may cover the roadway, making escape impossible and rescue difficult (▶Figures 14-51, 14-52, 14-54, and 14-55). Many people delay their evacuation because of the inconvenience and cost or because they spend time purchasing and installing materials to help protect their homes from the storm. Others delay evacuation until the last minute, thinking it will take them only an hour or two to drive inland. However, only single two-lane bridges to the mainland serve most barrier islands. Those bridges and roads become snarled with traffic, and accidents can cause further dangerous delays.

A shift of as little as 80 kilometers in the Galveston 1900 hurricane could have resulted in far fewer deaths. That distance is typically the National Hurricane Center's current twenty-four-hour lateral forecast error for the path of a hurricane. Recall that a shift of 1992's Hurricane Andrew just 32 kilometers to the north would have caused two or three times as much damage. Sixty-four kilometers to the south would have resulted in little monetary loss in the minimally populated Florida Keys (see "Case in Point: Hurricane Andrew").

The most difficult area for evacuation is the Florida peninsula, which is only a few meters above sea level and everywhere less than 200 kilometers across. A hurricane can easily cross the whole state without losing much strength. Population growth has been rapid, especially along the coast, and it is difficult to predict the precise path of a hurricane.

The deadliest Atlantic hurricanes have been in the Greater and Lesser Antilles; those areas have had 64 percent of the total losses. Before Mitch in 1998, Mexico and Central America had 15 percent of the total deaths and the United States mainland 20 or 21 percent. Major catastrophes in Central America include large death tolls:

- some 10,000 in Nicaragua and Honduras from Hurricane Mitch from October 22 to November 5, 1998;

- approximately 8,000 in Honduras from Hurricane Fifi from September 15 to 19, 1974;

- 7,192 in Haiti and Cuba from Hurricane Flora from September 30 to October 8, 1963; and

- more than 6,000 at Pointe-a-Pitre Bay, Haiti, on September 6, 1776.

Texas Department of Transportation photo.

▶**FIGURE 14-54.** Some traffic still tried to evacuate after Tropical Storm Francis in 1999 began to move in.

Texas Department of Transportation photo.

▶**FIGURE 14-55.** Traffic was blocked in 1999 as a storm surge from Tropical Storm Francis began to move in.

August 24–27, 1992: Category 5

The low pressure cell that would become Andrew began as thunderstorms over western Africa that then moved west over the Atlantic. It strengthened to tropical storm status by August 17. On August 21 and 1,600 kilometers off Florida, the disrupting upper-level winds died down, allowing the growth in the cloud height and increased wind velocities. A deep high pressure system developed over the southeastern United States that extended eastward to north of the storm, permitting Andrew to gain strength. By August 22, it had grown into a hurricane with winds more than 119 kilometers per hour. Strengthening was rapid. Over the northern Bahamas on August 23, winds reached 240 kilometers per hour. On August 24, the storm reached Dade County, near Miami, Florida, with a central pressure of 922 millibars, the third lowest pressure recorded in the twentieth century (▶Figure 14-56). Sustained winds reached 281 kilometers per hour, but local vortices reached 322 kilometers per hour. This storm was originally thought to be a Category 4 but is now known to

have been a Category 5—and with a 5.2-meter surge in sea level.

As Andrew crossed west into the Gulf of Mexico, winds again picked up to 193 kilometers per hour before the storm again went onshore as a Category 3 hurricane in a sparsely populated part of the south-central coast of Louisiana and nearby Mississippi (▶Figure 14-56); 1,250,000 people evacuated low-lying areas of Louisiana. Fifteen deaths occurred before the storm dissipated as it moved farther inland to the north. Once again, Andrew's devastation would have been much worse if its course had shifted it toward New Orleans, where dikes protect much of this huge city below sea level.

Sixty-one people died, and 88,000 buildings and thousands of cars were destroyed. In 2002 dollars, the storm did $35.8 billion in damage, $28 billion of it in Dade County. It was the most expensive natural disaster in U.S. history, but it could have been worse. If Andrew's track had been 30 or 35 kilometers farther north, damages would have been two or three times as high. If it had been 65 kilometers south, in the Florida Keys, dollar losses would have been much lower.

▶**FIGURE 14-56.** Hurricane Andrew's eye passed over Miami and the Gulf of Mexico, then looped just west of New Orleans, Louisiana, on August 26, 1992, before turning inland to the northeast.

NOAA GOES satellite image.

Table 14-4 Hurricane Deaths in the United States, 1900–2004

Hurricane Location	Year	Category	Deaths
Galveston, Texas	1900	4	>8,000
Southeast Florida	1928	4	1,836
Florida Keys, south Texas	1919	4	600
New England	1938	3	600
Florida Keys	1935	5	408
Southwest Louisiana, north Texas (Audrey)	1957	4	390
Northeastern United States	1944	3	390
Grand Isle, Louisiana	1909	4	350
New Orleans, Louisiana	1915	4	275
Galveston, Texas	1915	4	275

The largest numbers of deaths in U.S. hurricanes from 1900 to 2004 are shown in Table 14-4. Note that no recent hurricanes make the list. Part of the reason is better forecasting and warning systems, as well as better transportation facilities. Another part of the reason may be the lesser hurricane activity between 1970 and 1995.

The costliest U.S. hurricanes from 1900 to 2004 are shown in Table 14-5. Note that all but one of the costliest hurricanes have occurred since 1954. Costs reflect the rapidly growing populations along the coasts, more building in less suitable locations, and more expensive buildings.

It might seem that low-lying, subtropical islands such as Grand Cayman in the Caribbean and Guam in the western Pacific would be especially vulnerable to storm surge and wave damage. However, fringing coral reefs force storm waves to break well offshore, minimizing those effects. Water depths offshore of the reefs tend to be deep, and absence of a wide continental shelf prevents the storm surge from

rising as high. However, those characteristics do not make these islands safe. Because many of them are low-lying, they are especially vulnerable to severe wind damage.

Poor Countries: Different Problems

In contrast to developed countries, poor countries have dramatically different problems, many of which seem almost insurmountable because of poverty, culture, and disastrous land use practices. Much of Central America, for example, is mountainous with fertile valley bottoms that are mostly controlled by large corporate farms. Many people have big families that survive on little food with marginal shelter; unable to afford land in the valley bottoms, they live in poorly built dwellings. They decimate the forests for building materials and fuel for cooking, leaving the slopes vulnerable to frequent landslides. Others migrate in search of work to towns in valley bottoms where they crowd into marginal conditions on floodplains close to rivers that are subject to flash floods and torrential mudflows fed from the denuded slopes. Hurricane Mitch, which was Category 5 offshore but weakened rapidly to tropical storm status onshore, provides a dramatic and tragic example of what can happen (see "Case in Point: Hurricane Mitch").

Poor countries in which low-lying coastal areas attract and provide food for large populations also have severe problems. The delta areas of large rivers in Southeast Asia support large numbers of people because the land is kept fertile by the frequently flooding rivers. As those populations grow, and as heavy land use in the drainage basins of those rivers leads to more frequent and higher flood levels, people clamor for protection from the ravages of the floods. Both governments and local groups build more and higher

Table 14-5 Costliest Hurricanes in the United States, 1900–2004

Rank	Hurricane	Date	Location	Insured Loss*	Total Estimated Loss†
1	Andrew	1992, Aug.	Florida, Louisiana	20.807	35.82
2	Charley	2004, Aug.	Florida, Carolinas	6.755	11.82
3	Hugo	1989, Sept.	Georgia, Carolinas, Puerto Rico, Virgin Islands	7.69	10.55
4	Ivan	2004, Sept.	Alabama, Florida, Georgia, 12 others	6.0	10.5
5	Agnes	1972, June	Florida, northeast U.S.		9.32
6	Betsy	1965	Florida, Louisiana		9.22
7	Frances	2004, Sept.	Florida, Georgia, Carolinas, New York	4.4	7.7
8	Camille	1969, Aug.	Mississippi, Louisiana, Virginia		7.57
9	Georges	1998, Sept.	Alabama, Florida, Puerto Rico, Virgin Islands, Louisiana, Mississippi	3.35	6.53
10	Diane	1955	Northeast U.S.		6.0
11	Jeanne	2004, Sept.	Florida, Pennsylvania, Georgia, S. Carolina, Puerto Rico	3.245	5.68
12	Floyd	1999, Sept.	Carolinas, New Jersey, Virginia, Florida	2.226	4.778

* In billions of 2004 dollars. Figures from Insurance Institute of America and NOAA.
† Estimated loss assumed as 1.75 times insured loss. Older losses recalculated using U.S. Dept. of Commerce Price Deflator.

levees (see "Case in Point: Cyclones of Bangladesh and Calcutta, India").

Building Restrictions

Significant destruction of homes and other structures in historic hurricanes has caused some states such as North Carolina to dictate a setback line based on the probability of coastal flooding. Seaward of this line, insurance companies will not insure a building against wave damage. In spite of challenges, the courts have so far upheld the building prohibition.

The National Flood Insurance Reform Act of 1994 required federal mapping of areas that are subject to river or coastal floods. The purpose was to implement a Flood Insurance Rate Map (FIRM) provided by FEMA. The special flood hazard areas that were designated have a 1 percent chance of being flooded in any given year, that is, they are comparable to the 100-year floodplain of streams. Coastal communities wishing to participate in the NFIP must use maps of these flood hazard areas when making development decisions. Coastal areas subject to significant wave action in addition to flooding are more vulnerable to damage. Significant waves are considered those higher than 1 meter. Therefore, National Flood Insurance Program construction standards are to be more stringent in such coastal zones. National flood insurance premiums are to be higher for those who live in more vulnerable areas.

Flood insurance costs in 2003 for single-family dwellings built after 1981, with no basement and raised at least 4 feet above mean sea level, were generally $450 per year for the first $50,000 of replacement value and $145 per year for an additional $50,000. As with stream flooding, there is a thirty-day waiting period after purchase before the insurance takes effect. Given that hurricanes are not predictable for a specific location for more than a day or two in advance, it would seem prudent to maintain flood insurance if you live in an area of possible storm surge.

Some undeveloped areas were protected by the Coastal Barrier Resources Act of 1982, which prohibited federal incentives to development and prohibited the issuance of new flood insurance coverage. However, in areas not covered by that act, flood insurance remained available for elevated structures located as far seaward as the mean high-water line, regardless of local erosion rates.

Coastal building standards for the National Flood Insurance Program require the following:

■ All new construction be landward of mean high tide.

■ All new construction and major improvements must be elevated on piles so that the lowest floor is above the base flood elevation for a 1 percent chance of being flooded in any year.

■ Areas below the lowest floor must be open or have breakaway walls. Fill for structural support is prohibited. Raising the lowest floor above 100-year flood level

does nothing, of course, to prevent erosion. Piers may be undercut, and the eroding beach will eventually move landward from under the structure, causing its collapse.

Human alterations of dunes and mangrove stands (▶Figure 14-57) that would increase potential flood damage are prohibited. Even walking on dunes is generally prohibited because it disturbs the sand and any vegetation covering it, thereby making the dune more susceptible to erosion by both wind and waves. People build elevated walkways to cross the dunes to the beach (▶Figure 13-37, page 344, and Figure 14-58). (Text continues on page 391.)

▶**FIGURE 14-57.** Mangroves live in the brackish water of lagoons, anchoring their roots in limestone bedrock. These dense thickets slow the inland advance of storm surges.

▶**FIGURE 14-58.** Walkways and stairs cross a former dune to minimize erosion and reduction of dune height. However, this South Carolina dune was completely eroded away by a small spring storm.

October 24 to November 4, 1998: Nicaragua and Honduras

A major hurricane impacts the poorest countries the hardest. Mitch was the fourth strongest Atlantic hurricane since 1900, with a minimum central pressure of 905 millibars. It formed as a tropical storm in the southwestern Caribbean Sea on October 21, moved slowly northwestward to reach hurricane strength on October 24, rapidly strengthening to Category 5 from October 26 through 28. Maximum sustained surface winds reached 290 kilometers per hour. Weakening somewhat, Mitch hovered near the north coast of Honduras before moving southwest and inland as a tropical storm on October 29 (▶Figure 14-59). Waves north of Honduras probably reached heights of more than 13 meters. For the next two days, the storm continued westward over Honduras, Nicaragua, and then Guatemala, producing torrential rains and floods. Some mountainous areas received 30 to 60 centimeters of rain per day, with storm-total rainfalls as high as 1.9 meters. After several days of wandering over Central America, a weakened Mitch turned northeastward across Yucatan and over the Gulf of Mexico, where it strengthened to tropical storm status before pounding Florida's Key West on November 4–5.

Larger towns throughout Central America concentrate in valleys near rivers, where businesses and homes were flooded, buried in mudflows, or washed away (▶Figure 14-60). Most of the damage and deaths were related to the rains rather than wind or tidal surge. Mitch was the deadliest hurricane in the western hemisphere since 1780; it killed more than 11,000 people in Central America, most in floods and mudflows. Whole villages disappeared in floodwaters and mudflows, and 18,000 people were never found. More than 3 million others lost their homes or were otherwise severely affected.

During and after the storm, there were critical shortages of food, medicine, and water. Dengue fever, malaria, chol-

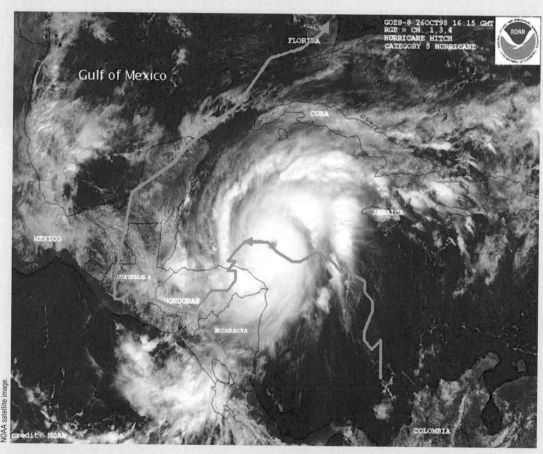

▶**FIGURE 14-59.** Hurricane Mitch, October 26, 1998, neared Honduras with winds at 180 miles per hour. Its cloudless eye is well defined. The storm track is shown as a green line that starts on the lower right and progresses to the upper right over this two-week period. Compare Figure 16-17.

USGS photo.

▶ **FIGURE 14-60.** Many homes in Honduras were buried by mudslides.

USGS photo.

December 16, 1998

▶ **FIGURE 14-61.** Washed out bridges crippled transportation and the distribution of food and fuel. This truck collapsed a bridge across the Manacal River—its underpinnings were weakened by incessant rains.

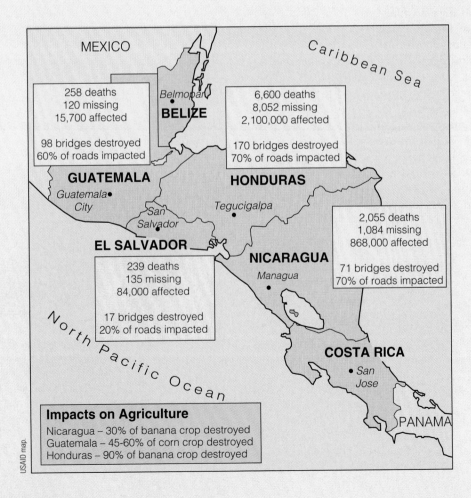

▶ **FIGURE 14-62.** This map details the deaths and damages from Hurricane Mitch following its path across Central America.

MEXICO

Caribbean Sea

Belmopan
BELIZE

258 deaths
120 missing
15,700 affected

98 bridges destroyed
60% of roads impacted

6,600 deaths
8,052 missing
2,100,000 affected

170 bridges destroyed
70% of roads impacted

GUATEMALA

Guatemala City

HONDURAS

Tegucigalpa

San Salvador

EL SALVADOR

239 deaths
135 missing
84,000 affected

17 bridges destroyed
20% of roads impacted

NICARAGUA

Managua

2,055 deaths
1,084 missing
868,000 affected

71 bridges destroyed
70% of roads impacted

North Pacific Ocean

COSTA RICA

San Jose

PANAMA

Impacts on Agriculture

Nicaragua – 30% of banana crop destroyed
Guatemala – 45-60% of corn crop destroyed
Honduras – 90% of banana crop destroyed

USAID map.

era, and respiratory illnesses were widespread. Roads were impassable (Figure 14-61), and helicopters for distributing relief supplies were in such short supply that some areas did not receive help for more than a week. Survivors surrounded by mudflows had to wait days for the mud to dry enough for them to walk to rescuers.

A large percentage of all crops were destroyed, including most of the banana and melon crops, as were shrimp farms in Honduras. Coffee and other crops in Nicaragua, Guatemala, and El Salvador suffered severe damage. Big agricultural companies that employed many of the people lost huge areas of farmland, so they had to lay off many employees. Damages came to more than $5.5 billion in 2002 dollars in Honduras and Nicaragua, two of the poorest countries. It will take decades for these nations to get back to where they were before Hurricane Mitch (Figure 14-62).

Many people live and work on the floodplains of large rivers and are used to flooding in hurricane season almost every year. When the river rises, they pull their furniture to the roof overhang, leave the house, and set up temporary shelters along a nearby raised highway until the water recedes in a couple of weeks. This time, Mitch took away nearly everything and left behind several feet of mud (Figure 14-60). With the help of government bulldozers and endless determination and hand labor, they try to put their lives back together.

Was the devastation of Hurricane Mitch a freak event that is not likely to repeat itself? Unfortunately, the combination of rapid population growth, widespread poverty, and lack of access to usable land makes Central America increasingly vulnerable to natural disasters. In 1995, 75 percent of Guatemala's and 50 percent of Nicaragua's populations were living in poverty, defined as living on less than $1 per day at 1985 prices. Those levels were worse after Hurricane Mitch. In 1974, 63 percent of Honduran farmers had access to only

6 percent of farmable land. Large corporate farms took over most of the fertile valley floors and gentle slopes for growing cotton, bananas, livestock, and irrigated crops. Peasants were forced onto steep slopes where they cleared forests for agriculture, building materials, and firewood; this caused increased soil erosion and the addition of sediment to rivers. Those who cannot survive by growing crops on those slopes move to cities in search of jobs. Lacking access to safe building sites there, they build shelters on steep, landslide-prone slopes or flood-prone riverbanks.

Improved hurricane forecasts are of limited use for these poverty-stricken people. Even if warned of approaching storms, most are reluctant to leave what little they own. In addition, they lack the resources to leave and have no way to survive when they get there. As a result, they are extremely vulnerable to such natural hazards.

The Bay of Bengal

Bangladesh and the northern coastal region of India around Calcutta have seen repeated terrible death tolls from tropical storms. These vast delta regions are low-lying and especially fertile—and one of the most densely populated regions on Earth, with some 130 million people in an area not much more than the size of New York state. People are exceptionally poor; their livelihoods are based directly on agriculture and therefore intimately tied to weather and its associated hazards. The delta lands on which people live and farm are virtually at sea level (▶Figure 14-63). Most people live on farms

NASA.

▶**FIGURE 14-63.** The delta of the Ganges River in Bangladesh is laced by numerous distributary channels and intervening land that is virtually at sea level (view to south, some 150 kilometers across).

rather than in the cities, and few rail lines connect the larger cities. Roads are narrow and crowded with heavy trucks, buses, and rickshaws. Bridges span only a few of the smaller river channels. People get around on bicycles, small boats, or ferries. Traffic jams in the cities are frequent on normal days. Imagine the chaos of trying to evacuate hundreds of thousands or millions of people with a couple of days' warning given such limited transportation.

Often there is adequate satellite warning to evacuate populations in the region, but most people do not leave because they believe that others will steal their belongings, they have lived through one major cyclone, and the average time between cyclones suggests to them that the next major one "will not come for many years," or if they die it is "God's will." The problem is enormous:

- In October 1737, 300,000 died in a surge that swept up the Hooghly River in Calcutta.

- In 1876, 100,000 died in another surge near the mouth of the Meghna River and thousands more in 1960 and 1965.

- In November 1970, cyclone winds of 200 kilometers per hour accompanied a 12-meter surge that swept across the low-lying delta of the Ganges and Brahmaputra rivers in Bangladesh. Four hundred thousand people died, many of them in the span of only twenty minutes. Whole villages disappeared, along with all of their people and animals.

- Again, on April 30, 1990, a cyclone with 145-mile-per-hour winds and a 6-meter surge swept into Bangladesh, drowning 140,000 people. The 1990 population of Bangladesh of 111 million is projected to double in thirty years, placing millions more on the delta.

- On June 6, 2001, nearly 100,000 people were stranded up to their chests in water following heavy monsoon rains. Villages were flooded when a flood protection embankment failed, and rail service was disrupted when the floods swept away a small bridge.

Catastrophic floods in Bangladesh are caused by several factors. The heavy monsoon rains from April through October are carried by warm, moist Indian Ocean winds from the southwest and magnified by the orographic effect of the air mass rising against the Himalayas. The rains are torrential and widespread in the drainage areas of the Ganges and Brahmaputra rivers, which come together in the broad delta region of Bangladesh. Rivers swell annually to twenty times their normal width. Those same drainage areas have been subjected to widespread deforestation and plowing of the land surface, which causes massive erosion and heavy siltation of river channels. The decrease in channel capacity causes the rivers to overflow more frequently.

In the delta region of Bangladesh, gradual compaction and subsidence of delta sediments coupled with gradual rises in sea level compound the problem. Relative sea level rise compared with the delta surface raises the base level of the rivers, reduces their gradient, and raises the water level everywhere on the delta. Cyclone-driven storm surges can put virtually the whole delta—75 percent of the country—underwater in a few hours (▶compare Figure 14-64).

Given that people live on the sea-level delta of two major rivers and in the path of frequent monsoon floods and frequent tropical cyclones, their options are few. The government plans to dredge and channelize more than 1,000 kilometers of riverbeds to improve navigation and build homes for flood victims on the new levees. Unfortunately, the impetus appears to be to improve food production rather than safety and will do little to prevent the problem. Ironically, it is those same floods that bring new fertile soils to the land surface. Dredging will increase the chances of flooding downstream; and the new levees, built from those fine-grained materials dredged from the delta's river channels, will be easily eroded during floods. Levee breaches will lead to avulsion of channels, widespread flooding, and destruction of fields. One useful suggestion is to construct high-ground refuges and provide flood-warning systems.

International Red Cross photo.

▶**FIGURE 14-64.** When flooded, the Mekong delta of Vietnam is underwater as far as you can see. The Ganges and Brahmaputra river deltas in Bangladesh also go underwater with the flooding and sea-level rise accompanying a tropical storm surge.

However, with still no national building code, state and local governments are responsible for enforcing their own codes. Some southeastern states have no statewide building codes at all. At the time that Hurricane Hugo hit South Carolina in 1989, half of the area of the state, including parts of the coast, had no building codes or enforcement at all. This lack of building codes results in much more widespread and severe destruction of property.

Most people agree that buildings should be built more strongly. Developers, builders, local governments, and many members of the public, however, often oppose such increased standards, especially those for winds. Their argument is that such regulations unnecessarily increase the cost of housing and limit economic development in the area. But studies by the research center of the National Association of Home Builders indicate that homes can be made much more resistant to hurricane damage at a cost of only a 1.8 to 3.7 percent increase in the total sales price of the property. Often increased standards fail to pass at all levels of government until a major disaster and huge losses make the need obvious to everyone. Even then, the issue of whether the state or counties are required to foot the cost of enforcing new regulations often thwarts new laws. When adequate building standards are not enacted and enforced, the general public is eventually forced to pay for the unnecessary level of damage. People's federal and local taxes could be lower if it were not for such unnecessary costs.

In the 1980s, the state of Florida designated a coastal construction control line (CCCL), seaward of which habitable structures are permitted only with adherence to certain standards of land use and building construction. The CCCL designates the zone that is subject to flooding, erosion, and related impacts during a 100-year storm. These standards are more rigorous than those of the National Flood Insurance Program and standard building codes. Hurricane Opal on October 4, 1995, provided a Category 3 test of the system. None of the 576 major habitable structures built to CCCL standards and seaward of the CCCL suffered substantial damage. Of the 1,366 preexisting structures in that zone, 768 (56 percent) were substantially damaged.

Hurricane Prediction and Warnings

Hurricane predictions include time of arrival, location, and magnitude of the event. The time of arrival is moderately predictable, but the specific location of landfall and the storm strength at landfall are poorly known more than a day or two in advance. We know that a hurricane's energy is drawn from the heat in tropical ocean water. But newly recognized large eddies of warm water 100 or more kilometers across can apparently spin off larger oceanic currents and boost the energy of hurricanes passing over

them, sometimes dramatically. Satellite sensors can monitor those thermal anomalies. Hurricane prediction, like any weather prediction, has significant uncertainty. Storms are monitored by weather satellites and by "hurricane hunter" aircraft that make daily flights into the storms to collect data on winds and atmospheric pressures. Within two to four days of expected landfall, they drop twenty to thirty 1-pound dropsonde sensing instruments into the storm from 30,000 to 40,000 feet. These transmit, wind, temperature, pressure, and humidity information, their fall being slowed by small parachutes. The path of the storm is commonly controlled by nearby high and low pressure systems.

Early warning in the United States, using weather satellites, allows people to evacuate by road or to reinforced high-rise buildings, such as those in Miami Beach. The National Hurricane Warning Center tries to give twelve hours warning of the hurricane path, though a modest storm can develop into a hurricane in less than that time and the storm path is often unpredictable. Most people believe it would take them less than one day to evacuate, but studies show that because of traffic jams and related problems most evacuations take up to thirty hours. With recent heavy development in coastal areas, 80 percent of the 40 million people living in areas subject to hurricanes have never been involved in an evacuation. The large seasonal tourist populations make matters even worse. Municipalities also view hurricane publicity as bad for tourism and property investment. Some people who have lived through one hurricane do not leave, believing that if they have done it before, they can do it again. Prospects for a calamity are genuine.

The costs of evacuations are also large. The usual estimate is that it costs roughly $1 million per coastal mile exclusive of any damages from the hurricane. Because forecasters refer to the statistical likelihood that a storm will strike a given length of the coast and there is significant uncertainty in the direction of the storm track, many more people are warned to evacuate than those in the final path of the storm. This is an unfortunate but necessary cost for the safety of the coastal population.

On average, 640 kilometers (400 miles) of coastline is warned of hurricane landfall within twenty-four hours. Of that, 200 kilometers (125 miles) may actually be strongly affected by the storm. Thus, some $275 million in costs were borne by people ultimately not in the storm's path. Clearly, more accurate forecasting could save significant costs.

The 2004 hurricane season showed that, even with our greatly improved prediction abilities, hurricanes still behave unpredictably. This was the first time four hurricanes came onshore in a single state since 1886 when Texas was the unfortunate victim. The first storm, Charley, approached the Gulf Coast of Florida as a Category 2 storm and was not expected to do severe damage. Much to the surprise of forecasters, Charley grew to a Category 4 storm in the few hours before it came ashore on August 13. Frances arrived on September 5 as a Category 2, Ivan (▶ Figure 14-6) made landfall

NOAA data.

▶**FIGURE 14-65.**
Four major hurricanes decimated Florida in August and September 2004, striking many areas more than once.

Map shows hurricane tracks in the North Atlantic Ocean with labels:

Charley Aug. 9 to 15
Ivan weakens
Jeanne Sept. 13 to 29
Frances Aug. 25 to Sept. 10
Ivan Sept. 2 to Sept. 18

North Atlantic Ocean

Geographic labels: Canada, United States, Cuba, Central America, South America, Africa

on September 15 as a Category 4, and then finally Jeanne dealt the final blow as a Category 3, following virtually the same track through Florida as Frances. In less than a month and a half, these storms wreaked havoc across Florida (▶Figures 14-65). The level of damage inflicted on coastal communities by Ivan should leave no doubt that people need to evacuate when warned (▶Figures 14-66 and 14-67). In some places, later storms hammered areas destroyed from the earlier storms as they were in the middle of repairs.

Categories of Hurricane Alerts

HURRICANE WATCH "A hurricane is possible in the watch area within 36 hours. Stay tuned to NOAA weather radio or a local station for additional advisories." This is the warning given during a hurricane watch. You should have prepared your property for high winds before this stage, but if you have not, then you should do the following immediately:

USGS.

▶**FIGURE 14-66.** Many buildings in Orange Beach, Alabama, were completely destroyed by Hurricane Ivan on September 15, 2004. Note that the front half and right side of the large five-story resort on the left collapsed into rubble. Anyone planning to ride out the storm there could not have survived. The one well-built house, near the center of both photos, survived, but the houses around it did not. Use red arrows to compare buildings.

▶FIGURE 14-67. Many beachfront homes in Orange Beach, Alabama, vanished in Hurricane Ivan in September, 2004. Even some behind the first row of buildings were destroyed; note the ground-level house to the right and behind the larger building on the right.

■ Fill your vehicle's gas tank. Gas station attendants may evacuate before you do.

■ Because winds cause the greatest damage, bring in anything that can become a damaging projectile in a high wind. Remove damaged or weak limbs on trees, along with extra branches to permit the wind to blow through. Remove outside antennas. Board up windows securely and reinforce garage doors.

■ Move a boat to a safe place, preferably above storm surge height. Secure a boat to its trailer with a rope or chain. Anchor the trailer to the ground or house with secure tiedowns.

■ Stock up on any prescription medications, food, and water—a one-week supply at home in addition to a three-day supply for evacuation.

■ Store as much clean drinking water as possible in plastic bottles, sinks, bathtubs. Public water supplies and wells often become contaminated, and electric pumps do not work without electricity. A small generator can be helpful, but stores sell out quickly when such crises arise.

■ Turn the refrigerator to the coldest setting. The power may be out for a long time.

■ Turn off appliances. The power surge while electricity is restored may damage them.

■ Turn off propane tanks and anchor them securely. Unplug small appliances; they may be damaged by a power surge.

■ Evacuate when authorities say to do so (▶Figures 14-54, 14-55).

■ Close and secure storm shutters or board up windows with plywood.

HURRICANE WARNING "A hurricane is expected in the warning area within twenty-four hours. If advised to evacuate, do so immediately." This message is typically given during a hurricane warning. The American Red Cross has a great deal of experience in planning for such warnings and suggests the following:

■ Plan to leave if you live on the coast, on an offshore island, in a floodplain, or in a mobile home. Tell a relative outside the storm area where you are going.

■ If you evacuate, take important items: identification, important papers (e.g., passports, insurance papers), prescription medicines, blankets, flashlights, battery radio and extra batteries, first aid kit, baby food and diapers, and any other items for a week or two in a shelter. Lock your house.

■ Listen to the radio for updates. Use telephones only for emergency calls.

■ Do not drive through floodwaters. The road may be washed out, and 2 feet of water can carry away most cars. Stay away from downed or dangling power lines; report them to authorities.

■ If trapped at home before you can safely evacuate, be aware that flying debris can be deadly. Stay in a small, strong, interior room without windows; stay away from windows even if they are shuttered. Hurricanes can spawn tornadoes. Close and brace exterior and interior doors. Use flashlights for emergency light; candles and kerosene lamps cause many fires.

■ If the eye of the hurricane passes over you, the storm is not over; high winds will soon begin blowing from the opposite direction, often destroying trees and buildings that were damaged in the first winds.

As recently as during Hurricane Floyd in 1999, the following became apparent:

1. People need to evacuate as soon as ordered rather than waiting to the last minute.

2. The huge increase in population along the coast in recent years has outgrown such infrastructure as highways. The roads are not able to cope with evacuating populations.

3. Changing freeways to single direction traffic away from the coast merely moves the bottleneck inland. It also creates severe safety problems at offramps where people traveling opposite from the normal direction of traffic try to get off to find lodging.

4. Lodging is completely inadequate to cope with the whole coastal population.

5. The latest approach is to move people 32 kilometers from the coast, beyond the limit of surge, to temporary protection, not hundreds of kilometers from the coast as was done in the past.

Hurricane Modification?

If we could somehow reduce the energy of a hurricane or dissipate it over a broad area, we could presumably reduce the hazard. Various schemes have been suggested, but none have proved feasible. For more than half a century, one suggestion has been to blow them apart using a nuclear bomb. Aside from the obvious problem of radioactive contamination, a moderate-size hurricane releases the energy equivalent of 400 twenty-megaton hydrogen bombs in one day. Clearly, that is not a viable possibility. Another suggestion is to spread a chemical film over the surface of the ocean under the hurricane to dramatically reduce evaporation from the sea surface. That would be similar to having the hurricane weaken as it passes over land or over cold water. Unfortunately, waves would immediately break up the chemical film and destroy its effectiveness. A third scheme is to spread minute particles of black carbon particles into the boundary layer around the hurricane's cloud shield. The black would absorb solar radiation, warm the boundary layer, increase convection in the outer part of the hurricane and in turn reduce the velocity of the hurricane's strongest winds. Dispersal of the carbon particles would be a technological challenge, the cost would be high, and the continent would lose precipitation that it normally receives.

One experiment at hurricane modification was actually attempted in the 1960s and '70s. The idea was to seed clouds in the outer edge of the eye wall with silver iodide. This would cause water droplets to freeze, releasing heat in the process. This would widen the radius of the eye and cause slowing of the maximum winds. The experiments did, in fact, appear successful in that maximum wind velocities decreased by 15 to 30 percent for some hours after the seeding. However, it was never really demonstrated that the seeding decreased the wind velocity; the costs of obtaining and maintaining suitable aircraft for monitoring the effects at high altitude were expensive. Further experiments were thwarted by the lack of hurricanes with well-developed eyes.

A More Damaging Hurricane than Any in Historic Time?

Hurricane damage is amplified by the following circumstances:

- heavy or prolonged rain;
- surge height (see Sidebar 14-2) amplified by funneling into a bay, wedging into shallow water of a gently sloping continental shelf, high northeast-quadrant winds pushing water up against the coast, and coinciding with high tide as in the Galveston hurricane of 1900;
- higher winds;
- higher storm waves pushed by higher winds (larger radius storm, longer fetch);
- warm eddies in the Gulf Stream adding energy to the storm;
- storm stalled by other weather systems such as nearby stationary high pressure systems; and
- following shortly after another hurricane when the ground is already saturated and when barrier islands and their beaches and dunes have been already stripped of sand (Hurricane Floyd in mid-September 1999 came only two weeks after Hurricane Dennis had saturated the ground and stripped sand from the beaches).

What would it take to have a still more devastating event? The highest category hurricanes are extremely damaging because of their high winds and because their low core atmospheric pressure leads to high surge heights. Lower-category hurricanes, however, often have a larger diameter, leading to longer fetch with higher storm waves and broader areas affected. A one-two punch could be provided by two or more large, back-to-back hurricanes. The effect could be amplified by hurricanes stalled by nearby stationary high pressure systems, by coinciding with an especially high tide, and by additional energy from huge warm-water cells in the Gulf Stream sweeping up the East Coast of the United States. If several of these influences were to overlap in time (see Figure 1-4, page 5), and this storm were to hit a major city such as Miami or New Orleans, we might have truly incredible damages.

KEY POINTS

✓ Hurricanes and other major storms cause severe coastal damage. Beaches and dunes are eroded, and buildings are severely damaged by wind, waves, and flooding. Damage is most severe where local zoning does not require raising buildings on posts and storm-resistant construction. **Review pp. 351–355.**

✓ Hurricanes, typhoons, and cyclones are all major storms that circulate counterclockwise with winds from 120 to more than 260 kilometers per hour. **Review pp. 355–357 and 360.**

✓ The strongest hurricanes, Category 5 on the Saffir-Simpson Hurricane Scale, have the lowest atmospheric pressure of less than 920 millibars and the highest wind speeds of more than 249 kilometers per hour. **Review pp. 355–357; Table 14-1.**

✓ Hurricanes that affect the southeastern United States form as atmospheric lows over warm subtropical water. The central core or eye of the hurricane is 20 to 50 kilometers in diameter, out of a total 160 to 800 kilometers. Clear, calm air in the low pressure eye is surrounded by the highest winds and stormy skies. **Review pp. 357 and 360–361; Figures 14-6, 14-11, and 14-12.**

✓ Hurricanes rotate counterclockwise, as in any low pressure area, but track clockwise, as with ocean currents. They grow off the west coast of Africa, then move westward with the trade winds. Most hurricanes are from August through October because it takes until late summer to warm the ocean sufficiently. They strengthen over warmer water and weaken over cool water or land. **Review pp. 360–361; compare Figures 14-14 to 14-16.**

✓ The greatest impacts from hurricanes are felt from North Carolina to Florida to Texas. Annual damages run from hundreds of millions to billions of dollars. **Review p. 360; Figure 14-17.**

✓ Nor'easters and other extratropical cyclones are similar to hurricanes except that they form in winter, lack a distinct eye, are not circular, and spread out over a large area. Like hurricanes, they are characterized by high winds, waves, and storm surges. **Review pp. 362–364.**

✓ Storm surges, as much as 7.3 meters high and 160 kilometers wide, result from a combination of low atmospheric pressure that permits the rise of sea level and prolonged winds pushing the sea into a broad mound. **Review pp. 368–371; Sidebars 14-1 and 14-2.**

✓ Surge hazards include significant rise of sea level and waves on top of the higher sea level. The surge and associated winds are concentrated in the northeast forward quadrant of the hurricane because of wind directions. **Review p. 369; Figures 14-24 and 14-25.**

✓ Damages from the hurricane depend on its path compared with shore orientation, presence of bays, forward speed of the hurricane, the height of dunes and coastal vegetation. **Review pp. 369–370.**

✓ The nature and quality of building construction has a major effect on damages. Floors, walls, and roofs need to be well anchored to one another, and buildings should be well attached to deeply anchored stilts. Damages from wind are greatest. **Review pp. 370–371.**

✓ High waves have much more energy and erode the beach lower and to a flatter profile, as well as erode dunes, especially if the waves are not slowed by shallow water offshore. **Review pp. 374–375; Figures 14-36 to 14-38.**

✓ Wind damages include blowing in windows, doors, and walls; lifting off roofs; blowing down trees and power lines; and flying debris. **Review pp. 375–378; Figures 14-43 to 14-48.**

✓ Rain and flooding from hurricanes can be greater with large-diameter, slower-moving storms. **Review pp. 378–381.**

✓ Many people think they can evacuate quickly and leave too late. Storm surges and downed trees and power lines often close roads. In addition to buildings and bridges, damages include the deaths of farm animals, agricultural damage, contaminated drinking water, and landslides. **Review p. 382.**

✓ Thousands of people in poor mountainous countries such as those in Central America die in floods and landslides triggered by hurricanes because they live in poorly constructed houses on floodplains and unstable steep slopes. **Review pp. 384–390.**

✓ Thousands of people living on low-lying deltas of major rivers die in floods and storm surges **Review pp. 384 and 389–390.**

✓ The National Flood Insurance Program requires that buildings in low-elevation coastal areas be landward of mean high tide and raised above heights that could be impacted by 100-year floods, including those imposed by storm surges. Restrictions are similar to those for streams. **Review pp. 385 and 391.**

✓ Hurricane predictions and warnings include the time of arrival, location, and magnitude of the event. The National Hurricane Warning Center tries to give twelve hours of warning, but evacuations can take as many as thirty hours. **Review pp. 391–394.**

IMPORTANT WORDS AND CONCEPTS

Terms

cyclone, p. 351

extratropical cyclone,
p. 362

eye, p. 357

fetch, p. 363

hurricane, p. 351

hurricane warning, p. 393

hurricane watch, p. 392

Nor'easter, p. 362

storm surge, p. 352

typhoon, p. 351

QUESTIONS FOR REVIEW

1. What causes a tropical cyclone or hurricane? Where does a hurricane get all of its energy?

2. Where do hurricanes that strike North America originate? Why there? Why do they track toward North America?

3. Why are coastal populations so vulnerable to excessive damages (other than the fact that they live on the coast)?

4. Where in a hurricane is the atmospheric pressure lowest and approximately how low might that be?

5. When is hurricane season (which months)? Why then?

6. What two main factors cause increased height of a storm surge?

7. What effects does the wind have on buildings during a hurricane?

8. What effects do the higher waves of hurricanes have on the coast?

9. If the forward speed of a hurricane is greater, what negative effect does that have? What positive effect does it have?

10. Which part of a hurricane does the greatest damage? (In other words, if the eye of a west-moving hurricane were to go right over Charleston, South Carolina, where would the greatest damage be?)

11. What is the difference in damage if a hurricane closely follows another hurricane—for example, by a week? Why?

12. Why is there more coastal damage if the sand dunes are lower?

13. What shape of roof is most susceptible to being lifted off by a hurricane? Why?

14. Why is it so important to cover windows and doors with plywood or shutters?

15. Why do developers, builders, local governments, and many members of the public oppose higher standards for stronger houses?

FURTHER READING

Assess your understanding of this chapter's topics with additional quizzing and conceptual-based problems at:

 http://earthscience.brookscole.com/hyndman.

CHAPTER 15

J. Pack, FEMA.

Seven people died in the Mossy Grove, Tennessee, tornado on November 12, 2002.

Thunderstorms

Thunderstorms, as measured by the density of lightning strikes, are most common in latitudes near the equator, such as central Africa and the rain forests of Brazil (▶ Figure 15-1). The United States has an unusually large number of lightning strikes and severe thunderstorms for its latitude. These storms are most common from Florida and the southeastern United States through the Midwest because of the abundant moisture in the atmosphere that flows north from the Gulf of Mexico (▶ Figure 15-1).

Thunderstorms form as unstable, warm, and moist air rapidly rises into colder air and condenses. As water vapor condenses, it releases heat. Because warm air is less dense than cold air, this added heat will cause the rising air to continue to rise in an updraft. This eventually causes an area of falling rain in an outflow area of the storm when water droplets get large enough through collisions. If updrafts push air high enough into the atmosphere, the water droplets freeze in the tops of **cumulonimbus clouds;** these are the tall clouds that rise to high altitudes and spread to form

THUNDERSTORMS AND TORNADOES

This worldwide map shows the average density of annual lightning flashes per square kilometer.

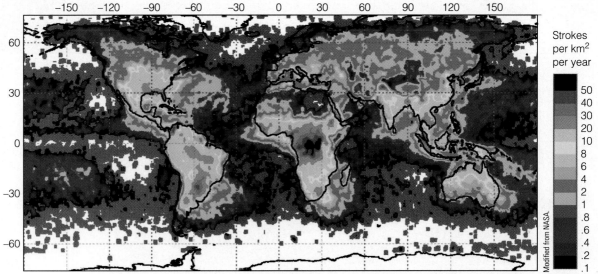

wide, flat. anvil-shaped tops (►Figure 15-2). This is where lightning and thunder form.

Cold air pushing under warm moist air along a cold front is a common triggering mechanism for these storm systems, as the warm humid air is forced to rapidly rise over the advancing cold air. Isolated areas of rising humid air from localized heating during the day or warm moist air rising against a mountain front or pushing over cold air at the surface can have similar effects. Individual thunderstorms average 24 kilometers across, but coherent lines of thunderstorm systems can travel for more than 1,000 kilometers. Lines of thunderstorms commonly appear in a northeast-trending belt from Texas to the Ohio River valley. Cold fronts from the northern plains states interact with warm moist air from the Gulf of Mexico along that line so the front and its line of storms moves slowly east.

Thunderstorms produce several different hazards. Lightning strikes kill an average of eighty-six people per year in the United States and start numerous wildfires. Strong winds can down trees, power lines, and buildings. In severe thunderstorms, large damaging hail and tornadoes are possible (see "Up Close: Jarrell Tornado, Texas, 1997").

Lightning

Lightning results from a strong separation of charge that builds up between the top and bottom of cumulonimbus clouds. Atmospheric scientists commonly believe that this **charge separation** increases as water droplets and ice particles are carried in updrafts toward the top of cumulonimbus clouds and collide with the bottoms of downward-moving ice particles or hail. The smaller upward-moving particles tend to acquire a positive charge, while the larger downward-moving particles acquire a negative charge. Thus, the top of the cloud tends to carry a strong positive charge, while the lower part of the cloud carries a strong negative charge (►Figure 15-3). This is a much larger but similar effect to static electricity that you build up by dragging

►**FIGURE 15-2.** A huge stratocumulus cloud spreads out at its top to form an "anvil" that foretells a large thunderstorm.

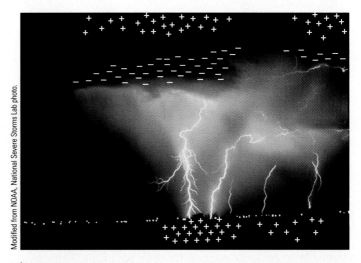

►**FIGURE 15-3.** In a thunderstorm, lighter positive-charged rain droplets and ice particles rise to the top of a cloud while the heavier negative-charged particles sink to the cloud's base. The ground has a positive charge. In a lightning strike, the negative charge in the cloud base jumps to join the positive charge on the ground.

On May 27, 1997, around 1 P.M., a tornado watch was issued for the area of Cedar Park and Jarrell, 65 kilometers north of Austin, Texas. Many people heard the announcement on the radio or on television, but most went on with their daily work. Storms are common in the hill country. This case seemed familiar: A cold front from the north had collided with warm, water-saturated air from the Gulf Coast to generate a line of thunderstorms. A tornado warning was issued at 3:25 P.M.

Just before 4 P.M., a tight funnel cloud swirled down from the dark clouds 8 kilometers west of Jarrell, a community of roughly 450 people. This tornado moved south–southeast along Interstate 35 at 32 kilometers per hour rather than taking a more typical easterly track. A local warning siren sounded ten to twelve minutes before the funnel struck.

When trained spotters saw a tornado on the ground, the alarm was sounded and everyone who could took shelter. Some sought protection in interior rooms or closets; few homes have basements because limestone bedrock is usually close to the surface. People in this area are advised to take shelter in closets and bathtubs with a mattress for cover, but in this case it did not matter. Within minutes, the F5 tornado wiped fifty homes in Jarrell completely off their foundation slabs. Hail the size of golf balls and torrential rain pounded the area. Wind speeds were 400 to 435 kilometers per hour for the twenty to twenty-five minutes the twister was on the ground. At least thirty people died.

One woman had hidden under a blanket in her bathtub. Her house blew apart around her, and both she and the tub were thrown more than 100 meters. She survived with only a gash in her leg. Some people watched the tornado approach and decided to outrun it by car. They survived, but in other tornadoes people have died doing this when they would have survived at home. Eyewitnesses reported that the Jarrell tornado lifted one car at least 100 meters before dropping it as a crumpled, unrecognizable mass of metal.

This was the second tornado to strike Jarrell; the first was only eight years previously on May 17, 1989.

One of several tornadoes during the same event moved south through the town of Cedar Park, demolishing a large Albertson's supermarket, where twenty employees and shoppers huddled in the store's cooler. One of us happened to be a few kilometers south of Cedar Park playing golf that hot and humid Texas morning. Thunderstorms began to build on the horizon, and the sky took on a greenish gray cast. Early in the afternoon, golf course attendants quickly drove around the course warning players that there were two spotted tornadoes in the area. Because thunderstorms and tornadoes are fairly common in the area, many people become complacent; several people thought about finishing their golf rounds. Reaching the car in a drenching downpour, we realized that there was no safe place to go. Our cell phones were useless because all circuits were busy. Fortunately, the tornadoes were north of us, so we drove south into Austin to wait out the storm.

---

your feet on carpet during dry weather, a charge that is discharged as a spark when you get near a conductive object.

The strong negative charges near the bottom of the clouds attract positive charges toward the ground surface under the charged clouds, especially to tall objects such as buildings, trees, and radio towers. Thus, there is an enormous electrical separation or potential between different parts of the cloud and between the cloud and ground. This can amount to millions of volts; eventually, the electrical resistance in the air cannot keep these opposite charges apart, and the positive and negative regions join with an electrical lightning stroke (▶ Figures 15-3 and 15-4).

Because negative and positive charges attract one another, a negative electrical charge may jump to the positive-charged cloud top or to the positive-charged ground. Air is a poor conductor of electricity, but if the opposite charges are strong enough they will eventually connect. Cloud-to-ground lightning is generated when charged ions in a thundercloud discharge to the best conducting location on the ground.

C. Clark photo, NOAA.

▶**FIGURE 15-4.** The return stroke on the left side of this photo is much brighter than both the small leader coming up from the ground and the cloud-to-cloud stroke on the right.

Negatively charged **step leaders** angle their way toward the ground as the charge separation becomes large enough to pull electrons from atoms. When this occurs, a conductive path is created that in turn creates a chain reaction of downward-moving electrons. These leaders fork as they find different paths toward the ground; as they move closer, positive leaders reach upward toward them from elevated objects on the ground (see the lower right side of Figure 15-5). If you ever feel your hairs pulled upward by what feels like a static charge during a thunderstorm, you are at high risk of being struck by lighting. When one of the pairs of leaders connects, a massive negative charge follows the conductive path of the leader stroke from the cloud to the ground. This is followed by a bright return stroke moving back upward to the cloud along the one established connection between the cloud and ground (▶ Figure 15-4). The enormous power of the lightning stroke instantly heats the air in the surrounding channel to extreme temperatures approximating 50,000°F or 28,000°C. The accompanying expansion of the air at supersonic speed causes the deafening boom that we hear as **thunder.**

In fewer cases, lightning will strike from the ground to the base of the cloud; this can be recognized as an upwardly forking lighting stroke (▶ Figure 15-5) rather than the more common downward forks observed in cloud-to-ground strokes. Lightning also strikes from cloud to cloud to equalize its charges, although there is little hazard associated with such cloud-to-cloud strokes (visible in Figures 15-3 and 15-4).

Lightning is visible before the clap of thunder because of the difference between the speed of light and the speed of sound. Sound travels a kilometer in roughly three seconds, while light will travel this distance almost instantaneously. Thus, the time between seeing the lightning and hearing the thunder is the time it takes for the sound to get to you. If the time difference is twelve seconds, then the lightning is about 4 kilometers away. It is generally recommended that you take cover if you hear thunder within thirty seconds of the lightning and stay in a safe place until you do not see lightning flash for at least thirty minutes.

Danger from lightning strikes can be minimized by observing the following:

- Take cover in an enclosed building. Do not touch anything that is plugged in. Do not use a phone with a cord; cordless phones and cell phones are okay. One of us was struck by lightning through a corded phone—not something you want to experience.

- Do not take a shower or bath or wash dishes.

- Stay away from high places or open fields or open water. Water conducts electricity.

- Stay away from tall trees. If there are tall trees nearby, stay under low bushes or areas of small trees.

- If trapped in the open, crouch on the balls of your feet, away from other people. Keep your feet touching to minimize the chance that a lightning strike will kill you as it goes up one leg, through your body, and down the other. Do not lie down because that increases your contact with the ground. You can be burned many meters away from the site of a strike.

- Stay away from metal objects, such as fences, golf clubs, umbrellas, and farm machinery (▶ Figure 15-6). Avoid tall objects such as trees or areas of high elevation such as a hill or mountain. Rubber-tired vehicles do not provide insulation from the ground because water on the tires conducts an electric charge.

Stay inside a car with the windows rolled up and do not touch any metal. Pull over and stop; do not touch the steering wheel, gearshift, or radio. The safety of a car is in the metal shield around you, not in any insulation from the tires.

Of the more than 100,000 thunderstorms in the United States each year, the National Weather Service classifies 10,000 as severe. Those severe storms spawn up to 1,000 tornadoes each year. The weather service classifies a storm as severe if its winds reach 93 kilometers per hour, spawns a tornado, or drops hail larger than 1.9 centimeters in diameter. Flash flooding from thunderstorms causes more than 140 fatalities per year (floods are reviewed in Chapter 11).

▶**FIGURE 15-5.** This ground-to-cloud lightning stroke was observed near East Lansing, Michigan, in spring 2004.

▶**FIGURE 15-6.** Reality can be gruesome. These cows were probably spooked by thunder and ran over against the barbed wire fence, where they were electrocuted by a later lightning strike. Note that they were at the base of a hill but out in the open.

Dr. Theodore Fujita photo; courtesy of Dr. Kaz Fujita.

Downbursts

Several airplane accidents in the 1970s spurred research into the winds surrounding thunderstorms. This research demonstrated that small areas of rapidly descending air, called **downbursts,** can develop in strong thunderstorms. Downburst winds as fast as 200 kilometers per hour and microburst (small downbursts with less than 4 kilometers radius) winds of up to 240 kilometers per hour are caused by a descending mass of cold air, sometimes accompanied by rain. These severe downdraft winds pose major threats to aircraft takeoffs and landings because they cause **wind shear,** which results in planes plummeting toward the ground as they lose the lift from their wings. Once Dr. Tetsuya (Ted) Fujita proved this phenomenon and circulated the information to pilots and weather professionals, the likelihood of airline crashes because of downbursts was greatly reduced.

When these descending air masses hit the ground, they cause damage that people sometimes mistake as having been caused by a tornado. On close examination, downburst damage will show evidence of straight line winds: Trees and other objects will lie in straight lines that point away from the area where the downburst hit the ground (► Figure 15-7). This differs from the rotational damage that is observed after tornadoes, where debris lies at many angles due to the inward flowing winds.

Hail

Hail causes $2.9 billion in annual damages to cars, roofs, crops, and livestock (► Figure 15-8). **Hailstones** appear when warm humid air in a thunderstorm rises rapidly into the upper atmosphere and freezes. Tiny ice crystals waft up and down in the strong updrafts, collecting more and more ice until they are heavy enough to overcome updrafts and

fall to the ground. The largest hailstones can be larger than a baseball and are produced in the most violent storms. Hailstorms are most frequent in late spring and early summer, especially April to July, when the jet stream migrates northward across the Great Plains. The extreme temperature drop from the ground surface up into the jet stream promotes the strong updraft winds. Hailstorms are most common in the plains of northern Colorado and southeastern Wyoming but rare in coastal areas. Hail suppression using supercooled water containing silver iodide nuclei has successfully been used to reduce crop damage; however, this practice was discontinued in the United States in the early 1970s because of environmental concerns.

Tornadoes

Tornadoes, the narrow funnels of intense wind, typically have rapid counterclockwise rotation (► Figure 15-9), though 1 percent or so rotate clockwise. They descend from the cumulonimbus cloud of a thunderstorm to wreck havoc on the ground. They form in certain large convective thunderstorms. Tornadoes are nature's most violent storms and the most significant natural hazard in much of the midwestern United States. They often form in the right-forward quadrant of hurricanes, in areas where the wind shear is most significant. Even weak hurricanes spawn tornadoes, sometimes dozens of them.

The United States has an unusually high number of large and damaging tornadoes relative to the rest of the world. The storms that lead to tornadoes are created through the collision of warm humid air moving north from the Gulf of Mexico with cold air moving south from Canada. Because there is no major east–west mountain range to keep these air masses apart, they collide across the southeastern and

Fred Phillips photos.

(a)

(b)

▶**FIGURE 15-8.** **(a)** A violent storm over Socorro, New Mexico, on October 5, 2004, unleashed hailstones, many larger than golf balls and some 7 centimeters in diameter. **(b)** Most cars caught out in the open suffered severe denting and broken windows. In some cases, hailstones went right through car roofs and fenders.

midwestern United States. These collisions of contrasting air masses cause intense thunderstorms that sometimes turn into deadly tornadoes.

A tornado path on the ground is generally less than 1 kilometer wide but up to 30 kilometers long. They rarely last more than thirty minutes. Typical speeds across the ground are in the range of 50 to 80 kilometers per hour, but their internal winds can be as high as 515 kilometers per hour, the most intense winds on Earth. The severity of a tornado is classified by those internal wind speeds and linked to their associated damage using the Fujita Tornado Scale (▶Table 15-1 and Figure 15-10).

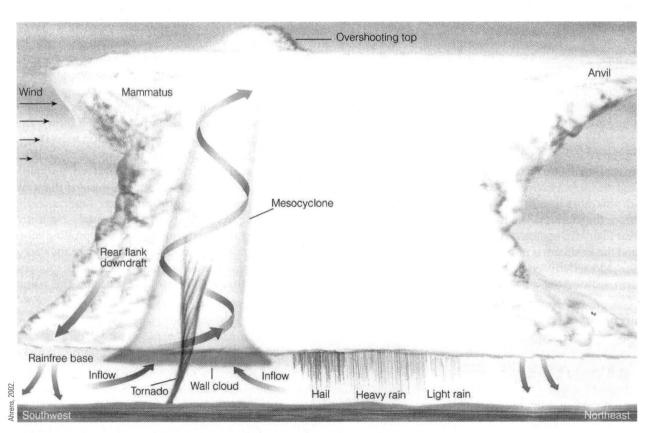

▶**FIGURE 15-9.** In this lateral view of a classic supercell system, the system is moving to the right.

Table 15-1 The Fujita Scale of Tornado Categories

Fujita Scale Value	Wind Speed		Number of Tornadoes (1985–93)	% per Year	% of Deaths	Damage
	Kilometers per Hour	Miles per Hour				
F0	64–118	40–73	478	51	0.7	Light: Some damage to tree branches, chimneys, signs.
F1	119–181	74–112	318	34	7.5	Moderate: Roof surfaces peeled, mobile homes overturned, moving autos pushed off roads.
F2	182–253	113–157	101	10.8	18.4	Considerable: Roofs torn off, mobile homes demolished, large trees snapped or uprooted. Light objects become missiles.
F3	254–332	158–206	28	3	20.5	Severe: Roofs and some walls torn off well-constructed houses, trains overturned, most forest trees uprooted, heavy cars lifted and thrown.
F4	333–419	207–260	7	0.8	36.7	Devastating: Well-constructed houses leveled, cars thrown, large missiles generated.
F5	420–513	261–318	1	0.1	16.2	Incredible: Strong frame houses lifted and carried considerable distance to disintegrate. Auto-size missiles fly more than 100 yards; trees debarked.
F6	>514				0	Winds are not expected to reach these speeds.

Modified from Ted Fujita; courtesy of Kaz Fujita.

The Fujita Scale

The **Fujita Tornado Scale** was devised by Dr. Ted Fujita at the University of Chicago. He separated probable tornado wind speeds into a six-point nonlinear scale from F0 to F5, where F0 has minimal damage and F5 has strong frame homes blown away (Table 15-1). In addition, Dr. Fujita compiled an F-scale damage chart and photographs corresponding to these wind speeds. Reference photographs of damage are distributed to National Weather Service offices to aid in evaluating storm intensities (▶Figure 15-10). Wind speeds and damages to be expected in different-strength buildings are shown in Tables 15-2a and 15-2b. Note that walls are likely to collapse in an F3 tornado in even a strongly built frame house; and in an F4, the house is likely to be blown down. Brick buildings perform better. In an F5 tornado, even concrete walls are likely to collapse.

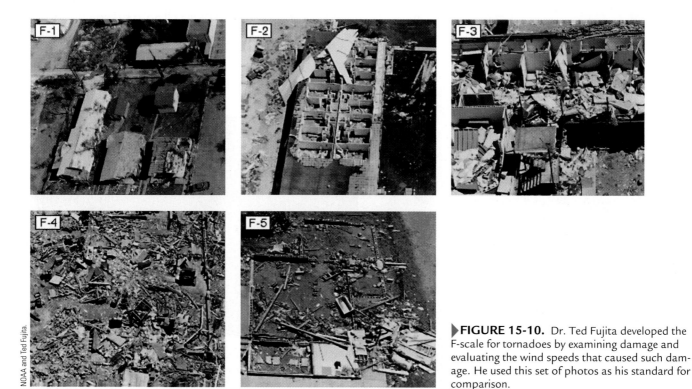

NOAA and Ted Fujita.

▶**FIGURE 15-10.** Dr. Ted Fujita developed the F-scale for tornadoes by examining damage and evaluating the wind speeds that caused such damage. He used this set of photos as his standard for comparison.

Table 15-2a Fujita Wind Scale

Wind Strength	Fujita Wind Scale					
	F0	F1	F2	F3	F4	F5
Miles per hour	40–73	74–113	114–158	159–207	208–261	262–319
Kilometers per hour	64–117	118–182	183–254	255–333	334–420	421–513

Table 15-2b Expected Damages for Different Types of Buildings Dependent on Tornado Strength*

Type of Building	Expected Damage by F-Scale Tornado					
	F0	F1	F2	F3	F4	F5
Weak outbuilding	Walls collapse	Blown down	Blown away			
Strong outbuilding	Roof gone	Walls collapse	Blown down	Blown away		
Weak frame house	Minor damage	Roof gone	Walls collapse	Blown down	Blown away	
Strong frame house	Little damage	Minor damage	Roof gone	Walls collapse	Blown down	Blown away
Brick structure	OK	Little damage	Minor damage	Roof gone	Walls collapse	Blown down
Concrete structure	OK	OK	Little damage	Minor damage	Roof gone	Walls collapse

*Simplified from Fujita, 1992.

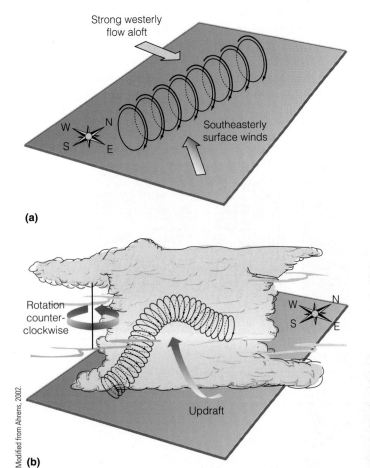

(a)

(b)

Modified from Ahrens, 2002.

▶ **FIGURE 15-11. (a)** Wind shear, with surface winds from the southeast, and winds from the west aloft. **(b)** This slowly rotating vortex can be pulled up into a thunderstorm, which can result in a tornado.

Tornado Development

Tornadoes generally form when there is a shear in wind directions, such as surface winds approaching from the southeast with winds from the west higher in the atmosphere. Such a shear can create a roll of horizontal currents in a thunderstorm as warm humid air rises over advancing cold air (▶ Figure 15-9). These currents, rolling on a horizontal axis, are dragged into a vertical rotation axis by an updraft in the thunderstorm to form a rotation cell up to 10 kilometers wide (▶ Figure 15-11). This cell sags below the cloud base to form a distinctive slowly rotating **wall cloud,** an ominous sight that is the most obvious danger sign for the imminent formation of a tornado (▶ Figure 15-12). **Mammatus clouds** can

NOAA, National Severe Storm Lab photo.

▶ **FIGURE 15-12.** A slowly rotating wall cloud descends from the base of the main cloud bank, an ominous sign for production of a tornado near Norman, Oklahoma, on June 19, 1980.

FIGURE 15-13. Mammatus clouds are a sign of the unstable weather that could lead to severe thunderstorms and potentially tornadoes. These formed over Tulsa, Oklahoma, on June 2, 1973.

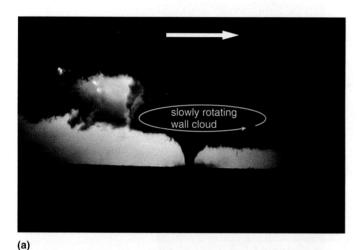

(a)

(b)

FIGURE 15-14. These two tornadoes are associated with slowly rotating prominent wall clouds. In **(a)**, a tornado descends from a wall cloud south of Dimitt, Texas, on June 2, 1995. In **(b)**, a tornado forms above this wall cloud and reaches the ground outside the wall cloud near Lakeview, Texas, on April 19, 1977. In both photos, the storm is moving from left to right.

be another potential danger sign, where groups of rounded pouches sag down from the cloud (▶Figure 15-13).

Strong tornadoes commonly form within and then descend from a slowly rotating wall cloud. A smaller and more rapidly rotating funnel cloud may form within the slowly rotating wall cloud or less commonly adjacent to it (▶Figure 15-14). If a funnel cloud descends to touch the ground, it becomes a tornado.

Tornadoes generally form toward the trailing end of a severe thunderstorm; this can catch people off guard. Someone in the path of a tornado may first experience wind blowing out in front of the storm cell along with rain, then possibly hail, before the stormy weather appears to subside (▶Figures 15-11 and 15-12). But then the tornado strikes. In some cases, people feel that the worst of the storm is over once the strong rain and hail has passed and the sky begins to brighten, unless they have been warned of the tornado by radio, television, or tornado sirens that have been installed in some urban areas that have significant tornado risk. Some tornados are invisible until they strike the ground and pick up debris. If you do not happen to have a tornado siren in your area, you may be able to hear an approaching tornado as a hissing sound that turns into a strong roar that many people have characterized as the sound of a loud oncoming freight train.

Conditions are favorable for tornado development when two fronts collide in a strong low pressure center (▶Figure 15-15). This can often be recognized as a **hook echo,** or

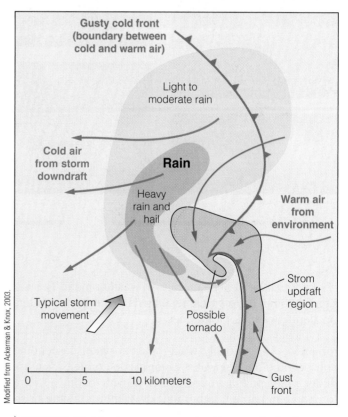

FIGURE 15-15. A common situation for tornado development is the collision zone between two fronts, commonly in the hook or "bow echo" of a rainstorm. A pair of curved arrows indicates horizontal rotation of wind in the lower atmosphere.

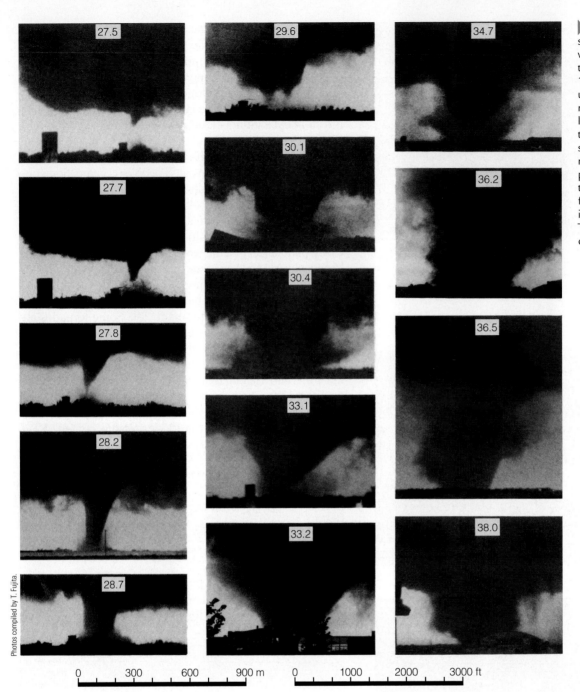

FIGURE 15-16. This series of fourteen photos was taken of the Fargo tornado on June 20, 1957. The times, in minutes, show that the funnel cloud descended in less than thirty seconds; the tornado then rapidly strengthened for the next minute. Just before the photo at 29.6 minutes, the funnel sheared off before strengthening again into a much wider funnel. This whole sequence took only ten minutes.

hook-shaped band of heavy rain on weather radar. This is a sign that often causes weather experts to put storm spotters on alert to watch for tornadoes.

Typically forming toward the rear of a thunderstorm, tornadoes are generally white or clear when descending and become dark as water vapor inside condenses in updrafts, which pull in ground debris. Growth to form a strong tornado can happen rather quickly, within a minute or so (▶ Figure 15-16), and last for ten minutes to more than an hour. Comparison of the winds of tornadoes with those of hurricanes (compare Table 15-1 with Table 14-1, page 356)

shows that the maximum wind velocities in tornadoes are twice those of hurricanes. Wind forces are proportional to the wind speed squared, so the forces exerted by the strongest tornado wind forces are four times those of the strongest hurricane winds. In many cases, much of the localized wind damage in hurricanes is caused by embedded tornadoes.

As a tornado matures, it becomes wider and more intense. In its waning stages, the tornado then narrows, sometimes becoming rope-like, before finally breaking up and dissipating (▶ Figures 15-17 and 15-18). At that waning stage, tightening of the funnel causes it to spin faster, so the tornado can still be extremely destructive.

Prediction and identification of tornadoes by the National Weather Service's Severe Storms Forecast Center in Kansas City, Missouri, uses Doppler radar, wind profilers, and automated surface observing systems. A **tornado watch** is issued when thunderstorms appear capable of producing tornadoes and telltale signs show up on the radar. A **tornado warning** is issued when Doppler radar shows strong indication of vorticity or rotation, or if a tornado is sighted. Before the warning stage, tornado spotters are alerted to watch for tornadoes. Warnings are broadcast on radio and television, and tornado sirens are activated if they exist in the potential path of tornadoes.

Tornado Damage and Risks

People are advised to seek shelter underground or in specially constructed shelters in their homes whenever possible. If no such space is available, people should at least go to some interior space with strong walls and ceiling and away from windows. People have been saved by going to an interior closet, or even lying in a bathtub. Unfortunately,

(a)

(b)

(c)

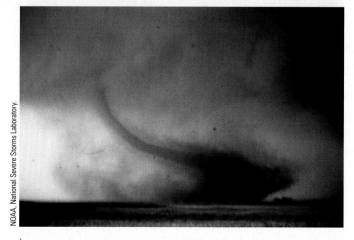

▶**FIGURE 15-17.** A big tornado south of Dimmitt, Texas, on June 2, 1995, sprays debris out from its contact with the ground **(a** and **b).** The storm dissipates slightly **(c).** This tornado tore up 300 feet of the highway where it crossed.

Harold Richter photos, NOAA National Severe Storms Laboratory.

NOAA, National Severe Storms Laboratory.

▶**FIGURE 15-18.** This thin, ropelike tornado was photographed at Cordell, Oklahoma, on May 22, 1981, just before it broke up and dissipated.

FIGURE 15-19. A basement, or at least an interior room without windows, would be a better choice for protection than this kitchen, which was destroyed by a tornado in Oklahoma.

A. Booher photo, FEMA.

A. Lamarre photo, U.S. Army Corps of Engineers.

FIGURE 15-20. The 1977 Birmingham, Alabama, tornado shows how selective the damage of tornadoes can be. The homes in the top part of this photo are completely demolished, while the home in the lower left mainly has roof damage.

in some cases a strong tornado will completely demolish houses and everything in them (▶Figure 15-19).

When Dr. Ted Fujita examined damage patterns from tornadoes, he noticed that there were commonly swaths of severe damage adjacent to areas with only minor damage (▶Figure 15-20). He also examined damage patterns in urban areas and cornfields, where swaths of debris would be left in curved paths (▶Figure 15-21). This led him to hypothesize that smaller vortices rotate around a tornado (▶Figure 15-22), causing intense damage in their paths but allowing some structures to remain virtually unharmed by

the luck of missing one of the vortices (Figure 15-20). Such vortices were later photographed on many occasions, supporting this hypothesis.

Those in unsafe places are advised to evacuate to a strong building or storm shelter if they can get there before the storm arrives. It is yet unclear whether vehicles provide more protection than mobile homes or lying in a ditch. FEMA still recommends that you lie in a ditch and cover your head, if you cannot get to a safe building; that will provide some protection from flying debris. Mobile homes are lightly built and are easily ripped apart—certainly not

▶**FIGURE 15-21.** Six 700-pound I-beams were pulled from an elementary school in Bossier City, Louisiana, and carried by a tornado along these paths. Other objects such as a diving board and a car were also carried significant distances.

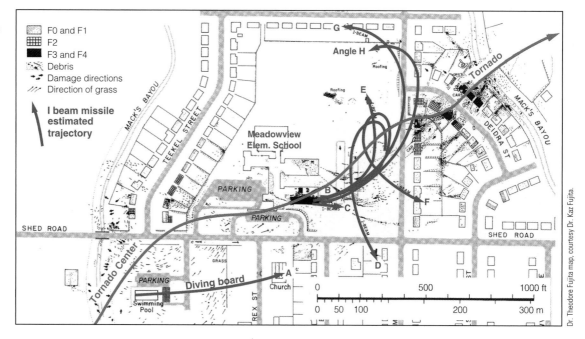

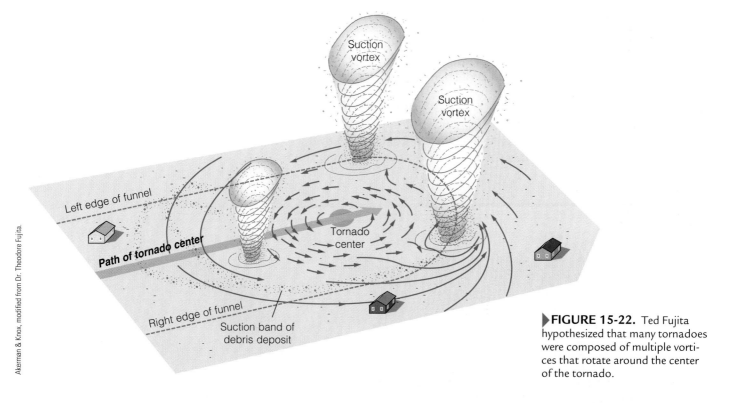

▶**FIGURE 15-22.** Ted Fujita hypothesized that many tornadoes were composed of multiple vortices that rotate around the center of the tornado.

a place to be in a tornado. Car or house windows and even car doors provide little protection from high-velocity flying debris such as two-by-fours from disintegrating houses.

Although cars are designed to protect their occupants in case of a crash, they can be rolled or thrown or penetrated by flying debris. If you are in open country and can tell what direction a tornado is moving, you may be able to drive to safety at right angles from the storm's path. Recall that the path of a tornado is often from southwest to northeast, so being north to east of a storm is commonly the greatest danger zone. Remember also that the primary hazard associated with tornadoes is flying debris, and much to peoples' surprise, overpasses do not seem to reduce the winds associated with a tornado. Do not get out of your car under an overpass and think that you are safe. In fact, an overpass can act like a wind tunnel that focuses the winds. Once a

THUNDERSTORMS AND TORNADOES **409**

▶**FIGURE 15-23.** The beam labeled "D" in Figure 15-21 ended up stuck in the ground at an angle.

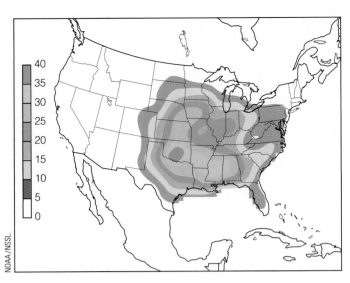

▶**FIGURE 15-24.** The areas of greatest tornado risk include much of the eastern half of the United States.

few people park under an overpass, this can cause the additional problem of a traffic jam, where helpless people may be stuck in the storm's path.

Although many people believe that the low pressure in a tornado vacuums up cows, cars, and people and causes buildings to explode into the low pressure funnel, this appears to be an exaggeration. Most experts believe that the extreme winds and flying debris cause almost all of the destruction. Photographs of debris spraying outward from the ground near the base of tornadoes suggest the same

(Figure 15-17b). However, even large and heavy objects can be carried quite a distance. The Bossier City tornado in Louisiana ripped six 700-pound I-beams from an elementary school and carried them from 60 to 370 meters away. Another I-beam was carried to the south, where it stuck into the ground in someone's backyard at an angle of 23 degrees from the horizontal (▶ Figures 15-21 and 15-23). In another documented case, several empty school buses were carried up over a fence by a tornado before being slammed back to the ground.

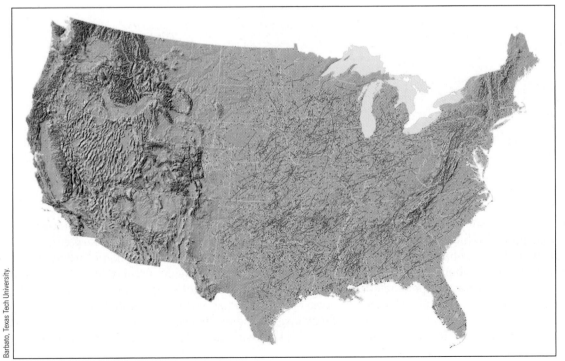

▶**FIGURE 15-25.** In this map of the paths for all recorded tornadoes in the United States from 1950–1995, the paths in yellow and blue are for smaller tornadoes (F0 to F2), while the paths in red are for larger tornadoes (F3 to F5).

Tornadoes
- Weak (F0-F1)
- Strong (F2-F3)
- Violent (F4-F5)
- Weak tracks
- Strong tracks
- Violent tracks

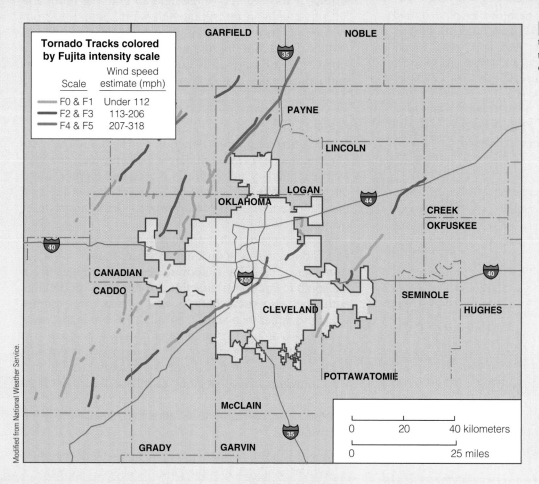

Tornado Tracks colored
by Fujita intensity scale

Scale	Wind speed estimate (mph)
F0 & F1	Under 112
F2 & F3	113-206
F4 & F5	207-318

Modified from National Weather Service.

0   20   40 kilometers

0   25 miles

FIGURE 15-26. This map of the May 3, 1999, tornadoes shows their paths and intensities around Oklahoma City.

FIGURE 15-27. An Oklahoma tornado on May 4, 1999, threw these cars into a crumpled heap.

A. Booher photo, FEMA.

One of the most severe tornado outbreaks in recent years was that of May 3, 1999, in central Oklahoma (▶ Figures 15-26 and 15-27). Eight storms producing fifty-eight tornadoes moved northeastward along a 110-kilometer-wide swath through Oklahoma City. Eighteen more tornadoes continued up through Kansas. Tornado strengths ranged from less than F2 to F5. Individual tornadoes changed in strength as they churned northeast. Fifty-nine people were killed and damages reached $800 million.

The largest known tornado outbreak to date started just after noon on April 3, 1974. A total of 148 tornadoes scored tracks from Mississippi all the way north to Windsor, Ontario, and New York state, with an overall storm path length of 4,180 kilometers. This **superoutbreak** lasted more than seventeen hours, killed 315 people, and injured 5,484 others. The map of the storm tracks (▶ Figure 15-28) shows that several of these tornadoes ended in downbursts.

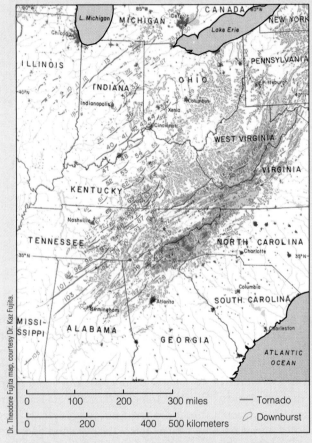

▶**FIGURE 15-28.** This map of the 148 tornado paths from the superoutbreak of April 1974 was compiled by Dr. Ted Fujita; his team of graduate students at the University of Chicago, including Dr. Greg Forbes; as well as others from the National Severe Storms Lab and other institutions.

The average number of tornadoes is highest in Texas and Oklahoma, followed by Kansas, Nebraska and adjacent states, Florida, and Louisiana. **"Tornado Alley,"** covering parts of Texas, Oklahoma, Arkansas, Missouri, and Kansas, marks the belt where cold air from the north collides frequently in the spring with warm, humid air from the Gulf of Mexico to form intense thunderstorms and tornadoes. Tornadoes are rare in the western and northeastern states (▶ Figures 15-24 and 15-25). An individual **tornado outbreak**—that is, a series of tornadoes spawned by a group of storms—has killed as many as several hundred people and covered as many as thirteen states (Table 15-3).

Tornado season varies, depending on location. The number of tornadoes in Mississippi reaches a maximum in April with a secondary maximum in November. Farther north, the maximum is in May, and in Minnesota it is in June. At these northern latitudes, tornadoes are virtually absent from November to February.

Most, though not all, tornadoes track toward the northeast. Storm chasers, individuals who are trained to gather storm data at close hand, know to approach a tornado from the south to southwest directions so they will not be in its path. They also know that it is safer to chase them on the flat plains rather than along the Gulf Coast, where the lower cloud base can hide the funnel from their view.

So what can we nonspecialists do to survive a tornado? A radio or television tuned to NOAA's weather radio network provides severe weather warnings. Typically, these warnings can provide up to ten minutes of lead time before the arrival of a tornado. General guidelines include the following:

- Move to a tornado shelter, basement, or interior room without windows. In some airports, such as Denver International, the tornado shelters are the restrooms.

- Flying debris is extremely dangerous, so if your location is at all vulnerable, protect your head with a bicycle or motorcycle helmet.

- In spite of television videos, a highway overpass is not a good location. Do not get out of your car and think you are safe. An overpass acts as a wind tunnel that can amplify the danger.

- Although cars can overturn, and flying debris can penetrate their windows and doors, they still provide some protection—especially below the window line.

Table 15-3 Deadliest Tornadoes on Record*

Name or Location	Date	Number of Tornadoes (and Number of States Affected)	Deaths	Estimated damage in Millions (1980 $)	Estimated damage in Millions (2002 $)
Tri-state: MO, IL, IN	March 18, 1925	7 (6)	689	18	39
Tupelo-Gainesville (MS, GA)	April 5-6, 1936	17 (5)	419	18	39
Enigma	February 19, 1884	60 (8)	420	3	6.5
Northern Alabama	March 21-22,1932	33 (7)	334	5	11
Super (see Fig. 15-28)	April 3-4, 1974	148 (13)	315	?	
Louisiana-Georgia	April 24-25,1908	18 (5)	310	1	2.2
St. Louis, Missouri	May 27, 1896	18 (3)	306	15	33
Palm Sunday	April 11-12,1965	51 (6)	256	200	438
Dierks, Arkansas	March 21-22,1952	28 (4)	204	15	33
Easter Sunday	March 23, 1913	8 (3)	181	4	9
Pennsylvania-Ohio	May 31, 1985	41 (3)	75		985
Carolinas	March 28, 1984	22 (2)	57		438
Oklahoma-Kansas (F5)	May 3-4, 1999	76 (2)	49		800
Southeastern United States	March 27, 1994	2 (2)	42		234
Jarrell, Texas (F5)	May 27, 1997	1 (1)	27		

*From FEMA, 1997, and other sources.

KEY POINTS

✓ Thunderstorms are most common at equatorial latitudes, but the United States has more than its share for its latitude. Storms form most commonly at a cold front when unstable warm, moist air rises rapidly into cold air and condenses to form rain and hail. Cold fronts from the northern plains states often interact with warm, moist air from the Gulf of Mexico to form a northeast-trending line of storms. **Review pp. 397–398.**

✓ Collisions between droplets of water carried in updrafts with downward-moving ice particles generate positive charges that rise in the clouds and negative charges that sink. Because negative and positive charges attract, a large charge separation can cause an electrical discharge—lightning—between parts of the cloud or between the cloud and the ground. If you feel your hairs being pulled up by static charges in a thunderstorm, you are at high risk of being struck by lightning. **Review pp. 398–400; Figure 15-3.**

✓ Thunder is the sound of air expanded at supersonic speeds by the high temperatures accompanying a lightning bolt. Because light travels to you almost instantly and the sound of thunder travels 1 kilometer in roughly three seconds, if the time between seeing the lightning and hearing the thunder is three seconds, then the lightning is only 1 kilometer away. **Review p. 400.**

✓ You can minimize danger by being in a closed building or car, not touching water or anything metal, and staying away from high places, tall trees, and open areas. If trapped in the open, minimize contact with the ground by crouching on the balls of your feet. **Review p. 400.**

✓ Larger hailstones form in the strongest thunderstorm updrafts and cause an average of $2.9 billion in damage each year. **Review p. 400.**

✓ Tornadoes are small funnels of intense wind that may descend near the trailing end of a thunderstorm; their winds move as fast as 515 kilometers per hour. They form most commonly during collision of warm, humid air from the Gulf of Mexico with cold air to the north. They are the greatest natural hazard in much of the midwestern United States. The greatest concentration of tornadoes is in Oklahoma, with lesser numbers to the east and north. **Review pp. 401–402; Figures 15-24 and 15-25.**

✓ The Fujita tornado scale ranges from F0 up to F5, where F2 tornadoes take roofs off some well-constructed houses, and F4 tornadoes level them. **Review pp. 402–404; Tables 15-1 and 15-2.**

✓ Tornadoes form when warm, humid air shears over cold air in a strong thunderstorm. The horizontal rolling wind flexes upward to form a rotating cell up to 10 kilometers wide. A wall cloud sagging below the main cloud base is an obvious danger sign for formation of a tornado. **Review pp. 404–405; Figures 15-9 and 15-12 to 15-14.**

✓ On radar, a hook echo enclosing the intersection of two fronts is a distinctive sign of tornado development. **Review pp. 405–406; Figure 15-15.**

✓ The safest places to be during a tornado are in an underground shelter or an interior room of a basement. Even being in a strongly built closet or lying in a bathtub can help. If caught in the open, you may be able to drive perpendicular to the storm's path. If you cannot get away from a tornado, your car may provide some protection, or lying in a ditch and covering your head will help protect you from debris flying overhead. **Review pp. 407–409.**

IMPORTANT WORDS AND CONCEPTS

Terms

charge separation, p. 398
cumulonimbus cloud, p. 397
downburst, p. 401
Fujita tornado scale, p. 402
hailstones, p. 401
hook echo, p. 405
lightning, p. 398
mammatus clouds, p. 404
step leader, p. 400

superoutbreak, p. 412
thunder, p. 400
thunderstorm, p. 397
tornado, p. 401
Tornado Alley, p. 410
tornado outbreak, p. 410
tornado warning, p. 407
tornado watch, p. 407
wall cloud, p. 404
wind shear, p. 401

QUESTIONS FOR REVIEW

1. When is the main tornado season?
2. How are electrical charges distributed in storm clouds and why? What are the charges on the ground below?
3. What process permits hailstones to grow to a large size?
4. Why do you see lightning before you hear thunder?
5. List the most dangerous places to be in a lightning storm.
6. What should you do to avoid being killed by lightning if caught out in the open with no place to take cover?
7. In what direction do most midcontinent tornadoes travel along the ground?
8. How fast do tornadoes move along the ground?
9. What is a wall cloud, and what is its significance?
10. Why does lying in a ditch provide some safety from a tornado?
11. How do weather forecasters watching weather radar identify an area that is likely to form tornadoes?
12. What is the greatest danger (what causes the most deaths) from a tornado?

FURTHER READING

Assess your understanding of this chapter's topics with additional quizzing and conceptual-based problems at:

 http://earthscience.brookscole.com/hyndman.

CHAPTER 16

John McColgan photo, U.S. Forest Service.

Fires in western Montana on August 6, 2000, chased these two elk into the relative safety of the Bitterroot River's upper drainage.

Wildland fires are a natural part of the evolution and renewal of nature. They benefit ecosystems by thinning forests, reducing understory fuel, and permitting the growth of different species and age groups of trees. But out of control, they also pose hazards to humans. They affect wildlands and often encroach on suburban or built-up areas. In some cases, the fuel and conditions foster a firestorm that is virtually impossible to suppress without a change in the conditions. Fires can be naturally started by lightning strikes, intentionally set for beneficial purposes, or set accidentally or maliciously. Lightning-started fires account for only 13 percent of forest fires, according to U.S. Forest Service figures. Prescribed debris burns that ran out of control accounted for 24 percent; amazingly, arson accounted for 26 percent. Various unintentional causes make up the remainder.

Investigators searching for clues to a fire's origin look for evidence left behind, including remnants of matches, cigarettes, and accelerants such as gasoline. They use wind

WILDFIRES
Fanned Flames

The afternoon of August 12, 2003, residents of the O'Brien Creek drainage on the southwestern fringes of Missoula, Montana, were ordered to evacuate. U.S. Forest Service firefighters on Black Mountain had been battling a large lightning-caused forest fire, one of many in the northern Rockies that summer (▶Figure 16-1). The fire was still largely uncontained, and the firefighters were trying to channel it away from dozens of homes in that forested urban interface. That afternoon, brisk winds from the west picked up and drove the fire into a frenzy.

Normally a wildfire will run quickly uphill (▶Figure 16-2), consuming fuel ahead and above it. When it reaches a ridge crest, the fuel ahead is downhill, slowing the fire. In this case, a 100-meter high ball of fire pushed by the wind crested a ridge and roared downhill toward O'Brien Creek. Burning embers flying ahead of the main fire ignited a new fire downslope. The new fire moved rapidly back upslope toward the advancing

Bureau of Land Management photo.

▶**FIGURE 16-2.** A forest fire races upslope through a pine forest.

fire, pulled by air rising in the main fire plume. That slowed the advance of the main fire, which then quickly turned laterally up the main channel of O'Brien Creek, consuming houses along with the forest.

In anticipation of the potential problem, firefighters had foamed houses in the area with Class A **foam,** a spray-on gel that can protect them from fire. This time, the fire was so hot that neither the foam nor fire-resistant building materials were sufficient. The only remains of some of the homes were their concrete foundations.

A month later, new grasses began sprouting, but the steep slopes remain vulnerable to rapid surface runoff from any heavy rain (▶Figure 16-3). After two or three years, sufficient ground vegetation will provide increasing soil protection from the direct impact of raindrops. In this way the post-fire erosion problem will gradually decrease.

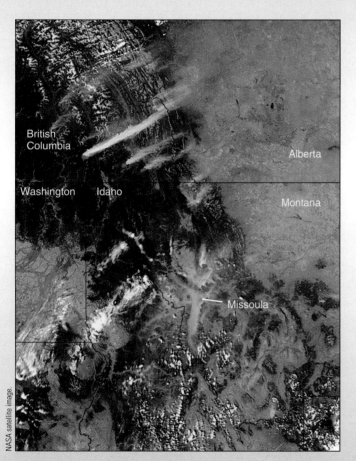

NASA satellite image.

British Columbia

Alberta

Washington   Idaho

Montana

Missoula

▶**FIGURE 16-1.** Forest fires spread throughout the northern Rockies on July 15, 2003, as shown by the red areas on the image. Fires in Glacier National Park are just south of the Alberta–Montana border. Fires in the mountains farther south surround Missoula, Montana. Smoke plumes, clearly visible as bluish gray areas, are quite different from the white clouds across the region.

Donald Hyndman photo.

▶**FIGURE 16-3.** This steep slope on the O'Brien Creek drainage at the west edge of Missoula, Montana, had been ravaged by fire one month earlier.

Lawrence-Berkeley Lab photo.

▶ **FIGURE 16-4.** The more severely burned white side of this tree indicates the direction from which the fire came; in this case, the fire moved upslope. Oakland-Berkeley hills fire, 1991.

direction when the fire began and indications of the direction in which the fire moved (▶ Figure 16-4).

Fire Process and Behavior

Trees and dry vegetation **burn** at high temperature by reaction with oxygen in the air. The main **combustible** part of wood is **cellulose**, a compound of carbon, hydrogen, and oxygen. When it burns, cellulose breaks down to carbon dioxide, water, and heat. A **fire** requires three **components**— fuel, oxygen, and heat—and cannot progress without all three (▶ Figure 16-5). Shrubs and trees also contain natural oils or saps that add to the combustibles. The large relative surface area of dry grasses, tree needles, and to a lesser extent shrubs leads to their greater ease of burning relative to the wood of trees. Fires in dry grass spread rapidly. With a small total mass of burnable matter on the ground—that is, grass, shrubs, and litter—a ground fire may move through fast enough to clear the understory but spare the trees. Some tree species such as ponderosa pine have evolved with time to survive fires that burn the ground vegetation.

Factors that affect the behavior of wildfires include fuel, weather, and topography. Fuel loading indicates the amount of burnable vegetation. The energy released also depends on the type of vegetation, its distribution, and its moisture

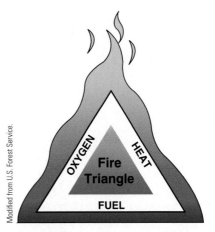

Modified from U.S. Forest Service.

▶ **FIGURE 16-5.** The "fire triangle": A fire requires fuel, oxygen, and heat. Without any one of these, a fire cannot burn.

National Interagency Incident Management Team.

▶ **FIGURE 16-6.** Where fire can easily reach up from ground vegetation through lower tree branches, it may crown. From there, embers may blow to start another fire downwind.

content. Fuels such as grass release less energy but can allow fire to spread rapidly. Whether fire on the ground can easily reach tree crowns can have a major effect on the rate of spread and intensity of a wildfire (▶ Figure 16-6). Wind carries burning embers above the treetops for considerable distances to start new fires.

People who live in the woods do so to get away from urban congestion and for the beauty of nature. They like to be surrounded by trees and shrubs (▶ Figures 16-7 and 16-8). In wildfires, trees and shrubs are fuel. People also tend to build with natural materials, especially wood, for siding and decks. Many even use wood roofing materials and stack firewood next to the house—more fuel for wildfires. Often houses will burn more readily than nearby trees because their wood is completely dry (▶ Figure 16-8). People

▶**FIGURE 16-7.** A wind-driven fire spreads over the hills east of Okanagan Lake near Kelowna, British Columbia, on August 20, 2003.

who insist on living in the forest can minimize the danger by clearing a broad area around their homes of all types of fuel and building with flame-resistant materials. Metal roofs, though not as natural, are flame-resistant. Burning embers falling on a metal roof will not ignite it, although if the roof gets sufficiently hot the wood framework supporting it will ignite. Even those who live not in the forest per se but in the urban interface near the forest are in danger, particularly in dry, hot weather where a fire is driven by high winds.

Weather fronts and thunderstorms can not only start wildfires but also dramatically affect their spread and direction. The worst **firestorms** are typically pushed by high winds, but extreme amounts of dry fuel can initiate firestorms that

▶**FIGURE 16-8.** Most homes on this ridge near Los Alamos, New Mexico, burned in the Cerro Grande fire on May 4, 2000, even though most of the trees did not.

CASE IN POINT
Storm King Mountain, Colorado

July 1994

Near Glenwood Springs, Colorado, the South Canyon fire of July 1994 burned 8.1 square kilometers of pinyon pines, junipers, and mountain shrubs on Storm King Mountain. Fourteen firefighters died when the fire blew up and trapped them upslope and downwind. Bedrock in the area is mostly sandstone, pebbly sandstone, and shale. Hill slopes are steep, some layers dip subparallel to the slope, and 2.3 square kilometers of the area are part of an old landslide. The formation weathers easily to a silty and sandy soil that has a history of producing debris flows.

In the weeks that followed the fire, large amounts of ash and mineral soil moved downslope into drainage channels. Some moved by **dry ravel,** the downslope slippage of loose grains, and some by the impact of raindrops and overland flow. Loose, silty sand and ash accumulated to as much as a meter deep along the sides of most drainages, adding to coarser fragments of colluvium collected by slow downslope movements over the years. This set the stage for erosional events that were to follow.

On the night of September 1, 1994, torrential rains flushed all of this loose material downslope into larger canyons to mobilize debris flows. With no vegetation to slow the flow over the ground surface, heavy rain led to rapid runoff. Rills and gullies formed on the bare hillsides, indicating rapid overland flow. These efficiently carried surface runoff to larger channels. They came down across Interstate Highway 70 as mud, boulders, and debris to engulf thirty vehicles and push some into the adjacent Colorado River. Fortunately, no one died in these flows. Some debris fans blocked almost half of the river. Every major drainage in which the headwaters burned supplied debris flows that reached canyon mouths. The flows show characteristics of both debris flows and hyperconcentrated flows. Calculated peak velocities for many flows were between about 4.6 and 8.5 meters per second. Calculated discharges were from 73 to 113 cubic meters per second.

generate their own winds as air rushes in to replace air rising convectively in the heat of the fire.

On a hillside, especially a steep one, flames and sparks rise with their heat, promoting the upslope movement of fires into new fuel. Thus, fires generally move rapidly upslope but not down. Although flaming branches and trees can fall downslope, advance in that direction tends to be slow. To be upslope from a fire is to be in serious danger. For the same reason, a fire moves downwind (see "Case in Point: Storm King Mountain, Colorado"). Winds accelerate

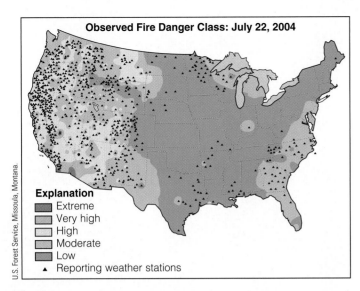

Observed Fire Danger Class: July 22, 2004

Explanation
- ■ Extreme
- ■ Very high
- ■ High
- ■ Moderate
- ■ Low
- ▲ Reporting weather stations

U.S. Forest Service, Missoula, Montana.

▶ **FIGURE 16-9.** This fire-danger rating map was prepared from NOAA satellite sensors for July 22, 2004.

fires both by directing the flames at new fuels and bringing in new oxygen in the air to aid in burning. High winds can carry burning embers for a kilometer or more to start new spot fires well ahead of the main fire. The **convective updraft** of heat from a fire draws in new air from the sides of the fire, helping to fan the flames. Canyons can funnel air as in a chimney and cause more rapid spreading; however, ridge tops often dramatically slow further spread. A fire reaching a ridge crest can create an updraft of air from the far side to fan the flames, but the same updraft can help prevent the fire from moving down the other side.

Regionally, fires tend to advance in the direction of the prevailing winds so that fires in the northern United States and southern Canada tend to burn toward the northeast with the westerly winds (see smoke drift in Figure 16-1). In the trade winds belt of southern United States and northern Mexico, fires tend to burn toward the southwest. Thus, homes in the prevailing downwind direction from fire-prone or high-fuel areas are more at risk than those upwind from such areas.

The risk of wildfires in a given region is estimated by the national fire-danger rating system, which uses National Weather Service data collected by means of National Oceanic and Atmospheric Administration (NOAA) weather satellites. A high-resolution infrared sensor is used to measure greenness of vegetation and interfuel moisture content (▶ Figure 16-9). This greenness is compared with the typical value for each map area. Such estimates are used to infer fire danger and to inform the public through news media and roadside signs.

Weather most directly affects **fuel moisture.** A few years of less than normal moisture dehydrates the soil and lowers the water table, providing less moisture for the growth of trees and undergrowth and less moisture in all of the vegetation. It may take several wet seasons to replenish the water

shortfall below the surface. Thus, surface vegetation during a drought may green up in the rainy season but dehydrate quickly as dry weather returns. Fires start more easily and spread rapidly. High humidity makes it more difficult for fires to start and to continue burning, whereas high temperatures and winds progressively reduce moisture over time and increase the fire hazards. Drought maps provide a record of longer-term moisture deficits (▶ Figure 16-9).

Hazards to the public increase with the number of people who choose to live within or next to forest, range, and other wildlands. Not only do they put themselves and their property in danger but also they endanger firefighters and encourage fire management officials to expend limited resources in protection of individuals' properties rather than managing the overall fire. From 1985 to 1994, an average of 73,000 fires per year burned on federal lands, consuming more than 3,000,000 acres and costing $411 million for suppression. Additional millions were lost in timber values, and indirect costs resulting from landslides and floods are unknown. Between 1990 and 1993, fires destroyed more than 4,200 homes in California alone. In ten days in Southern California in October 2003, catastrophic fires destroyed 3,452 homes (see "Case in Point: Southern California Firestorm, 2003").

Individual fires can be catastrophic. On October 20, 1991, a firestorm in the hills upslope from Oakland, California, destroyed or damaged 3,354 single-family homes and 456 apartments (▶ Figure 16-10). It killed twenty-five people, and damage amounted to $2.2 billion. Eight years after the fire, residents replaced 80 percent of their homes in the same locations with their views of San Francisco Bay and the Golden Gate Bridge. However, new restrictions dictate that roofs be fire-resistant, decks and sheds be built with heavier materials that would not burn as quickly, and landscaping be done using fire-resistant plants.

Lawrence-Berkeley Lab photo.

▶ **FIGURE 16-10.** Little remained of expensive houses on the steep slopes of the Oakland–Berkeley hills but their chimneys and foundations.

The first of eleven major fires of the season began in Southern California on October 21, 2003. Hot, dry Santa Ana winds that blow westward out of the desert Southwest gusted to almost 100 kilometers per hour, spreading the fires rapidly out of control (Figure 16-11). Ten days later, the fires had burned more than 300,000 hectares or 3,000 square kilometers, about the area of Rhode Island. They leveled 3,452 homes and killed twenty-two people. Nine hundred of the homes were in the San Bernardino Mountains east of Los Angeles in a fire likely started by arson on October 25. The largest, the Cedar fire in San Diego County, burned 2,207 homes and caused fifteen deaths (Figure 16-12). A lost hunter trying to signal rescuers started that fire.

Many of those who died had ignored evacuation orders, waiting until the last minute to leave. By then the fast-moving flames overtook their lone evacuation road, cutting off all escape.

The worst fires were in San Diego County, where the high cost of available land encouraged developers and individuals to build in areas vulnerable to brush fires. Some neighbor-

Carol Jandrall photo, California Department of Forestry.

 FIGURE 16-12. Many homes were closely surrounded by trees and brush, which made it impossible for firefighters to save them from the Cedar fire east of San Diego.

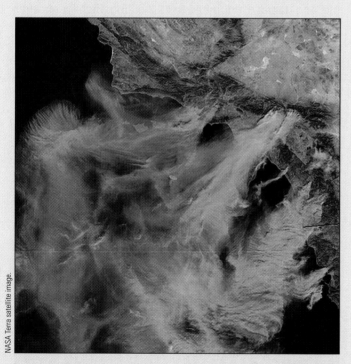

NASA Terra satellite image.

 FIGURE 16-11. The Southern California fires on October 26, 2003, can be seen in this satellite photo. The largest area of fire at the right edge of the image is on the eastern outskirts of San Diego. The largest area of fires in the north-central part of the image is on the northern outskirts of Los Angeles. The Santa Ana winds blow the fires and smoke toward the southwest.

hoods were closely spaced wood houses surrounded by pine trees, with some trees wedged against flammable roofs and wood decks. California state law now requires strict standards for building with fireproof materials and clearing brush around homes. However, hundreds of thousands of people ignore the rules and enforcement has been limited (Figure 16-12).

Four years of drought in the San Bernardino Mountains northeast of Los Angeles left the trees vulnerable to bark beetle infestation and killed large numbers of them. Although there are tight controls on fire-resistant building materials and brush removal, the cost of tree removal can be as high as $850 per tree, so few people removed the dead trees. As one resident put it, "Those dead pine trees are just gasoline on a stick."

Fifteen thousand firefighters fought the fires, along with helicopters equipped with giant water buckets, and air-tanker aircraft that drop massive volumes of fire retardant. Insured losses from these fires amounted to more than $1.25 billion. A change in weather on November 1 calmed the Santa Ana winds and brought moisture from the Pacific. Two or three centimeters of white, powdery snow fell in Big Bear Lake resort area near San Bernardino, slowing the fires and permitting firefighters to build fire lines. The only drawback was that the rain and snow caused mudslides that closed highways.

Two months later, on December 25, torrential rains fell, 4.39 inches on the San Bernardino Mountains and 8.57 inches on the San Gabriel Mountains. On the south side of Cajon Pass between them, the route of Interstate Highway 15, the unprotected soils of the burned area unleashed heavy mudslides that raced down canyons and through a trailer camp and recreation center. Fifteen people died in mud as much as 4 to 5 meters deep.

Costs of firefighting and cleanup are estimated to reach another $2 billion. Fire insurance is still available; however, insurance companies build the heightened risk into their premiums. In some areas, insurers are becoming less willing to write coverage for fire. In others, homeowners are told to replace roofs and clear brush if they want to be covered.

Who pays for the costs of fighting fires and trying to save the homes of those who choose to live in fire-danger zones? Fire-insurance rates, though rising, pay for only a small part of the cost because federal and state governments pay most of the costs of fighting the fires and cleanup and generally pay part of the cost to help people rebuild. Some houses are built in "**indefensible locations**" such as narrow canyons that are too dangerous for firefighters. In Hook Canyon on the southeast side of Lake Arrowhead, 350 homes burned. Some homeowners' associations in scenic areas banned property owners from cutting trees, as in the same area in the San Bernardino Mountains. They lifted the ban less than a year before the fire, but by then it was too late to do much even for those inclined to remove trees.

Population increases in outlying areas compound the problem. In recent years as the population in the Sierra Nevada nearly doubled, so did the number of fires and acres burned. Property damage in the same period went up 5,000 percent.

In this region's Mediterranean climate, more than 90 percent of the rainfall comes during the winter and early spring (between November and April). Chaparral or scrub brush that covers many Southern California hillsides, burns in wildfires every thirty-five to sixty-five years, primarily during late summer or early fall when hot, dry Santa Ana winds blow off the deserts to the east. The hydrophobic soil layer produced by fires (▶Figure 16-13; also see Chapter 8, page 213), and the lack of vegetation, lead to greater overland flow and high stream flows. Studies suggest that erosion increases by a factor of 10 following chaparral fires. The frequency of fires and the damages produced are likely to increase dramatically as more people move to the woods.

Thomas Spittler photo, California Division of Forestry.

▶**FIGURE 16-13** In hydrophobic soil such as this produced by the Banning fire in Southern California in 1993, the dry layer lies beneath the dark gray hydrophobic layer, which is composed of soil and ash.

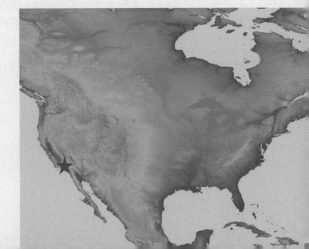

From October 25 to November 3, 1993, Santa Ana winds fanned twenty-one major fires in Southern California, burning 76,500 acres and 1,171 buildings. Three people died. Fires in summer 2000 and again in 2003 burned hundreds of thousands of acres of forest in Idaho and Montana.

Fire Suppression and Prevention

A variety of approaches are typically used for fire suppression. Highly trained smokejumpers parachute into remote areas of lightning-caused spot fires to exterminate them before they spread out of control. Bulldozers cut firebreaks to remove vegetation down to bare ground to slow the spread of fire on the ground. Aircraft drop giant loads of water or borate fire retardant along the lateral and advancing edges of a fire to help direct and contain it (▶ Figure 16-14). Helicopters scoop huge buckets of water from nearby streams or lakes and dump them on the edge of fires. Firefighters on the ground cut fire lines to prevent fires from spreading laterally. For the largest fires that threaten towns or critical facilities, firefighters sometimes deliberately set back burns. In a **back burn,** a fire is set near a road or facility to burn back toward the main fire; the convective rise of hot air in the main fire draws the new flames toward it. In a **burn-out**, a large area in the path of a big fire is burned under controlled conditions and suitable weather, generally beginning at a road or river; again it burns back toward the main fire, eliminating fuel and stopping the progress of the main fire. Before lighting a burnout, the forest or range between the burnout and area to be protected is thoroughly drenched using planes and helicopters. The technicalities of a burnout are complex and dangerous, but they can work remarkably well. In July 2003, firefighters set a large burnout that saved the town of West Glacier at Glacier National Park from a huge fire that was bearing down on the area.

The best protection for homes in forests is fuel reduction before the fires start. Homeowners should use flame-resistant roof materials, remove underbrush and tree limbs, and move firewood away from houses (▶ Figures 16-15 and 16-16).

Aggressive fire suppression by federal and state agencies since the 1920s has thwarted fire in most of the forests of the United States. Tree densities and fuel loads have risen to critical levels, ideal conditions for catastrophic fires. Finally realizing this, federal agencies recently shifted to mitigation and management of fires, including prescribed burns, and now permit wildfires to burn in wilderness areas.

Erosion Following Wildfires

Major fires often result in severe slope erosion in the few years following the fire. In normal, unburned areas, vegetation has a pronounced effect in slowing the infiltration of rainfall and snowmelt into the ground and in minimizing

▶**FIGURE 16-14.** An aerial tanker drops a load of water on the edge of an Idaho fire in August, 1996. The house, sitting upslope from the fire, is in severe danger.

▶**FIGURE 16-15.** This is all that remained of a forest home in Colorado after a June 6, 2000, fire.

▶**FIGURE 16-16.** In the Cerro Grande fire at Los Alamos, New Mexico, in May 2000, houses on one side of a street in a forested community burned to the ground whereas those on the other side remained unscathed.

surface runoff. Evapotranspiration by trees and other vegetation decreases the amount of water reaching the ground. Tree needles, leaves, twigs, decayed organic material, and other litter on the ground soak up water and permit its slow infiltration rather than turning it into surface runoff.

Intense fires burn all of the vegetation, including litter on the forest floor. The organic material burns into a **hydrocarbon residue** that soaks into the ground. This material fills most of the tiny spaces between fine soil grains in the top few centimeters of soil and sticks them together (▶ Figure 16-17). In some cases, water can no longer infiltrate the surface soils, thus most of it runs off. Such impervious soils are called **hydrophobic.** The hydrophobic nature of soils is often attributed to intense burning, but the relationship to fires is not entirely clear. What is clear is that water more readily runs off the surface of the charred soil.

Big rain droplets beat directly on the unprotected ground. When rainfall is heavy and vegetation and ground litter have been removed by fire, the water cannot infiltrate fast enough to keep up with the rainfall. Rain drops splash fine grains off the surface, and the runoff carries them downslope as sheetwash or overland flow. Unprotected soils are easily gullied (▶ Figure 16-18). The local channels lead to formation of rills and gullies. All of this surface runoff quickly collects downslope in larger gullies and small valleys. Water collecting in a valley with a large enough watershed quickly turns into a flash flood or debris flow. These destructive and dangerous floods rip out years of accumulated vegetation, soil, and loose rock, carrying everything down valley in a water-rich torrent. (Debris flows are covered in Chapter 8 on landslides, pages 212–220; see also "Case in Point: Bitterroot Valley Fires, Summer 2000.")

Erosion following a fire cannot be prevented, but it can be minimized. Federal and state agencies and individuals can plant vegetation, grass, shrubs, and trees that prevent

▶**FIGURE 16-18.** These gullies were eroded by a short-lived rainstorm one year after the 2000 fires in the southern Bitterroot Valley, Montana.

Donald Hyndman photo.

the direct impact of raindrops on bare soil. Over large areas, slopes are often seeded soon after a fire. Straw can be packed into tubes or laid out in bales to provide a barrier to overland flow, thus reducing slope erosion. Falling dead trees across the slope can have a similar effect. Drains can direct water flow laterally to valleys to minimize surface gullying. On small steep patches of particularly vulnerable soil, sheets of plastic can be spread to prevent water from reaching the soil.

Rapid accumulation of water because of overland flow following a fire creates a severely shortened and heightened stream hydrograph. This can lead to flash floods that wash out bridges, roads, and buildings. After a region has burned, it is unsafe to drive or walk in small valley bottoms during intense rainstorms because swift runoff funnels most of the precipitation into adjacent gullies and then quickly into sequentially higher-order streams. Even where the storms are several kilometers away in the headwaters of a burned drainage, a flash flood downstream can appear under a clear blue sky. Homes that may have survived a fire in usually dry valley bottoms are vulnerable to debris flows from all of the new sediment washed downslope after a fire.

Some Concerns

One needs to ask whether public funds should be used to help people rebuild in places where wildfires and the associated floods and erosion are ever-present. Our conscience

M. Rieger photo, FEMA.

▶**FIGURE 16-17.** This mudslide followed the Hayman, Colorado, fire on July 5, 2002.

The fire season began in June with a scattering of lightning-caused fires in the mountains flanking the Bitterroot Valley just east of the Montana–Idaho border. On July 15, the Little Blue Fire near Painted Rocks Lake burned more than 23 square kilometers. It forced the evacuation of people from twenty-five homes near the spectacular Bitterroot Range. By the end of July, five major fires were burning in the mountains around the southern Bitterroot Valley.

Then on the night of July 31, a dry lightning storm started seventy-eight new fires. Within a week, some merged into giant forest fire complexes (see chapter opener photo). Many homes in the forest or on its fringes burned. Eventually, seventy homes and ninety-four vehicles were lost, and more than 1,500 people had to evacuate their homes. The Forest Service closed the Bitterroot National Forest to public use and soon thereafter the state closed all public lands until the fire danger subsided.

The forest consists of areas of thick-barked ponderosa pine trees that live as open stands of large trees that are resistant to fire. Other areas are thicker stands of Douglas fir and grand fir trees that do not survive most fires. Forest management practices that harvested the large ponderosas and vigorous fire suppression more than doubled the percent of fire-vulnerable trees. Fires then burn more intensely and over larger areas than before modern forest management practices.

Contributing to the extreme conditions was the weather. The strong El Niño of 1998 caused dry weather in the Pacific Northwest in addition to the southwestern United States. Lower than normal precipitation continued through the summer and early fall of 1999, and the snowpack for much of the mountain West, including this area, was less than 70 percent of normal. Lower elevations lost their snow cover in early February, causing vegetation to green up early. Spring showers contributed to growth of abundant fuel for fires. Then the normally active storm track across the Pacific Northwest in June weakened. A high pressure system in the Southwest channeled a series of weak low pressure systems into western Montana. These provided little moisture but spawned numerous lightning storms. By late July, drought conditions were severe to extreme. Vegetation moisture dropped to critical levels in the forests. Leafy plants were turning to fall colors, and pine trees began dropping needles. August saw only 10 percent of its normal precipitation.

In addition to lightning-caused fires, human-caused fires increased in number. The Blodgett fire immediately west of Hamilton, Montana, began near a campground and popular trailhead at the end of July. It rapidly spread both north and south in steep terrain so that by August 3 it had burned 5.3 square kilometers, and by August 8, 23 square kilometers.

It finally consumed more than 46 square kilometers and required the evacuation of people from 900 homes.

The proximity of the Northern Region Forest Service Headquarters and Smoke Jumpers Center in Missoula helped suppression efforts. Initial attack crews managed to control 86 percent of all of the fires within five days. Extremely dry conditions permitted fires to spread rapidly with breezes as little as 5 kilometers per hour. On the afternoon of August 6, with a temperature of 92°F and a relative humidity of 11 percent, the winds were 32 kilometers per hour, promoting rapid expansion of the fires.

Several large fires in central Idaho and western Montana remained uncontrolled into September and burned a total of more than 1,335 square kilometers. In total, 1,185 square kilometers of forest land burned in the southern Bitterroot Valley area. The fires generated their own strong winds, and smoke columns reached as high as 9 kilometers. Burning twigs drifted 2 to 3 kilometers to ignite new fires. Firefighting crews were recruited from across the United States, and 600 soldiers and crews came from Australia and New Zealand. A total of 3,000 firefighters and twice that many support people worked on the fires at the height of the disaster. With insufficient resources, the highest priority was to protect firefighters and the public, then protect communities, then homes and other structures, and finally critical natural resources.

Land use and settlement also changed in recent years, with consequences for both forests and people. Between 1990 and 1998, the population of most of the Bitterroot Valley increased by 40 percent. People are drawn to the natural beauty of the area and the rural environment. Much of the growth is in unincorporated areas within or next to fire-vulnerable forests. Many of the newcomers are not familiar with the risks of forest fires or floods on seemingly peaceful little streams. They build their homes in heavily treed areas right at the edge of a little stream with almost no floodplain, or on a tiny alluvial fan right at the mouth of the gully feeding it. In spite of the unsuitability of these building sites, the owners expect firefighters to concentrate on saving their homes at the expense of the rest of the forest.

Some people refused to leave despite mandatory evacuation orders. Several stayed to protect their property; they viewed the evacuation orders as "typical government overreaction." But when they saw homes burning in other areas, most agreed to leave.

Fire's reputation in leading to floods and mudslides in the following few years held true in the Bitterroot Valley. A series of intense thunderstorms in mid-July 2001 did not lead to many new fires because sufficient rain in June kept the forests green. But those same storms unleashed short-lived but torrential rains onto unprotected soils. Intense fires had burned not only trees and ground vegetation but also the litter, including pine needles, twigs, and other dry vegetation on the forest floor. Organic material burned into a resinous material that glued together near-surface grains in the soil, making it mostly impervious to rainwater.

Rain from the torrential downpours, instead of striking tree leaves and needles and collecting in the forest litter before slowly soaking into the ground, struck the ground directly. The fine grains and resinous burned nature of the soil made it almost impervious to water; even without the effect of the fires, the soil could not accommodate the downpour. Water poured off the surface as sheetflow, collecting into gullies and tributary channels where it roared downslope as flash floods and debris flows. The effect was most extreme where fires burned much of a large drainage basin; water ran off the surface over a large area and collected rapidly in the main channels. In places, a twenty- to thirty-minute rainfall was intense enough to severely erode gullies in hillsides, even where there was almost no upslope drainage basin.

After the fires, the U.S. Forest Service placed erosion-prevention devices on hill slopes to slow future runoff and reduce sediment in streams. They draped **straw wattles,** looking like six-inch diameter snakes of netting full of straw, across vulnerable slopes near streams and felled trees across slopes (▶ Figure 16-19). Most proved ineffective in the brief but torrential downpours of the following summer. They completely disappeared in the deluge. The Forest Service improved drainage with larger culverts across many kilometers of roads and trails. The floods quickly filled them with boulders and branches, and washed out the roads. The work temporarily reassured people living in vulnerable areas, but much of it proved futile.

One woman in a house in the 50-meter-wide floodplain of Laird Creek recounted a frightening scene in a little tributary that emptied onto a small alluvial fan almost directly opposite her house. A wall of water 5 meters high came down the small channel with giant turbulent waves coming in rapid pulses. It carried sand, gravel, and boulders up to a half-meter across, leaving sand grains stuck to tree bark as much as 4 meters above the resulting deposit. The whole event lasted fifteen minutes. Surprisingly, the house survived, though the carefully tended yard around it was trashed both by this side channel and by high water in the main stream.

Donald Hyndman photo.

▶**FIGURE 16-19.** These straw wattles were placed below the McClain Creek landslide in the Bitterroot Range, Montana.

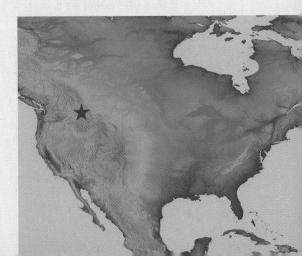

tells us that we should help those in need, but such help merely encourages them and others to live in such vulnerable areas. Perhaps, as FEMA has finally learned, we should help people who are willing to relocate to more suitable and less vulnerable places. At the same time, we need to zone such hazardous areas as off-limits to building.

With the fiercely independent attitudes of many rural residents, zoning restrictions are considered unacceptable infringements on their freedoms. They insist on their right to do as they wish with their own property. An alternative to zoning is to make clear that if people do build there, they must not expect public help in times of crisis. Such people should not expect help in fighting fire, stabilizing streams or hillsides before or after floods, or in rebuilding after catastrophes. If people still insist on living in fire-prone areas, they could create special fire-prevention districts that would tax their members—for example, a few thousand dollars per year—to create a pool of funds to pay the costs of future fire protection for their homes.

Insurance companies generally set premiums at levels based on risk and replacement costs. However, when governmental agencies spend millions of dollars fighting fires, and most of those efforts are expended in protecting homes, much of the real cost is on the shoulders of the general public rather than on the people who choose to live in wooded areas.

So do we fight fires or let them burn? Until a few years ago, the U.S. Forest Service and its Smokey Bear mascot maintained that fire was bad and all fires should be extinguished as quickly as possible. In the 1990s, that policy changed to permit fires to burn in uninhabited wilderness areas. It is now recognized that fire is a natural part of wildland evolution, that it is necessary for the health of rangelands and forests. Because the United States spends upward of $1 billion per year on fighting fires, you might expect some significant long-term benefit. However, many experts argue that we get no long-term benefit. In fact, preventing fires leads to buildup of wildland fuels that ultimately leads to worse fires.

KEY POINTS

✓ Fires can be naturally started by lightning strikes, intentionally set for beneficial purposes, or set accidentally or maliciously. Lightning strikes cause only 13 percent of all wildfires, while arson accounts for double that. **Review p. 415.**

✓ Fire requires fuel, oxygen, and heat. Lacking any one of these, fire cannot burn. The higher surface area of dry grass and needles makes them burn faster and more easily. **Review p. 417.**

✓ People who live in the woods are surrounded by fuel for fire. Many build their homes of wood, which provides more fuel for any wildfire. The worst firestorms are pushed by high winds that both bring in oxygen and blow the flames into more fuel. **Review pp. 417–419.**

✓ Prolonged dry weather reduces fuel moisture and increases fire danger. Thus, satellite imaging for greenness compared with normal conditions provides an indication of regional fire danger. **Review p. 419; Figure 16-9.**

✓ The catastrophic October 2003 fires in Southern California were aggravated by drought-weakened forests killed by bark beetles, numerous homes built from wood in forests, trees closely surrounding the houses, and the strong Santa Ana winds that fanned the flames. **Review pp. 420–421.**

✓ Hydrophobic soils sealed by fire cause water to run off the surface rather than soak in; this can lead to flash floods and mudflows. **Review pp. 422–423.**

✓ Techniques for fire suppression include cutting firebreaks down to bare ground, helicopters dumping large buckets of water, and borate bombers dumping huge loads of fire retardant. In out-of-control fires that threaten critical facilities, firefighters sometimes set back burns to burn back toward the advancing fire and thus deprive it of fuel. **Review p. 422.**

✓ Some concerns involve costs to the public of fighting fires. Firefighters are directed to first protect human lives, then people's homes, and after that the forest. Thus, much or most of the cost of protecting a few who choose to live in dangerous places is borne by the vast majority of the public who do not. **Review pp. 423 and 426.**

IMPORTANT WORDS AND CONCEPTS

Terms

back burn, p. 422	firestorms, p. 418
burn, p. 417	foam, p. 416
burnout, p. 422	fuel moisture, p. 419
combustible, p. 417	hydrocarbon residue,
cellulose, p. 417	p. 423
convective updraft, p. 419	hydrophobic, p. 423
dry ravel, p. 418	indefensible locations,
fire components,	p. 421
p. 417	straw wattles, p. 425

QUESTIONS FOR REVIEW

1. Wildfires are beneficial to forests in what two ways?

2. What are the two main causes of forest fires?

3. What two conditions lead to more fire-prone forests?

4. What can people living in the woods or forest do to minimize fire danger to their houses?

5. Why do winds accelerate fires? Give two specific reasons.

6. Why do fires advance faster upslope than downslope?

7. What natural conditions and processes of a fire lead to new fires well beyond the burning area of a large fire?

8. On which side of a forest is a home at greater risk to fire and why?

9. In addition to pouring water or retardant on fires, what techniques are used to fight fires? Name two specific techniques.

10. Why is hill slope erosion more prevalent after severe wildfires? Be specific.

11. What main techniques are used to minimize post-fire erosion? Name two specific and quite different techniques.

12. What are the main differences in a stream hydrograph after a large fire?

FURTHER READING

Assess your understanding of this chapter's topics with additional quizzing and conceptual-based problems at:

 http://earthscience.brookscole.com/hyndman.

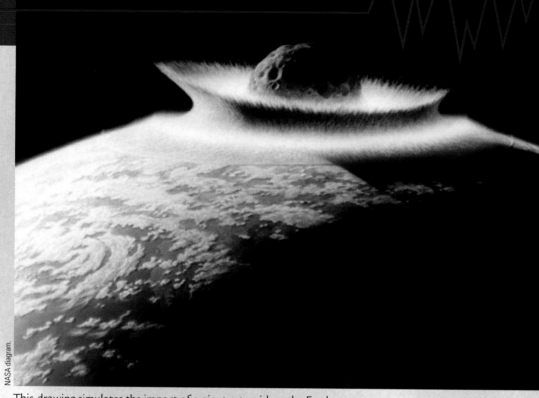

This drawing simulates the impact of a giant asteroid on the Earth.

IMPACT OF ASTEROIDS AND COMETS

Projectiles from Space: The Ultimate Catastrophe?

Asteroid impacts are not significant natural hazards on most people's horizon; however, no other physical hazard has such a dire potential. Although the odds of a huge asteroid colliding with Earth are tiny, the consequences of such an impact would be truly catastrophic. A large impact could wipe out civilization on Earth.

Meteor showers appear in the night sky when small pieces of rock from space enter the atmosphere at high speeds. When this occurs, friction with the surrounding air molecules heats the surrounding air to white-hot incandescence. Earth's atmosphere shields us from the impact of most meteorites, because small ones burn up in the upper atmosphere. The air around a large meteorite heats to become incandescent, but the cores of such meteorites typically do not get especially hot on entry into the atmosphere. Many fall on buildings or dry grass without starting a fire. A meteorite that fell in Colby, Wisconsin, on July 14, 1917, was cold enough to condense moisture in the air and become coated with frost.

As with most hazards, there are innumerable small meteorites, fewer large ones, and fortunately only a rare giant one. A fireball is a larger rock entering the Earth's atmosphere that glows for a longer period of time before it either disintegrates or survives to strike the Earth. Because meteorites travel at speeds much greater than the speed of sound, we hear only those that are relatively close. If we hear no sound, the meteorite is probably more than 100 kilometers away.

Relatively few falls are witnessed so the meteorite fragments can be collected; fewer than 1,000 total have been witnessed in the United States. Only twenty or thirty witnessed falls lead to meteorite finds worldwide each year. Sometimes, large meteorites break up in the Earth's atmosphere and fall as a **strewn field.** The Allende meteorite that fell in Chihuahua, Mexico, in 1968 scattered fragments over an oval-shaped area 50 kilometers long and up to 10 kilometers wide. Stony asteroids from 10 to 100 meters in diameter generally disintegrate on collision with the atmosphere.

The terms that refer to various space objects that impact Earth can be a bit confusing:

Meteors are objects that produce light in the sky as they streak through Earth's atmosphere.

Meteorites are the pieces of rock that survive passage through the atmosphere to collide with Earth. Most begin in the asteroid belt between Mars and Jupiter. The planets of our solar system lie in a mathematical progression of distance from the sun except that one planet is missing where the asteroid belt is found.

Asteroids are chunks of space rock orbiting the sun just like Earth. The asteroids appear to be the remnants of material that had not coalesced into a planet at about the time the other planets formed around our sun. Collisions between asteroids and the gravitational influence of the sun and planets pull some asteroids out of their normal orbits. The majority of these asteroids are less than 3 kilometers in diameter and most are 100 meters to 1 kilometer in diameter (▶ Figure 17-1).

Comets are similar to asteroids but consist of ice and some rock, essentially "dirty snowballs" (▶ Figure 17-2). They do not come from the asteroid belt but range far beyond our solar system where they make up the *Oort*

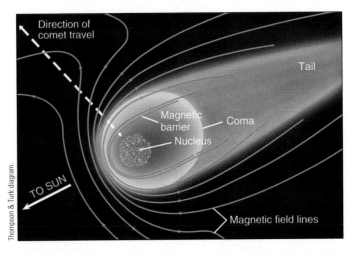

▶**FIGURE 17-2.** A comet consists of a solid nucleus of a rock and ice mixture surrounded by a "coma" of dust and gas. The tail is a mixture of water, other volatiles, and dust that the solar wind sprays away from the direction of the sun.

cloud. The Oort cloud forms a vast spherical region around the sun; it extends to a distance of more than 100,000 times the distance of Earth from the sun. It contains billions of comets. Most comets have heads less than 15 kilometers in diameter, but they travel at speeds up to 60 or 70 kilometers per second. At those velocities, impact with Earth would be a catastrophe. Some comets traverse our solar system as frequently as once every 10 years. They are the ones of greatest concern because they have the highest chance of coming close to Earth. See "Case in Point: Comet Hale-Bopp."

An inner doughnut-shaped zone of trillions of comets, the Kuiper comet belt, lies in the plane of the solar system and extends to 20,000 times the distance from Earth to the sun. Because comets spray off water, other volatiles, and dust to form their glowing tails when they come within the influence of the solar wind from the sun, they gradually become dehydrated. Eventually, they lose virtually all of their water, leaving only their rocky material. At that point, they are not easily distinguished from asteroids. In fact, there may be a continuous gradation between rocky comets and icy asteroids. The tail of a comet does not indicate its direction of

▶**FIGURE 17-1.** This photo compilation shows Asteroid 460 from four different sides.

The spectacular Hale-Bopp comet, with a diameter of approximately 40 kilometers, was seen by most people on Earth between January and May 1997. It came within Earth's orbit but on the other side of the solar system. Its closest approach to the Earth on March 22 was still 200 million miles away. If it had collided with the Earth, the energy expended would be tens to hundreds of times larger than that of the dinosaur-killing asteroid of 65 million years ago. It is a long-period comet that spends most of its orbit far beyond our solar system before blowing through our solar system after thousands of years. Whether one of these objects annihilates us is a matter of where Earth is in this shooting gallery when one of these objects passes through.

travel; rather, the tail points away from the sun. The tail is the spray of volatiles and dust blown off its surface by the solar wind (▶Figures 17-2 and 17-3).

The general term **bolide** includes both asteroids and comets.

Meteorites come in several types, all of which are somewhat similar to rocks thought to make up the deeper interior of the Earth:

Iron meteorites make up 6 percent of all meteorites. They consist mostly of a nickel-iron alloy that probably crystallized slowly in the deep interior of a large, solid body in our solar system. Collision between such bodies, and their breakup, leads to some collisions with Earth. Iron meteorites are extraordinarily heavy, with densities of 7.7 to 8 grams per cubic centimeter. This compares with most terrestrial rocks, which have densities of 2.6 to 3.3 grams per cubic centimeter; and with water, which has a density of 1 gram per cubic centimeter.

Stony-iron meteorites make up less than 1 percent of all meteorites. They consist of nearly equal amounts of magnesium and iron-rich silicate minerals such as

▶**FIGURE 17-3.** "Linear comet."

olivine and pyroxene in a nickel-iron matrix. They probably come from a zone between the deeper iron-rich parts of a large asteroid and the outer stony parts.

Chondrites and Achondrites

Chondrites are stony meteorites that make up 93 percent of all meteorites. They consist primarily of olivine and pyroxene, magnesium-iron-rich minerals, along with a little feldspar and glass. They have densities of approximately 3.3 grams per cubic centimeter. Millimeter-scale silicate spheres called **chondrules** enclose nickel-iron inclusions within or surrounding the chondrule.

Achondrites are stony meteorites that are similar to basalt, a common rock on Earth. They consist of variable amounts of olivine, pyroxene, and plagioclase feldspar.

Identification of Meteorites

Iron meteorites are distinctive. They are different from other rocks nearby and are almost three times as dense as most common rocks. They will generally be black in color unless oxidation over many years has turned their surface brown. Iron meteorites are extremely hard, virtually impossible to break with a hammer. With time out in the weather, they rust to iron oxides. Pieces of manufactured iron are more abundant and may be similar but can be distinguished by polishing a surface of the rock. Iron meteorites show intersecting sets of parallel lines marking the internal structure of the nickel-iron minerals. When these lines are accentuated by acid etching, they show the distinctive patterns that are diagnostic of iron meteorites.

Even stony meteorites are distinctively heavy. They are made of peridotite that is 15 to 20 percent more dense than most common rocks. Their heft is distinctive. Most contain enough metallic iron to be still heavier. Stony meteorites may be broken, exposing the fresh interior of the meteorite. If the broken surface is unaltered, you may see small inclusions of metallic silver-gray-colored nickel-iron that are strongly suggestive of a meteorite.

Some meteorites are rounded by their passage through the atmosphere. Iron meteorites are commonly more angular and sometimes twisted-looking. Some show rounded thumb-sized depressions. Others have an orientation related to their direction of travel, with a smooth leading end and pitted rear end.

Collision Course?

On December 6, 1997, astronomers of the University of Arizona Spacewatch program spotted a huge chunk of space rock, an asteroid some 1.5 kilometers in diameter, apparently on a near-collision course with Earth. Asteroids spotted in telescopes are tracked with time until astronomers have enough information to determine how close their trajectory will take them to Earth. In March 1998, the astronomer who spotted this particular rock was able to plot the path of Asteroid 1997 XF11 as coming dangerously close to Earth. The nearest approach to Earth was to be on October 26, 2028. However, in his excitement to announce the event to the press, he neglected to check earlier measurements on the same object. The corrected results indicate that the asteroid would come no closer than 2.5 times the distance to the moon.

An asteroid of that size traveling at more than 30 to 40 kilometers per second and striking the Earth would cause unbelievable damage. The energy expended would be that of 2 million Hiroshima-size atomic bombs. According to Jack G. Hills of the Los Alamos National Laboratory, if it struck the Atlantic Ocean, it would create a tsunami more than 100 meters high that would obliterate most of the coastal cities around that ocean. If it hit on land, the crater formed would be 30 kilometers across and darken the sky for weeks or months with dust and vapor. For comparison, the infamous asteroid that exterminated the dinosaurs and as much as 75 percent of all other species on Earth 65 million years ago was 10 to 15 kilometers in diameter.

Most asteroids stay in the **asteroid belt** between the orbits of Mars and Jupiter, where they pose no danger to Earth. The dangerous few are difficult to spot because their trajectories toward Earth leave them nearly stationary in the night sky. They are recognized on sequential images as changing position over several weeks, and their approximate paths are then calculated.

To understand the dangers and consequences of a large asteroid colliding with Earth, we need to consider the evidence for past impacts and their environmental effects.

Evolution of an Impact Crater

One might imagine that an asteroid striking the Earth would create a big hole in the ground that has a shape dependent on the incoming angle of the object. However, the incredible velocity and therefore the energy of the impactor requires that it explode violently on impact. The effect is more like a missile being fired into a surface of sand. It blasts a nearly round hole regardless of the impact angle (▶Figures 17-4 and 17-5). If the impactor is large enough, the explosion violently compresses material in the bottom of the crater, accelerating it to speeds of a few kilometers per second and ejecting material outward at hypervelocity (▶Figure 17-6). The center of the crater rebounds rapidly to form a central cone; that cone and the outer rim almost immediately collapse inward to form a wider but shallower final crater.

Most impacts arrive obliquely to the Earth's surface, a significant proportion at 5 to 15 degrees from horizontal. Such bolides ricochet and often break up in the atmosphere into five to ten fragments that still have roughly 50 percent of the original velocity. That scenario spreads devastation over a much larger area of the Earth. Numerous smaller masses still traveling at hypervelocities would provide the heat and energy to cause widespread reaction between nitrogen and oxygen in the atmosphere to generate nitrates that would combine with water and form nitric acid. The resulting acidic rain can be quite damaging to buildings, as well as to crops and natural vegetation.

In 1992, the Shoemaker-Levy 9 comet broke up into twenty-one fragments during a close approach to Jupiter as it was pulled apart by the huge planet's intense gravitational field. Then in July 1994, the fragments, all less than 1 kilometer in diameter, impacted one after another over a period of six days in an arc across that planet. Comets typically travel

David Roddy photo, USGS.

▶**FIGURE 17-4.** This panoramic view of Meteor Crater in Arizona shows its relatively small size for an impact crater, 1.2 kilometers across.

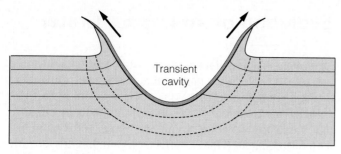

(a) End excavation

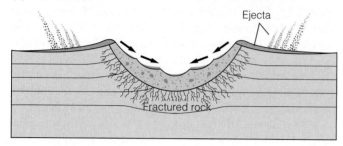

Ejecta

Fractured rock

(b) Modification

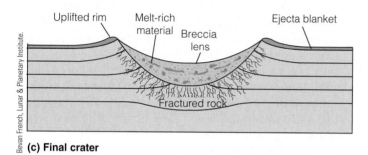

Uplifted rim

Melt-rich material

Breccia lens

Ejecta blanket

Fractured rock

Bevan French, Lunar & Planetary Institute.

(c) Final crater

▶**FIGURE 17-5.** The evolution of a moderate-size impact crater: **(a)** A transient crater is excavated, compressed, and fractured, and the base of the cavity melts with the rim raised. **(b)** The ejecta blanket spreads around the cavity, and the rim slumps back into the cavity. **(c)** Fallback material partly fills the cavity, along with some melt-rich material.

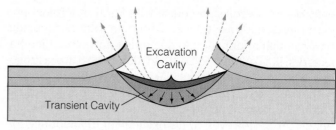

Excavation Cavity

Transient Cavity

(a) Excavation and compression under cavity

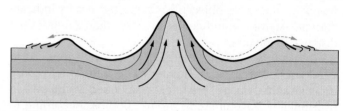

(b) Uplift and excavation

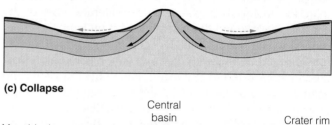

(c) Collapse

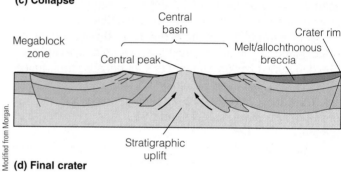

Central basin

Megablock zone

Central peak

Crater rim

Melt/allochthonous breccia

Stratigraphic uplift

Modified from Morgan.

(d) Final crater

▶**FIGURE 17-6.** The evolution of a large impact crater: **(a)** A transient crater is excavated and compressed, and the base of the cavity melts. **(b)** The base of the transient cavity rebounds as excavation continues. **(c)** The raised rim of the transient crater and central uplift both collapse to form a larger and shallower crater basin partly filled with inward-facing scarps, large blocks, smaller fragments, and melt rocks. **(d)** The final crater is much broader and shallower than the initial transient crater. *Autochthonous* refers to materials more or less in their original position.

at higher velocities than asteroids; this one was traveling at roughly 60 kilometers per second. Apparently simultaneous impacts at different sites on Earth suggest that these may be fragments of larger asteroids or comets. In other cases, multiple impacts may be fragments that broke up after ricocheting off the atmosphere.

Some Key Points and Evidence for an Impact

All impacts produce craters. Relatively small impacts, such as the one that created the 50,000-year-old Meteor Crater in Arizona (Figure 17-4) or recent smaller craters on the moon (▶Figure 17-7) remain as open craters rather than collapsing (see "Case in Point: Meteor Crater, Arizona"). The largest identified open crater is the 100-kilometer-diameter Popigai Crater of Siberia (see "Case in Point: Popigai Cra-

ter in Northern Siberia"). The older Manicouagan Crater of eastern Canada is preserved as a striking ring of lakes in the basement rocks of the Canadian Shield (▶Figure 17-8). By contrast, complex craters form when the walls of a broad, deep crater collapse inward to create a wider but shallower crater (▶Figure 17-6).

Because most asteroids travel at velocities of 15 to 25 kilometers per second, these velocities translate into incredible energies. Recall that the energy of a moving object is equal to its mass times the square of its velocity (see Sidebar 17-1).

NASA Mars Global Surveyor image.

1500 m

▶**FIGURE 17-7.** This simple, small, bowl-shaped crater on Mars is a smoothly rounded depression with a raised rim.

NASA.

▶**FIGURE 17-8.** The giant Manicouagan Crater of the eastern Canadian Shield shows as a dark ring of lakes in this NASA Earth Observatory image. The ring of lakes is 65 kilometers in diameter.

CASE IN POINT
Meteor Crater, Arizona

This is the classic open-crater impact site, 65 kilometers east of Flagstaff, Arizona, that is so well known to the general public. It is small as impact craters go, only 1.2 kilometers across and 180 meters deep, but it is nicely circular and has distinctly raised rims (▶ Figure 17-4). Being only 50,000 years old, it is also well preserved. The projectile was an iron meteorite with a diameter of some 60 meters. It came in at 15 kilometers per second to explode with the energy of 20 million tons of TNT, equivalent to that of the largest nuclear devices. The target rock, Coconino sandstone, shows good evidence for shock features, including shocked quartz and lechatelierite, a fused silica glass. A shepherd found a piece of iron in 1886, and a prospector found many more in 1891. One piece found its way to a mineral dealer in Philadelphia who recognized it as an iron meteorite. The dealer visited the site and found numerous fragments of iron meteorites. Unfortunately, no craters were then known to be formed by meteorite impacts, and even the great G. K. Gilbert, chief geologist of the U.S. Geological Survey, misinterpreted the crater as a volcanic crater or limestone sinkhole around which the meteorite fragments had fallen by coincidence.

Daniel Barringer, a mining engineer interested in the iron as an ore, filed claim to the site in 1903 and began intensive exploration of it with many drill holes. He found meteorites under rim debris and under boulders thrown out from deep in the crater. Barringer presented scientific papers in which he concluded that the crater had formed by meteorite impact, but few scientists were convinced. A few years later, two astronomers separately visiting the crater concluded that it had been caused by a meteorite, but that given its size and velocity the meteorite would have vaporized or disintegrated completely. The scientific community remained unconvinced until Eugene Shoemaker, then a graduate student, studied the crater and its materials in detail, finding shock-melted glass containing meteoritic droplets and the extremely high-pressure minerals coesite and stishovite. Stishovite also requires temperatures greater than about 750°C.

Thus, the energy doubles for every doubling of the mass of the asteroid but quadruples for every doubling of the velocity. Because comets are mostly ice, with a density of 0.9 grams per cubic centimeter, their overall densities including their rock component tend to be similar to that of water, 1.0 grams per cubic centimeter. Doubling the size of a comet doubles the energy. However, comets tend to travel at much higher velocities than asteroids—for example, 60 to 70 kilometers per second. Because the energy quadruples

Donald Hyndman photo.

▶**FIGURE 17-9.** Robert Hargraves, who discovered the Beaverhead impact site in Medicine Lodge Valley southwest of Dillon, Montana, is pictured here with a shatter cone cluster in an outcrop. Note that the apex of each shatter cone points upward.

for every doubling of the velocity, comets can have extremely high energy in spite of their lower densities.

On impact, the kinetic energy of the incoming object is converted to heat and vaporization of the asteroid and the target materials. This melts more rock, excavates a crater, and blasts out rock and droplets of molten glass. The result is a huge fireball that heats and melts rock and burns everything combustible.

Rocks on the receiving end of an impact also show distinctive features, especially **shatter cones** (▶see Figures 17-9 and 17-10). These cone-shaped features, with rough striations radiating downward and outward from the shock effect, range from roughly 10 centimeters to more than a meter long and are considered diagnostic of bolide impact. They form most readily in fine-grained massive rocks. Apexes of the shatter cones point upward toward the shock source. Those cones directly under the impactor should be vertical; those off to the sides flair down and outward from the source in "horsetail" fashion. Thus, the distribution of orientations of shatter cones provides evidence for the location of the center of the impact site. Individual cones are often initiated at a point of imperfection, sometimes a tiny pebble as in Figure 17-10. The Sudbury structure north of Toronto, Ontario, has some of the best-known shatter cones (see "Case in Point: The Sudbury Complex, Ontario"). They are distributed over an area 50 kilometers by 70 kilometers around the intrusion.

Fragments and dust sprayed out from a large impact site can drift around the Earth. The enormous impact at the end of the Cretaceous period (see Appendix 1, Geologic Time) blew out enough material to deposit a thin, dark layer called the Cretaceous-Tertiary **boundary clay (K-T boundary)** (▶Figure 17-11). The dust and elemental carbon soot layer from fires ignited by the impact fireball 65 million years ago is generally only a centimeter or so thick and chocolate brown to almost black in color. The large amount of soot is related to worldwide fires that burned much of the vegetation on the planet. The clay in this boundary layer contains grains of both shocked quartz and other minerals and tiny,

Donald Hyndman photo.

▶**FIGURE 17-10.** Note the tiny pebble at the apex of the cone (arrow) in this close view of a shatter cone at the Beaverhead impact site.

FIGURE 17-11. The Cretaceous-Tertiary (K-T) boundary layer is exposed at Bug Creek in northeastern Montana. The inconspicuous lowest dark layer (arrow) marks this important boundary.

Donald Hyndman photo.

now hollow, microtektite-like spherules, the original glass droplets sprayed out during impact. Those droplets of glass, generally a millimeter or two in diameter, are typically the most obvious signatures of an impact feature that are found in sediments (▶ Figure 17-12). (See "Case in Point: Ries Crater in Germany.") The abundance of quartz in the boundary clay strongly suggests that the dinosaur-killing impact was on quartz-rich continental rocks, probably in an area rich in granite, gneiss, or sandstone.

Sometimes melting of the asteroid produces the glass; sometimes it forms by melting the target material (▶ Figure 17-13). The spherules are often altered to green clay. In silicate target rocks, the impact melt may also form sheets, dikes, and ejecta fragments, and be disseminated in breccias. The melt develops under extremely high impact pressures. Impact-melt compositions are a mixture of the compositions of the target rocks that were shocked above their melting temperatures. Because the impacting meteorite is often melted as well, the impact melt may contain extraordinary amounts of nickel, iridium, platinum, and other metals that are abundant in iron meteorites.

Also found in the boundary clay of the 65 million-year-old event are anomalous amounts of iridium and other plat-inum-group elements, a phenomenon called the **iridium anomaly.** The anomalous amounts are tiny, some 0.5 to 10 parts per billion, but those elements are essentially absent in most rocks except for meteorites and highly mafic rocks such as peridotite and other rocks derived directly from the Earth's mantle. Some iridium is vaporized during the eruption of large basaltic volcanoes such as those in Hawaii, but the amounts are too small to account for the amounts found at the K-T boundary clay. The boundary clay from the initial find contains 30 times as much as would be expected from the normal fallout of meteoric dust. Elsewhere, as in Denmark, the clay contains up to 340 times as much. That boundary in sedimentary rocks marks the demise of the dinosaurs and 60 to 75 percent of all species. Vaporization of the impacting bolide is thought to spread the iridium worldwide, where it has now been found in roughly 100 sites.

Where an asteroid impacts into an ocean, tsunami would be formed by water flowing into and back out of the crater. Some of these tsunami could be extremely high. For example, the wave produced by the Chicxulub impact is calculated to have moved as a wave 200 meters high. The runup likely averaged more than 150 meters in height, with a maximum of 300 meters.

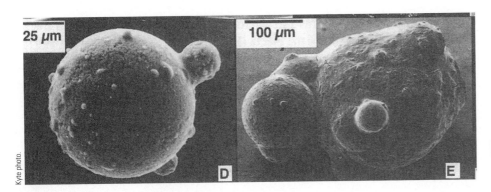

Kyte photo.

25 µm

100 µm

FIGURE 17-12. Tiny impact spherules are droplets of molten glass sprayed out from the impact site.

IMPACT OF ASTEROIDS AND COMETS **435**

The Precambrian-age Sudbury intrusion is 140 kilometers in diameter. Although it is not an open crater, its widespread shatter cones lead to essentially universal acceptance that it was caused by asteroid impact. Sudbury is also the largest nickel deposit in the world. It has been suggested that the nickel originated in the impacting meteorite, but there is considerable disagreement on that point.

If the target rocks contain quartz, the extreme shock pressures produced by impact deform the quartz to produce so-called **shocked quartz grains** that show multiple sets of shock lamellae imposed by more than 60,000 atmospheres of pressure (▶Figure 17-13). These are pressures found at depths of 200 kilometers in the Earth, or deep in the planet's mantle. Bohor and others (1984) of the U.S. Geological Survey discovered shocked quartz in the 65-million-year-old Cretaceous-Tertiary boundary clay. Some workers suggested that shocked quartz had formed by particularly violent volcanic explosions, but those suggestions have since been dispelled. Volcanic grains show only a single set of deformation lamellae that can be formed at much lower deformation pressures. Still higher-pressure forms of silica, stishovite and coesite, can also form under some circumstances, such as those at Meteor Crater, Arizona.

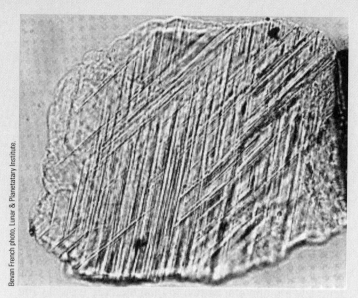

Bevan French photo, Lunar & Planetary Institute.

▶**FIGURE 17-13.** This impact-shocked grain of quartz is less than 1 millimeter across. Multiple sets of thinly spaced deformation planes are imposed by a high-intensity impact.

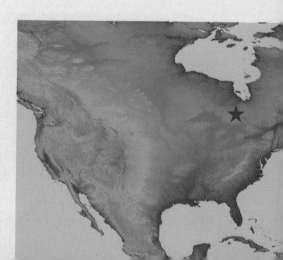

The 15.1-million-year-old, 24-kilometer-diameter Ries Crater formed in limestone, shale, and sandstone over crystalline basement rocks. The impact ejected a blanket of sedimentary rock fragments that were cold when they landed to the east. These fragments contain droplets of frozen melt derived from the explosive melting of underlying crystalline basement rocks. Shock features are widespread. The impact ejected glass droplets or **tektites** to the east of the impact site for 260 to 400 kilometers. Passage through the atmosphere aerodynamically shaped the glass, but it landed cold.

The interpreted sequence of events begins with high-speed shock waves and vapor blown out at shallow angles to the Earth's surface, quickly followed by high-speed ejection of target material. The tektites formed as melted target material, probably ejected above the atmosphere before falling back to Earth. The crater is circular, but the eastward fallout pattern of the tektites suggests that the impactor arrived from the west.

Chances of a Large Asteroid Impact on Earth

As with other hazards, small impacts are quite common, while giant events are rare (▶Figure 17-14). On average, a 6-meter-diameter bolide (either a rock asteroid or an ice or rock comet) collides with Earth every year; one 200 meters in diameter collides every 10,000 years on average. Only 1,500 or so asteroids larger than 1 kilometer in diameter are known in Earth-crossing orbits, those that pass through Earth's orbit around our sun. The largest is 41 kilometers in diameter. Most of these cross the Earth's orbit only at long intervals, so the chances of a collision with Earth are fortunately extremely small. The orbits of some asteroids vary from time to time because of the gravitational pull of various planets. Innumerable smaller asteroids also cross Earth's orbit. We live in a cosmic shooting gallery with no way to predict when one will hit the bull's-eye. We can only estimate the odds.

Earth is constantly sweeping up stray asteroids, so their number should be decreasing with time. However, collisions between asteroids in the asteroid belt create new asteroids that leave that belt; some of those fall into Earth-crossing orbits.

Also dangerous are comets, balls of ice and rock that have orbits outside our solar system but become visible when they pass close to the sun or Earth. Some 10 to 12 percent of impacts on the Earth and the moon are from comets. There is some evidence that comet showers have a periodic

component of 26 million to 32 million years, corresponding to paleontological extinctions, but some scientists argue against that assertion. Although the number of Earth impacts with time was thought to have slowed some 3.5 billion years ago, a recent study of the ages of glass-melt spherules in lunar soils indicates that after 500 million years ago, the impact rate increased again to previously high levels.

The periodicity of major impacts, based on the ages of impact craters and on theoretical aspects of interacting orbits and other movements within our galaxy, suggests an average interval of 33 million years. A periodicity of genus extinction events has been suggested as between 26 and 31 million years.

Likely Consequences of Impacts with Earth

Impact of a Large Asteroid

The demise of the dinosaurs 65 million years ago was associated with the death of between 40 and 70 percent of all species (see "Case in Point: Chicxulub Impact Crater"). If a bolide of that size were to impact Earth today, it would annihilate virtually everyone on the planet and a large proportion of other species. Almost immediate flash incineration near ground zero would accompany strong shock waves. Nitrogen, which makes up 78 percent of the atmosphere, burns with oxygen, which makes up 21 percent of the atmosphere, to form nitrous and nitric oxide. Those oxides combine with water to form nitric acid, which then becomes strongly acid rain. Sufficient nitrogen oxide and therefore acid rain were likely produced with a 10-kilometer-diameter impactor thought to be the primary cause of mass extinction 65 million years ago.

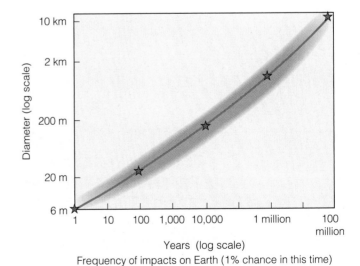

▶**FIGURE 17-14.** This log graph charts the approximate chance of an impact of an asteroid of a given size hitting the Earth.

Yucatan Peninsula, Mexico: 65 Million Years Ago

An asteroid 10 to 15 kilometers in diameter opened an initial temporary crater estimated to be 80 to 110 kilometers in diameter in the Yucatan peninsula of eastern Mexico (▶Figure 17-15). Its vertical walls collapsed immediately inward to form a shallower and broader crater basin 195 kilometers in diameter, with a pronounced central uplift. The Chicxulub is widely thought to be the impact site that killed the dinosaurs and the majority of other species on Earth at that time. The energy released from such an impact should have been equivalent to that of 100 trillion tons of TNT, or a million 1980 eruptions of Mount St. Helens. An impact large enough to have this effect should theoretically create an initial crater 200 kilometers in diameter and a much larger final diameter, but Chicxulub is the largest crater of the right age that has been found so far. The crater was slowly buried later by the quiet accumulation of limestones on a continental shelf, so it is not exposed at the surface (▶Figure 17-16). However this burial also preserved some features that tend to be eroded away with time. It has been studied through drilling and geophysical methods.

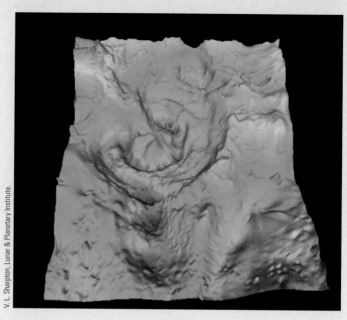

▶**FIGURE 17-16.** This map of the Chicxulub Crater was created using geophysical imaging because the crater is buried below the ocean and partly filled with sediment.

Aside from the final crater shape, evidence for such an impact includes worldwide distribution of shocked quartz grains, the extremely high pressure silica mineral coesite, and glass spherules near the K-T boundary in Mexico and Haiti. Huge tsunami waves formed in the Gulf of Mexico. At Brazos River, Texas, these waves were 50 to 100 meters high. Massive submarine slope failures were common around the Gulf of Mexico and along the East Coast of North America at that time.

The Manson impact structure in central Iowa is also 65 million years old, but at 35 kilometers in diameter it is much too small to be the main impact site for the event. Because concurrent impacts are known for other events, breakup of the asteroid may have caused multiple impacts. This impact structure is buried under ice age glacial deposits but has been studied by drilling and geophysical methods. Basement granite and gneiss under the center of the crater have been uplifted at least 4 kilometers above their original position. As with most other well-documented asteroid impact sites, shocked mineral grains are present in the target rocks.

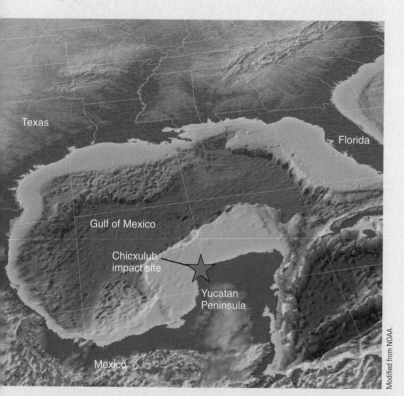

▶**FIGURE 17-15.** The Chicxulub impact site, 65 million years ago, is at the northern edge of the Yucatan Peninsula of eastern Mexico. The low-lying areas around the current Gulf of Mexico were a shallow continental shelf 65 million years ago.

Mass extinctions, for those species that were lucky enough to be far away from the impact, would occur after the impact from any of a number of indirect causes. Acid rain would kill vegetation and sea life all around the planet. Dust, soot from fires, and nitrogen dioxide would blot out the sun so animals that were not incinerated would freeze and starve to death. Plants would also die because of the drop in temperature and sunlight. With the impact of a 10-kilometer asteroid, land temperatures would, depending on assumptions, likely drop worldwide to freezing levels within a week to two months. Because of the large heat capacity of water, sea-surface temperatures would drop only slightly. Widespread fires would ignite from lightning strikes after much of the vegetation had died from either freezing in late spring or from toxic atmospheric effects.

The vaporization of a 10-kilometer-diameter chondritic bolide on impact with Earth would probably not only generate strong acid rain but also spread nickel concentrations of between 130 and 1,300 parts per million, even when diluted with 10 to 100 times the amount of target Earth material. A nickel-iron impactor would generate even more. This is many times the toxic level for chlorophyll production in plants. Seeds and roots would likely not recover for a long time.

A Modest-Sized Asteroid Impact

Impact of a more modest-sized asteroid 1.5 to 2 kilometers in diameter is considered enough to kill perhaps one-quarter of the people on Earth and threaten civilization as we know it. The chance of that smaller-sized asteroid striking Earth in any one year is estimated at one in several hundred thousand (Figure 17-14). This probably seems like such a minute chance as to be irrelevant. People do believe, however, in their chances of being dealt a royal flush in poker (1 chance in 649,739) or of winning the multimillion dollar lottery jackpot (1 chance in 10 to 100 million). The chance that the Earth will be struck by a civilization-ending asteroid next year is greater than either of those. As with other natural hazards, the low odds of such an event does not mean that it will be a long time before it happens. It could happen at any time.

The consequences of such an impact would be brutal for life as we know it. All other hazards and disasters pale in comparison. The fireball or ejecta from the impact would ignite fires within hundreds of kilometers of the impact site. A heavy plume of smoke would linger for years in the atmosphere. Recall that energy expended near the impact site would burn atmospheric nitrogen and oxygen to form nitric oxide. Mixing of the nitric oxide with water in the atmosphere would produce nitric acid and therefore acid rain that would add to the problem. Sulfate aerosols and water would be added to the atmosphere as well. A large portion of the ozone layer, which protects us from the sun's ultraviolet rays, probably would be destroyed.

At about the same time, dust blown into the stratosphere would block sunlight to that equivalent to an especially cloudy day almost worldwide. Large particles would settle out quickly, but dust particles smaller than 1/1,000 millimeter would remain in the stratosphere for months. Any temporary increase in temperature from widespread fires would be quickly replaced by cooling because of less solar radiation. The dust would be distributed worldwide because much of it would be blown out of the atmosphere before settling back into it. All agriculture would probably be wiped out for a year, and summertime freezes would threaten most agriculture after that. Many specialists view global-scale wars over food as inevitable. Such desperate conflicts would likely kill a large percentage of the world's population. Others argue that humans are more resourceful than that. If there were, say, a decade of warning before the event, sufficient food supplies could be grown and stored to outlast the period of darkness. However, that is a pretty big "if"! To add insult to injury, earthquakes would be generated within hundreds of kilometers of the impact site; and if the impact were into an ocean, tsunamis could inundate coastal areas for tens of kilometers inland, areas inhabited by a large proportion of the world's population.

Doomsday?

What if astronomers were to discover a 10-kilometer asteroid or comet, the size of the one that did in the dinosaurs 65 million years ago and they determined that it was on a collision course with Earth? What would we see without a telescope and when would we first see it? By the time it reached inside the moon's orbit, it would be three hours from impact. It would first appear like a bright star, becoming noticeably brighter every few minutes. An hour from impact, it would be as bright as Venus. Fifteen minutes from impact, it would appear as an irregular mass, rapidly growing in size. Three seconds from impact, it would enter Earth's atmosphere traveling at perhaps 30 kilometers per second. After that, the impact would be as recorded in the extinction event 65 million years ago.

What Could We Do About an Incoming Asteroid?

Clearly we could not survive the scenario outlined above. So what, if anything, can we do about it—that is, other than burying our heads in the sand and waiting for the inevitable? Suggestions include blasting it into pieces with a nuclear bomb or attaching a rocket engine to it to deflect it away from the Earth. Unfortunately, blasting it into smaller pieces might just pepper a large part of the Earth with thousands of smaller pieces, not a comforting scenario. Conventional explosives on one side of the asteroid might deflect it rather than shattering it. Another possibility is to deflect the asteroid by changing the amount of heat radiated from one side; for example, we could coat one side with white paint. The effect is weak, but over a long time, it could change its orbit enough to narrowly miss the Earth.

NASA catalogs **near-Earth objects** that are larger than 1 kilometer in diameter. The Jet Propulsion Laboratory in Pasadena, California, can detect objects down to 10 to 20 meters or so in diameter. Smaller ones are not important; most flame out before they reach Earth's surface. However, the impact of a large one would be catastrophic. Sometimes scientists tracking asteroids do not get much warning before one comes close. On December 26, 2001, they spotted an object 0.3 kilometer across. Twelve days later, it came within 800,000 kilometers of Earth, roughly twice the distance to the moon, a frighteningly small distance. If it had collided with the Earth, it would likely have destroyed an area the size of Texas or all of the northeastern states and southern Ontario. Another object 60 meters across came within 460,000 kilometers of the Earth on March 12, 2002. Astronomers did not detect it until four days after it passed because it came from the direction of the sun and thus could not be seen. Another 100-meter diameter object was also first spotted in June 2002, three days after it missed us by only 120,000 kilometers.

Falling Rocks: Your Personal Chance of Being Hit by a Meteorite?

Asteroid impacts have been recognized worldwide. The largest proportion presumably fell into the oceans that cover two-thirds of the Earth's surface. Most of those have remained undetected because details of the ocean floor are not well known. Because older parts of the ocean floors have been subducted into oceanic trenches, early impacts into the oceans will forever remain undetected. On continents, impact sites are broadly distributed but concentrated in areas of greater population where people find them or in areas of good exposure because of lack of vegetation (▶Figure 17-17).

Meteorites fall from the sky daily, but has anyone been struck? The only well-documented case was on November 30, 1954, in Sylacauga, Alabama, when a 3.8-kilogram meteorite crashed through the roof of a house, bounced off a radio, and hit Mrs. Hulitt Hodge, who was sleeping on a sofa. She was badly bruised but otherwise okay.

More recently, on Saturday, June 12, 2004, at 9:30 A.M., a 1.3-kilogram stony meteorite crashed through the roof of a home in Auckland, New Zealand. The rock was 7 by 13 centimeters, gray with a black rind, and quite rounded (▶Figure 17-18). It was especially hot when it landed on the Archer's family sofa in the living room, but no one was hurt.

On June 8, 1997, a 24-kilogram meteorite crashed into a garden 90 kilometers northeast of Moscow, Russia, where it excavated a 1-meter-deep crater. In Wethersfield, Connecticut, on April 8, 1971, a 0.3-kilogram meteorite came through the roof of a house into the living room at night. No one was injured. By incredible coincidence, in 1982 a 2.7-kilogram meteorite struck a house just 3 kilometers away. In another unusual case, on October 9, 1992, a bright fireball seen by thousands of people from Kentucky to New Jersey came down in Peekskill, New York, where it mangled the trunk of Michelle Knapp's car. Hearing the crash, she went outside to find the 11.8-kilogram meteorite in a shallow pit under the car. It was still warm and smelled of sulfur. In 1860, a meteorite falling on a field near New Concord, Ohio, killed a colt. Another killed a dog in Nakhia, Egypt, in 1911. There are no records of a person having been killed—at least

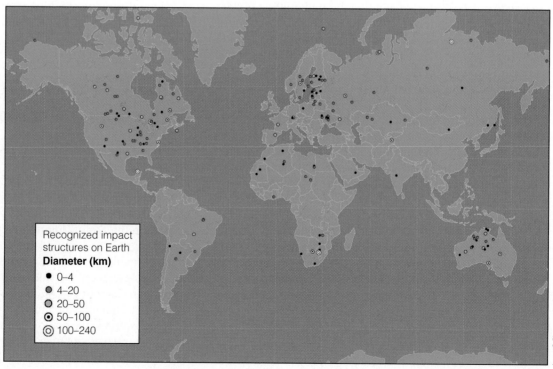

▶**FIGURE 17-17.**
Impacts of various sizes have peppered Earth for billions of years. The uneven distribution of these structures is mainly related to how easily they are identified because of population density and land cover.

Recognized impact structures on Earth
Diameter (km)
● 0–4
◉ 4–20
◎ 20–50
◉ 50–100
◎ 100–240

FIGURE 17-18. This 2004 meteorite that crashed through the roof of a home in Auckland, New Zealand, is approximately 13 centimeters long.

Meteorite Quarterly photo.

so far! Clearly, you need not stay awake at night worrying about the possibility of being hit by a meteorite.

There have also been cases of a bolide that involved grazing contact with Earth but no actual impact. See "Case in Point: Tunguska, Siberia."

Other Consequences of a Major Asteroid Impact

Lunar Maria

The large, essentially circular features that you see on a dark night that cover most of the surface of the moon are *lunar maria* or "seas." Most are filled with basalt. These are universally recognized as the products of major impacts, mostly during the early evolution of the moon and Earth. Many are enormous, much larger than any known impact sites on Earth. Why have no such giant impact sites been found on the Earth, especially given that the much greater gravitational attraction of the Earth should draw in many more asteroids than the moon?

Flood basalts on Earth have a curious habit of forming at about the time of major extinctions of life, so a relationship between them seems possible. The immense Deccan basalts of western India, for example, erupted at about the time of the demise of the dinosaurs and a large proportion of other species, 65.7 million years ago. If there is a cause-and-effect relationship, what is it? Most of the other handful of major flood basalt fields on the Earth erupted at the times of most other major extinctions of animal species. One imaginative but controversial suggestion is that the impacts of giant asteroids caused not only major extinction events but also triggered mantle melting and eruption of the flood basalts by sudden relief of pressure in the Earth's mantle.

One problem with this proposal is that a layer containing shocked quartz grains has been found recently—not at the base of the 65 million-year-old Deccan flood basalts in India, but between two major periods of basalt eruption.

The Inevitable Event

The impacts of asteroids and comets with Earth only rarely affect individuals and have not killed groups of people in historic time, at least none that have been recorded. A long time without an event, and in this case an especially long time compared with human life spans, leads to the widespread belief that it will never happen, at least not to us. However, we now have enough information on objects in Earth-crossing orbits that we have a pretty good idea of the odds of a significant impact with Earth. It is, of course, quite clear that it will eventually happen—we just do not know when.

Are we prepared for that inevitable event? The short answer is, clearly not. In one sense, there is no point in staying awake at night worrying about being hit by a stray space rock. Certainly, the odds of a person being hit in a human lifetime are rather small. In addition, worrying about the possibility would do no good. For a small, though deadly rock, we can neither predict such an event nor see it coming before being struck. That conclusion holds for even larger impactors that could kill thousands of people. For still larger bolides, the difficulty of spotting incoming objects heading

directly for Earth creates a major predicament. For a larger doomsday object that is somehow spotted as it grows larger on Earth approach, we have neither decided on a formal plan of action nationally or internationally nor set up the mechanism for implementation of that action. We just do not want to think about the possibility.

If you were in a position of power or influence, what would you do?

KEY POINTS

✓ Although the chances of a giant asteroid striking Earth are quite small, such an impact could wipe out civilization on Earth. Numerous small meteorites, far fewer large ones, and only rarely a giant one strike Earth. **Review pp. 428 and 441.**

✓ Asteroids are pieces of space rock orbiting the sun like the Earth does. Comets are ice with some rock that occasionally loop through our solar system. Meteors are light-producing objects streaking through the sky. Meteorites are the pieces of rock that actually collide with Earth. **Review pp. 429–430.**

✓ Meteorites include those consisting of iron, chondrites composed of the dark minerals olivine and pyroxene, and achondrites that are similar to the basalt found on Earth. Proportions of the three types are similar to such rocks in the interior of the Earth. **Review p. 430.**

✓ Occasionally, asteroids come relatively close to Earth. An asteroid with a 1.5-kilometer diameter that was spotted in 1997 is expected to come as close as 2.5 times the distance to the moon in October 2028. If calculations are incorrect and it were to strike the Atlantic Ocean, a tsunami more than 100 meters high would obliterate most of the Atlantic coast cities. **Review pp. 430–431.**

✓ The high velocity of a large incoming asteroid, at tens of kilometers per second, requires that it blast a deep, round crater that compresses and melts the rock below, spraying it outward in all directions. **Review pp. 431–432.**

✓ Distinctive features at and around the site of a major impact include shatter cones of rock radiating downward and outward, droplets of molten glass, quartz grains showing shock features, and for giant impacts, a layer of carbon-rich clay containing the metal iridium. **Review pp. 434–435.**

✓ The average chance of impact from a 200-meter-diameter asteroid is 1 in 10,000 years. Impact of a 10-kilometer-diameter object that could wipe out most or all of civilization is expected to occur

once every 100 million years on average. **Review pp. 437–439; Figure 17-14.**

✓ The asteroid that annihilated the dinosaurs and most other life forms 65 million years ago was likely 10 to 15 kilometers in diameter and struck in the Yucatan peninsula of eastern Mexico. **Review pp. 437–438.**

✓ Hazards from a large impact include a firestorm, soot that would block the sun and cause prolonged freezing and death of plants, strong acid rain, and nickel poisoning of plants. **Review pp. 437 and 439.**

✓ Spotting an asteroid heading directly for Earth can be almost impossible because it may hardly move in the night sky; coming from the direction of the sun, it would not be seen at all. The chance of an individual being struck by a small meteorite are insignificant. **Review p. 440.**

IMPORTANT WORDS AND CONCEPTS

Terms

achondrites, p. 430
asteroid belt, p. 431
asteroids, p. 429
bolide, p. 430
chondrites, p. 430
comets, p. 429
iridium anomaly, p. 436
iron meteorites, p. 430

K-T boundary, p. 434
meteorites, p. 429
meteors, p. 429
near-Earth objects, p. 440
shocked quartz grains, p. 436
stony-iron meteorites, p. 430
tektites, p. 437

QUESTIONS FOR REVIEW

1. What is the relationship between the size of meteorites and the number of a given size?
2. Where do asteroids originate in space?
3. What is the path of comets around the sun?
4. How is the tail of a comet oriented? (Which way does it point?)
5. Name two general kinds of meteorites and describe their composition.

6. How fast do asteroids travel in space?

7. Why do asteroids create a more-or-less semicircular hole in the ground, regardless of whether they come in perpendicular to the Earth's surface or with a glancing blow?

8. What is the sequence of events for the impact of a large asteroid immediately after it blows out of the crater?

9. What evidence is there that some large comets or asteroids break up on close encounter with a planet?

10. List five quite different direct physical or environmental effects of impacts of a large asteroid in addition to the excavation of a large crater.

11. Roughly what proportion of Earth's human population would likely be killed by impact of an asteroid 1.5 to 2 kilometers in diameter?

12. Why would astronomers have difficulty in recognizing a large incoming asteroid headed for direct impact on Earth?

13. Sixty-five million years ago, a large asteroid struck the Earth. Where did it apparently happen?

FURTHER READING

Assess your understanding of this chapter's topics with additional quizzing and conceptual-based problems at:

 http://earthscience.brookscole.com/hyndman.

CHAPTER 18

Steep, unstable slopes are not a good place for a house. This became clear to a resident in Laguna Canyon, Southern California, in March 1998 when El Niño storms released torrents of mud that engulfed many homes.

Dave Gatley photo, FEMA.

THE FUTURE
Where Do We Go from Here?

Those who cannot remember the past are condemned to repeat it.

—George Santayana, 1905

We Are the Problem

In the introductory chapter of this book, we emphasized that a hazard exists where a natural event is likely to harm people or property. Similarly, we noted that a disaster is a hazardous event that affects humans or property. A major natural event—an earthquake, a volcanic eruption, a landslide, a flood, or a hurricane or a tornado—in a remote area is merely a part of the ongoing process of nature. Problems arise when people place themselves in environments where they can be impacted by such major events. It makes no sense to blame nature for natural events that have been going on for hundreds of millions of years. The problem is not the natural event, but humans.

The solution then would seem simple. Keep people from locations that are hazardous. Unfortunately, that is easier said than done. Nomadic cultures learned to go with the flow; they learned to live with nature rather than fight it. Impacted by major events, they merely moved to safer ground, leaving a floodplain when the water began to rise, moving away from a volcano when it began to rumble.

Part of the problem now is that there are billions of us humans. As civilization evolved, we moved to where we could live comfortably and where we had the resources to meet our needs for food, water, energy, shelter, and transportation. We settled along coastlines and rivers where water was available, crops would grow, the climate was hospitable, and we could trade goods with other groups of people. We settled near volcanoes where soil was fertile for crops. As time went on, we concentrated in groups for defensive reasons; towns grew larger but essentially in the same locations that were most suitable for our basic needs.

As our cities grew, our infrastructure became entrenched. Our shelters, our transportation systems, and our communications networks became permanent—houses, stores, factories, roads, bridges, dams, water supply systems, and power lines. When natural events impacted us, we fought back. We built levees to hold back rising rivers, not realizing that with our towns built on floodplains, we were living in the natural high-water paths of many rivers. We should not have built in these places to begin with, but back then we did not know any better. We should have built our towns on higher terraces above the level of floods. We could still use floodplains to grow crops but abandon them temporarily whenever the water rose. Unfortunately, that is all in hindsight. Our towns grew up where they did, and we have to live with the consequences.

We continue to make poor choices. As the people of industrialized countries became more affluent and as free time and efficient transportation became widely available, we began making choices based on leisure, recreation, and aesthetics. We began building homes close to streams, on the edges of sea cliffs (▶Figure 18-1), on offshore barrier islands, in the "shelter" and view of spectacular cliffs and majestic volcanoes. When the larger natural events have come along, we have complained that nature is out of control. But nature is not the problem; we are the problem. We should live in safer places.

People take risks all the time. Most people understand that there are risks in all aspects of life; we accept many through lack of choice. There is little we can do about a stray meteorite unexpectedly nailing us, and we know that the odds of it happening are remote. Walking in the forest, we know that trees sometimes fall; we know the odds of one hitting us are remote so we do not think about it. Every time we get in a car and head out on the road, we under-

▶**FIGURE 18-1.** A block of ten cliff-top houses in Pacifica, California, began to lose the battle of cliff collapse in March 1998. By March 2004, only two houses remained.

stand that some other driver may cause a collision by being drunk, by falling asleep, by running a red light, or by merely being inattentive. Yet we accept the risk because, although accidents happen, we feel it is not likely to happen to us.

The same feelings cloud our judgment when we decide to build or buy a house in a location that suits our fancy. That spectacular view over the ocean, that place on a beautiful sandy beach, that tranquil site at the edge of a scenic stream—these are all desirable places to live. Those attractive locations are also subject to **natural hazards** that endanger our houses and perhaps also our lives.

Even people who are knowledgeable of particular natural hazards sometimes succumb to the lure of a great place. Yes, catastrophes have happened to others, "but it won't happen to me." Our wanting "something special" clouds our judgment. There are those who are well aware of a particular hazard but are willing to live with the consequences, both physical and financial. In an area prone to earthquakes, they incur the expense of building earthquake-resistant homes and purchasing expensive earthquake insurance. In a landslide-prone area, they install expensive drainage systems and may even pay for rock bolting the bedrock under their houses. Because landslide insurance is not generally available, they are willing to risk losing the complete value of their houses. For most of us, however, our home is the largest investment we will ever make; we borrow heavily, paying for it over fifteen to thirty years. Unfortunately, if the house is destroyed in a landslide, we are still liable for paying off the mortgage, even though the house no longer exists.

Hazard Assessment and Mitigation

"Soft" solutions for hazardous areas (including zoning to prevent building in certain areas and strict building codes to minimize damage to buildings and their occupants) are much less expensive than the **"hard" alternatives** (including levees on rivers, concrete barriers and riprap along coasts, catchment basins below debris-flow channels, and installing drains on a landslide-prone slope).

Dramatic advances in science and technology in the last century have led the public to believe that we can control **nature's rampages.** Many people believe that if a river levee fails or houses and roads are washed away on a barrier island, we just did not build the levee high enough or emplace sufficient protection before the storm. They do not realize that the more we hold back the effect of a large natural event, the worse the effect will be eventually. The natural outcome is inevitable; it is just a matter of time.

The availability of federal funds in the last several decades to repair and rebuild after a natural disaster has led many states, local governments, and individuals to believe that they are entitled to such help. However, since 1988, federal policy has gradually shifted to emphasize **mitigation**. Federal funds are still available for rebuilding but in a safer way or safer location. Almost as important has been the increasing role of insurance companies. After Hurricane Andrew in 1992 and the Mississippi River floods of 1993, many companies have either refused to renew policies or have dramatically increased insurance premiums to cover anticipated losses. These changes—coupled with land use regulations, strictly enforced building codes, fines, and various financial incentives—are all important in reducing losses.

Societal Attitudes

Even where lives are lost, collective memories of a disaster go back only a couple of years. People quickly forget that it could happen again.

Homes badly damaged by some natural disaster are typically ordered abandoned or removed. Even where such homes are condemned because they are too dangerous to inhabit, individuals and real estate groups sue for permits to rebuild and reoccupy. Often they are successful, and the problem escalates.

Homeowners commonly blame, and often sue, others for damages to property that they purchase. Who should be held responsible—the seller, the purchaser, the developer or real estate agent, the government? A good rule of thumb is the old adage of *Caveat emptor,* or "**Buyer beware.**" In many aspects of society, however, the seller is held responsible if he or she is aware of some aspect of a property that is damaged or endangered but the problem is not obvious. The issue can be a leaky roof, a cracked foundation, high radon levels in the home, previous flood damage, or the house is resting on a landslide. If the property is damaged but the buyer is not aware of it, the buyer is not generally held responsible. The same responsibility rests on developers or real estate agents who represent that a property is undamaged or safe from a particular hazard. If a developer purchases a property and then finds out that it rests on a landslide, he or she cannot represent to potential buyers that the property is safe from landsliding (▶ see Figure 18-2).

Homeowners who face losing major investments such as their homes often try to pin the blame on someone else, especially someone with "deep pockets." If the developer has disappeared or declared bankruptcy, then the next in line is often a government entity. Did the government provide a permit to the developer to build on a piece of property? Should a county or city be to blame for permitting building on a site that was prone to landsliding or flooding? Many areas have few restrictions on building in hazardous sites. In other areas, local ordinances require that a professional geologist or geotechnical engineer certify that a site is safe from sliding or flooding. Unfortunately, as in any profession, there are those who are less competent or less ethical than others. There are instances where a developer has fired a local expert who provided an assessment that would not permit building on a piece of property for proposed devel-

▶**FIGURE 18-2.** According to the realtor involved, this house, one of the last two remaining at the north end of a block of ten houses atop the sea cliff in Pacifica, California, sold for $450,000 in March 2004. How long will it be before its part of the cliff also collapses? This house is at the far end of the block of houses shown in Figure 18-1.

(a)

(b)

▶FIGURE 18-3. (a) In 1988, the newly remodeled Atlantic House restaurant and gift shop just south of Charleston, South Carolina, sat on the beach on numerous sturdy pilings. (b) After Hurricane Hugo in September 1989, the site was almost unrecognizable. Compare the bend in the road and riprap in the upper right and the parking lot. As for the restaurant and its walkway, only the pilings remain.

opment and then hired another "expert" who would provide a more positive view.

In many areas, **zoning restrictions** concerning natural hazards are limited. In many counties in South Carolina, for example, there are few restrictions on building on active barrier islands (▶Figure 18-3) or on the quality of construction to minimize wind damage during a hurricane. Although there have been dramatic advances in hurricane forecasting, increases in coastal populations leave roads jammed with traffic and too little time for evacuation. In Seattle and many other hilly West Coast cities, real estate agents are not required to disclose the fact that landslides have been occurring in a particular area for decades.

In some cases, however, the damage is heavily influenced by the homeowner's own behavior. If a home begins breaking up on a landslide, broken water-supply pipes may leak water into the slope. Did the pipes leak before movement began, thereby making the slope more prone to failure? Or did the slope fail, causing breakage of the pipes? Did a homeowner's roof and gutter system, heavy watering, and septic drain field feed water into the ground to make a slope less stable? Was the slope destabilized by the cutting and filling used to provide a house site on a slope?

Did the building of levees along a river cause flood levels to rise higher than they would have without the levees? Was a debris-flow dam not built high enough or a debris basin not built large enough to protect homes downslope? Or should the homes simply not have been built in such a hazardous site in the first place?

Should landowners be permitted to do whatever they wish with their property? **Property rights** advocates often say so. If a governmental entity permits building on land within its jurisdiction, should the taxpayers in the district shoulder the responsibility if there is a disaster? Should a landowner be prevented from developing a piece of prop-erty that might be subjected to a disaster? If so, has the government effectively taken the landowner's anticipated value without compensation, a taking characterized in the courts as **reverse condemnation?**

Who should bear the cost of hazard mitigation—the developer or landowner who wishes to build on a parcel of land or all taxpayers in the form of local, state, or federal taxes? If one person is permitted to "protect" his or her property from a flood or hurricane, is he or she transferring that hazard to someone else downstream or down the coast?

Such questions are not merely hypothetical; they arise all the time. Answers are not simple. They are the subject of lawsuits, countersuits, and court cases at all levels.

Education

Given the grim assessment above, it would be easy to despair that not much can be done—that the public will always end up paying for someone else's greed or poor judgment. Given the antitaxation feelings of a significant proportion of the public, most people should be receptive to reducing costs in futile efforts to control nature. Billions of dollars are expended in protecting property in areas that are inherently high risk, efforts that are doomed to ultimate failure. Once people realize that many major projects to protect the property of selective groups entail large costs to great numbers of people but benefit only a few, then the support for such projects should diminish. If so, the solution should be to educate the public about natural hazards and the processes that control them.

Many of these problems are political. Politicians are elected to serve their constituents. Their decisions may be made to better the lives of those they serve or for purposes of reelection based on their performance. People at all levels of government have direct control over expenditures for major publicly funded projects, so they need to be educated as to the effects and consequences of such projects. However, because officials serve at the wishes of their constituents, it is the general public that actually holds the cards.

Often the issue revolves around bringing money and jobs to a community or state, regardless of the long-term advantages or disadvantages. Members of a community are often happy to have a multimillion-dollar flood-protection or shore-protection project even if the state's share of the costs is spread to others throughout the state or throughout the country. The dollars brought in provide significant short-term financial benefit to a broad spectrum of the community. The community may also be happy to get such a project even if its members do not clearly understand the long-term effects. Sometimes the lasting effects, such as loss of a beach, are detrimental to the characteristics that drew them to the area in the first place.

Clearly, we need to educate people at all levels, from the general public to developers and realtors, business people, banks and insurance companies, and those who hold political office. Education can be a struggle because people do not want to hear that they should not or cannot live in a location that strikes their fancy. The opportune time to effectively convey this message is within a year or so after a major natural catastrophe. Then the cause and effect are clear in people's minds; we merely need to help clarify cause and effect for them and specify what individuals can do to prevent future similar occurrences. Unfortunately, even then people's memories of the trauma and costs fade quickly with time. They gradually begin to believe that it was a rare occurrence that is not likely to affect them again.

If we can make people responsible for their own actions, then losses to both property and life can be minimized. The **personal responsibility** must be passed not only to individuals but also to all involved groups and organizations. We all stand to gain if people can be dissuaded from living in hazardous areas. For people to accept the **consequences**, we must help them understand the processes of natural hazards and consequences of dealing with them.

Instead of addressing natural hazards on a short-term basis, both for gains and losses, we need to consider and deal with the long-term issues. We need to design for the **long-term future**.

Different Ground Rules for the Poor

In countries where poverty is widespread, the forces that drive many people's behavior are different than those in prosperous countries. In much of Central America and parts of southeastern Asia, for example, millions of people lack the resources to choose to live in certain places for reasons of aesthetics. They live where they can provide food and shelter. In Guatemala and Nicaragua, giant corporate farms now control most of the fertile valley bottoms, leaving the peasants little choice but to work for them in the fields at minimal wages and to provide their own shelter in the steep landslide-prone hillsides. Others who cannot find such work migrate to urban areas in search of work, again relegated to living in steep surrounding hills. Both groups of people clear forests to grow food, provide building materials, and gather firewood for cooking. Their choice of steep hillsides as a place to live is a hazardous one, but they have little choice in order to survive. Compounding the problem is that fertility rates and population growth in desperately poor countries are among the highest in the world, so more people are forced to live in less suitable areas. Large proportions of those populations do not minimize family sizes for religious and cultural reasons. For such poor societies, the reduction of vulnerability to natural hazards does not depend much on strengthening the zoning restrictions against living in dangerous areas or improving warning systems, though education can help. More so, it depends on cultural, economic, and political factors. Because more affluent individuals and corporations control many of those factors, the poor are left to fend for themselves. The environmental spin-offs, however, affect everyone.

The catastrophic December 2004 earthquake and tsunami in Sumatra and nearby areas bring into focus some of the problems faced by poor people living on low-lying coasts in much of the tropics. Large numbers of people with big families survive on fishing, staffing resorts for more affluent vacationers, and providing a broad range of spin-offs from these businesses. Earnings are very low and people live virtually on beaches at the edge of normal high tide, in homes poorly built from mud bricks or light-weight timbers and plaster. Roofs are typically sheets of corrugated iron. Even without a major earthquake or tsunami, any large storm will rip off roofs and parts of walls to destroy what few belongings people have.

The magnitude 9 earthquake crumbled many of the coastal homes in Sumatra, apparently killing a lot of people. The giant tsunami that followed a few minutes later crushed most of the remaining homes within hundreds of meters of the beach (▶ Figure 18-4) and swept up the debris and survivors in a churning mass that only a fortunate few survived. Hundreds of kilometers away in Thailand, Sri Lanka, India, and nearby countries, thousands of people living in similar circumstances died because they were unaware of the incoming tsunami (Figure 18-5).

What could have prevented the catastrophe or at least minimized its effects? Most of the victims lived in the coastal areas because of proximity to their means of livelihood or because the rugged mountainous terrain provided few other suitable building sites. The small number of buildings left standing were generally two-story reinforced concrete structures built by the few who could afford such luxury. Clearly most of the people had little choice of a better location. In hindsight, building a tsunami warning system would have helped, provided that warnings could be circulated automatically and very quickly to the populace. Since such

Tyler Clements photo, U.S. Navy.

▶**FIGURE 18-4.** Virtually all of the low-lying coastal area around Banda Aceh, Sumatra, was wiped clean; houses and many roads disappeared, especially near coastal estuaries.

events generally come at lengthy intervals, the public need to be aware and reminded regularly of the possibility and what to do when the warning comes. Education on the nature of earthquakes, their relationship to tsunami, and what to do in such an event could have saved tens of thousands.

The prospect of global warming adds an additional dimension. Most scientists agree that our temperatures are increasing, with more rapid erosion of coastlines, along with more extremes in weather that cause more landslides, floods, hurricanes, and wildfires. Some small islands in the

Indian Ocean and far from the earthquake epicenter were completely overwashed by the 2004 tsunami waves. As sea level continues to rise, such low-lying islands will gradually succumb to the sea, even without a catastrophic event.

Finally, we remind the reader that developed or prosperous countries lose economic value; underdeveloped or poor countries lose lives. Both eventualities are undesirable, but neither seems likely to change in the near future. Both depend on **societal attitudes** and people's behavior. We need to begin there.

▶**FIGURE 18-5**. In Matara, Sri Lanka, more than 1,500 kilometers from the epicenter, small coastal villages were destroyed. Only a few homes, set well back from the beach and surrounded by trees, survived.

Asitha Koggala photo.

KEY POINTS

✓ A disaster is a hazardous event that affects humans or property. We are the problem. The problem is not the natural event, but humans. **Review pp. 444–445 and 447–449.**

✓ "Soft" solutions for hazardous areas include zoning to prevent building in certain areas and strict building codes minimize damage and are much less expensive. "Hard" alternatives including levees on rivers and riprap along coasts are expensive and create other problems. **Review p. 446.**

✓ The public believe that we can control nature's rampages. However, most are short term solutions that create other problems. **Review p. 446.**

✓ Federal policy has gradually shifted to emphasize mitigation. **Review p. 446.**

✓ In many aspects of society, the seller is held responsible if he or she is aware of some aspect of a property that is dangerous or damaged. Homeowners commonly blame and often sue others for damages to property but should remember the old adage of *Caveat emptor*, or "Buyer beware." **Review p. 446.**

✓ In many areas, zoning restrictions concerning natural hazards are limited. **Review p. 447.**

✓ In some cases, the damage is heavily influenced by the homeowner's own behavior. **Review p. 447.**

✓ Who should bear the cost of hazard mitigation—the developer or landowner who wishes to build on a parcel of land or all taxpayers in the form of local, state, or federal government taxes? **Review p. 447.**

✓ We need to educate people at all levels, including the general public, developers and realtors, business people, banks and insurance companies, and politicians at all levels. **Review pp. 447–448.**

✓ If we can make people responsible for their own actions, then losses to both property and life can be minimized. **Review p. 448.**

✓ In poor societies, the reduction of vulnerability to natural hazards does not depend on zoning restrictions or improving warning systems, but more on cultural, economic, and political factors. Affluent individuals and corporations commonly control many of those factors, so the poor are left to fend for themselves. **Review pp. 448–449.**

✓ Developed or prosperous countries lose economic value; underdeveloped or poor countries lose lives. **Review p. 449.**

IMPORTANT WORDS AND CONCEPTS

Terms

buyer beware, p. 446
consequences, p. 448
"hard" alternatives (solutions), p. 446
long-term future, p. 448
mitigation, p. 446
natural hazards, p. 445
nature's rampages, p. 446

personal responsibility, p. 448
property rights, p. 447
reverse condemnation, p. 447
societal attitudes, p. 449
"soft" solutions, p. 446
zoning restrictions, p. 447

QUESTIONS FOR REVIEW

1. If someone's house on a floodplain is damaged or destroyed in a flood, who is to blame? Why?

2. Why so people live in dangerous places like offshore barrier bars?

3. What is the main distinction between "soft" solutions to natural hazards and "hard" alternatives? Provide examples.

4. What is meant by hazard mitigation?

5. If I sell a house that is later damaged by landsliding, who is responsible, that is, what are the main considerations?

6. When the federal government provides funds to protect people's homes from floods or wildfires, individual homeowners benefit at the expense of taxpayers. What are the two main types of alternatives to elimination of taxpayer expense for natural hazard losses?

7. What is meant by the expression, "Those who ignore the past are doomed to repeat it"? Provide an example related to natural hazards.

8. When we note that people need to take responsibility for their own actions in living in a hazardous environment, what is different about the behavior of poor people living in underdeveloped countries?

FURTHER READING

Assess your understanding of this chapter's topics with additional quizzing and conceptual-based problems at:

 http://earthscience.brookscole.com/hyndman.

GEOLOGICAL TIME SCALE

Era	Period	Epoch	Began (Millions of Years Ago)
Cenozoic	Quaternary	Holocene	0.01 (10,000 years ago)
		Pleistocene	1.8
	Tertiary	Pliocene	5
		Miocene	24
		Oligocene	34
		Eocene	55
		Paleocene	65
Mesozoic	Cretaceous		145
	Jurassic		213
	Triassic		248
Paleozoic	Permian		286
	Pennsylvanian		325
	Mississippian		360
	Devonian		410
	Silurian		440
	Ordovician		505
	Cambrian		544
PRECAMBRIAN			
Proterozoic			2,500
Archean			4,500

MINERAL AND ROCK CHARACTERISTICS RELATED TO HAZARDS

The following section outlines the characteristics of many common rocks, especially as they relate to geological hazards. Earthquakes and landslides are among the processes that break down rocks to smaller particles, which are then transported downstream in flowing water. The strength of rocks, therefore, affects when a fault will break and how strong the associated earthquake will be, when a hillside will slide and in what way, and how easily big boulders are broken down to be transported in streams. Whether rocks are at or near Earth's surface or deep beneath the surface also matters.

Rocks at least a few kilometers below the surface are warm or even hot. They are under high pressure because of the load of rocks and sediments above them. Like ice in a glacier, they may be brittle at the surface but deform plastically deep below under such loads. Hard rocks at the Earth's surface may break during an earthquake or collapse in a landslide. Those same rocks, warmer and persistently stressed at depth, may slowly flow. Some hard taffy seems rigid or brittle if you bite it hard; if you bite more gently and slowly, it bends and flows. Like warm tar, rocks may flow when stressed by tectonic movements. Many rocks deform more easily if they are soaked with water. We discuss these differences in more detail in the separate chapters in which they are most relevant.

Geologists classify rocks into three basic groups: igneous, sedimentary, and metamorphic. We focus on their descriptive characteristics here.

Igneous Rocks

Igneous rocks are those that solidify from molten magmas. Those that crystallize slowly deep underground are called **plutonic** rocks; they generally show distinct mineral grains that are easily visible to the unaided eye. Their grains interlock with one another because some grains crystallize first at high temperature within the magma; others crystallize later at somewhat lower temperatures, filling in the spaces between earlier-formed grains so the grains are in direct contact with no spaces in between. Most igneous rocks are massive—that is, they show little or no orientation of their grains. In a few cases, where the magma was moving as it crystallized, the grains may show parallel-oriented grains. If the magma crystallized within a kilometer or so of the Earth's surface, gases dissolved in the magma separate as steam bubbles that are trapped between the straight sides of the grains. Such rocks contain cavities bounded by the straight sides of the mineral grains that grew in the magma.

Volcanic rocks are formed when a magma reaches the Earth's surface. If the magma has low viscosity and contains little gas, it may pour out as a lava flow. If it is highly viscous and contains significant dissolved gas, it is likely to blast out as loose particles of volcanic ash. If the ash settles while still hot, its particles may fuse together to form a solid rock. Alternatively, loose ash that sits around for hundreds or thousands of years may become cemented into solid rock because water dissolves the surfaces of particles and then precipitates along their contacts, cementing them together. Blasting ash out of the vent typically chills the shreds of magma so quickly that minerals cannot crystallize into grains; the particles are glass. Volcanic rocks and processes are described in some detail in Chapters 6 and 7 on volcanic rocks.

Sedimentary Rocks

Sedimentary rocks form at the Earth's surface by deposition or precipitation from water or sometimes wind. Weathering processes break down rocks into particles that are transported downslope into streams, lakes, and ultimately the ocean. Loose particles are described as sediments. Chemicals dissolved from the surfaces of sediment grains that are in contact with water for hundreds to millions of years again precipitate to form cements that bind the grains together to form solid sedimentary rocks. Individual rocks of sand-sized grains, for example, become sandstone, and mud-sized grains become mudstone or shale. Where water in a lake or ocean contains large amounts of dissolved constituents, those chemicals may concentrate further by evaporation in a warm environment and precipitate to form chemical sedimentary rocks such as limestone. Marine animals such as clams and coral may take in chemicals dissolved in the water to precipitate them in their shells. Tiny corals often form colonies that precipitate enough calcium carbonate to build whole coral reefs that fringe coastlines in warm-water environments.

Metamorphic Rocks

Metamorphic rocks begin as sedimentary or igneous rocks but have been buried well beneath the Earth's surface, where they are subjected to higher temperatures and pressures. Heated mineral grains react with one another and with water along their boundaries to form new metamorphic minerals. If sediment grains were not interlocking, the new grains grow to interlock with one another. If the rock undergoing metamorphism is sufficiently deep where it becomes soft and plastic, it often flows in response to tectonic stresses and its mineral grains become oriented. This is one of the distinctive characteristics of many metamorphic rocks. Because a few igneous rocks such as some granites, show such orientation, there can be uncertainty and debate as to whether such a rock is actually igneous or metamorphic. Where sedimentary or igneous rocks are locally heated by contact with an intruding magma, the resulting rock becomes a contact metamorphic rock. Because most contact metamorphic rocks are not buried deeply enough to flow and take on mineral orientation, they are most commonly massive.

Weak Rocks and Strong Rocks

Most people think of shale as weak and granite as strong, and in a general way that is quite true. Shale, which is a sedimentary rock that forms when clay-rich muds compact and dry out, splits easily into thin sheets that may crumble in your hand. Shale in a hillside may crumble and slide when the load of the slope above pushes on a shale layer. If the platy sheets of the shale are oriented parallel to the slope, then the shale may split easily and slide on those sheets. Pieces of shale eroded by a stream crumble to tiny fragments as other rocks tumble against them. The tiny scraps are easily picked up by the swirling currents and carried far downstream. In a fault zone, a large mass of shale is so easily broken that it cannot build up a large enough stress to cause a large earthquake.

Granite, on the other hand, consists of interlocking grains of hard minerals, quartz, and feldspars. Granite in a hillside is so strong that it rarely fails unless it is exposed in a vertical cliff. Then it may collapse in a rockfall if the cliff is oversteepened or if cracks in the granite fill with water that freezes and expands. Boulders of granite that fall into a stream may tumble along the bottom in high water, but as the water level falls the stream is not energetic enough to move them. They often collect in place to form a steeper gradient and rapids. A large mass of granite in a fault zone does not easily break. Stress on the rocks on opposite sides of the fault needs to build up to a high level before the fault slips. When it does it may produce a large earthquake.

Many igneous rocks such as granite, and many metamorphic rocks such as quartzite and gneiss, which are formed by heating at high temperatures far below the Earth's surface, can be considered strong. Many sedimentary rocks such as shale or sands in which the grains are not cemented together are weak. As used here, strong rocks are hard to break. They do not collapse as easily in landslides or rockfalls; they do not erode as easily in streams. Rock strength also depends on whether the rock contains weak layers or many fractures.

Common Rock-Forming Minerals

To understand the makeup and behavior of rocks, we need to keep clear the distinction that **elements** such as oxygen, silicon, iron, and aluminum combine chemically to form **minerals** such as quartz, feldspars, and micas. Groups of minerals go together to form **rocks** (▶Figure A2-1). Some of the more common rocks are discussed below.

Rocks are agglomerations of grains of individual minerals. Mineral grains are individual crystals of chemical compounds (Figure A2-1). Each type of mineral has a crystalline structure controlled by the size and electrical charge on the atoms of its elements. Although there are literally thousands of different minerals, only a handful are common rock-forming minerals, and those are formed from only a few elements. Some grew to form igneous rocks by crystallization from high temperature melts or magmas. Shale, sandstone, and most other sedimentary rocks were formed by the disintegration of other rocks, followed by erosion and deposition of the sediment. In most cases, the grains are later cemented together to form a solid rock. Some sedimentary rocks grow by precipitation from cold, watery solutions. After deep burial under other rocks, some sedimentary rocks are subjected to the high pressure of other rocks on top and heat up to form metamorphic rocks. Elements in some minerals in the hot rocks diffuse into nearby minerals and react with them to form new metamorphic minerals.

Most minerals are recognizable by their visible properties. The abundant elements (and their standard chemical abbreviations) in most rocks are: oxygen (O), silicon (Si), aluminum (Al), iron (Fe), magnesium (Mg), calcium (Ca), sodium (Na), and potassium (K).

Rocks dominated by light-colored minerals are themselves light-colored or **felsic.** Rocks dominated by dark-colored minerals are themselves dark-colored or **mafic.** See ▶Figure A2-2. Distinctive properties of the common minerals are listed in Table 2-1. Figure A2-2 shows that igneous rocks range from pale to dark. The rock names are arbitrary divisions that permit us to talk about them efficiently without full descriptions.

We use various terms to describe the characteristics of rocks and minerals. Colors are described, as you might expect, in terms of gray, yellow, brown, red, and green, but most colors are imposed by small amounts of iron oxides that are hydrated to varying degrees. Those shades of color

FIGURE A2-1. Elements combine chemically to form minerals. Combinations of minerals make rocks.

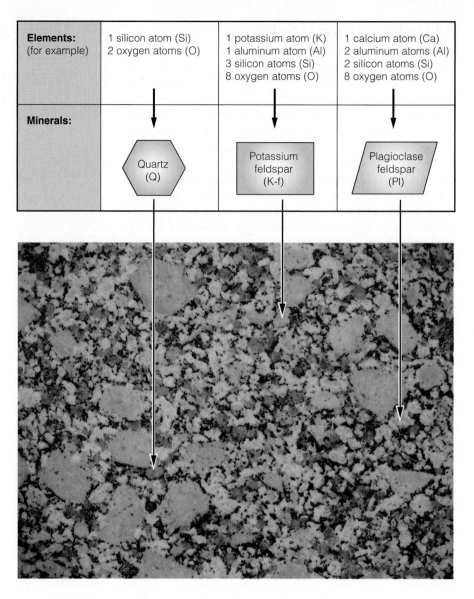

Elements: (for example)	1 silicon atom (Si) 2 oxygen atoms (O)	1 potassium atom (K) 1 aluminum atom (Al) 3 silicon atoms (Si) 8 oxygen atoms (O)	1 calcium atom (Ca) 2 aluminum atoms (Al) 2 silicon atoms (Si) 8 oxygen atoms (O)
Minerals:	Quartz (Q)	Potassium feldspar (K-f)	Plagioclase feldspar (Pl)

are less important than how light or dark the color is. The darkness of the color often reflects the amount of iron in the rock; that also often relates to other aspects of rock composition as well. The **cleavage** of a mineral refers to the flat breakage surfaces that are controlled by the mineral's internal arrangement of atoms. The most obvious example is the prominent cleavage of mica flakes. Other minerals such as feldspar have two different orientations of cleavages. Rocks can also have cleavage because of the parallel orientation of flat or elongate mineral grains. Prominent examples include slate or schist in which small to larger grains of micas control the platy breakage of the rock. The **hardness** of a rock is technically the ease with which you can scratch it, such as with the tip of a pocket knife. Hard rocks are those that contain large amounts of hard minerals such as quartz and feldspars that cannot easily be scratched with a knife. Those that contain large amounts of soft minerals

such as micas or calcium carbonate minerals can be easily scratched and are described as soft. The ease of breakage of a rock or mineral is not hardness but brittleness. A poorly cemented sandstone, for example, might be hard because it contains lots of grains of quartz; but it also might be brittle or crumbly because the grains are not well cemented together. Most limestones are soft but not brittle or crumbly.

The characteristics and strength of a rock depends in large part on the minerals that it contains, and the nature of a mineral depends on the chemical elements and their arrangement within the mineral.

Four oxygen atoms surround each silicon atom in a tetrahedral arrangement; six oxygen atoms surround each aluminum atom to make an octahedral array (▶ Figure A2-3). Those sheets stacked like the pages of a book and more or less weakly bonded together are the basic components of the

Table A2-1 The Common Rock-Forming Minerals and Their Main Properties

Mineral	Main Elements*	Properties
Quartz		Generally, medium to pale gray as grains in rocks; commonly white in veins that fill fractures in rocks.
		Typically glassy looking.
		Especially hard (cannot scratch with a knife); no flat cleavage surfaces.
		Abundant in granite, rhyolite, sandstone, schist, gneiss.
Potassium feldspar	K, Al	Pale colors: white, pinkish, beige.
		Hard, with flat cleavage surfaces in two directions at 90 degrees.
		Abundant in granite and rhyolite.
Plagioclase feldspar	Na, Ca, Al	Pale colors: white, pale gray, sometimes a bit greenish.
		Hard with flat cleavage surfaces in two directions at 90 degrees.
		Abundant in granite, gabbro, and most other igneous rocks.
Biotite (dark mica)	K, Al, Mg, Fe	Dark brown or black flakes; almost golden if weathered.
		Soft (can easily scratch with a knife); has one shiny, smooth cleavage direction.
		Biotite can amount from 5% to 20% of granite and more than 50% in metamorphic rocks such as mica schist.
Muscovite (white mica)	K, Al	Yellowish white flakes, soft.
		Has one shiny, smooth cleavage direction.
		Generally less abundant than biotite but can make up more than 10% of a granite and more than 50% of a schist.
Clay	Al and often Mg, Ca, Na, K	Microscopic flakes similar to micas but much softer and weaker.
		Make up highly variable amounts of soils.
		Some clay minerals take on water when wet and make a soil susceptible to landsliding (see descriptions of important clay minerals below).
		A major component of shale. See Figures A2-3 to A2-5.
Hornblende (an amphibole)	Ca, Mg, Fe, Al	Dark green or black.
		Short rod-shaped grains.
		Two shiny cleavage surfaces at 60-degree directions to one another.
		Can amount to 10% to 20% in granite and similar rocks; commonly ~50% in a metamorphosed basalt rock called *amphibolite*.
Pyroxene	Ca, Mg, Fe, Al	Dark green, dark brown, or black.
		Short stubby grains.
		Two less prominent cleavage surfaces at 90 degrees.
		Occurs primarily in dark-colored rocks; most abundant in gabbro.
Olivine	Mg, Fe	Green, glassy looking, hard.
		Generally occurs in dark-colored rocks, especially in basalt and gabbro.

*In addition to silicon and oxygen.

clay minerals. Between these negatively charged combination sheets, some clays have positively charged atoms or ions such as sodium or calcium that hold the sheets together.

Kaolinite is the simplest clay. It consists of alternating sheets of silicon tetrahedra and aluminum octahedra (▶ Figure A2-4). Although they have no overall charge, the layers are weakly positive on one side and weakly negative on the other. These positive and negative charges weakly bind the structure together, resulting in extremely soft crystals. Kaolinite does not absorb water between its layers and does not expand and contract as it wets and dries. It generally forms through weathering in especially wet and warm environments. Kaolinite commonly forms deep soils, some considerably more than 100 meters deep. Because it is so weak, it is frequently involved in landslides.

Smectite is a group of complex clay minerals that has silicon, aluminum, and magnesium sheets bound together to make combination layers. Loosely bound atoms of water

▶**FIGURE A2-2.** Percentages of minerals in common igneous rocks. Aphanitic rocks have grains that are too small to see without a magnifier; phaneritic rocks are distinctly grainy looking.

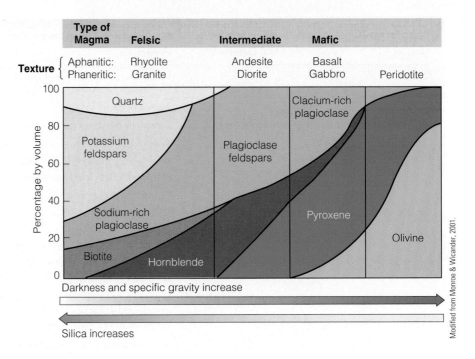

Type of Magma	Felsic	Intermediate	Mafic	
Texture { Aphanitic:	Rhyolite	Andesite	Basalt	
Phaneritic:	Granite	Diorite	Gabbro	Peridotite

Darkness and specific gravity increase

Silica increases

Modified from Monroe & Wicander, 2001.

Tetrahedron   **Octahedron**

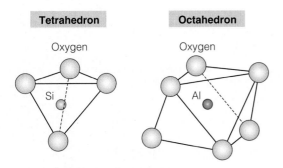

▶**FIGURE A2-3.** Atomic arrangement of the main building blocks in clay minerals. Oxygen atoms surround either silicon or aluminum atoms.

and sodium or calcium bind the sheets together (▶Figure A2-5). When the clay gets wet, water seeps between the layers. Then the clay swells to become plastic and structurally extremely weak. Smectite generally forms through weathering in a dry environment, especially in volcanic ash.

Many of the sedimentary formations that lie beneath the High Plains just east of the Rocky Mountains are rich in smectite. Soils developed on them become extremely slippery when wet; they are highly unstable and likely to slide. If this slimy mineral did not exist, landslides would be much less abundant. Because it swells when it gets wet, smectite also wreaks havoc with houses that are built on soils containing this mineral (see Chapter 9's section on swelling soils).

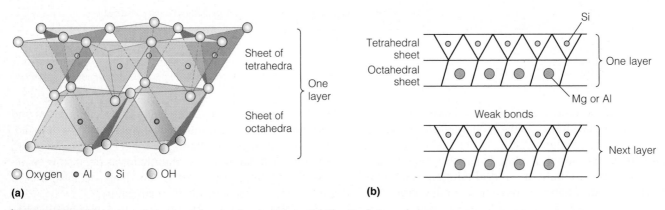

(a)   ○ Oxygen • Al ∙ Si ◎ OH

(b)

▶**FIGURE A2-4.** **(a)** Kaolinite-type structure in some clays. **(b)** The Kaolinite-type structure in some clays consists of sheets of tetrahedra and octahedra. Compare the tetrahedra (inverted triangles here) and octahedra. Oxygen atoms are at each corner.

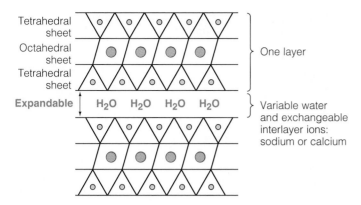

▶**FIGURE A2-5.** In the smectite structure, the interlayer cations (positive-charged atoms) are sodium or calcium that can be replaced by water. This causes the interlayer space to expand and the clay to swell.

(a)

(b)

Common Rocks

Granite

Granite, a common igneous rock (▶Figures A2-1 and A2-6), is found primarily in mountainous areas such as the Rockies and Appalachians and in deeply eroded "basement" rocks such as those of eastern Canada and of the Adirondacks of New York. It is light-colored, grainy, and strong. It breaks to form sharp edges but typically weathers to rounded corners on outcrops. It crystallized from molten magma well below the Earth's surface to form strongly interlocking grains. Its minerals include abundant plagioclase and sometimes potassium feldspar, moderately abundant quartz, and small amounts of micas or hornblende. Diorite is a darker, less common rock that is otherwise similar to granite. When granite magma reaches the surface to erupt in a volcano, it solidifies as fine-grained rhyolite.

Gabbro

Gabbro (▶Figure A2-7) is a coarse-grained igneous rock that crystallizes underground in large masses from the same magma that erupts at the surface as basalt lavas. It is medium to dark gray in color and strong. Early crystallizing grains of dark pyroxene often settle in the magma to form black layers as in the photo. Gabbro is not common.

(c)

▶**FIGURE A2-6.** Granite is a light-colored, grainy, massive, strong, and hard igneous rock with few fractures.

(a)

(b)

FIGURE A2-7. (a) Gabbro is a darker-colored, grainy, massive, strong, and hard igneous rock with few fractures. (b) In some cases, early crystallized dark grains of pyroxenes settle out to form dark layers.

Rhyolite

Rhyolite (▶ Figure A2-8) is a common, light-colored, fine-grained volcanic rock that has the same composition as granite but which solidified at the surface. It can be white, pink, yellow, or pale green, the color provided by superficial amounts of various iron oxides. Most rhyolite is ash and rather easily eroded, though vertical cliffs are not uncommon. Vertical fractures are well spaced and permit large slabs to break off as dangerous rockfalls. The sample at the bottom shows fragments of pumice.

(a)

(b)

(c)

▶FIGURE A2-8. Rhyolite is a light-colored, fine-grained, massive, generally soft igneous rock with few fractures. (a) If crystallized at shallow depths below the surface, rhyolite may be hard. (b) Pieces of pumice (slightly darker) may be surrounded by fine-grained ash. (c) Pieces of pumice are frothy with gas holes formed in the rising magma as pressure on it decreased.

(a)

(b)

▶**FIGURE A2-9.** Andesite is a medium-colored, fine-grained, massive igneous rock with few fractures. Its composition and characteristics are intermediate between those of basalt and rhyolite.

(a)

(b)

▶**FIGURE A2-10.** **(a)** Basalt is a dark-colored, fine-grained, massive igneous rock with few fractures. Its composition is equivalent to that of gabbro. **(b)** Basalt with gas holes and green crystals of olivine.

Andesite and Dacite

Andesite (▶Figure A2-9) is a volcanic rock with color and characteristics that are intermediate between those of rhyolite and basalt. Its color ranges from gray to pinkish gray, green, or brown. Some andesites contain larger grains of white plagioclase (as in the photo) or greenish black hornblende or pyroxene. Andesite forms strong, hard lava flows but weak ash.

Dacite is a paler volcanic rock, related to andesite, and intermediate in composition and appearance between andesite and rhyolite.

Basalt

Basalt, the most common volcanic rock, is low in silica and typically forms lava flows. It is usually black but can be slightly brownish or greenish. Most basalts are fine-grained and massive (▶Figure A2-10a); some are full of steam holes

▶ **FIGURE A2-11.** Some shales or mudrocks are marked by mud-cracks that formed as wet mud dried and shrank.

▶ **FIGURE A2-12.** Sandstone is made of grains of sand cemented together to make a solid rock.

◖ see Figure A2-10b. Some basalts contain larger dark grains of black pyroxene or green olivine. Basalt forms strong, widespread lava flows but less abundant and weak cinder piles.

Shale and Mudstone

Shale and **mudstone** are dried mud or mudrocks; some even show ripple marks or mudcracks as in ▶ Figure A2-11. These rocks range widely in color from medium to dark gray, red, green, yellow, and brown. Shale is almost as soft and flaky as the clay-rich mud that forms it. Mudstone is more massive and much stronger; it contains lots of tiny grains of quartz and feldspar. Shale may form thick layers on gentle slopes or thin layers between stronger sedimentary rocks such as sandstone or limestone. Shale and its lightly metamorphosed equivalent, slate, form thin plates that break easily and make weak slopes and weak layers between rocks that easily fail in landslides.

Sandstone

As its name implies, **sandstone** (▶ Figure A2-12) is made of sand grains—generally quartz but sometimes with feldspar. Its color ranges from white to yellow, brown, red, or green. It may be rather weak if in thin layers with weak rocks in between or lightly cemented. It may be strong if well-cemented or if metamorphosed to interlocking grains, in which case it is called *quartzite*.

Limestone

Limestone (▶ Figure A2-13) is calcium carbonate, soft to scratch but strong. It dissolves slowly in rainwater to form ragged rock surfaces and underground caverns but does

▶ **FIGURE A2-13.** Light-colored limestone with a rough surface.

not erode easily. Its color is commonly pale gray, often weathering to a soft-looking but a rough and sandy-feeling surface. Dissolution of limestone under soil is the most common cause of sinkholes.

Slate

Slate (▶ Figure A2-14) is lightly metamorphosed, strongly compressed shale. It is somewhat harder but still splits fairly easily into flat sheets. Those sheets lie flat on steep slopes and facilitate sliding. Its color is most commonly dark gray or green.

(a)

▶**FIGURE A2-15.** Shiny micas in this schist are slightly wavy because of deformation after metamorphism. The little lumps are dark red garnets.

(b)

▶**FIGURE A2-16.** Gneiss is similar to schist but contains pale layers that are dominated by light-colored feldspars alternating with dark mica-rich layers. Some were heated to high enough temperatures to melt and thus produce the pale granite layers.

▶**FIGURE A2-14.** **(a)** Slate, a lightly metamorphosed but strongly compressed mud rock, breaks into thin plates or flakes that make weak slopes that are subject to landsliding. **(b)** An outcrop view from the Sierra Nevada northwest of Sacramento, California.

Schist

Schist (▶Figure A2-15) is a strongly metamorphosed shale or slate. The grains range from 1 to several millimeters across. Mica-rich varieties are weak and prone to sliding on steep slopes. Schist splits into flat sheets; those and individ-

ual mica grains lie flat and facilitate sliding in translational landslides. The sample in Figure A2-15 contains dark red garnets.

Gneiss

Gneiss (▶Figure A2-16) is a generally hard and strong highly metamorphosed rock. It consists of white grains of feldspar, glassy quartz, generally some micas, and sometimes hornblende or garnet. It is characterized by distinct thin layers dominated by dark grains separated by layers of light-colored grains. Gneiss forms bold cliffs that break into slabs or blocks like granite.

Serpentinite

Serpentinite (▶Figure A2-17) is a low-grade metamorphic rock that forms by hydration of rocks from the Earth's mantle, dark rocks originally consisting of olivine and some pyroxene. It generally rises toward the Earth's surface along fault zones, where it can be found in a wide range of shades from medium green to black. Serpentinite is weak, soft, slippery, and creates high risk of landslides. Its weakness and presence in fault zones leaves it strongly deformed. Often only larger blocks of serpentinite remain in a thoroughly sheared matrix, as in the outcrop photo. Serpentinite is found in mountainous areas of western and eastern North America and to a minor extent in areas of basement metamorphic rocks.

(a)

Rocks, Landscapes, and Hazards

In summary, hard, strong rocks such as granite, gabbro, gneiss, much basalt, and some andesite and sandstone do not erode easily. In cool climates, they form steep slopes, rocky cliffs, and narrow canyons. In dry climates, weathering and formation of clays on the surfaces of feldspar grains in granites causes gradual disintegration of the rocks to form sandy soils that erode easily when it rains. Cliffs of these rocks are prone to rockfalls. In warm, moist climates, they may weather to soft clays. Fractures in these rocks are typically spaced at decimeters to meters; those fractures help drain water, keeping slopes relatively dry.

Rhyolite can be quite hard and resistant to erosion if the rock solidified at high temperature to fuse the grains together. More often it settles out of the air as cold ash that is not well stuck together; if it is not later cemented, it erodes easily.

Limestone is soft in the sense that it is easy to scratch with a knife. However, where it forms thick layers, limestone tends to be strong and forms prominent cliffs. It dissolves slowly in rainwater near the water table to form underground caverns. Where cavities above these grow sufficiently large near the surface, the ground may collapse to form sinkholes, sometimes taking roads or houses down with them.

Especially soft and crumbly rocks such as shale, slate, and mica schist break up and erode easily to form gentle slopes. They are prone to landsliding because water does not drain out easily and the flat sheets slide easily past one another. Where layers of shale separate layers of stronger rocks, their flat planes form prominent zones of weakness. Where those weak zones slope nearly parallel to a hillside, large masses of rock can be prone to sliding. Serpentinite is simply a slope-failure disaster waiting to happen.

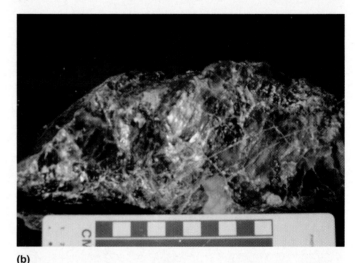
(b)

(c)

▶**FIGURE A2-17.** **(a)** Dark, thoroughly sheared serpentinite typically contains somewhat bigger but still weak lumps shown in this 5-meter-wide outcrop. **(b,c)** Small samples that show the typical smooth, curving surfaces formed by shear.

CONVERSION FACTORS

Length

1 millimeter (mm)	=	**0.03937**	**inches**
1 centimeter (cm)	=	**0.3937**	**inches**
1 meter (m)	=	**3.281**	**feet**
1 kilometer (km)	=	**0.6214**	**miles (~5/8 mile)**
1 inch (in)	=	*2.540*	*cm*
1 foot (ft)	=	*30.48*	*cm*
1 mile (mi)	=	*1609*	*m = 1.609 km*

Area

1 square centimeter (cm^2)	=	**0.1550**	**square inches (in^2)**
1 square meter (m^2)	=	**10.76**	**square feet (ft^2)**
1 square kilometer (km^2)	=	**0.3861**	**square miles**
1 hectare (=10,000 m^2)	=	**2.471**	**acres**
1 in^2	=	*6.452*	*cm^2*
1 ft^2	=	*929.0*	*cm^2*
1 mi^2	=	*2.590*	*km^2*

Volume

1 cubic centimeter (cm^3)	=	**0.06102**	**cubic inches (in^3)**
1 cubic meter (m^3)	=	**35.31**	**cubic feet (ft^3) = 1.308 cubic yards (yd^3)**
1 liter	=	**61.02**	**in^3 = 1.057 qt. = 0.2642 gallons**
1 ft^3	=	*0.02832*	*m^3*
1 yd^3	=	*0.7646*	*m^3*
1 gallon U.S.	=	*3.785*	*liter = 0.1337 ft^3*

Mass

1 gram (gm)	=	**0.03527**	**oz = 0.002205 pounds (lb)**
1 kg	=	**2.205**	**lb**
1 oz	=	*28.35*	*gm*
1 pound (= 16 oz.)	=	*453.6*	*gm = 0.4536 kg*

Velocity

1 meter per second (m/sec.)	=	**3.281**	**feet per second = 3.6 km/hr**
1 mile per hour	=	*1.609*	*kilometer per hour (km/hr)*

Temperature

°C	=	**5/9 (°F–32)**
°F	=	*9/5 °C + 32*

Prefixes for SI units

Giga	$= 10^9$	$= 1{,}000{,}000{,}000$	times
Mega	$= 10^6$	$= 1{,}000{,}000$	times
Kilo	$= 10^3$	$= 1000$	times
Hecto	$= 10^2$	$= 100$	times
Centi	$= 10^{-2}$	$= 0.01$	times
Milli	$= 10^{-3}$	$= 0.001$	times
Micro	$= 10^{-6}$	$= 0.000{,}001$	times
Nano	$= 10^{-9}$	$= 0.000{,}000{,}001$	times
Pico	$= 10^{-12}$	$= 0.000{,}000{,}000{,}001$	times

GLOSSARY

100-year flood A flood magnitude that comes along once every 100 years on average.

100-year floodplain The area of a river's floodplain that is covered by water in a 100-year flood.

aa Blocky basalt with a ragged, clinkery surface.

acceleration The rate of increase in velocity. During an earthquake, the ground accelerates from being stationary to a maximum velocity before slowing and reversing its movement.

achondrite A stony meteorite that is similar to basalt in composition.

acid rain Rainfall containing acidic components such as sulfur or nitrogen oxides or high concentrations of carbon dioxide that are detrimental to the health of plants or people.

active fault A fault that is likely to move again, especially those that have moved in the last 10,000 years.

active margin A convergent continental margin—for example, at a subduction zone.

adiabatic Change in volume and temperature without change in total heat content.

adiabatic cooling. *See* adiabatic.

adiabatic lapse rate (dry and wet) The rate of change of temperature as an air mass changes elevation.

aftershocks Smaller earthquakes after a major earthquake that occur on or near the same fault. They may occur for weeks or months after the main shock.

air mass A more or less coherent body of air with about the same temperature and humidity.

alluvial fan A fan-shaped deposit of sand and gravel at the mouth of a mountain canyon, where the stream gradient flattens at a main valley floor.

alluvium Sand and gravel deposited by running water.

amplitude The size of the back-and-forth motion on a seismograph.

andesite An intermediate-colored, fine-grained volcanic rock. It has a silica content of approximately 60 percent, has intermediate viscosity, and forms slow-moving lavas, fragments, and volcanic ash.

angle of repose The maximum stable slope for loose material; for dry sand it is 30 to 35 degrees.

anticline An arch or convex upward bend in a layer of rock.

aquifer A permeable formation saturated with enough water to supply a spring or well; a water-bearing area of rock that will provide usable amounts of water.

arc volcano One of a chain of andesitic volcanoes that parallels an oceanic trench or subduction zone and lies on the plate overlying the descending slab. A continental arc lies on the edge of a continent; an island arc lies on oceanic crust.

artificial levee An artificial embankment of loose material placed at the edge of a channel to prevent floodwater from spreading over a floodplain.

aseismic fault A fault on which no earthquakes have been recorded.

ash Loose particles of volcanic dust or small fragments.

ash fall Volcanic ash falling through the air or collecting loosely on the ground.

ash-fall tuff Volcanic ash from an ash fall that has compacted or cemented to form a solid rock.

ash flow A mixture of hot volcanic ash and steam that pours at high velocity down the flank of a volcano. Also called *nuée ardenté* or *pyroclastic flow*.

ash-flow tuff The solid rock formed by compaction or cementation of ash from an ash flow.

asteroid A chunk of rock orbiting the sun in the same way that Earth orbits the sun.

asteroid belt The distribution of asteroids in a disk-shaped ring orbiting the sun.

asthenosphere Part of Earth's mantle below the lithosphere that behaves in a plastic manner. The rigid and brittle lithosphere moves over it.

atom The smallest particle of matter that takes part in an ordinary chemical reaction.

average The usual or ordinary amount of something. Specifically, the sum of a group of values divided by the number of values used.

avulsion The permanent change in a stream channel, generally during a flood when a stream breaches its levee to send most of its flow outside of its channel.

back burn or back fire A fire deliberately lit by firefighters just ahead of a wildfire to use up fuel that the wildfire needs to continue burning. Rising heat in the wildfire draws in air from all directions, thereby pulling the backfire toward the wildfire.

backshore The landward part of the beach that is horizontal or slopes landward; also called *backbeach*.

backwash After water from a wave moves up onto the beach (swash), much of that water moves back to the ocean (backwash).

barrier bar A natural ridge of sand built by waves a few meters above sea level that is parallel to the shoreline and just offshore. The term is commonly used for a sand bar across the mouth of a bay or inlet.

barrier island A near-shore, coast-parallel island of sand built up by waves and commonly capped by wind-blown sand dunes. Also called a *barrier bar*.

barrier reef A long coral reef, typically parallel to shore and separated from the land by a lagoon.

basalt A black or dark-colored, fine-grained volcanic rock. It has a silica content of approximately 50 percent and low viscosity, and its flows downslope rapidly.

base isolation A mechanism to isolate a structure from earthquake shaking in the ground; often flexible pads between a building and the ground.

base surge The high energy blast of steam and ash that blows laterally during the initial stages of a volcanic eruption.

bay mouth bar A natural sand bar across the mouth of a bay or inlet.

beach The deposit of loose sand along the shore and deposited by waves. It includes sand seaward of either cliffs or dunes and extends offshore to a depth of perhaps 10 meters.

beach hardening The placement of beach "protection" structures: riprap, groins, and related features.

beach nourishment Addition of sand to a beach to replace that lost to the waves. Also called *beach replenishment*.

beach scarp A nearly vertical seaward face on a berm that was eroded by waves.

bedding Layers in a sedimentary rock.

bedload Heavier sediment in a stream that is moved along the stream bed rather than in suspension.

Benioff zone A zone of earthquakes near the top of a descending oceanic plate where it descends under an adjacent plate. The earthquakes extend as deep as 600 kilometers and can form by slip along the descending fault zone or by bending or extension in the downgoing plate.

bentonite A soft "swelling" clay that forms by alteration of volcanic ash, swells when wet, and becomes extremely slippery. It is prone to landsliding and can deform houses built on it.

berm The uppermost accumulation of beach sand and other sediment left by waves.

blind thrust A thrust fault that does not reach Earth's surface. It is not evident at the surface.

block A mass of cold, solid rock ejected from a volcano and larger than 25 centimeters across.

body wave A seismic wave that travels through the interior of Earth. P and S waves are body waves. Surface waves travel near Earth's surface and are not body waves.

bolide A general term that includes both asteroids and comets.

bomb A 6- to 25-centimeter diameter mass of cold, solid rock that is ejected from a volcano.

bond The attraction energy of positive and negative charges that holds two atoms together. Also called a *chemical bond*.

brackish Water intermediate in salinity between sea water and fresh water. Often true of the water in lagoons connected to the ocean.

braided stream A stream characterized by interlacing channels that separate and come together at different places. The stream has more sediment than it can carry, so it frequently deposits some of it.

breach Failure of part of a river's levee, leading to some flow outside the main channel.

breakwater An artificial offshore barrier to waves constructed to create calm water for a beach or for anchoring boats.

buoyancy The ability of an object to float.

burn Combustion of cellulose and other materials by combination with oxygen to produce carbon dioxide and water vapor.

buyer beware The old saying that warned a buyer to be wary that what he or she was purchasing might not be as good as it appears. The buyer pays the consequences.

caldera A large depression, generally more than 1 kilometer across, in the summit of a volcano; formed by collapse into the underlying magma chamber.

caldera collapse Collapse of the ground over a magma chamber to form a caldera.

capillary fringe The zone above the water table where moisture seeps upward between soil grains.

carbon dioxide A gas (at ordinary surface conditions) making up less than 1 percent of the atmosphere and which is toxic to humans at concentrations of several percent.

catastrophe Disaster, calamity.

catastrophism The idea that Earth's features were formed quickly by catastrophic forces. The evidence is derived from a literal interpretation of creation as outlined in the Bible. Recently, many geologists have recognized that although Earth is extremely old and that its surface features have evolved over a long time, much of the change occurs in relatively few major events separated by long periods with little action.

cause Whatever makes something happen.

cavern A large, natural underground cave or tunnel, most commonly in limestone. Also used for a soil cavern developed over a limestone cavern.

cellulose The woody parts of trees and plants.

Celsius (°C) The temperature scale based on 0 for the freezing point of water and 100 for the boiling point of water at Earth's surface and under normal conditions. Degrees Celsius are 5/9 the size of degrees Fahrenheit, the scale used by the public in the United States.

chance The likelihood of an event. For example, if the chance of an event is 1 percent, then the event will occur one time in 100 tries.

channelization Straightening and confining stream flow within artificial barriers along the sides of the stream.

chaos theory The theory that seemingly random events can unexpectedly lead to major, seemingly unrelated events. Chance effects can multiply and ultimately lead to major events.

charge The electrical energy, either a positive or a negative charge, stored in a thundercloud.

charge separation The separation distance or potential between positive and negative charges in a thundercloud. The greater the separation, the more violent the electrical discharge when positive and negative charges connect.

Chinook winds Winds that warm by adiabatic compression as they descend from high elevations of a mountain range to low elevations on the plains to the east.

chondrite A stony meteorite consisting primarily of the minerals olivine and pyroxene; the most common kind of meteorite.

cinder cone A small steep-sided volcano consisting of basaltic cinders.

cinders Small fragments of basalt full of gas holes that were blown out of a cinder cone.

clay Particles of sediment smaller than 0.004 millimeter. *See also* clay minerals.

clay minerals Exceptionally fine-grained (often clay-sized), soft, hydrous aluminum-rich silicate minerals with layered molecular structures.

climate The weather of an area averaged over a long period of time.

clockwise The direction of rotation of the hands of a clock.

cloud A cluster of condensed fine droplets of water in the atmosphere.

coast That part of the continental margin adjacent to the ocean. It is of variable width depending on context but is generally the zone affected by ocean-controlled processes.

coastal bulge The region above a subduction zone in which the overlying continental plate flexes upward before slip on the subduction zone causes a major earthquake.

coastal inlets Map-view indentations in a coastline; a bay or estuary.

cohesion The attraction between small soil particles that is provided by the surface tension of water between the particles.

cold front The line of boundary between a large mass of cold air advancing under an adjacent large mass of warm air.

collision zone The zone of convergence between two lithospheric plates.

columnar jointing Basalt lava flows often cool and shrink, breaking into five- or six-sided columns oriented perpendicular to the surface of the flows.

combustible Burnable.

comet A mass of ice and some rock material traveling at high velocity in the gravitational field of the sun but travel-ing outside our solar system before passing through it on occasion.

compaction The reduction in volume of clay, soil, or other fine-grained sediments in response to an overlying load.

components of stress Stress, such as the load of a landslide mass, can be resolved into directional components. For example, one component is the amount of the load that tends to pull the mass parallel to the slope.

composite volcano A stratovolcano consisting of layers of ash, lava, and assorted volcanic rubble.

compound An inorganic substance formed by chemical combination of two or more other substances.

compressional stress A force that squeezes or compresses something.

consequences The result or effect of an action.

constructive interference Where independent highs overlap highs and lows overlap lows; the highs amplify highs and the lows amplify lows.

continental arc. *See* arc volcano.

continental crust The upper 30 to 60 kilometers of the continental lithosphere that has an average density of 2.7 grams per cubic centimeter and an average composition similar to that of granite.

continental drift The gradual movement of continents as oceans spread and separate.

continental margin The boundary between continental lithosphere and oceanic lithosphere.

continental shelf The shallowly submerged edges of the continent to a depth of approximately 100 meters below sea level.

continent–continent collision Collision between two continental plates.

convection The slow circulation and overturn of an unevenly heated fluid or plastic mass driven by the rise of lower density hot fluids relative to higher density cooler fluids. Earth's mantle below the lithosphere may circulate by convection, rising under spreading ridges and sinking below subduction zones.

convective updraft The rapid rise of a heated, expanded air mass driven by its lower density. The removal of air from below draws outside air into the rising "chimney."

convergent boundary A tectonic plate boundary along which two plates come together by either subduction or continent–continent collision.

coral reef An offshore ridge of limestone built by the actions of coral and calcareous algae.

core Innermost part of Earth consisting of nickel and iron. The inner part is solid, the outer part liquid.

Coriolis effect Rotation of Earth from west to east under a fluid such as the atmosphere or oceans permits that fluid to lag behind Earth's rotation. Fluid flows shift to the right in the northern hemisphere and to the left in the southern hemisphere as a result of this effect. Thus a southward-moving fluid appears to curve off to the west in the northern hemisphere.

counterclockwise The direction of rotation opposite to that of the hands of a clock.

cover collapse Collapse of the roof material over an underground cavity, often a soil cavern over a limestone cavern.

creep (1) Surface layers of soil on a slope move downslope more rapidly than subsurface layers. (2) Slow, more-or-less continuous movement on a fault. Creeping faults either lack earthquakes or have only small ones.

cross-bedding Internal layers inclined to the overall layers in a sediment or volcanic ash deposit.

crowning The rise of heat that draws fire to the tops of trees.

crust. *See* continental crust, oceanic crust.

crystal settling Denser early-formed mineral grains in a magma sometimes settle to the bottom of a magma chamber, leading to magmatic differentiation.

cumulonimbus cloud A thundercloud with a flat anvil-shaped top; a sign of a nearby thunderstorm.

current The continuous movement of a mass of fluid in a definite direction.

cutoff When adjacent meanders in a stream come close together, the stream may cut through the narrow neck between them to bypass and abandon the intervening meander loop. The cut through the narrow neck is a cutoff.

cyclic events Events that come at evenly spaced times, such as every so many hours or years.

cyclone A large low pressure weather system that circulates counterclockwise in the northern hemisphere and clockwise in the southern hemisphere. It is equivalent to a hurricane or a typhoon.

dacite A light-colored volcanic rock intermediate in composition and appearance between andesite and rhyolite.

daylighted surface A layer inclined less steeply than the slope so that it is exposed at the surface down lower on the slope.

debris avalanche A fast-moving avalanche of loose debris that flows out a considerable distance from its source.

debris flow A slurry of rocks, sand, and water flowing down a valley. Water generally makes up less than half the flow's volume.

deepwater waves A wave in water deep enough that its water movement does not touch bottom—that is, a wave in water that is deeper than half the wave's length.

deformation The change in shape of an object.

density The mass (e.g., grams) of a material per area (e.g., cubic centimeter). Water is 1 g/cm³; granite is 2.7 g/cm³.

deposition The accumulation of sediment previously carried by a stream.

desert A particularly dry climate in which evaporation exceeds precipitation. An area with less than 25 centimeters of precipitation annually. A desert does not need to be hot.

destructive interference Where independent highs overlap lows and lows overlap highs; the lows tend to cancel out the effects of the highs.

developed countries Industrial countries that are wealthy and have a large middle class.

dike A mass of magma that has injected into a fracture and solidified.

dip The angle of inclination of a plane as measured down from the horizontal.

disaster Catastrophe, calamity.

discharge Measured volume of water flowing past a cross section of a river in a given amount of time. Usually expressed in cubic meters per second or cubic feet per second.

displacement The distance between an initial position and the position after movement on a fault.

dissolution Minerals or rocks dissolving in water.

divergent boundary A spreading plate boundary such as a midoceanic ridge.

dome A bulge on a volcano that is pushed up by molten magma below.

downburst A localized, violent, downward-directed wind below a strong thunderstorm.

drainage basin The area upslope from a point in a valley that drains water to that point.

dredging Scooping up loose sediment such as gravel from a stream to separate any gold it may contain.

dripstone Calcium carbonate that precipitates in a cavern from percolating carbonate-rich water. Stalactites, stalagmites, and columns are types of dripstone.

dry climate Sometimes described as an area with evaporation greater than precipitation, sometimes as an area with too little precipitation to support significant plant life.

dry ravel Dry, loose sand and other soil materials that trickle down postfire gullies to collect in larger canyons. The material accumulates until a rainstorm flushes it downslope in a debris flow.

duff Loose, combustible twigs, leaves, and grass on the forest floor.

dune A ridge or mound of sand deposited by the wind, most commonly at the head of a beach.

dynamic equilibrium The condition of a system in which the inflow and outflow of materials is balanced.

earthquake The ground shaking that accompanies sudden movement on a fault, movement of magma underground, or a fast-moving landslide.

earthquake forecast The statement that a future event will occur in a certain area in a given span of time (often decades) with a particular probability. *Compare with* earthquake prediction.

earthquake prediction The statement that a future event will occur at a certain time at a certain place. *Compare with* earthquake forecast.

Earth's magnetic field reversals Magnetic stripes on ocean floor.

ebb current The seaward drainage of water from an estuary or from a lagoon through a gap between segments of a barrier island as high tide recedes.

effluent stream. *See* gaining stream.

elastic A type of deformation in which a rock deforms without breaking. If the deforming stress is relaxed, the rock returns to its original shape. *Contrast* with plastic.

elastic rebound theory The theory applied to most earthquakes in which movement on two sides of a fault leads to bending of the rocks until they slip to release the bending strain during an earthquake.

element One of ninety-two naturally occurring materials such as iron, calcium, and oxygen that cannot be separated chemically. Elements combine to make minerals.

El Niño Elevated sea-surface temperatures that lead to dramatic changes in weather in some areas every few years—for example, rains in coastal Peru and western Mexico and southwestern California. *See also* La Niña.

emissions Gases given off by a volcano.

energy The capacity for doing work. The kinetic energy of a large wave can move huge boulders. The heat energy of a rising mantle plume can melt the overlying Earth crust.

ENSO El Niño-southern oscillation; the alternation between El Niño and its opposite extreme La Niña.

epicenter The point on the Earth's surface directly above the focus; the initial rupture point on a fault.

equator The imaginary line around Earth equidistant between the north and south poles or 90 degrees from each pole.

equilibrium The chemical balance of reactions between minerals or minerals and water.

equilibrium profile (1) The longitudinal profile or slope of an equilibrium stream (*see* graded stream); (2) the seaward slope of a beach that is adjusted so that the amount of sediment that waves bring onto the beach is adjusted to the amount of sediment that the waves carry back offshore.

equinox The two times of the year that the sun is directly above the equator; approximately March 21 and September 21.

erosion The wearing down and transport of loose material at the Earth's surface, including removal of material by streams, waves, and landslides.

eruption The processes by which a volcano expels ash, steam and other gases, and magma.

eruption cloud The cloud of ash, steam, and other gases blown out of a volcano during an eruption.

eruption column The more-or-less cylindrical mass of ash and gases forcefully blown up from a volcanic vent during an eruption.

eruptive rift A crack in the ground from which lava erupts.

eruptive vent The point of eruption of material from a volcano.

estuary The submerged mouth of a river valley where fresh water mixes with seawater.

evaporation The change of a substance from a liquid to a vapor.

evapotranspiration Water returned to the atmosphere by a combination of evaporation of water from the soil and leaves, and transpiration of moisture from the leaves. *See also* evaporation, transpiration.

expansive soil Clay-rich soil that expands when it absorbs water. Most expansive soil contains the swelling clay smectite.

explosion crater A depression formed by the forceful blasting out of gases and fragments from a volcano.

exsolve The process by which gas or other material separates from a solid rock.

extratropical cyclone A cyclone formed outside the tropics and commonly associated with weather fronts. Nor'easters are an example.

extrusive rocks Igneous rocks that extrude onto Earth's surface. Also called *volcanic rocks*.

eye The small-diameter core of a hurricane; characterized by lack of clouds and little or no wind.

far-field tsunami A tsunami generated by an earthquake far from the point of impact of the wave.

fault An Earth fracture along which rocks on one side move relative to those on the other side. A strike-slip fault (generally vertical) has relative lateral movement of the two sides. A normal fault (generally, steeply inclined) has the upper block of rock moving down compared with the lower block. Reverse faults (generally, steeply inclined) and thrust faults (generally, gently inclined) have the upper block of rock moving up compared with the lower block.

fault creep The slow, more-or-less continuous movement on a fault, in contrast to the sudden movement of a section of fault during an earthquake.

fault gouge Ground up or pulverized rock along a fault zone; caused by friction during fault movement.

felsic Light-colored igneous rocks.

fetch The distance over which wind blows over a body of water. The longer the fetch, the larger the waves that will be developed.

fire The heat, light, and flame of something burning.

fire components Fuel, oxygen, and heat are all needed to have a fire. A fire cannot burn without all three components.

firestorm A large, extremely hot fire generating a large updraft that pulls in vast amounts of air from the sides to fan the flames.

fissure A fracture at the surface that has opened by widening. *See also* rift.

flash flood A short-lived flood that appears suddenly, generally in a dry climate, in response to an upstream storm.

flood A stream flow high enough to overtop the natural or artificial banks of a stream.

flood basalt A broad expanse of basalt lava that cooled to fill in low-lying areas of the landscape.

flood channel A channel occupied during a flood.

flood frequency The average time between floods of a given height or discharge.

flood fringe The fringe or backwater areas of a flood that store nearly still water during a flood and gradually release it downstream after the flood.

floodplain Relatively flat lowland that borders a river, usually dry but subject to flooding every few years.

floodway Area of floodplain regulated by FEMA for federal flood insurance.

fluidization The process of changing a soil saturated with water to a fluid mass that flows downslope.

foam A fire-retardant gel sprayed on a combustible house or other building to prevent it from burning in a wildfire.

focus The initial rupture point on a fault. The epicenter is the point on Earth's surface directly above the focus. Also called *hypocenter.*

foliation The planar aspect of a rock that is controlled by parallel orientation of mica grains or other closely spaced parallel features.

forearc Area between the volcanic arc and the subduction zone.

foreshocks Smaller earthquakes that precede some large earthquakes on or near the same fault. They are inferred to mark the initial relief of stresses before final failure.

fractal Features that look basically the same regardless of size; for example, the coast of Norway looks serrated on the scale of a map of the world or on a map of only part of that coast.

fracture Any crack in a rock.

frequency The number of events in a given time, such as the number of back-and-forth motions of an earthquake per second.

frictional resistance The resistance to downslope movement of a flow or landslide.

front The boundary between one air mass and another of different temperature and moisture content.

fuel Burnable cellulose-rich material that includes wood, trees, shrubs, grass, and other dry vegetation.

fuel moisture Amount of moisture in a fuel. The lower the moisture content, the more readily the fuel will burn and at a higher temperature.

Fujita scale The scale of tornado wind speed and damage devised by Dr. Tetsuya Fujita.

fumarole Vent that emits steam or other gases.

funnel cloud Narrow, rapidly spinning funnel-shaped "cloud" that descends from a violent storm cloud. *See also* tornado.

g Acceleration of gravity—9.8 meters per second squared. *See also* acceleration.

gaining stream A stream, typically in a climate with abundant rainfall, that lies below the water table and gains water from groundwater.

gas emissions Gas given off by a volcano.

geophysics Study of Earth's physical properties.

geotechnical Study and interpretation of engineering properties of soil and rock.

geotextile fabric Strong permeable cloth that permits water to pass but not soil and rock particles.

geyser Intermittent eruption of hot and boiling water from a vent in an area of volcanic rocks.

giant debris avalanche Giant avalanche of loose debris formed by collapse of a large oceanic volcano such as Hawaii or the Canary Islands.

glacial outwash Sand and gravel deposited by a meltwater stream from a glacier.

global warming Warming of Earth's surface temperatures, especially over the past 100 years. Much of the recent warming has been attributed to increases in greenhouse gases.

graded river. *See* graded stream.

graded stream Stream in equilibrium with its environment; its slope is adjusted to accommodate the amount of water and sediment amounts and grain size provided to it.

gradient Slope along the channel of a streambed, typically expressed in meters per kilometer or feet per mile.

granite Light-colored, grainy igneous rock consisting of the minerals quartz, the two feldspars, orthoclase and plagioclase, and sometimes a dark mineral, either mica or amphibole.

gravity Attraction or pull between two masses, such as a mass of rock on a hillside and Earth below. *See also* g.

greenhouse gases Atmospheric gases such as carbon dioxide and methane that retain heat much like a greenhouse; solar energy can get in, but the heat cannot easily escape.

groin Wall of boulders, concrete, or wood built out into the surf from the water's edge to trap sand that moves in longshore drift. The intent is to hinder loss of sand from the beach.

groundwater Water in the saturated zone below the ground surface.

Gulf Stream Warm oceanic current that moves north from the Gulf of Mexico, past the East Coast of North America, and across the North Atlantic to maintain mild temperatures in northern Europe.

gullying Cutting of channels into a slope by running water.

gunite. *See* shotcrete.

Hadley cell Large-scale atmospheric circulation involving rise of warm air near the equator and sinking of cooler air near 30° north (and south) of the equator.

hailstone Single pellet of hail formed when droplets of water freeze before falling to the ground.

harbors Deep, calm-water shelter areas for ships. A port or loading and unloading area for ships.

hardening. *See* beach hardening.

"hard" solutions Solutions to a problem that involve building protective structures such as riprap and levees or continuous pumping of water from a landslide-prone slope.

harmonic tremor Rhythmic shaking of the ground that often accompanies magma movement.

Hawaiian type lava Fluid basalt lava of the type that erupts from Hawaiian volcanoes.

hazard An environment that could lead to a disaster if it affects people.

headland An erosionally resistant point of land jutting out into the sea (or a large lake) and often standing high.

heat Energy produced by the collisions between vibrating atoms or molecules. The faster the vibration, the greater the heat and the higher the temperature.

heat capacity The capacity to hold heat. Metals such as iron heat and cool easily; they have low heat capacity. Water has a high heat capacity.

Hertz The number of cycles of an earthquake per second, a measuring unit of frequency of shaking.

high pressure system An area characterized by descending cooler dry air and clear skies. Winds rotate clockwise around a high pressure system (in the northern hemisphere). *See also* right-hand rule.

high tide High sea level, often twice a day, and controlled by the gravitational pull of the sun and moon.

hook echo A curved, hook-shaped pattern on weather service radar that is indicative of a supercell thunderstorm. The hook shape often indicates development of a tornado.

hot spot The area over a rising hot magmatic plume in Earth's mantle.

hotspot volcano An isolated volcano, typically not on a lithospheric plate boundary, but lying above a plume or hot column of rock in Earth's mantle.

hot spring A place at which hot water reaches the ground surface.

hurricane A large tropical cyclone in the North Atlantic or east Pacific ocean, with winds greater than 118 kilometers per hour. Similar storms elsewhere are called typhoons or cyclones.

hurricane warning A hurricane watch is upgraded to a warning when dangerous conditions of a hurricane are likely to strike a particular area within twenty-four hours.

hurricane watch An indication that a hurricane may affect a region within thirty-six hours.

hydraulic mining A mining method that used high-pressure water hoses to wash gold-bearing gravels down into sluice boxes where gold could be separated from the gravel.

hydrograph A graph that shows changes in discharge or river stage with time.

hydrologic cycle The gradual circulation of water from ocean evaporation to the continents, where it falls to the ground as rain or snow and soaks into the ground to feed vegetation, groundwater, and streams. From there, some water evaporates and some returns to the oceans.

hydrophobic Soil sealed by hydrocarbon resins. These soils will not permit water to soak in.

hydrothermal Pertaining to hot water, usually of volcanic origin.

hypothesis A proposal to explain a set of data or information, which may be confirmed or disproved by further study.

ice age A period of low temperature lasting thousands of years and marked by widespread ice sheets covering much of the northern hemisphere. Two to four ice ages are recorded in North America, and as many as twenty in deep sea sediments of the last 2 million years.

igneous rocks Rocks that crystallize from molten magma, either within the Earth as plutonic rocks (e.g., granite and gabbro) or at Earth's surface as volcanic rocks (e.g., rhyolite, andesite, and basalt).

indefensible location A home or building location that is especially vulnerable to wildfire. Particularly important are dry vegetation and other fuels near the building and a location high on a hill, one that fire can easily reach upslope.

influent stream. *See* losing stream.

inlet (1) A gap in a coastline that connects the ocean with a bay, harbor, or coastal lagoon. (2) A small, narrow embayment or arm of the sea into a coastline.

insurance A service by which people pay a premium, usually annually, to protect themselves from a major financial loss they cannot afford. The insurance company that collects the premiums pays for covered losses.

intensity scale The severity of an earthquake in terms of the damage that it inflicts on structures and people. It is normally written as a Roman numeral on a scale of I to XII.

intertidal zone The beach zone between low and high tides.

intrusive rocks Igneous rocks that intrude into other rocks within the Earth.

ion A charged atom.

iridium anomaly A thin clay-bearing layer of sediment 65 million years old that has an abnormally high content of the platinum-like metal iridium.

iron meteorite A meteorite consisting of a nickel and iron alloy.

island arc volcano. *See* arc volcano.

isostacy Lower-density crust floats in Earth's higher density mantle. Also called *isostatic equilibrium.*

jet stream The high speed air current traveling from west to east across North America at an approximate altitude of 10 to 12 kilometers.

jetties Walls of boulders, concrete, or wood built perpendicular to the coast at the edges of a harbor, estuary, or river mouth. The intent is to prevent sediment from blocking a shipping channel.

joint One of a group of fractures with similar orientation.

karst The ragged top of limestone exposed at the surface, resulting from dissolution by acidic rainfall and groundwater. The ground is often marked by sinkholes and caverns.

K-T boundary The boundary between the Cretaceous and Tertiary geological time periods, from approximately 65 million years ago. The boundary is marked by a thin layer of sooty clay that contains the iridium anomaly.

Kyoto Protocol The 1997 agreement between many countries to reduce their emissions of greenhouse gases into the atmosphere.

ladder fuels Fuels of different heights that permit fire to climb progressively from burning ground material to brush, small trees, low branches, and finally to the tops of tall trees.

lagoon A narrow, shallow body of seawater or brackish water that parallels the coast just inland from a barrier island.

lahar A volcanic mudflow.

landslide Downslope movement of soil or rock.

La Niña The opposite of El Niño.

lapilli A particle of volcanic ash between 2 millimeters and 6 millimeters across.

lapse rate. *See* adiabatic lapse rate.

latent heat The heat added to a gram of a solid to cause melting or to a liquid to cause evaporation. Also the heat produced by the opposite changes of state.

lateral spreading The movement of a landslide that spreads out laterally as it moves downhill.

latitude Imaginary lines on Earth's surface that parallel the equator. The equator is 0 degrees latitude; the poles are 90 degrees north and south latitude.

lava Magma that flows out onto the ground surface.

lava dome A large bulge or protrusion, either extremely viscous or solid on the outside, and pushed up by magma.

lava flow. *See* lava.

lava lake A large, deep, pool of lava in a crater or caldera.

lava tube An underground tunnel left when lava flowing downslope pours out of a lava flow after the surrounding lava has solidified.

left-lateral One sense of movement on a strike-slip fault; if you stand on one side of the fault, the opposite side has moved to the left.

levee. *See* natural levee.

lightning The visible electrical discharge that marks the joining of positive and negative electric charges between clouds or between a cloud and the ground.

liquefaction A process in which water-saturated sands jostled by an earthquake rearrange themselves into a closer packing arrangement. The expelled water spouts to form a sand boil.

lithosphere Rigid outer rind of Earth approximately 60 to 100 or so kilometers thick; it forms the lithospheric plates.

lithospheric plates A dozen or so segments of the lithosphere that cover Earth's outer part.

Little Ice Age An especially cold period in Europe between the 1400s to the early 1800s. Widespread glaciers flowed down to low elevations, and many harbors froze where they never do today.

longitude Imaginary lines on Earth's surface oriented north–south and perpendicular to the equator.

longshore current A current parallel to the shore caused by refraction of waves coming in to the beach at an angle.

longshore drift The gradual migration of sand or gravel along the shoreline resulting from waves repeatedly carrying the sediment grains obliquely up onto shore and then straight back to the water's edge.

long-term future Consideration of long-term effects on our environment (that is, many decades out).

losing stream A stream, typically in a dry climate, that lies above the water table and loses water to an aquifer.

low-energy coast A coast not affected by large waves.

low pressure system An area of low atmospheric pressure that is characterized by rising warmer and humid air and cloudy skies. Winds rotate counterclockwise around a low pressure cell (in the northern hemisphere). *See also* right-hand rule.

low tide Low sea level, often twice a day, and lacking the gravitational pull of the sun and moon.

low-velocity zone Zone of low seismic velocity that marks the boundary between the lithosphere and asthenosphere. Presumed to contain a small percentage of partial melt.

maar A shallow crater with low rims, often on nearly flat topography, and formed by gas-rich volcanic eruptions.

mafic A dark-colored igneous rock.

magma Molten rock. When it flows out on the ground surface, it is called *lava*.

magmatic differentiation The process by which a magma separates into rocks of different compositions—for example, by settling of early formed minerals or giving off gases.

magnetic field The area around a magnet in which magnetism is felt.

magnetic stripes Strips of ocean floor, parallel to an oceanic ridge, that alternate between weak and strong magnetism parallel to Earth's current magnetic field.

magnitude The relative size of an earthquake, recorded as the amplitude of shaking on a seismograph. The range of amplitudes is so large that the number (open-ended but always less than 10) is recorded as a logarithm of the amplitude. Different magnitudes are based on amplitudes of different seismic waves: M_L = Richter or local magnitude; M_s = surface wave magnitude; M_b = body wave (P or S wave) magnitude; M_w = moment magnitude (most important for extremely large earthquakes).

mammatus clouds Downward rounded pouches protruding from the base of the overhanging anvil of a thundercloud.

mangrove Large leafy bushes or trees that form dense thickets in shallow brackish waters of coastal lagoons and estuaries.

mantle The thick layer of material below the thin crust and above Earth's core. Mostly peridotite in composition in its upper part. Its density approximates 3.2 or 3.3 g/cm³ in the upper part and 4.5 g/cm³ in the lower part.

mass A measure of the quantity of matter.

meander One of a series of pronounced back-and-forth bends in a stream channel.

mean sea level The average height of the sea, averaged over a long time, specifically nineteen years. Inferred to be midway between average high and low tides.

melting temperature The temperature at which a rock melts. Granite commonly melts between 700° and 900° Celsius. Basalt commonly melts near 1200 to 1400° Celsius.

Mercalli Intensity Scale. *See* intensity scale.

metamorphic rocks Original rocks of any kind that have been changed by heat or pressure; sometimes changes occur in chemical composition.

meteor A piece of space rock that heats to a white hot incandescence when it streaks through Earth's atmosphere.

meteorite An asteroid that passes through the atmosphere to reach Earth.

methane A gaseous hydrocarbon emitted during volcanic eruptions. It can be burned to generate heat; if released into the atmosphere, it is a greenhouse gas.

microtektite Translucent droplets of glass that have come through Earth's atmosphere.

migrating earthquakes Earthquakes that occur in a sequential manner along a fault over time.

millibar A measure 1/1000 of 1 bar, which is the normal atmospheric pressure at sea level.

mineral A naturally occurring inorganic compound with its atoms arranged in a regular crystalline structure. *Compare with* rock.

mining groundwater Removal of water in the ground without replacement.

mitigation Changes in an environment to minimize loss from a disaster.

moho Abbreviated colloquial name for the Mohorovičic discontinuity.

Mohorovičic discontinuity Boundary between Earth's crust and mantle. It is detected from the contrast between the slower seismic velocities of the crust (generally 5 to 7 km/sec.) and the upper mantle (approximately 8 km/sec.).

molecule The smallest combination of two or more atoms held together by balanced atomic charges.

moment magnitude (M_w) The magnitude of an earthquake based on its seismic moment; depends on the rock strength, area of rock broken, and amount of offset across the fault.

monsoon A seasonal wind that blows from the southeast for half of the year bringing warm, moist air and heavy rains from the Indian Ocean onto the south Asian continent.

mudflow A flow of mud, rocks, and water dominated by clay or mud-sized particles.

multipurpose dam A dam justified by perceived multipurpose benefits—for example, flood control, hydroelectric power, water supply, and recreation.

narrowing of a harbor A harbor that becomes narrower toward the land.

National Flood Insurance Program The flood insurance program guaranteed and partly funded by the U.S. government.

natural hazards Hazards in nature that endanger our property and physical well-being.

natural levee Natural embankment of sediment at the edge of a stream, where sediment is deposited as floodwaters slow and spread over an adjacent floodplain.

nature's rampages The perception of many people that a natural catastrophe involves nature going on an abnormal rampage. Actually, nature is not doing anything that it has not done for millions of years; we humans have merely put ourselves in harm's way.

near-Earth object A space object such as an asteroid or comet that narrowly misses the Earth.

Nor'easter A strong extratropical winter storm that moves up the east coast of North America with high winds and high waves. These can be as damaging as hurricanes.

normal fault. *See* fault.

normal stress The component of stress perpendicular to (normal to) one of Earth's planar surfaces.

North Atlantic Oscillation The winter atmospheric pressure pattern over the North Atlantic Ocean that brings major storms every few years.

nucleation The process by which atoms come together to form molecules that are large enough to grow or by which gas dissolved in magma separates to bubbles that are large enough to grow.

nuée ardenté. *See* ash flow.

obsidian Volcanic glass, usually of rhyolite composition. It is generally a water-poor magma that solidified before it could nucleate and crystallize.

ocean An extremely large body of saltwater that surrounds the continents.

oceanic crust The upper part of Earth's lithosphere under the oceans. It consists of basalt and gabbro and is typically 7 kilometers thick.

oceanic ridge A high-standing rift or spreading zone in an ocean—for example, the mid-Atlantic Ridge or the East Pacific Rise.

oceanic trench A long depression of valley in the seafloor where the ocean floor descends into a subduction zone.

oceanic volcano A basalt volcano in the ocean.

offset The distance of movement across a fault during an earthquake.

orographic effect The effect created when moisture-bearing winds rise against a mountain range: They condense, form clouds, and rain.

overburden Soil or other material above the bedrock. Often used in reference to waste material that must be removed to get at desired material below.

overland flow Surface runoff of water in excess of that which is able to infiltrate the ground.

oxbow lake A small lake left when a meander of a stream is cut off and left after a flood.

ozone O_3, the form of the oxygen molecule that has three atoms of oxygen rather than the normal two that is found in air. A layer rich in ozone in the upper atmosphere protects animals from much of the harmful ultraviolet radiation from the sun.

pahoehoe Basalt lava with a ropy or smooth top.

paleo– A prefix referring to events in the past.

paleoflood data Information on previous floods gathered from erosional and depositional features left by such a flood.

paleomagnetism The study of past characteristics (orientation and strength) of Earth's magnetism as preserved in rocks formed at various times in the past.

paleoseismology The study of former earthquakes from examination of offset rock layers below the ground surface.

paleovolcanology The study of former volcanic events from examination of offset rock layers below the ground surface.

pali The Hawaiian term for giant cliffs; these are now known to be the headscarps of giant submarine landslides.

Pangaea Supercontinent that began to break up to form today's continents 225 million years ago.

passive margin A coastline that is not a plate boundary—for example, the east coast of Canada and the United States.

Peléan eruption A large ash-rich eruption that typically produces ash flows.

perforated pipes Pipes full of holes that are pushed into a water-saturated landslide. Water seeps through the holes into the pipes and then down through the gently sloping pipes to the surface.

period The time between seismic waves or the time between the peaks recorded on a seismograph.

permeability The ease with which water can move through soils or rocks.

personal responsibility The idea that people should be responsible for their own actions rather than blaming someone else for permitting them to do something that was to their detriment.

phreatic A volcanic eruption dominated by steam.

phreatomagmatic Similar to a phreatic eruption but with a higher proportion of magma.

pillow basalt Basalt that flows into water or wet mud chills on its outside surfaces to form elongate fingers of basalt a meter or so across.

piping Water seeping through a dam or levee can begin to carry sand and cause erosion.

plastic A type of deformation in which a rock deforms gradually without breaking. If the deforming stress is relaxed, then the rock does not return to its original shape. Contrast with *elastic*.

plate. *See* lithospheric plates.

plate tectonics The theory that lithospheric plates that move relative to one another collide in some places, pull apart in others, and slide past one another in still others. These movements cause earthquakes and volcanic eruptions as well as build mountain ranges. The theory is supported by a wide range of data.

plateau basalt A large area of basalt flows that forms an elevated plateau when its thin edges are eroded away.

Plinian eruption Extremely large eruptions that involve continuous blasts of ash. Examples include Vesuvius in 79 A.D. and Mount St. Helens in 1980.

plume A rising, upward-flaring zone of hot rock in Earth's mantle.

plutonic A distinctly grainy igneous rock that solidified below Earth's surface.

point bar Deposits of sand and gravel on the inside (concave side) of a meander bend of a stream.

poor sorting A sediment with a mixture of different grain sizes. *Compare with* sorted.

popcorn clay Clay that swells when wet to take on the surface characteristics of a layer of popcorn.

pore pressure The pressure of water in pore spaces between soil or sediment grains tends to push the grains apart and reduce the contact force between the grains. That can facilitate landsliding.

porosity The percentage of pore space in a sediment or rock.

precipitate Crystallization of chemical constituents from solution.

precipitation Water that falls as rain, hail, or snow.

predictions. *See* earthquake prediction.

pressure The force per unit area; a continuously applied force spread over a body.

property rights The right to do what one wishes with one's personal property.

pumice Frothy volcanic rock dominated by gas bubbles enclosed in glass; typically pale in color and floats on water.

P wave The compressional seismic wave that shakes back and forth along the direction of wave travel.

pyroclastic material Fragmental material blown out of a volcano—for example, ash, cinders, and bombs.

pyroclastic flow. *See* ash flow.

quick clay Water-saturated mud deposited in salty water tends to consist of randomly oriented flakes of clay with large open spaces between the flakes. If the salt is flushed out, the flakes are unstable and may easily collapse and flow almost like water.

quicksand Loose water-saturated sand; if a mass is placed on quicksand, it will sink in the sand and water is pushed to the surface.

radiometric The measurement of the age of a rock by the analysis of radioactive constituents in the rock and their products, along with their known rates of decay.

rain Atmospheric precipitation in the form of falling water droplets.

rain shadow Drier climate on the downwind side of a mountain range.

recurrence interval The average number of years between an event of a certain size in a location—for example, floods, storm surges, and earthquakes. Also known as a *return period*.

refraction. *See* wave refraction.

regional metamorphism Metamorphism spread over a large region.

relative humidity The percentage of moisture in air relative to the maximum amount it can hold (at saturation) under its given temperature and pressure.

resisting force The force (such as friction or a load pushing in the opposite direction) that resists downslope movement.

resurgent bulge A rejuvenated bulge in the floor of a caldera.

resurgent caldera A huge collapse depression at the Earth's surface that sank into a near-surface magma chamber during eruption of the magma.

return period. *See* recurrence interval.

reverse condemnation If the government restricts a property owner from using his or her property for some potential purpose that has effectively taken the value of the property for that purpose.

reverse fault. *See* fault.

rhyolite A light-colored, fine-grained volcanic rock. It has a silica content of approximately 70 percent and high viscosity, and it generally erupts as volcanic ash and fragments.

Richter magnitude scale The scale of earthquake magnitude invented by Charles Richter. *See also* magnitude.

riffles The part of a stream with shallow rapids—generally, the straighter section between meander bends.

rift A spreading zone on the flank of a volcano from which lavas erupt. *See also* rift zone.

rift zone An elongate spreading zone in Earth's lithosphere.

right-hand rule The rule used to remember the direction of wind rotation in a low or high pressure cell in the northern hemisphere. With your right thumb pointing in the direction of overall air movement, your fingers point in the direction of wind rotation.

right lateral One sense of movement on a strike-slip fault; if you stand on one side of the fault, the opposite side moves to the right during an earthquake.

ring of fire The zone of volcanoes and arc volcanoes above subduction zones that surround the Pacific Ocean.

rip current A short-lived surface current that flows directly off a beach and through the breaker zone. It carries water back to the sea that had piled up onto the beach.

riprap Coarse rock piled at the shoreline in an attempt to prevent wave erosion of the shore or destruction of a near-shore structure or a streambank.

risk The chance of an event multiplied by the cost of loss from such an event.

river stage Water height in a river, measured relative to some arbitrary fixed point.

rock A mass of interlocking or cemented mineral grains. Compare with *mineral*.

rockbolts Long bolts drilled into and expanded in an unstable rock mass to help keep it from landsliding.

rockfall A rock mass that falls from a steep slope.

rotational landslide A landslide in which the mass rotates as it slides on a basal slip surface. Also called a *slump*.

runoff The portion of precipitation that flows off the ground surface.

runup The height to which water at the leading edge of a wave rushes up onto shore. Also used for the height to which a tsunami wave rushes up onshore.

sackung A natural spreading fracture across a steep slope; caused by gravitationally driven downslope movement.

Saffir-Simpson scale The commonly used five-category scale of hurricane intensity based on wind speeds and sea-water damage.

sand Particles of rock that range in size from 1/16 to 2 millimeters in diameter.

sand boil A pile of sand brought to the surface in water expelled from the ground by liquefaction at shallow depth or during flood when the water pressure under a channel forces groundwater to the surface outside a levee.

sand dune An accumulation of wind-blown sand, most commonly along the upper beach above high-tide level.

sand sheets Layers of nearly homogeneous, unlayered sand laid down by a tsunami wave that sweeps sand inland from a beach.

sand spit A wave-deposited area of sand jutting into the sea or a lake and often parallel to the coast.

Santa Ana winds Southern California trade winds that flow southwest, bringing dry air from the continental interior.

scientific method The method used by scientists to solve problems. They analyze facts and observations, formulate hypotheses, and test the validity of the hypotheses with experiments and additional observations.

scoria A lava full of gas bubbles; fewer gas bubbles and heavier pumice. Most commonly basalt composition.

sea cliff A seaward-facing cliff produced by wave erosion.

seafloor spreading The spreading of oceanic lithosphere at oceanic ridges.

seasons Spring, summer, fall, and winter and their changes in temperature, humidity, and other aspects of weather.

seawall Walls of boulders, concrete, or wood built along a beach front and intended to protect to prevent shore structures from wave erosion.

sedimentary rock A rock deposited from particles in water, ice, or air—or precipitated from solution—and then cemented into a solid mass. Examples are sandstone, shale, and limestone.

seiche The back-and-forth sloshing of a lake or other closed body of water. It can be caused by an earthquake, landslide, or changes in atmospheric pressure.

seismic gap A section of an active fault that has not had a recent earthquake. Earthquakes elsewhere on the fault suggest that the gap may have an earthquake in future.

seismic sea wave. *See* tsunami.

seismic wave A wave sent outward through Earth in response to sudden movement on a fault. *See also* P wave, S wave, surface wave.

seismogram The record of seismic waves from an earthquake or other ground motion as recorded on a seismograph.

seismograph The instrument used to detect and record seismic waves.

shallowing continental shelf The area where water over the continental shelf becomes shallower near shore.

shallow-water wave A wave in water shallow enough that its water movement touches bottom at a depth shallower than 1/2 of its wavelength.

shield volcano An extremely large basalt-lava volcano, such as those in Hawaii, with gently sloping sides.

shocked quartz Grains of the mineral quartz (silicon dioxide) that show parallel sets of deformation planes that are indicative of hypervelocity shock stresses.

shore The narrow strip of land next to the ocean or a lake between low tide and the upper limit of wave erosion.

shore profile The slope of the beach.

shotcrete A fluid cement-type material that is sprayed on a slope to prevent water penetration. Also called *gunite*.

silica tetrahedra An arrangement of four oxygen atoms in a four-cornered pyramid around a single silicon atom.

sinkhole A ground depression caused by collapse into an underground cavern.

slip plane, slip surface The sliding surface at the base of a landslide. Also called a *slide plane*.

slope angle The angle of a slope as measured down from the horizontal.

slump. *See* rotational landslide.

societal attitudes The attitudes of people—for example, the lack of acceptance of personal responsibility for their own actions.

"soft" solutions Solutions to a problem that involve avoiding the problem through restrictive zoning and building codes that minimize damage.

soil For engineering purposes, all of the loose material above bedrock.

soil creep Slow downslope movement of near-surface soil or rock; caused by numerous cycles of heating and cooling, freezing and thawing, burrowing animals, and trampling feet.

solstice The two times of the year that the sun overhead is farthest from the equator; approximately June 21 and December 21.

soluble Able to be dissolved, typically in water.

solution The process of dissolving rocks of minerals in water.

sorted A sediment is well sorted if its grains are all about the same size. *Compare with* poor sorting.

southern oscillation. *See* ENSO.

spreading zone or rift zone Boundary along which lithospheric plates spread apart or diverge.

stage The height of water above an arbitrary datum in a channel during a flood.

stalactite A cylindrical deposit of calcium carbonate that grows down from the roof of a cavern by condensation of carbonate-rich water.

stalagmite A cone-shaped deposit of calcium carbonate that grows up from the floor of a cavern by evaporation of carbonate-rich water.

step leader Electrical charges that advance downward from a thundercloud but do not manage to reach the ground.

Stokes Law The formula that expresses the rate of settling of spherical particles in a fluid.

stony-iron meteorite Essentially a chondrite that contains some nickel-iron.

storm A large atmospheric disturbance involving rain and wind.

storm surge The rapid sea-level rise caused by both low atmospheric pressure of a major storm and the strong winds that accompany the storm and push water forward.

strain Change in size or shape of a body in response to an imposed stress.

stratovolcano A large, steep-sided volcano consisting of layers of ash, fragmental debris, and lava.

straw wattle A net in the shape of a large hose, filled with straw, and used to hinder surface runoff and erosion on a slope denuded by fire or landslide.

streambed mining Excavation of sand and gravel from a stream bed.

stream order A way of numbering streams in a hierarchy that designates a small unbranched stream in a headwaters as first order, the stream it flows into as second order, and so on.

stress The forces on a body. These can be compressional, extensional, or shear.

strike The compass direction of a horizontal line on a plane such as a fault or a rock layer.

Strombolian eruption Frequent mild eruptions of basalt or andesite scoria, typically forming a cinder cone.

subduction The process in which one lithospheric plate (usually oceanic) descends beneath another.

subduction zone Convergent boundary along which lithospheric plates come together and one descends beneath the other; often ocean floor descending beneath continent.

submarine canyon A deep canyon in the continental shelf and slope offshore and extending about perpendicular to the shore.

submarine collapse. *See* submarine landslide.

submarine landslide Subsea level collapse of the flank of an oceanic volcano such as in Hawaii or the Canary Islands.

submarine volcano A volcano that does not rise above the ocean surface.

subsidence Settling of the ground in response to extraction of water or oil in subsurface soil and sediments, drying of peat, or formation of sinkholes. Subsidence of the ocean floor occurs by cooling of hot lithosphere.

sulfur dioxide A toxic volcanic gas consisting of two oxygen atoms attached to one sulfur atom.

supercell A particularly strong rotating thunderstorm that can spin off dangerous tornadoes.

superoutbreak A large group of tornadoes produced along a major storm front.

surf The near-shore zone of turbulent water, including breaking waves and swash as the water returns to the sea.

surface rupture length The length of a fault broken during an earthquake.

surface wave The seismic wave that travels along and near Earth's surface. These waves include Rayleigh waves (which move in a vertical, elliptical motion) and Love waves (which move with horizontal perpendicular to the direction of wave travel).

surge. *See* base surge, storm surge.

suspension Sediment grains carried within the water column and buoyed up by turbulent eddies—in contrast to sediment grains transported along a stream bed.

swash Water from a wave moves up onto the beach (swash), then much of that water moves back to the ocean (backwash).

S wave The seismic shear wave that shakes back and forth perpendicular to the direction of wave travel. S waves do not pass through liquids.

swelling soil A soil that expands when wet; generally, a soil that contains the swelling clay smectite.

syncline A trough or convex downward bend in a layer of rock.

talus Coarse, angular rock fragments that fall from a cliff to form a cone-shaped pile banked up against the slope.

tectonic. *See* plate tectonics.

tektite Droplets of molten rock—glass—that may have formed by superheated splash from a hypervelocity impact on either the moon or Earth.

tensional stress Stress involving pulling apart of rock masses.

tephra. *See* pyroclastic material.

texture The arrangement of minerals in a rock. Flat grains of mica, for example, may be arranged parallel to one another to form a rock that breaks into sheets. Equidimensional grains may form a massive rock.

thalweg The line connecting the deepest parts of the channel along the length of the stream bed.

theory A scientific explanation for a broad range of facts that have been confirmed through extensive tests and observations.

thrust fault. *See* fault.

thunder The violent boom caused by superheating and supersonic expansion of the air along a lightning stroke.

thunderstorm A storm accompanying clouds that generate lightning, thunder, rain, and sometimes hail.

tidal current The flow of water through a narrow passage, such as between the ocean and a lagoon, through segments of a barrier island, in response to sea level change between high and low tides.

tidal wave A colloquial but incorrect name for tsunami.

tide The change in sea level, generally once or twice a day, in response to the sun and moon's gravitational pull.

tiltmeter A device like a carpenter's level that records any change in slope on the flank of a volcano.

tornado A near-vertical narrow funnel (generally only a kilometer or so in diameter) of violently spinning wind associated with a strong thunderstorm.

tornado alley The region of the central United States, between Texas and Kansas, that is noted for frequent tornadoes.

trade winds Regional winds that blow from northeast to southwest between latitudes 30 degrees north (or south) and the equator and centered near 15 degrees north (or south).

transform fault Boundary along which lithospheric plates slide laterally past one another.

translational slide A landslide that moves approximately parallel to the slope of the ground.

transpiration The passing of water vapor through the pores of vegetation to the atmosphere.

tree mold A cylindrical hole in a lava flow that is left when lava surrounded and burned a tree.

trench An elongate depression in the ocean floor at a subduction zone between two tectonic plates and most commonly at the edge of an active continental margin. Most are at the margins of the Pacific Ocean.

trimline The line along a mountainside along which tall trees in the forest upslope are bounded by distinctly shorter trees downslope.

triple junction A junction between three lithospheric plates.

tropical cyclone A large rotating low pressure cell that originates over warm tropical ocean water. Depending on location, they are called hurricanes, typhoons, or cyclones.

tropical storm A storm at tropical latitudes.

tropics The warm climate area between the Tropic of Cancer (33 degrees north of the equator) and the Tropic of Capricorn (33 degrees south).

tsunami An abnormally long wavelength wave most commonly produced by sudden displacement of water in response to sudden fault movement on the seafloor. Can also form when a landslide, volcanic eruption, or asteroid impact displaces water.

tsunami warning When a significant tsunami is identified, officials order evacuation of endangered low-lying coastal areas.

tsunami watch The alert is issued when a magnitude 7 or larger earthquake is detected somewhere around the Pacific Ocean or some other ocean that may see dangerous tsunamis.

tuff A rock formed by consolidation of volcanic ash.

turbulence Eddies and irregular current flow.

typhoon. *See* cyclone.

undercurrent A subsurface current of water moving in a different direction or velocity than that at the surface—for example, the current at the west end of the Mediterranean Sea at Gibraltar.

underdeveloped countries Poor countries with a large proportion of the population living in poverty and small proportions of both middle class and wealthy.

Uniform Building Code A national code for construction standards to provide safety for people in high hazard zones for earthquakes, hurricanes, and other natural hazards.

uniformitarianism The theory that Earth's features today have been generated by processes that we see going on today and have been going on for millions of years—the "present is the key to the past."

upwelling The rise of deep, cold, nutrient-rich ocean water, especially in areas of the trade winds.

urbanization Change of an area with addition of buildings and pavement such as in a city.

VEI. *See* Volcanic Explosivity Index.

velocity The speed or rate of movement of something. It can refer, for example, to the rate of ground motion in an earthquake, rate of water flow in a stream, or the rate of flow of air from one place to another (wind).

vent The eruptive site of a volcano.

viscosity The resistance to flow of a fluid because of internal friction.

volcanic arc The chain of volcanoes that lies above an active subduction zone.

volcanic ash Any small volcanic particles; formally, particles less than 2 millimeters across.

Volcanic Explosivity Index A scale of volcanic eruption violence based on volume, height, and duration of an eruption.

volcanic gases Gases such as steam, carbon dioxide, sulfur oxides, chlorine compounds, and methane that are emitted by a volcano.

volcanic weather Weather generated during a volcanic eruption. The high temperature above an erupting volcano draws in outside air that rises and cools. Moisture in the air condenses to form rain and stormy weather.

volcano A mountain formed by the products of volcanic eruptions.

Vulcanian eruption A style of eruption that is more violent than Strombolian and less violent than Peléan.

wall cloud The rotating area of cloud that sags below the main thundercloud base and from which a tornado may develop.

warm front The boundary between a large mass of cold air advancing under an adjacent large mass of warm air.

water pressure The pressure of water spaces between grains in the ground.

watershed That part of a landscape that drains water down to a given point on a stream.

water table Nearly the level of the top of the saturated zone, the level to which water would rise in a shallow well.

water vapor The invisible gaseous form of water.

wave (1) The elongate undulation on the surface of a body of water that generally moves perpendicular to the crest of the undulation. (2) The front of a pulse of energy that moves outward from a disturbance such as an earthquake.

wave base The greatest depth of wave motion that stirs sediment on the bottom—generally, 10 meters or so.

wave crest The highest elongate part of a wave.

wave energy The capacity of a wave to do work—that is, to erode the shore. Because wave energy is proportional to the square of wave height, high waves do almost all coastal erosion.

wave frequency The number of waves that pass a point per second.

wave height The vertical distance between a wave crest and an adjacent trough.

wavelength The distance from crest to crest of a wave—for example, in an earthquake wave or a water wave.

wave period The time between passage of two wave crests.

wave refraction The bending of wave crests where one end of the wave drags bottom and slows down in shallower water.

wave train A group of waves more or less the same size that are moving in the same direction.

wave trough The lowest elongate part of a wave.

weather The conditions—temperature, humidity, air motion, and pressure—of the atmosphere at a particular place and time.

weather fronts. *See* cold front, warm front.

weathering The gradual destruction of rock materials through physical disintegration and chemical decomposition by exposure to natural agents such as moisture in the atmosphere.

welded ash. *See* welded tuff.

welded tuff Ash from a hot ash flow that is hot enough when deposited that the particles fuse together to form a solid rock. If it is hot enough and under sufficient load of overlying ash, it may completely fuse to form obsidian.

westerlies. *See* westerly winds.

westerly winds Regional winds that blow from southwest to northeast and centered 45 degrees north (or south) of the equator.

wind The movement of air from an area of higher to lower atmospheric pressure.

wind duration The extent of time the wind blows over a particular location.

wind shear The short-distance change in wind direction or velocity.

wing dam Segments of walls built to protrude into the current of a river to increase the depth of water in the channel to facilitate shipping.

zone of accumulation The lower end of a landslide in which material moving downslope accumulates.

zone of depletion The upper end of a landslide from which material moves downslope.

zone of saturation A subsurface zone in which all of the pore spaces are filled with water.

zoning restrictions Laws that prohibit building in certain areas.

INDEX

CIP indicates Case in Point boxes.